TRANSACTIONS
OF THE
INTERNATIONAL
ASTRONOMICAL UNION

VOLUME XIVB – PROCEEDINGS

OTTO HECKMANN

PRESIDENT OF THE INTERNATIONAL ASTRONOMICAL UNION
1967–1970

INTERNATIONAL COUNCIL OF SCIENTIFIC UNIONS
INTERNATIONAL ASTRONOMICAL UNION
UNION ASTRONOMIQUE INTERNATIONALE

TRANSACTIONS
OF THE
INTERNATIONAL ASTRONOMICAL UNION
VOLUME XIVB

PROCEEDINGS
OF THE FOURTEENTH
GENERAL ASSEMBLY
BRIGHTON 1970

Edited by

C. DE JAGER

General Secretary of the Union

A. JAPPEL

Executive Secretary

PUBLISHED FOR THE INTERNATIONAL COUNCIL OF SCIENTIFIC UNIONS
WITH THE FINANCIAL ASSISTANCE OF UNESCO

D. REIDEL PUBLISHING COMPANY / DORDRECHT-HOLLAND
1971

Published on Behalf of
the International Astronomical Union
by
D. Reidel Publishing Company
Dordrecht, Holland

Printed in The Netherlands by D. Reidel, Dordrecht

PREFACE

The changing character of the IAU General Assemblies becomes most clear from a comparison of the agenda of the Brighton meeting with that of one of the earlier meetings. The fourth General Assembly (Cambridge Mass., 1932) had about 240 participants, registered guests included, the Brighton meeting had about 2300 people attending. The Cambridge meeting lasted 5½ working days, of which, however, three half days were exclusively devoted to excursions, leaving four real meeting days. At that time the nearly 30 commissions had each only one meeting, during part of a morning or afternoon; some commissions did not meet at all. There was one public lecture, by Sir Arthur Eddington, on 'The Expanding Universe'. Most of the small European countries were represented by two or three delegates only, but the delegations of France and Great Britain were composed of 20 and 18 scientists respectively; at that time there were only two delegates from Germany.

After the Brighton General Assembly, with about 200 commission meetings – one commission met eleven times! – six Joint Discussions, four Invited Discourses, a Special Meeting and hardly time for excursions, there were a few complaints about too many overlapping meetings. In my opinion this reflects the encouraging intensity and sparkling tension of the modern General Assemblies and of astronomical life in general! In organizing future General Assemblies utmost care will be taken, of course, to possibly avoid overlap of meetings, and to secure enough time for discussions, but a simple computation shows that with 2400 participants, ten meeting rooms, and 8 meeting days there will be an average speaking time of 12 minutes per participant! This explains the tendency, nowadays, of shifting the very detailed specialist discussions towards the symposia and colloquia which more and more tend to cluster around the General Assemblies – there were *eleven* such meetings scattered in time and place around the Brighton General Assembly. This tendency will certainly continue.

The present volume, *Transactions* **XIVB**, contains the report of the Executive Committee for the triennial period 1967–69; the reports of the Inaugural Ceremony and the General Assembly at Brighton, the reports of the Commission Meetings, and further what is customary called the Astronomer's Handbook; the most important part of that section is the new Statutes, By-laws and Working Rules of the Union, as accepted at the XIVth General Assembly. We found it also necessary to publish a new version of our Style Book which has changed in several respects after its first publication in 1966. Finally there are the usual lists of commission officers and members, and the general list of Union members.

I wish to express my sincere thanks to miss G. Drouin for translating various parts of this volume into French.

C. DE JAGER
General Secretary

CONTENTS

Part 3

Reports of Meetings of Commissions
Comptes Rendus des Séances des Commissions

Part 4

Astronomer's Handbook

PART 1

REPORT OF THE EXECUTIVE COMMITTEE
1967–1969

EXECUTIVE COMMITTEE
1967–1970

PRESIDENT

Professor Dr O. Heckmann, Director, The European Southern Observatory, Bergedorfer Strasse 131, 205 Hamburg 80, D.B.R.

VICE-PRESIDENTS

Professor M. K. V. Bappu, Director of the Astrophysical Observatory, Kodaikanal, India
Professor W. N. Christiansen, School of Electrical Engineering, University of Sydney, Sydney, N.S.W., Australia
Professor L. Gratton, Laboratorio di Astrofisica, Casella Postale 67, Frascati, Rome, Italy
Professor J. Sahade, Observatorio Astronómico, La Plata, Argentina
Professor M. Schwarzschild, Princeton University Observatory, Peyton Hall, Princeton, New Jersey 08540, U.S.A.
Professor A. B. Severny, Director of the Crimean Astrophysical Observatory, USSR Academy of Sciences, P/O Nauchny, Crimea, U.S.S.R.

GENERAL SECRETARY

Dr L. Perek, Director, Astronomical Institute of the Czechoslovak Academy of Sciences, Budečská 6, Praha 2, Czechoslovakia

ASSISTANT GENERAL SECRETARY

Professor C. de Jager, Director, The Astronomical Institute, Zonnenburg 2, The Netherlands

Introduction

The present report covers the period of three years from 1 January 1967 to 31 December 1969. It follows closely the pattern as set by previous reports, especially that for the years 1964–1966, formally approved by the XIIIth General Assembly and published in volume XIIIB of the *Transactions of the IAU*.

The period under report includes the first eight months of the year 1967 (until 31 August) which fell under the responsibility of the former Executive Committee, but it does not include the first seven and a half months of the year 1970 still within the responsibility of the present Executive Committee.

Decisions of the Executive Committee related to the XIIIth General Assembly and printed in *Transactions* volume XIIIB or published in the *Information Bulletin of the IAU* will not be repeated here. Thus, the present report is actually restricted to formal matters and deals with facts necessary for permanent record. The reports on the activity of IAU Commissions, Inter-Union Commissions, Astronomical Services, etc., have been included in the *Reports of Astronomy*, 1970, to be published as volume XIVA of the *Transactions*.

The Activities Report deals with the administration of the Union, meetings of the Executive Committee, IAU Membership, adherence of countries, Commissions of the IAU, colloquia and symposia organized by the IAU, IAU publications, and relations to other organizations.

The Report on IAU Finances gives a rounded off picture of the receipts and payments of the Union during the period under report, and presents an outline of the financial problems the Union has to face.

The Report as a whole is supplemented by an appendix listing data on Adhering Organizations, IAU Members and IAU Publications.

Activities Report

The President of the Union summoned, with the authority of the Executive Committee, an Extraordinary General Assembly which was held in Prague, Czechoslovakia, on 15 May 1968. This Extraordinary General Assembly modified articles 4a and 11a of the Statutes of the International Astronomical Union in that it added to the six existing categories of adherence, categories 7 and 8 with 20 and 30 units of contribution respectively. The proceedings of this Extraordinary General Assembly have been published in *Information Bulletin* No. 20.

Administration

The Executive Committee and the officers responsible for the administration of the Union have had to deal with an ever increasing number of problems arising from the natural growth of the IAU, its complex relationship with other organizations and from the rapid progress in astronomy. The organization of symposia and colloquia, and the preparation of IAU publications, have now permanently become the responsibility of the Assistant General Secretary. Another development of apparently institutional character is the meetings of executive officers (President, General Secretary and Assistant General Secretary), regularly held at least twice a year in order to co-ordinate the Union's activities in its manifold fields of interest.

The administrative office in Prague had to be re-inforced by a short-hand typist over the periods of intensified work, especially from 1 September 1969 on. Moreover, the President, the General Secretary and the Assistant General Secretary benefited from the services of several assistants and secretarial help in Hamburg, Prague and Utrecht, generously provided by their respective home institutions. It is hoped that the financial situation of the Union will permit one of the future General Secretaries to assemble an IAU staff large enough to cope with the administration of the Union without external help.

Meetings of the Executive Committee

The Executive Committee, in its former composition, held the 29th meeting in Prague during the thirteenth General Assembly, having had sessions on 20, 21, 27 and 29 August 1967. The meeting was presided over by the President of the Union Prof. P. Swings. All members and both counsellors were present.

The 30th meeting of the Executive Committee in its present composition, was held in Prague on 31 August 1967 under the presidency of Prof. O. Heckmann. All members and both advisers were present. Further, the Executive Committee held the following meetings:

The 31st meeting in Hamburg-Bergedorf from 26 to 30 August 1968. Excused were: Prof. A. B. Severny, Vice-President, and Prof. P. Swings, adviser.

The 32nd meeting in Frascati (Rome) from 21 to 26 September 1969, when all members and both advisers were present.

The Executive Committee formed, at its 31st meeting, a Statutes Committee to prepare a revised version of the IAU Statutes and By-laws. This committee, which consisted of the President Prof. O. Heckmann, the Vice-President Prof. M. Schwarzschild, the General Secretary Dr L. Perek, the adviser Prof. J.-C. Pecker, the former General Secretary of long standing Dr D. H. Sadler, and the

Executive Secretary Dr A. Jappel, met in Frankfurt a/M in December 1968 and worked out a draft of the revised Statutes and By-laws for submission to the Executive Committee. The approved draft has been published in *Information Bulletin* No. 23.

Between meetings of the Executive Committee, the business of the Union was, where necessary, conducted by correspondence between members and advisers.

Adhering Countries

Lebanon withdrew from the Union in 1967, and Colombia has adhered to it since that year. The 32nd meeting of the Executive Committee admitted Iran to the Union from 1969 on. Cuba has applied for admission, and its application will be submitted to either the 33rd meeting of the Executive Committee or the XIVth General Assembly, according to whether the present or revised Statutes and By-laws will be in force at the time of submission.

The list of Adhering Countries and corresponding Adhering Organizations, together with the approximative number of individual Members in each country, is given in the appendix on pages 25–26.

Members of the IAU

The number of IAU Members was 1572 on 1 January 1967. A further 450 new Members have been admitted at the XIIIth General Assembly since then. On the other hand, the General Secretary has been informed of the decease of 29 Members, so that the Union had 1993 Members on 28 February 1970.

The list of Members whose decease has been notified since 1 January 1967 is given in the appendix on page 26.

Commissions of the IAU

The *Transactions of the IAU*, volume XIVA, subtitled *Reports of Astronomy* 1970, include the reports prepared by the Presidents of each of the 38 Commissions of the Union, assisted by their Vice-Presidents, Organizing Committees, those responsible for the various working groups or committees formed within the Commissions, and by the Commissions themselves.

The Executive Committee approved, at its 31st meeting, to change the title of Commission No. 5 from Commission for Abstracts and Bibliography to Commission for Documentation; in French from Commission des Analyses de Travaux et de Bibliographie to Commission de Documentation.

In the period under report, Commissions have co-opted a number of new members from among the Members of the Union. On the other hand, Members of the IAU complied with the wish of the Executive Committee not to adhere to more than three Commissions, and resigned their membership in Commissions in excess of three.

Prof. A. I. Lebedinskij has been succeeded by Prof. A. A. Mihajlov, nominated as Vice-President of Commission No. 17 on the Moon.

The Presidents of Commissions will prepare proposals for the new composition of Commissions including co-opted members and new IAU Members to be admitted at the General Assembly.

The Executive Committee formed, at its 32nd meeting, a Working Group on Numerical Data in Astronomy and Astrophysics with the following membership: Prof. Ch. Moore-Sitterly (Chairman), Prof. C. W. Allen, Prof. M. Migeotte, Prof. W. Fricke.

The purpose of the Working Group is to deal with problems of collecting and disseminating numerical data in astronomy and astrophysics, prepare a programme of procedure, act as intermediary between the IAU and CODATA, and prepare a meeting of interested Members at the XIVth General Assembly.

Between the General Assemblies, Commissions organized several colloquia and other projects financed by the Union.

Colloquia of IAU Commissions
(held between 1 January 1967 and 31 December 1969):

IAU Colloquium on Spectroscopy, Bombay, India, 9–18 January 1967 (Commission 14).

IAU Colloquium on The Zodiacal Light and the Interplanetary Medium, Honolulu, Hawaii, 30 January–2 February 1967 (Commissions 21 and 22).

IAU Colloquium on Space Spectroscopy, Evanston, Ill., U.S.A., 3–5 May 1967 (Commissions 9, 14, 29, 44).

IAU Colloquium on The Gravitational *n*-Body Problem, Paris, France, 16–18 August 1967 (Commissions 17, 28, 33, 37).

IAU Colloquium on Non-Periodic Phenomena in Variable Stars, Budapest, Hungary, 5–9 September 1968 (Commissions 29, 42).

IAU Colloquium on Mass-Loss from Stars, Trieste, Italy, 12–16 September 1968 (Commissions 29, 42).

IAU Colloquium No. 1 on the Problem of Variation of Geographical Coordinates in the Southern Hemisphere, La Plata, Argentina, 4–5 November 1968 (Commission 19).

IAU Colloquium No. 2 on Spectrum Formation in Stars with Steady State Extended Atmospheres, Munich, FRG, 16–19 April 1969 (Commission 36).

IAU Colloquium No. 3 on Interstellar Dust, Jena, GDR, 22–26 August 1969 (Commission 34).

IAU Colloquium No. 4 on Stellar Rotation, Columbus, Ohio, U.S.A., 8–11 September 1969 (Commissions 29, 35, 36, 45).

IAU Colloquium No. 5 on Visual Double Stars, Nice, France, 8–10 September 1969 (Commission 29).

IAU Colloquium No. 6 on Mass-Loss and Evolution in Close Binaries, Elsinore, Denmark, 15–19 September 1969 (Commission 42).

Specific Projects of Commissions

The list of Specific Projects of Commissions as approved by the XIIIth General Assembly (or, during the period under report, by the Executive Committee) is given, together with the corresponding expenditure, in the report on the financial situation of the IAU (page 17). These projects are, as a rule, continuations of the previous projects and thus need no further comment.

Symposia

IAU Symposia

During the period under report, the following symposia were organized by the Union:

No.	Subject	Place	Date
32	Continental Drift, Secular Motion of the Pole and Rotation of the Earth	Stresa	March 1967
33	Physics and Dynamics of Meteors	Tatr. Lomnica	September 1967
34	Planetary Nebulae	Tatr. Lomnica	September 1967
35	The Structure and Development of Solar Active Regions	Budapest	September 1967
36	Ultraviolet Stellar Spectra and Related Ground-based Observations (jointly with COSPAR)	Lunteren	June 1969
37	Non-Solar X- and Gamma-Ray Astronomy (jointly with COSPAR)	Rome	May 1969
38	Spiral Structure of the Galactic System	Basel	Aug./Sept. 1969
39	Cosmical Gas Dynamics	Crimea	September 1969
40	Planetary Atmospheres	Marfa/Texas	October 1969

Symposium No. 32 was organized jointly with IUGG. Symposia Nos. 33, 34 and 35 were held shortly after the XIIIth General Assembly.

The following IAU Symposia will be held shortly before or immediately after the XIVth General Assembly:

No.	Subject	Place	Date
41	New Techniques in Space Research (jointly with COSPAR)	Munich	10–14 August 1970
42	White Dwarfs	St Andrews	10–14 August 1970
43	Solar Magnetic Fields	Paris	31 August – 4 September 1970
44	External Galaxies and Quasi Stellar Sources	Uppsala	10–14 August 1970
45	Motion, Evolution of Orbits and Origin of Comets	Leningrad	4–11 August 1970
46	The Crab Nebula	Jodrell Bank	6–8 August 1970

Symposia with IAU participation

Symposium on Solar Flares and Space Research, Tokyo, 9–11 May 1968. Organized by COSPAR, cosponsored by IAU, IUCSTP, URSI and IUGG.

Symposium on Meteorites, Vienna, 7–13 August 1968. Organized by UNESCO, cosponsored by IUG, IAEA and IAU.

Colloquium on Planetary Atmospheres and Surfaces, Woods Hole, U.S.A., 11–15 August 1969. Joint Symposium of IAU and URSI.

International Symposium on Solar-Terrestrial Physics, Leningrad, 12–20 May 1970. Organized by IUCSTP, cosponsored by IAU, IUGG and URSI.

The IAU was, in addition, represented at numerous symposia and meetings of many ICSU Organizations, as announced in the *IAU Information Bulletin.*

IAU Publications

Budgetary considerations and the fact that the Union cannot afford facilities of a publishing house prompted the Executive Committee radically to reformulate the publication policy of the IAU. Thus it was resolved that

—The old contract with Academic Press and Willmer Brothers should be terminated.

—A new publisher should be found.

—The Union should only finance the production of such publications as are meant for free distribution to Members.

—The distribution of IAU publications on a subscription basis should be discontinued.

—Only institutions of developing countries should obtain IAU publications free of charge.

—All symposium proceedings should be published for the account and risk of the publisher.

—For such publications, the publisher should pay the Union a royalty.

—Such royalties should primarily be used for purchasing the volumes to be given free of charge to institutions of developing countries.

These principles were embodied in the contract made with D. Reidel Publishing Company, Dordrecht, Holland.

Transactions of the IAU

The *Transactions of the XIIIth General Assembly* were published in two volumes, A and B. Volume A included the corrected Draft Reports, distributed free to Members of the Union. Volume B, subtitled *Proceedings of the Thirteenth General Assembly, Prague 1967* was distributed free of charge to those who were Members of the IAU at the time of the thirteenth General Assembly.

The Invited Discourses, Joint Discussions and the results of Special Meetings held during the XIIIth General Assembly were published in the form of a symposium volume under the title *Highlights of Astronomy as presented at the XIIIth General Assembly of the IAU 1967.*

The Draft Reports and Transactions volumes were financed from the budget of the Union. The *Highlights of Astronomy* already went to the charge of D. Reidel.

The Transactions volumes sold as follows:

By Academic Press in	1967–1969	1966	1965 and/or earlier
vol. XIA	32	18	46
vol. XIB	34	32	9
vol. XIIA	91	29	423
vol. XIIB	170	367	
vol. XIIC	327	197	
By D. Reidel	1968–1969		
vol. XIIIA	451		
vol. XIIIB	451		

IAU Symposia

IAU symposia up to and including No. 31 were still the responsibility of the Union under the terms of the old contract. It should be noted that Symposium No. 29 only appeared in Russian, and is not available in English.

IAU symposia up to and including No. 35 already were published by D. Reidel, without any financial responsibility of the IAU.

Academic Press effected the following sales:

Symposium No.	Number of copies sold 1967–1969	
24	140	(319 sold in 1966)
25	142	(322 sold in 1966)
26	155	(309 sold in 1966)
27	769	
28	660	
30	476	
31	658	

The sales by D. Reidel were as follows:

Symposium No.	Number of copies sold 1968–1969
32	1363
33	863
34	856
35	921
Highlights	829

IAU Information Bulletin

The *IAU Information Bulletin* was prepared and edited by the General Secretary, and printed, published and distributed by D. Reidel. It appeared, as a rule, twice a year and went, free of charge, to IAU Members and institutions on a special distribution list.

Publication Policy

The Executive Committee examined in detail several proposals for lower publication costs with a view of obtaining lower selling prices. Unfortunately, all possibilities examined proved impracti-

cable, and thus it was resolved not to change the present arrangements, which are as follows:
—The Draft Reports have been abolished.
—The Reports of Commissions will be published in *Transactions* volume A, to be distributed free of charge to Members at the General Assembly. The production costs of this volume are the responsibility of the Union.
—The Proceedings of the General Assembly will be published in *Transactions* volume B, to be distributed free to those having been Members of the IAU at the time of the General Assembly. Production costs to be borne by the Union.
—*Information Bulletin* of the IAU to be financed by the Union.
—All other IAU publications to be financed by the publisher, on his account and risk, but the Union to obtain a royalty of 5 % of the selling price. Free distribution only to the Organizing Committee of the Symposium in question and to Institutions of developing countries on a special distribution list.
—*Information Bulletin* financed by the IAU and distributed free of charge to IAU Members and selected institutions.

The financial aspects of this publication policy are commented on in the Report on the Financial Situation.

Relations to other Organizations

IAU Representation

1. *International Council of Scientific Unions.* The General Secretary represented the Union at the seventh meeting of the ICSU Executive Committee in Rome (October 1967). At the twelfth General Assembly of ICSU held in Paris, in September 1968, the General Secretary was elected Vice-President of ICSU for two years. The General Secretary also represented the Union at the meeting of the ICSU Executive Committee held in Yerevan, in October 1969.

The General Secretary is the present representative of the Union on the Executive Committee of ICSU.

2. *ICSU Organizations.* The Union participates in the activity of a number of Special and Scientific Committees, and Inter-Union Commissions formed and sponsored by ICSU. For more details please see pages 13 onwards of volume XIIC of the *Transactions*.

The following list only gives the present state of IAU representation in these organizations:
(*a*) Committee on Space Research (COSPAR): C. de Jager (succeeded L. Gratton in 1968).
(*b*) International Abstracting Board (IAB): J. B. Sykes.
(*c*) Federation of Astronomical and Geophysical Services (FAGS): D. H. Sadler (President), B. Guinot.
(*d*) Comité International de Geophysique (CIG): G. Righini (Vice-President), R. Michard.
(*e*) Inter-Union Committee on Frequency Allocations for Radio Astronomy and Space Science (IUCAF): J. F. Denisse, V. A. Sanamian, F. G. Smith (Chairman), O. Hachenberg.
Inter-Union Committee on the Ionosphere (IUCI): discontinued.
Inter-Union Commission on Solar Terrestrial Physics (IUCSTP) created in 1966: Z. Švestka.
Inter-Union Commission on Science Teaching (IUCST): Miss Edith Müller.
Committee for Data on Science and Technology (CODATA): Mrs. Ch. Moore-Sitterly.
Inter-Union Commission on Spectroscopy: A. H. Cook, E. Edlén, J. G. Phillips, M. J. Seaton.

Other Organizations

The Union is represented in the following organizations:
(*a*) La Fondation Internationale du Pic-du-Midi: A. Lallemand.
(*b*) Le Comité Consultatif pour la Définition de la Seconde (CCDS) du Bureau International des Poids et Mesures: B. Guinot, H. M. Smith.
(*c*) Le Comité Consultatif pour la Définition du Mètre du Bureau International des Poids et Mesures: A. H. Cook.

(*d*) Le Comité Consultatif International des Radiocommunications: H. M. Smith.
(*e*) Various Services of FAGS:
 Bureau International de l'Heure (BIH): H. M. Smith, F. Zagar.
 International Polar Motion Service: Organizing Committee of IAU Commission 19.
 Quarterly Bulletin on Solar Activity: Organizing Committee of IAU Commission 10.
 International Ursigrams and World Day Service (IUWDS): R. Michard.
 Solar Particles and Radiation Monitoring Organization (SPARMO): C. de Jager.

Direct Relations with UNESCO

The IAU had the benefit of three direct contracts with UNESCO, which provided for grants to the International Summer Schools for Young Astronomers, jointly organized in 1967, 1968 and 1969. These summer schools proved a great success, and the Executive Committee wishes to express to UNESCO its heartfelt thanks for its generous financial help.

The 1967 grant of $ 8 000 according to contract UNESCO/NS/0134/67 was paid to the Organizers of the Summer School in Manchester direct, and is not included in the accounts of the Union.

The two other grants, each of $ 8 000, towards the summer schools in Arcetri (1968) and Hyderabad (1969) respectively are included in the financial report.

Moreover, the Union obtained from UNESCO a grant of $ 2 000 towards the studies of Mr. B. De of India at the University of Michigan. This grant was paid directly to Mr. De.

Relations with ICSU

Over the period under report, the Union enjoyed the most understanding support of ICSU in all its efforts. In addition to the annual UNESCO subvention, transmitted to the Union by ICSU, this organism paid the Union, jointly with UNESCO, further $ 5 000. A similar grant of $ 4 000 has been made for the year 1970.

The Executive Committee feels that it has to extend its special thanks to ICSU for the generous assistance given to the Union.

Report on IAU finances

In accord with the policy adopted in 1961, the Executive Committee presents to the Members of the Union a 'Summarized Account of Receipts and Payments' covering the period from 1 January 1967 to 31 December 1969. All amounts are given in U.S. dollars converted from other currencies at the following ICSU conversion rates, treated as exact:

£ sterling: $ 2.80 ($ 2.40 after devaluation)
U.S. $ = 3.60 Dutch guilders = 4.90 French francs = 14.36 Czechoslovak crowns.

The period under report was marked by the devaluation of the Pound sterling in 1967 and that of the French franc in 1969. This resulted in a total loss to the Union of $ 478.69 on its English and French bank accounts. The conversion rate of $ = 4.90 French francs, as used throughout the summarized account, was the official ICSU rate of exchange until 1 November 1969.

Certain bank operations, such as transfers from one account to another and reimbursements, were effected at current rates of exchange and thus resulted in exchange differences.

The summarized account has been derived from 14 (one for each currency and each year) individual accounts of receipts and payments audited and certified as correct by the Union's professional accountants. These accounts will be available for examination by the Finance Committee at the XIVth General Assembly.

SUMMARIZED ACCOUNT OF RECEIPTS AND PAYMENTS

The statement on pages 12–14 is a simple account of receipts and payments collected for convenience under a few main headings. The following explanatory notes provide more detailed information on each item of the accounts.

Receipts

1. *National Subscriptions.* The XIIIth General Assembly of the IAU, held in Prague, resolved to fix the amount of the unit of contribution at 900 gold francs, the gold franc being equivalent to 0.3266 U.S. dollars. Thus, the unit of contribution is 293.94 U.S. dollars.

The modification of articles 4a and 11a of the Statutes of the IAU, by adding to the 6 existing categories of adherence categories 7 and 8 with 20 and 30 units of contribution respectively (Extraordinary General Assembly held in Prague on 15 May 1968 – see Proceedings published in *Information Bulletin* No. 20), led two countries to adhere in category 7.

As on 31 December 1969, the IAU had 44 adhering countries paying annually a total of 184 units of contribution. Eighteen countries adhered in category 1, nine each in categories 2 and 3, three each in categories 5 an 6, and two countries adhered in category 7.

The contributions of 39 adhering countries had passed through the bank before 31 December 1969. Three further payments made in 1969 passed through the bank after 31 December 1969. One adhering country owes its dues to the Union since 1966, another one since 1965. One adhering country has already paid its contribution for 1970, and the account of another country shows a balance to its credit.

SUMMARIZED ACCOUNT OF RECEIPTS AND PAYMENTS (IN U.S. DOLLARS)
FOR THE PERIOD 1967–1969

Receipts

U.S. dollars

1. Subscriptions from adhering countries

	1966 (or before)	1967	1968	1969	1970	
1967	3	38	1 + 3 part.			35 011.48
1968	—	4	33 + 1 part.	1 + 1 part.		44 056.16
1969	—	—	10	39	1	52 614.57
1967–1969	3	42	44 + 4 part.	40 + 1 part.	1	131 682.21

2. Sales of publications   U.S. dollars
   1967..   ..   ..   13 391.69
   1968..   ..   ..   16 597.70
   1969..   ..   ..   13 129.93
                                     43 119.32

3. Subscriptions to IAU publications   0

4. Interest on savings and other accounts
   1967..   ..   ..   77.30
   1968..   ..   ..   6.65
   1969..   ..   ..   738.96
                                     822.91

5. Subvention from UNESCO, through ICSU
   1967..   ..   ..   14 000.00
   1968..   ..   ..   14 000.00
   1969..   ..   ..   23 987.03
                                     51 987.03

6. Contracts with UNESCO
   1967..   ..   ..   500.00
   1968..   ..   ..   8 000.00
   1969..   ..   ..   6 000.00
                                     14 500.00

7. Other receipts
   1967..   ..   ..   74 603.39
   1968..   ..   ..   31 136.29
   1969..   ..   ..   11 788.16
   Transfer from cancelled savings account-Prague 1.39
   Gain through difference of rate of exchange £ 1.11
                                     117 530.34

Total receipt   359 641.81

Bank accounts as on 1 January 1967:
—savings accounts 71 244.19
—current accounts 18 518.91

                                     89 763.10

Effective balance as on 31.12.1969 ..   ..   23 498.47

                                   113 261.57

SUMMARIZED ACCOUNT OF RECEIPTS AND PAYMENTS (IN U.S. DOLLARS)
FOR THE PERIOD 1967–1969

Payments

	U.S. dollars	U.S. dollars
1. Administrative office		
1967..	15 240.26	
1968..	10 729.56	
1969..	11 960.94	
		37 930.76
2. Subscriptions to ICSU		
1967..	708.09	
1968..	700.23	
1969..	881.12	
		2 289.44
3. Expenses of Commissions		
1967..	—	
1968..	63.67	
1969..	676.92	
		740.59
4. Specific projects of Commissions approved by the General Assembly		
1967..	11 992.67	
1968..	9 045.86	
1969..	7 784.07	
		28 822.60
5. General Assembly		
1967 (XIII GA and as per analysis)	21 309.14	
1968 (XIV GA Brighton)..	396.60	
1969 (XIV GA Brighton)..	684.92	
		22 390.66
6. Printing of Publications by IAU		
1967..	48 369.61	
1968..	28 260.97	
1969..	1 919.18	
		78 549.76
7. Meetings of the Executive Committee		
1967..	—	
1968..	6 124.61	
1969..	5 929.37	
		12 053.98
8. Organization of IAU Symposia and Colloquia and participation in other scientific meetings		
1967..	5 140.36	
1968..	2 522.87	
1969..	10 718.53	
		18 381.76
9. Publication of symposia not by IAU		0
	to carry over	201 159.55

				carried over	201 159.55

10. Inter-Union commissions, committees on the ICSU, etc.

1967..	.. ..	2 147.81	
1968..	.. ..	2 200.00	
1969..	.. ..	2 200.00	
			6 547.81

11. Other expenses of representation

1967..	.. ..	534.06	
1968..	.. ..	70.00	
1969..	.. ..	777.32	
			1 381.38

12. Specific projects approved by the Executive Committee

1967..	.. ..	1 100.00	
1968..	.. ..	—	
1969..	.. ..	—	
			1 100.00

13. Programmes under direct UNESCO contract

1967..	.. ..	1 500.00	
1968..	.. ..	6 000.00	
1969..	.. ..	8 629.92	
			16 129.92

14. Reimbursement of loan from ICSU

1969..	.. ..		4 000.00

15. Reimbursements to savings accounts

1969..	.. ..		38 763.89

16. Other Expenses

1967..	.. ..	3 162.62	
1968..	.. ..	30 111.19	
1969..	.. ..	20 395.07	
Loss through devaluation of £ in 1967	.. .. ..	134.82	
			53 803.70

Total payments	322 886.25
Excess of receipts over payments	36 755.56
Total ..	359 641.81

Bank Accounts as on 31 December 1969:

—savings accounts	57 987.10
—current accounts	55 274.47
	113 261.57

2. *Sales of publications.* The complexity of the accounts from the printers and publishers, and the many different arrangements for the sale of transactions and symposium volumes as well as reprints from them, make it impracticable to give separate statements for each volume and reprint sold. Generally, the revenue from IAU publications is made up of

(a) Receipts from Academic Press Inc. (London) Ltd., for sales of Transactions and Symposium volumes.

(b) Royalties from the same publishers.

(c) Royalties from Cambridge University Press.

(d) Royalties from D. Reidel Publishing Company.

(e) Receipts from organizations and individuals for sales of IAU publications by the Secretariat.

(f) Receipts from organizations and individuals from sales of reprints.

The following is a synopsis of the receipts, separately for 1967–1968 and 1969.

Receipts from sales of publications 1967–1968:

	$
—Academic Press, for sales of IAU *Transactions* up to and including volume XIIC and IAU Symposia up to and including volume No. 31	18 828.33
—Royalties from Academic Press	15.24
—Royalties from Cambridge University Press	310.01
—Sales of IAU publications by the Secretariat	496.10
—Sales of reprints	10 339.71
Total	29 989.39

Receipts from sales of publications 1969:

—Academic Press, for sales of IAU *Transactions* up to and including volume XIIC and IAU Symposia up to and including volume No. 31	5 262.81
—D. Reidel, royalties for sales of IAU *Transactions*, vol. XIIIA, XIIIB, *Highlights of Astronomy*, and IAU Symposia 32–35	7 711.26
—Cambridge University Press, royalties 1968	18.35
—Academic Press, royalties 1968	9.46
—Sales of IAU publications by the Secretariat	128.05
Total	13 129.93

Recapitulation:

Receipts 1967–1968	$ 29 989.39
Receipts 1969	$ 13 129.93
Total for 1967–1969	$ 43 119.32

3. *Subscriptions to IAU publications.* The system of distributing IAU publications on a subscription basis has been terminated in 1966. It is suggested to omit this heading in the budget for 1971–1973.

4. *Interest on reserves in savings banks.* The interest earned in 1967 on the savings accounts at the Dime Savings Bank, First National City Bank, Chemical Bank and Bowery Savings Bank were transferred to the current account of the IAU with the Chemical Bank when these savings accounts were closed. The interest on the savings account at the American Savings Bank has been allowed to accumulate on this account. The interest of $ 582.58 earned on the savings account at the Amsterdam-Rotterdam Bank in 1969 has been transferred to the current account of the IAU with this bank.

In addition, the current accounts in Dutch guilders and French francs attracted interest in the total amount of $ 240.33.

5. *UNESCO subvention through ICSU.* Detailed accounts of the expenditure of this subvention are submitted to UNESCO and ICSU each year. The distribution of the subvention is left to the discretion of the IAU within the permitted UNESCO categories.

In 1969, the normal UNESCO subvention of $ 14 000 was increased by a combined UNESCO-ICSU grant of $ 5 000. The amount of $ 23 987.03 was paid erroneously by ICSU, and the excess of $ 5 000 had to be reimbursed. $ 12.97 are bank charges.

6. *UNESCO contracts.* The Union obtained two contracts from UNESCO during the period under report:

(*a*) Contract UNESCO/NS/0847/68 providing a grant of $ 8 000 towards the organization of the International Summer School for Young Astronomers in Arcetri.

(*b*) Contract UNESCO/SC/1492/69 providing a grant of $ 8 000 towards the organization of the International Summer School for Young Astronomers in Hyderabad.

The last payment of $ 2 000 under the terms of contract SC/1492/69 will be made in 1970. $ 500 represent the balance payable under the terms of contract UNESCO/NC/3034/66 (see Agenda and Draft Reports, 1967, page xlii)

7. *Other receipts.* The entries under this heading largely misrepresent the actual receipts of the Union, as they include internal bank transfers, transient items, reimbursements, etc. The following is an analysis of the amounts of $ 74 603.39, 31 136.29 and 11 788.16 for the years 1967, 1968, and 1969 respectively:

Other receipts in 1967:	$
—transfer from savings accounts	56 564.79
—loan by ICSU	10 000.00
—transfer from current accounts	3 100.00
—various reimbursements	3 618.88
—reimbursement of loan to staff member	988.17
—response to appeal for funds	200.00
—transfer from Státní spořitelna	82.17
—personal deposit	49.38
Total	74 603.39

Other receipts in 1968:	
—transfer from current accounts	29 765.13
—various reimbursements	787.07
—response to appeal for funds	392.73
—transfer from ČSOB to Petty-Cash	100.00
—personal deposit	91.36
Total	31 136.29

Other receipts in 1969:	
—transfer from current accounts	6 693.07
—various reimbursements	1 567.41
—COSPAR funds deposited on ČSOB account	2 400.00
—contribution by Astron.-Meteorol. Anstalt Basel towards Symposium No. 38	934.03
—rectification of erroneous transfer to LOC of Symposium 36	159.00
—personal deposit	34.65
Total	11 788.16

Recapitulation:

1967	$ 74 603.39
1968	$ 31 136.29
1969	$ 11 788.16
Grand total	$ 117 527.84
+	2.50 (see page 12)
	117 530.34

Specific Projects of Commissions

Commissions, number	Project	Allocated in 1967 for 1968–70 $	Paid in 1967 $	Paid in 1968 $	Paid in 1969 $	Total 1967–69 $	Total in 1968–69 $	Balance for 1970 from allocation $
6	IAU Telegram Bureau	1 600	666.66	0	1 066.66	1 733.66	1 066.66	533.34
10	Cartes Synoptiques	0	882.00	326.66	0	1 208.66	326.66	
16	Centre de Documentation	1 400	665.33	450.00	450.00	1 565.33	900.00	500.00
20	Minor Planet Center	1 300	500.00	0	400.00	900.00	400.00	900.00
27	Catalogue of Variable Stars	3 500	0	771.82	771.81	1 543.63	1 543.63	1 956.37
29	Information Bulletin for the Southern Hemisphere	1 500	0	0	500.00	500.00	500.00	1 000.00
38	Exchange of Astronomers	22 000	9 278.68	8 269.20	5 167.41	22 715.29	13 436.61	8 563.39
42	Bibliography and Program Notes for Eclipsing Binaries	200	0	0	200.00	200.00	200.00	0
Total		31 500	11 992.67	9 817.68	8 555.88	30 366.23	18 373.56	13 453.10

It is seen that the bulk of this sum is made up of withdrawals from the savings banks, which were used for financing the production of IAU publications, transfers from one current account to another, the ICSU loan and various reimbursements. These operations cannot be considered effective receipts, as they are balanced by corresponding payments.

Payments

The headings have been numbered to correspond to those of the budget approved by the XIIIth General Assembly in Prague, August 1967.

1. *Administrative Office.* The expenses of the Administrative Office include the salaries, social security and pension payments of the IAU staff, and expenses for the regular meetings of the Executive Officers of the Union. It should be noted that thanks to the generosity of the Observatoire de Nice, Astronomical Institute of the Czechoslovak Academy of Sciences, the Astronomical Institute at Utrecht, the European Southern Observatory, and the Deutsche Forschungsgemeinschaft, the Union paid no rental, had the free use of all communication facilities, and only spent a negligible amount for stationery.

2. *Subscriptions to ICSU.* The IAU makes an annual contribution towards the administrative expenses of ICSU of an amount equal to 2 % of the income from subscriptions from Adhering Countries for the previous year.

3. *Expenses of Commissions.* The participation of the President of Commission No. 17 in the meetings of the Committee on Lunar Nomenclature and a visit of the Secretary of the International Summer School for Young Astronomers to Switzerland resulted in expenses exceeding the budgetary amount of $ 500 by $ 240.59.

4. *Specific Projects of Commissions.* Details of the allocations approved by the General Assembly are given in the report of the Finance Committee to the XIIIth General Assembly. (*Transactions IAU*, vol. XIIB, for 1967; *Transactions IAU*, vol. XIIIB, for 1968 and 1969.)

Actual payments on these projects were as in the list appended. This list requires the following comments:

The grants made towards the publication of the Cartes Synoptiques represent balances from previous years and are not covered by the total sum allocated in 1967.

The sum actually charged to Commission No. 38 is $ 4 968.92. The difference of $ 198.49 between this sum and that in the list is due to the devaluation of the French franc. It should be noted that thanks to an additional ICSU subvention of $ 5 000 to Commission No. 38, the Executive Committee could reduce the allocation from the original $ 22 000 to $ 19 000, so that the total amount at the disposal of that Commission for 1968–1970 is $ 24 000. The actual balance in favour of Commission No. 38 is therefore $ 10 563.39 + $ 198.49 = $ 10 761.88, out of which $ 2 225 have already been committed for 1970.

The total of $ 1 543.63 paid to Commission No. 27 for the publication of the Catalogue of Variable stars has not passed through the bank accounts and is therefore not included in the summary account.

The balance as on 31 December 1969 is made up of credits in favour of different projects which may, but need not be spent in 1970.

5. *Expenses of the General Assembly.* The amount of $ 21 309.14, as given in the summary account, had been withdrawn from the IAU current bank accounts in connection with the XIIIth General Assembly in Prague, 1967. This sum was used as follows:

Purpose:	$
(a) Members of the Executive Committee (travel and living expenses)	5 995.00
(b) Presidents of Commissions (travel expenses)	6 943.00
(c) Young astronomers (travel expenses)	3 093.20
to carry over	16 031.20

	carried over	16 031.20

(*d*) IAU staff and external assistance	1 578.25
(*e*) Bank expenses	213.52
(*f*) Opening of current account at Československá obchodní banka	1 000.00
(*g*) Symposia Nos. 33 and 35 (grants towards participation)	430.00
(*h*) Reimbursement of grants to General Secretary, Presidents of Commissions and Young Astronomers (reimbursed later)	2 056.17
Total	21 309.14

The costs of the General Assembly itself were $ 17 822.97, that is by about $ 2 200 more than in 1964.

6. *IAU Publications*. The production costs of IAU publications were as follows:

In 1967–1968	$
(*a*) Willmer Brothers Ltd., Birkenhead, England, for printing IAU symposium volumes up to and including vol. 31, Draft Reports 1967, sheets of *Transactions* vol. XIIIA and reprints from these publications	62 107.63
(*b*) D. Reidel Publishing Company, Dordrecht, Holland, for binding and distribution of IAU *Transactions*, vol. XIIIA, production and dispatch of IAU *Transactions* vol. XIIIB, printing and distribution of *IAU Information Bulletins* Nos. 19 and 20, and purchase of IAU symposium volumes Nos. 32 through 35 and *Highlights of Astronomy* for free distribution	13 222.96
(*c*) Etablissements Pierotti, Nice, France, for printing *IAU Information Bulletins* Nos. 17 and 18	1 210.78
(*d*) Purchase of other IAU publications resold to customers	89.21
Total	76 630.58

In 1969	
(*a*) D. Reidel Publishing Company, Dordrecht, Holland for printing and distribution of *IAU Information Bulletins* Nos. 21, 22	1 770.28
(*b*) Purchase of IAU publications for free and other distribution	148.90
Total	1 919.18

Recapitulation:

Cost of IAU publications in 1967–1968	$ 76 630.58
in 1969	$ 1 919.18
Grand total	$ 78 549.76

The amount spent for production of IAU publications in 1968–1969 was $ 30 180.15.

7. *Meetings of the Executive Committee.* The expenses arisen from the 1967 meeting in Prague are included in those for the General Assembly, heading 5, as above. Total expenditure was higher than that for the previous meetings of the Executive Committee: it has become more and more difficult to finance the administrative meetings of the IAU from sources others than its own. The expenses for the meetings of the Statutes Committee have been included in this heading, this committee having been an extension of the Executive Committee rather than one of its projects.

8. *Organization of Symposia and Colloquia.* The payments made were, as a rule, for travel grants towards participation in the meetings; organizational expenses were insignificant. The grants were allocated by the Organizing Committees of the meetings in cooperation with the Assistant General

Secretary; actual payment was made by the Administrative Office. The following table shows the payments in detail:

Symposium No.	Subject, Place, Date	Payment in $
32	Continental Drift, Secular Motion of the Pole and Rotation of the Earth, Stresa, Italy, 21–25 March 1967	1 270.56
33	Physics and Dynamics of Meteors, Tatranská Lomnica, Czechoslovakia, 3–9 September 1967	849.98 + (350)*
34	Planetary Nebulae, Tatranská Lomnica, Czechoslovakia, 3–9 September 1967	425.00
35	The Structure and Development of Solar Active Regions, Budapest, Hungary, 4–8 September 1967	1 466.00 + (80)*
36	Ultraviolet Stellar Spectra and Related Ground-based Observations, Lunteren, Netherlands, 24–27 June 1969	2 007.92
37	Non-Solar X- and Gamma-Ray Astronomy, Rome, Italy, 8–10 May 1969	572.40**
38	Spiral Structure of the Galactic System, Basel, Switzerland, 29 August – 4 September 1969	1 567.03
39	Cosmical Gas Dynamics, Crimea, U.S.S.R., 9–19 September 1969	1 750.00
40	Planetary Atmospheres, Marfa, Texas, U.S.A., 26–31 October 1969	1 450.00
	Total	11 358.89

Colloquium No.	Subject, Place, Date	Payment in $
–	On Spectroscopy, Bombay, India, 9–18 January 1967	500.00
–	On the Zodiacal Light and the Interplanetary Medium, Honolulu, Hawaii, U.S.A., 30 January – 2 February 1967	—
–	On Space Spectroscopy, Evanston, Ill., U.S.A., 3–5 May 1967	498.82
–	On the Gravitational n-Body Problem, Paris, France, 16–18 August 1968	500.00
–	On Non-Periodic Phenomena in Variable Stars, Budapest, Hungary, 5–9 September 1967	500.00
–	On Mass-Loss from Stars, Trieste, Italy, 12–16 September 1968	360.00
1	The Problem of Variation of Geographical Constants in the Southern Hemisphere, La Plata, Argentina, 4 and 5 November 1968	500.00
2	Spectrum Formation in Stars with Steady State Extended Atmospheres, Munich, FRG, 16–19 April 1969	407.22
3	Interstellar Dust, Jena, GDR, 22–26 August 1969	495.56
4	Stellar Rotation, Columbus, Ohio, U.S.A., 8–11 September 1969	500.00
5	Visual Double Stars, Nice, France, 8–10 September 1969	439.80
6	Mass-Loss and Evolution in Close Binaries, Elsinore, Denmark, 15–19 September 1969	475.00
7	Proper Motions, Minneapolis, Minnesota, U.S.A., from 21 to 23 April 1970 – advance payment	500.00
	Total	5 676.40

* See page 19.
** The original grant had amounted to $ 756.00, of which $ 183.60 were reimbursed.

In addition, the Union contributed financially to the following meetings:

	Contribution in $
IAEA Conference on Meteorite Research, Vienna, Austria, 7–13 August 1968	553.20
IUCST Congress on the Integration of Science Teaching, Varna, Bulgaria, 12–19 September 1968	409.67
International CODATA Conference, Arnoldshain, FRG, 30 June – 5 July 1968	200.00
Total	1 162.87

Recapitulation:	$
Symposia	11 358.89
Colloquia	5 676.40
Other Meetings	1 162.87
Total	18 198.16
Reimbursed	183.60
Grand Total	18 381.76

9. *Publication of Symposia not by the IAU*. During the term under report the Union did not subsidize publications issued by other organizations.

10. *Inter-Union Commissions*. The payments made consisted of contributions to the Inter-Union bodies concerned, and of a refund of an unspent UNESCO subvention. Expenditure was as follows:

	$
Reimbursement to ICSU of unspent UNESCO contribution to IUCSTR	447.73
Contribution to CIES (1967, 1968, 1969)	600.00
Contribution to IUCAF (1967, 1968, 1969)	3 000.00
Contribution to IQSY (for 1966)	500.00
Contribution to IUCSTP (1968, 1969)	2 000.00
Total	6 547.73

11. *Representation expenses*. The budgetary provisions permitted the President and General Secretary to participate in the Basel celebration of the 50th anniversary of the Union, to pay the travel expenses of the General Secretary to ICSU and other meetings, to purchase a wreath on the occasion of Professor Danjon's decease, etc.

12. *Projects authorized by the Executive Committee*. No such projects were initiated during 1968–1969. The contribution of $ 1 100.00 towards the publication of the Information Bulletin for the Southern Hemisphere was a payment which had become due in 1967.

13. *UNESCO Contracts*. The payments under the terms of the UNESCO contracts were as follows:

	$
Balance from contract NS/3034/66 (grant to Mr. De la Reza)	1 500.00
First instalment according to contract NS/0847/68 (Summer School Arcetri)	6 000.00
Last instalment under the terms of the same contract	2 000.00
First instalment according to contract SC/1492/69 (Summer School Hyderabad)	6 629.92
Total	16 129.92

The balance under the terms of contract SC/1492/69 is payable in 1970.

Details of payments in respect of the contracts are given in the reports to UNESCO. They are available to the Finance Committee for examination.

14. *Reimbursement of Loan from ICSU*. The ICSU loan to the Union of $ 10 000 has to be repaid as follows under the terms of contract: $ 2 000 in 1969; $ 3 000 in 1970; $ 3 000 in 1971 and $ 2 000 in 1972. Thus, the payment of $ 4 000 covers the instalment for 1969 and two thirds of that for 1970.

15. *Reimbursement to Savings Accounts*. Under the budgetary provisions, $ 38 500 had to be repaid to the IAU savings accounts until 31 December 1970. The amount actually reimbursed is $ 38 763.89. The excess of $ 263.89 is due to a difference between the ICSU and effective rate of exchange of the Dutch guilder. The difference between the $ 56 564.79 withdrawn from the savings accounts and the $ 38 763.89 repaid is $ 17 800.90.

16. *Other Expenses*. Reference is made to the comment on heading 7 under Receipts. The analysis of the amounts of $ 3 162.62 for 1967, $ 30 111.19 for 1968 and $ 20 395.07 for 1969 is as follows:

Other Expenses in 1967	$
Transfer from current account at Chemical Bank to that at ČSOB	2 002.75
Refund to Taiwan of erroneously paid contribution	195.96
Refund of excess payment for reprints	16.85
Discount	0.50
Bank Charges	11.91
Refund of personal deposit	52.13
Loan to staff member	882.52
Total	3 162.62

Other Expenses in 1968	$
Transfer from current account at Chemical Bank to that at ČSOB	6 002.95
Discount on Canadian cheques	29.48
Bank expenses	28.01
Transfer from current account at Société Générale to current accounts at ČSOB and Amsterdam-Rotterdam Bank	20 429.51
Transfer from current account at Barclays Bank to that at Amsterdam-Rotterdam Bank	3 604.53
Exchange losses	16.71
Total	30 111.19

Other Expenses in 1969	$
Transfer from current account at Chemical Bank to that at ČSOB	5 000.00
Advance payment towards 32nd meeting of Executive Committee (reimbursed)	282.23
Reimbursement to Taiwan of erroneously paid contribution	295.77
Reimbursement to ICSU of excess subvention	5 000.00
Transfer from current account at Amsterdam-Rotterdam Bank to that at ČSOB	6 063.33
Reimbursement to LOC Symp. 36 of payment to Glusheva	159.00
Travel and living expenses of General Secretary in connection with attendance at IAB General Assembly	263.20
Transfer from current account at ČSOB to Petty Cash	350.00
Transfer from current account at ČSOB of COSPAR deposit	2 414.00
Loan	134.00
Transfer from current account at Barclays Bank to that at Société Générale	362.64
Bank expenses and losses	70.90
Total	20.395.07

Recapitulation: $

Other Expenses 1967		3 162.62
	1968	30 111.19
	1969	20 395.07
	Total	53 668.88
Loss through devaluation		134.82
	Grand total	53 803.70

The reimbursement of the ICSU loan (partial) and that to the savings accounts is entered in headings 14 and 15 respectively under Payments. A transfer of $ 5 000 from the current account at the Chemical Bank has not yet been put to the credit of the IAU current account with the ČSOB.

AUDIT OF THE ACCOUNTS 1967–1969

Conformément à la mission que vous avez bien voulu nous confier, nous vous présentons le résultat de notre examen des comptes de l'Union Astronomique Internationale pour la période du 1ᵉ janvier 1967 au 31 décembre 1969.

L'examen, notamment, a porté sur les documents suivants:
—comptes des recettes et dépenses pour l'année 1967 vérifiés par nous-mêmes
—comptes des recettes et dépenses pour l'année 1968 vérifiés par nous-mêmes
—comptes des recettes et dépenses pour l'année 1969 vérifiés par nous-mêmes
—comptes des recettes et dépenses et situation des comptes bancaires pour la période de trois ans: 1967, 1968 et 1969.

Les comptes ont été comparés avec les cahiers de comptes) reçus et pièces comptables diverses.

Nous certifions que ces comptes sont en accord complet avec les informations et explications que nous avons demandées pour nous permettre d'effectuer notre examen.

Le compte, ci-inclus, des recettes et dépenses, résumant les comptes des années 1967, 1968 et 1969 est la représentation exacte des dépenses et recettes de l'Union Astronomique Internationale pour la période du premier janvier 1967 au trente et un décembre 1969.

A Paris, le 6 Mars 1970

ROGER BACLE m.p.
H.E.C. – Expert-Comptable

CONCLUSION

The assets of $ 113 261.57 as on 31 December 1969 (the $ 66.07 of the Nice petty cash included) compare favourably with those of $ 89 763.10 as on 1 January 1967; the active balance is $ 23 498.47.

The budget approved by the XIIIth General Assembly for the years 1968–1970 provided for an expenditure of $ 225 300 balanced by an income of $ 225 300. This budget, compared with the actual receipts and payments of the years 1968–1969, gives the balance for the year 1970, as follows:

	Income in $	Expenditure in $
Budget 1968–1970, as approved by the XIIIth General Assembly (UNESCO contracts not included)	225 300.00	225 300
Actual figures for 1968 and 1969 (adjusted and UNESCO contracts excluded)	163 103.53	146 538
Balance for 1970	62 196.47	78 762
Available for payments in 1970	16 565.35	
Total	78 762.00	

The budgetary estimates for 1970, on the other hand, are as follows:

Budget 1970 (UNESCO contracts not included)

Receipts in $			Payments in $	
1. National subscriptions	52 000		1. Administrative office	21 309.50
2. Sales of publications	4 200		2. Subscription to ICSU	1 040.00
4. Interest on reserves in savings banks	2 300		3. Expenses of Commissions	500.00
5. UNESCO subvention through ICSU	18 000		4. Specific projects of Commissions	19 151.59
			5. XIVth General Assembly	21 000.00
			6. IAU publications	17 000.00
	Total	76 500	8. Organization of symposia	14 500.00
			10. Inter-Union commissions	2 200.00
			11. Representation expenses	500.00
Excess of payments over receipts		21 801.09	14. Reimbursement of ICSU loan	1 000.00
			16. Bank losses	100.00
	Total	98 301.09	Total	98 301.09

Estimated excess of payments over receipts for 1970	$ 21 801.09
Still available from approved budget for payments in 1970	$ 16 565.53
Estimated excess of expenditure over income	$ 5 245.56

Explanatory Notes

Item 4, Specific projects of Commissions has been corrected for the actual balance in favour of Commission No. 38 and the UNESCO/ICSU subvention of $ 2 000 it is to receive in 1970, as well as for a balance of $ 1 500 due from 1967 for the publication of the *Card Catalogue of Clusters and Associations.* Item 5 of Payments (XIVth General Assembly) includes the amounts of $ 3 000 shifted from the budget of Commission No. 38 in favour of young astronomers wishing to attend the General Assembly, and of $ 2 000 to obtain for the same purpose from UNESCO. Item 6, IAU Publications: the sum of $ 17 000 is made up of the estimated payment of $ 14 000 for the production of *Transactions* volume XIVA and of that for printing the Agenda for the XIVth General Assembly, the Report of the Executive Committee, the Agenda of Commissions and *Information Bulletin* No. 24.

The Finance Committee at the XIVth General Assembly will be asked to endorse the supplementary budget for 1970, as proposed above, and to approve a comprehensive budget for the years 1971–1973.

APPENDIX

A. LIST OF COUNTRIES ADHERING TO THE UNION

The following is a list of the 44 countries that adhered to the Union in December 1969, giving also the year of admission, the approximate number of Members, and the Adhering Organizations.

Country	Year	Members	Adhering Organizations
Argentina	1927	16	Comité Nacional de Astronomía, La Plata.
Australia	1939	46	Australian National Committee for Astronomy, Sydney.
Austria	1955	12	Österreichische Akademie der Wissenschaften, Wien.
Belgium	1920	46	Académie Royale de Belgique, Bruxelles – Koninklijke Vlaamse Academie van België, Brussel.
Brazil	1961	10	Conselho Nacional de Pesquisas, Rio de Janeiro.
Bulgaria	1957	7	Bulgarian Academy of Sciences '7 November', Sofia.
Canada	1920	59	Canadian National Committee for the IAU, Department of Energy, Mines and Resources, Ottawa.
Chile	1947	5	Universidad de Chile, Santiago de Chile.
Colombia	1967	1	Academia Colombiana de Ciencias Exactas, Fisicas y Naturales, Bogotá.
Czechoslovakia	1922	38	National Committee of Astronomy of Czechoslovakia, Czechoslovak Academy of Sciences, Praha.
Denmark	1922	14	Danish National Committee for Astronomy, det Kongelige Danske Videskabernes Selskab, København.
Finland	1948	10	Finish Academy of Sciences and Letters, Helsinki.
France	1920	147	Comité National Français d'Astronomie, Académie des Sciences, Paris.
G.D.R.	1951	31	Nationalkomitee für Astronomie der DDR, Deutsche Akademie der Wissenschaften zu Berlin.
G.F.R.	1951	109	Rat Westdeutscher Sternwarten zu München.
Greece	1920	16	Greek National Committee for Astronomy, Academy of Athens, Athens.
Hungary	1947	10	Hungarian Academy of Sciences, Budapest.
India	1964	40	Indian National Committee for the International Astronomical Union, Indian National Science Academy, New Delhi.
Iran	1969	0	Tehran University, Tehran.
Ireland	1947	4	National Committee of Astronomy, Royal Irish Academy, Dublin.
Israel	1954	5	The Israel Academy of Sciences and Humanities, Jerusalem.
Italy	1921	54	Consiglio Nazionale delle Ricerche, Roma.
Japan	1920	73	National Committee of Astronomy of Japan, Science Council of Japan, Tokyo.
Korea (PDR)	1961	3	Academy of Sciences, Phyongyang.
Mexico	1921	10	Universidad Nacional de Mexico, Mexico D.F.
Netherlands	1922	48	Koninklijke Nederlandse Academie van Wetenschappen, Amsterdam.
New Zealand	1964	2	Royal Society of New Zealand, Wellington.
Norway	1922	10	Det Norske Videnskaps-Akademi, Oslo.
Poland	1922	38	Polska Akademia Nauk, Warsawa.

Country	Year	Members	Adhering Organizations
Portugal	1924	12	Seccao Portuguesa das Uniones Internacionais Astronómica, Geodésica e Geofisica, Lisboa.
Roumania	1928	13	Conseil Scientifique de l'Observatoire de Bucarest, Bucarest.
South Africa	1938	9	Council for Scientific and Industrial Research, Pretoria.
Spain	1922	15	Comision Nacional de Astronomia, Madrid.
Sweden	1925	37	Kungl. Vetenskaps-Akademien, Stockholm.
Switzerland	1923	20	Schweizerische Naturforschende Gesellschaft, Basel.
Taiwan (Republic of China)	1959	9	Academia Sinica, Nankang, Taipei.
Turkey	1961	11	Türk Astronomi Dernegi, Istanbul.
U.A.R.	1925	8	Astronomical Center at Helwan, Egyptian Region.
United Kingdom	1920	189	The Royal Society, London.
U.S.A.	1920	553	The National Academy of Sciences, Washington D.C.
U.S.S.R.	1920	261	Academy of Sciences of the U.S.S.R., Moscow.
Vatican City State	1932	6	Pontificia Academia delle Scienze, Città del Vaticano.
Venezuela	1953	1	Academia de Ciencias Fisicas, Matematicas y Naturales, Caracas.
Yugoslavia	1935	13	Conseil Fédéral pour la Coordination de la Recherche Scientifique, Belgrade.

B. LIST OF DECEASED MEMBERS

The Executive Committee regrets to record the loss by death of the following 29 Members of the Union (at the date of 28 February 1970).

Mrs. Z. N. Aksent'eva	9 April 1969	C. C. Kiess	16 October 1967
H. D. Babcock	8 April 1968	A. Kohlschütter	28 May 1969
A. Birkenmajer	30 September 1967	J. J. Kubikowski	11 November 1968
G. Demetrescu	15 June 1969	N. I. Kučerov	17 October 1964
A. J. Deutsch	11 November 1969	A. I. Lebedinskij	8 September 1967
E. J. Dijksterhuis	18 May 1965	Mrs. S. L. McDonald	6 December 1963
J. Dufay	5 November 1967	A. V. Markov	19 November 1968
J. H. Focas	3 January 1969	O. F. W. Mathias	4 February 1969
C. W. Cartlein	20 December 1965	R. W. Michie	27 March 1969
E. Gullón	9 July 1968	Miss R. J. Northcott	29 July 1969
S. E. Hajkin	30 July 1968	Mrs. S. V. Romanskaja	26 November 1969
L. G. Henyey	18 February 1970	L. I. Semenov	22 March 1965
E. Hertzsprung	21 October 1967	S. C. Venter	26 January 1968
C. Hoffmeister	2 January 1968	M. H. Wrubel	26 October 1968
J. Ikaunieks	27 April 1969		

C. LIST OF IAU PUBLICATIONS

Transactions of the IAU, volume XIIIA, pp. CVI + 1048, price $ 34.75 (Members $ 23.15)

Editor: L. Perek. Published by D. Reidel, Publishing Company, Dordrecht, Holland.

Transactions of the IAU, volume XIIIB, pp. IX + 309, price $ 9.75 (Members $ 6.50)

Edited and published as *Transactions*, volume XIIIA.

Highlights of Astronomy as presented at the XIIIth General Assembly of the IAU 1967, pp. X + 548, price $ 26 (Members $ 17.50)

Edited and published as *Transactions*, vol. XIIIA.

IAU Symposium Volumes

No. 27. *The Construction of Large Telescopes*, pp. XII + 246, price £ 3-10-0d (Members £ 2-7-0d)

Editor: D. L. Crawford. Published by Academic Press Inc. (London) Ltd.

No. 28. *Aerodynamic Phenomena in Stellar Atmospheres*, pp. XXVII + 483, price £ 7-0-0d (Members £ 5-0-0d)

Editor: R. N. Thomas. Published by Academic Press Inc. (London) Ltd.

No. 29. *Non-Stable Phenomena in Galaxies*, pp. 271 (in Russian only) price 2.40 roubles

Editor: M. A. Arakeljan. Published by the Publishing House of the Academy of Sciences of Armenian SSR, Yerevan.

No. 30. *Determination of Radial Velocities and their Applications*, pp. XIII + 262, price £ 4-4-0d (Members £ 2-16-0d)

Editors: A. H. Batten and J. F. Heard. Published by Academic Press Inc. (London) Ltd.

No. 31. *Radio Astronomy and the Galactic System*, pp. X + 501, price £ 7-0-0d (Members £ 5-0-0d)

Editor: H. van Woerden. Published by Academic Press Inc. (London) Ltd.

No. 32. *Continental Drift, Secular Motions of the Pole, and Rotation of the Earth*, pp. VI + 107, price $ 6.40, in U.S.A., and Canada $ 7.20 (Members $ 4.30 and 4.80 respectively)

Editors: Wm. Markowitz and G. Guinot. Published by D. Reidel Publishing Company, Dordrecht, Holland.

No. 33. *Physics and Dynamics of Meteors*, pp. XIII + 525, price $ 25.– (Members $ 16.70)

Editors: L. Kresák and P. M. Millman. Published by D. Reidel Publishing Company, Dordrecht, Holland.

No. 34. *Planetary Nebulae*, pp. XVI + 469, price $ 22.50 (Members $ 14.80)

Editors: D. E. Osterbrock and C. R. O'Dell. Published by D. Reidel Publishing Company, Dordrecht, Holland.

No. 35. *Structure and Development of Solar Active Regions*, pp. xii + 608, price $ 27.80 (Members $ 18.50)

Editor: K. O. Kiepenheuer. Published by D. Reidel Publishing Company, Dordrecht, Holland.

IAU Colloquium Volumes

Commission No. 14. Proceedings of the International Conference on Spectroscopy. Cosponsored by IUPAC and IUPAP

Chairman of Organizing Committee: R. K. Asundi. Two mimeographed volumes. Bhabha Atomic Research Centre, Modular Laboratories, Trombay, Bombay 74, India.

Commissions Nos. 14, 29, 35 and 36. Proceedings of the Colloquium on Late-type Stars

Editor: Mrs. M. Hack, Osservatorio Astronomico di Trieste, 1966.

Commission No. 26. Proceedings of the Colloquium on the Evolution of Double Stars

Editor: J. Dommanget, Observatoire Royal de Belgique, Communications, Série B. Nr. 17, 1967.

Commission No. 36. Proceedings of the Colloquium on the Blanketing Effect

Editor: K. H. Böhm. *Journal of Quantitative Spectroscopy and Radiative Transfer* 6, No. 5 (1966).

Commission No. 10. Proceedings of the Colloquium on the Fine Structure of the Solar Atmosphere

Editor: K. O. Kiepenheuer. *Forschungsberichte* 12, Deutsche Forschungsgemeinschaft, Franz Steiner Verlag, Wiesbaden, 1966.

Commission No. 33. Proceedings of the Colloquium on Les nouvelles méthodes de la dynamique stellaire

Editors: F. Nahon and M. Hénon. Editions du
Centre National de la Recherche Scientifique,
Paris, 1967.

Commissions Nos. 29 and 42. Proceedings of
the Second Trieste Colloquium on Astrophysics, Mass Loss from Stars

Editor: Mrs. M. Hack. D. Reidel Publishing
Company, Dordrecht, Holland, 1968.

Commissions Nos. 7, 33 and 37. Proceedings
of the Colloque sur les problèmes des n-corps

Editor: B. Morando. Editions du Centre National de la Recherche Scientifique, Paris, 1968.

Publications

by, or on behalf of, IAU Commissions and by Astronomical Services

Commission No. 6. IAU (Telegram Bureau)
Circulars Nos. 2001–2196

Issued by the IAU Central Bureau for
Astronomical Telegrams.

Commission No. 10. Heliographic Charts of
the Photosphere (Volume XIII, Nos. 1, 2, 3)

Published by the Eid. Sternwarte in Zürich,
under the direction of M. Waldmeier.

Commission No. 10. *Cartes Synoptiques de la
Chromosphère Solaire* (Volume V, fascicule 1)

Prepared by Observatoire de Paris at Meudon,
under the direction of R. Michard.

Commission No. 16. *Newsletter* No. 1–6

Published by the President of Commission.

Commission No. 20. Minor Planet Circulars
Nos. 2695–3022

Mimeographed circulars issued by the Minor
Planet Center, Cincinnati.

Commission No. 23. Information regarding
the different sections (Hyderabad, Perth-Edin-
burgh, Oxford-Potsdam, Melbourne, Catania,
Tacubaya, Uccle-Paris, etc.) may be obtained
from the respective Observatories

Prepared by the Observatories in question, in
co-operation with Commission No. 23.

Commission No. 26. *Circulaire d'Information*
Nos. 42–49

President: P. C. Couteau, Observatoire de
Nice, Nice.

Commission No. 27. *Information Bulletin on
Variable Stars* Nos. 166–396

Prepared by L. Detre, Konkoly Observatory,
Budapest.

Commission No. 31. Annual Report of the
BIH, 1967, 1968, 1969

Published by the Bureau International de
l'Heure, Paris.

Commission No. 35. *Circular letters* Nos. 1–7

Editor: A. Massevich, Astronomical Council
of the U.S.S.R. Academy of Sciences, Moscow.

Commission No. 41. *Bibliography of books
published in 1966 on the History of Astronomy,*
1967

Editor: P. G. Kulikovskij, Astronomical Council of the U.S.S.R. Academy of Sciences,
Moscow.

*Bibliography of books published in 1967 on the
History of Astronomy,* 1968

Commission No. 41. *Information Circular* Nos.
14–17

Published by the President of the Commission.

Monthly Notes of the IPMS Nos. 11 (1966) and
No. 12 (1969)

Published by International Polar Motion
Service, Mizusawa-shi.

IAU Information Bulletin Nos. 18–23

Prepared by the General Secretary.

PART 2

INAUGURAL CEREMONY
REPORT OF THE GENERAL ASSEMBLY

INAUGURAL CEREMONY

Tuesday, 18 August at 10^h 30^m

The Inaugural Ceremony of the XIVth General Assembly of the International Astronomical Union took place in The Dome at Brighton, in the presence of a distinguished gathering of representatives of the Government of Great Britain, the Town of Brighton, the University of Sussex, the Royal Society, the National Organizing Committee, and many other organizations.

The Union was honoured by the presence of Her Excellency, Her Majesty's Secretary of State for Education and Science, The Right Hon. Mrs Margaret Thatcher, M.P., The Mayor of Brighton, Alderman H. Nettleton, The Vice-Chancellor of the University of Sussex, Professor Asa Briggs, The President of the Royal Society, The Right Hon. The Lord Blackett, O.M., C.H., P.R.S., and the Chairman of the National Organizing Committee, Sir Bernard Lovell, O.B.E., F.R.S.

Dr D. H. Sadler, O.B.E., Chairman of the Local Organizing Committee, acted as chairman for the ceremony.

After a few words of welcome to Members and guests, the Chairman for the Ceremony called upon Alderman H. Nettleton, The Mayor of Brighton, to address the Assembly.

ADDRESS BY THE MAYOR OF BRIGHTON, ALDERMAN H. NETTLETON

'Mr. Chairman, Ladies and Gentlemen,

It is a great pleasure for me, as Mayor of Brighton, to have this opportunity of extending a warm welcome to such a large assembly of distinguished visitors.

I need hardly tell you that we greatly value the presence of a great conference of this kind in our town. I think it is true to say that Brighton has already established itself as one of the principal centres for conferences of all kinds. The visit of the General Assembly of the International Astronomical Union can only confirm that opinion.

It is customary on occasions such as this, for a Mayor to remind his visitor of the attractions and amenities of his town. I do not propose to take up your time in this way today. I am sure you will be eager to take a little relaxation between your Conference sessions and to discover these attractions for yourselves.

May I remind you that this great building and the Eastern palace nearby were at one time the property of our King George IV. The Royal Pavilion, as the former palace is called, was the seaside home and this great hall was constructed for use as the Royal Stables – can you imagine this hall with a large fountain in the centre and with horse and coaches passing in and out. How times have changed.

We in Brighton owe much to King George IV. He left us some remarkable buildings which attract a tremendous amount of attention all over the world.

However, I must return to my task of welcoming you all to Brighton. I do indeed welcome you most heartily.

The Mayoress and I much enjoyed meeting so many of you last evening. It was a very stimulating occasion.

To all of you, I express my very best wishes for a most successful conference at Brighton. Come again as soon as you can.'

French translation:

Allocution du Maire de Brighton

'Monsieur le Président, Mesdames, Messieurs,

C'est un grand plaisir pour moi de souhaiter, en ma qualité de Maire de Brighton, la bienvenue à une aussi grande assistance de visiteurs éminents.

Il est à peine utile de vous dire combien nous apprécions qu'une aussi vaste assemblée puisse se réunir dans notre ville. Il est vrai que Brighton s'est déjà révélée comme l'un des principaux centres de réunions de toutes sortes. La Quatorzième Assemblée Générale de l'Union Astronomique Internationale ne peut que confirmer cette renommée.

Il est d'usage, en de telles occasions, que le Maire de Brighton fasse valoir auprès de ses visiteurs les agréments de sa ville. Mon propos n'est pas de prendre sur votre temps pour le faire. Je suis persuadé que vous désirerez tous profiter des moments de détente entre vos sessions de travail pour découvrir par vous-mêmes les attraits de notre cité.

Puis-je vous rappeler que le grand édifice où nous nous trouvons, et le Palais voisin, furent autrefois la propriété du Roi George IV. Le Pavillon – ainsi s'appelle ce palais – était sa résidence d'été, et cette immense salle fut construite pour abriter les Écuries Royales – imaginez-la telle qu'elle était autrefois avec, en son milieu, une grande fontaine et l'incessant va-et-vient des montures et des carrosses. Les temps ont bien changé!

Notre ville de Brighton doit beaucoup au Roi George IV. Grâce à lui, elle offre à l'admiration de ses visiteurs de remarquables édifices dont le rayonnement s'étend loin dans le monde.

Il me faut toutefois revenir à mon aimable tâche de vous souhaiter à tous la bienvenue à Brighton.

Ma femme, et moi-même, avons pris grand plaisir hier soir à accueillir un grand nombre d'entre vous. Ce fut pour nous une réunion très stimulante.

A vous tous, j'adresse mes vœux les plus chaleureux pour le succès complet de votre conférence. Revenez dans notre ville aussi souvent que vous le pourrez.'

Dr Sadler expressed his thanks to Alderman Nettleton for his address and invited Professor Asa Briggs, Vice-Chancellor of the University of Sussex, to address the gathering.

ADDRESS BY THE VICE-CHANCELLOR OF THE UNIVERSITY OF SUSSEX, PROFESSOR ASA BRIGGS

'As Vice-Chancellor of the University of Sussex, I would like to offer the warmest of welcomes to the General Assembly of the International Astronomical Union.

We are a young university in terms of any time scale – astronomical or historical – young and, I believe, lively and creative – certainly keenly interested in the international dimensions of life and thought (not to speak of research) in the twentieth century.

We are proud, too, of our close association with the Royal Observatory at Herstmonceux, and of the fact that from the start astronomy has figured actively in the range of our academic preoccupations.

There is no subject in the world which is more international in its appeal and in its spirit of cooperation than astronomy, and what other subject could possibly draw us closer together not only as academic colleagues but as citizens of mankind?

There is also no town in the world that has a richer experience in the pleasure of entertaining than Brighton. We hope that your stay here will be very happy, as well as very rewarding, and that you will have the opportunity while here not only to contemplate or to speculate about the universe, but to see for yourself what is here on the ground – in the University, in the town, and in the surrounding and very beautiful Sussex countryside.

Brighton has not always been a centre of educational advance. An old nineteenth-century guide book told visitors to the town 'instead of poring through long cylinders at the stars, let them have licence to admire the eyes of the ladies.'

Today, all this has changed. We are profiting as a university from our newness to foster educational innovation – in the curriculum, in our teaching methods, and in academic and social organisation – while at the same time insisting upon the interaction of effective teaching and exploratory research.

From whatever part of the world you come, we hope that while you are with us you will feel that our University is yours.'

French translation:

Allocution du Vice-Chancelier

'En ma qualité de Vice-Chancelier de l'Université du Sussex, j'adresse mes vœux les plus chaleureux de bienvenue à l'Assemblée Générale de l'Union Astronomique Internationale.

Notre université est jeune par l'âge – que l'on utilise l'échelle historique des temps ou l'échelle astronomique – mais aussi, je veux le croire, vivante et créatrice, et sans aucun doute passionnément attirée par les dimensions internationales de la vie et de la pensée du 20ème siècle (pour ne pas parler de la recherche).

Nous sommes fiers également de l'étroite collaboration qui nous unit à l'Observatoire Royal de Herstmonceux, et du fait que l'astronomie figure en bonne place parmi nos préoccupations académiques.

Il n'est pas dans le monde de discipline plus internationale par son attrait et son esprit de coopération que l'astronomie; et quelle autre, en effet, pourrait nous rendre plus proches les uns des autres, non seulement en tant que collègues universitaires, mais comme citoyens du monde?

Il n'est pas non plus au monde une ville aussi riche que Brighton d'une longue expérience en matière de distraction. Nous espérons que votre séjour parmi nous sera heureux et profitable et que vous n'aurez pas seulement l'occasion de contempler l'univers et de spéculer à son sujet, mais aussi de voir ce qui est sur la terre – à l'Université, à Brighton, dans ses environs, et dans notre si belle campagne du Sussex.

Brighton n'a pas toujours été un centre universitaire. Un vieux guide du 19ème siècle vantait à ses visiteurs les agréments de la ville et, au lieu d'observer les étoiles à travers des lunettes, les incitait plutôt à admirer les beaux yeux des dames!

Aujourd'hui tout est changé. Notre université prend avantage de sa jeunesse pour innover dans le domaine de l'éducation – dans les programmes, dans nos méthodes d'enseignement, dans l'organisation sociale et universitaire – tout en insistant sur la nécessité d'une interaction entre la pratique de l'enseignement et cette exploration qu'est la recherche.

De quelque partie du monde que vous veniez, nous espérons que, pendant votre séjour parmi nous vous vous considererez notre Université comme la vôtre.'

The Chairman thanked Professor Briggs and then followed a musical interlude played on the organ by Lady Jeans. On the programme were: Sir William Herschel (1738–1822), Sonata V in D minor and Prelude for Diapasony in C major, and J. S. Bach (1685–1750) Prelude and Fugue in G major.

The Chairman, Dr D. H. Sadler, complimented Lady Jeans on her brilliant performance and then invited Lord Blackett, President of the Royal Society, to take the floor.'

ADDRESS BY THE PRESIDENT OF THE ROYAL SOCIETY, THE RIGHT HON. THE LORD BLACKETT, O.M., C.H., P.R.S.

'Secretary of State, Mr Mayor, Mr Vice-Chancellor, Mr President,

In the name of the Royal Society it is my pleasure to welcome you to this fourteenth General Assembly of the International Astronomical Union. During the 50 years of the history of the Union we have had this opportunity on only one previous occasion – in 1925 when the second General Assembly was held in Cambridge. At that moment our knowledge of the universe was being transformed as the great discoveries then in progress at Mt. Wilson were revealing the structure of the

Milky Way system and the nature of the extragalactic nebulae. In the 45 years which have intervened since that time there have been startling revolutions in astronomical techniques. The developments in radioastronomy and in astronomical observations from space vehicles have now given astronomers the ability to observe the universe over wide regions of the electromagnetic spectrum. The subject of astronomy has never before been in such an exciting phase where discoveries leave us spellbound in their breadth and scope.

I am impressed, too, by the extent of the technical developments now embraced in the astronomical researches. Indeed, the great advances in the subject today are the direct result of the close interplay between the astronomers, the technologists and the industries of our contemporary societies.

One of my distinguished predecessors in this office of President, Sir Joseph Banks, objected most strongly to the formation of the Royal Astronomical Society 150 years ago, on the grounds that it would be harmful to the Royal Society. He would, I am sure, be interested to see the situation today in which the relations are so cordial that the Royal Society is formally the host organisation for this assembly, having discharged its responsibility through the Royal Society National Committee for Astronomy, all the members being Fellows of the Royal Astronomical Society.

In recent years the Royal Society has given much attention to the international exchange and collaboration amongst scientists. Ever since its formation the International Astronomical Union has been outstanding in the extent of the co-operation between its members throughout the world; and this is one more reason why I am so glad to be able to extend a welcome to you today. Many of the fundamental aspects of astronomy – for example the astronomical and navigational ephemerides – are the results of international collaboration amongst your members. It is, indeed, encouraging to see the great extension of this world wide collaboration to many of the contemporary developments in astronomy and astrophysics.

In welcoming you today on behalf of the Royal Society, may I express the hope that your deliberations will continue to be in the great spirit which has pervaded your whole history and may I wish your Union continued success in the future'.

French translation:

Allocution du Président de la Royal Society

'Madame le Secrétaire d'Etat, M. le Maire, M. le Vice-Chancelier, M. le Président,

Au nom de la Royal Society, j'ai le grand plaisir de vous souhaiter la bienvenue à la Quatorzième Assemblée Générale de l'Union Astronomique Internationale. Depuis la fondation de l'Union, voici 50 ans, ce n'est qu'une seule fois que nous avons eu l'occasion de le faire précédemment, c'était en 1925, à Cambridge, lors de la Seconde Assemblée Générale. A cette époque notre connaissance de l'univers était bouleversée par les grandes découvertes faites au Mont Wilson et qui nous ont révélé la structure du système de la Voie Lactée et la nature des nébuleuses extragalactiques. Durant les 45 années qui ont suivi, d'étonnantes révolutions ont affecté les techniques astronomiques. Le développement de la radioastronomie, celui des observations faites à partir d'engins spatiaux ont permis aux astronomes d'observer l'univers dans les domaines extrêmement étendus du spectre électromagnétique. L'astronomie n'a jamais été dans une phase plus excitante que maintenant, et ses découvertes nous fascinent par leur amplitude et leur portée.

Je suis impressionné également par l'extension que prennent les développements techniques liés aux recherches astronomiques. Certes les grands progrès d'aujourd'hui sont le résultat direct de l'étroite interaction qui existe entre les astronomes, les technologues et les industries de notre société contemporaine.

Sir Joseph Banks, qui fut l'un de mes distingués prédécesseurs à la Présidence de la Société, s'élevait très vigoureusement il y a 150 ans contre la création de la Royal Astronomical Society, car il pensait qu'elle ferait du tort à la Royal Society. Je suis persuadé qu'il serait heureux de constater qu'aujourd'hui nos relations sont si cordiales que c'est la Royal Society qui officiellement accueille cette assemblée, mais qu'elle a délégué ses responsabilités à son Comité National d'Astronomie dont tous les membres sont également membres de la Royal Astronomical Society.

Ces dernières années, la Royal Society a accordé toute son attention aux échanges et à la collaboration internationale entre scientifiques. Depuis sa création, l'Union Astronomique Internationale a toujours fait prévaloir une politique de large coopération entre ses membres dans le monde entier; et c'est une raison supplémentaire pour moi d'avoir la joie de vous souhaiter la bienvenue aujourd'hui. Bien des aspects fondamentaux de l'astronomie – par exemple les éphémérides astronomiques et nautiques – résultent de la collaboration internationale entre vos membres. Il est vraiment encourageant de constater que cette vaste collaboration internationale s'étend aux progrès actuels de l'astronomie et de l'astrophysique.

En vous accueillant aujourd'hui au nom de la Royal Society, permettez-moi d'exprimer l'espoir que vos délibérations se poursuivront avec l'élévation d'esprit qui a toujours prévalu au cours de votre histoire, et de souhaiter à votre Union de continuer dans l'avenir son fructueux développement.'

Dr Sadler expressed his thanks to Lord Blackett for his address, and invited Sir Bernard Lovell, Chairman of the National Organizing Committee, to speak.

ADDRESS BY THE CHAIRMAN OF THE NATIONAL ORGANIZING COMMITTEE, SIR BERNARD LOVELL, O.B.E., F.R.S.

'Secretary of State, Mr Mayor, Mr Vice-Chancellor, Mr President,

Lord Blackett has explained that the Royal Society has discharged its responsibility as the host organisation for this assembly, through its National Committee for Astronomy; as the present Chairman of this Committee it is my pleasure to welcome you today. I should perhaps hasten to add that this National Committee quickly passed on its responsibilities to a National Organising Committee, but that nearly all the work has been done by the Local Organising Committee and the Finance and Policy Committee. To the Chairmen, Mr Sadler and Professor Tayler, of these two small committees and their members I am deeply grateful for dealing so admirably with the mammoth task of making the practical arrangements for this Assembly.

This is, indeed, a memorable year for British astronomy. In the Royal Astronomical Society we are celebrating the 150th anniversary of our foundation. In the 50 years history of this Union, Fellows of the Royal Astronomical Society have been prominent. We claim three Presidents – Dyson, Eddington, and Spencer Jones, and three General Secretaries – Fowler, Stratton and Sadler. Furthermore, it is surely a most happy circumstance that we have amongst us Lady Jeans, the widow of one of Britain's most eminent astronomers, who will be performing music composed by Sir William Herschel, the first President of the Royal Astronomical Society in 1820.

During our recent history there have been periods of gloom and despondency in British astronomy when it seemed that we would be unworthy of our heritage. On the occasion of the previous meeting of the Union in England in 1925 it seemed that the possibility of further significant contributions from British observational astronomers had vanished forever. Many influential voices were to be heard, asking that we should concentrate our efforts on theoretical work, and leave the observing to those who worked under clearer skies. It is our good fortune that the developments in radio astronomy removed the handicap of our location on Earth, and in recent years we have been able to make our own contributions to the tremendous advances in man's understanding of the universe.

These developing possibilities appropriately coincided with an entirely new outlook by the State on the support of the astronomical sciences. British astronomers, today, are full of appreciation and praise for the support which they receive from Her Majesty's Government through you, Secretary of State. We are now benefiting to the extent of more than £ 3 million per year for the support of ground based optical and radio astronomy alone – and that is equal to one half of this country's expenditure on all forms of research and development 35 years ago when I began my career. It is a remarkable change in our fortunes enabling us to plan our national and international researches as a major element in the development of world astronomy.

It is therefore in the spirit of hope and confidence shared by my colleagues whom I represent that

I welcome you today. We look forward eagerly to the proceedings of these days and promise our fullest co-operation in the future affairs of the Union.'

French translation:

Allocution de Sir Bernard Lovell

'Madame le Secrétaire d'Etat, M. le Maire, M. le Vice-Chancelier, M. le Président,

Lord Blackett vient de vous dire que la Royal Society avait dévolu à son Comité National d'Astronomie la responsabilité d'accueillir cette Assemblée. Aussi est-ce avec plaisir qu'en ma qualité de Président de ce comité je vous souhaite la bienvenue. Je me hâte d'ajouter que ce Comité National a bien vite passé ses pouvoirs à un Comité National d'Organisation, mais qu'en fait c'est le Comité Local d'Organisation qui a fait presque tout le travail, avec le Comité des Finances et d'Orientation. Je suis profondément reconnaissant à M. Sadler et au Professeur Tayler, Présidents de ces deux comités, ainsi qu'aux membres de ces comités, pour avoir su si admirablement mener à bien la tâche gigantesque que représente l'organisation pratique de cette Assemblée.

1970 est vraiment une année dont l'astronomie britannique se souviendra. La Royal Astronomical Society a célébré le 150ème anniversaire de sa fondation. Au cours des 50 années de son histoire, plusieurs membres de la Royal Astronomical Society ont tenu au sein de l'Union des rôles prépondérants. Trois d'entre eux ont été Présidents de l'Union – Dyson, Eddington et Spencer Jones – et trois autres Secrétaires Généraux – Fowler, Stratton et Sadler. De plus, un heureux concours de circonstances nous permet d'avoir parmi nous Lady Jeans, qui fut la compagne d'un de nos astronomes les plus éminents et qui interprétera ce soir pour nous des œuvres de Sir William Herschel qui fut en 1820 le premier Président de la Royal Astronomical Society.

Au cours de son histoire récente, l'astronomie britannique a connu des périodes d'abattement et de découragement: nous nous pensions indignes de cet héritage. Lors d'une précédente réunion de l'Union en Angleterre, en 1925, il semblait que l'espoir d'importantes contributions de la part des astronomes britanniques dans le domaine de l'Observation s'évanouissait pour toujours. Des voix autorisées nous demandaient d'abandonner les observations à ceux qui jouissaient d'un ciel plus limpide et de concentrer nos efforts vers les recherches théoriques. Heureusement pour nous les progrès de la radio-astronomie ont supprimé le handicap de notre position sur la Terre et récemment nous avons pu contribuer efficacement aux progrès considérables réalisés par l'homme dans sa compréhension de l'univers.

Cette période de développement coïncide heureusement avec l'attitude toute nouvelle de l'Etat en faveur des sciences astronomiques. Aujourd'hui, les astronomes britanniques apprécient vivement, avec toute leur gratitude, l'appui que le Gouvernement de Sa Majesté leur apporte par votre intermédiaire, Madame le Secrétaire d'Etat. Nous bénéficions maintenant de plus de 3 millions de livres sterling par an pour les seules recherches radioastronomiques et optiques – ce qui représente la moitié des crédits dont disposait toute la recherche pure et appliquée il y a seulement 35 ans, au début de ma carrière. Ce changement de notre fortune nous permet de considérer nos recherches nationales et internationales comme un élément primordial dans le développement mondial de l'astronomie.

Aussi est-ce avec un espoir et une confiance que partagent les collègues que je represente, que je vous souhaite la bienvenue aujourd'hui. Nous attendons avec impatience les réunions des prochains jours et nous vous promettons notre plus complète coopération à l'activité future de l'Union.

The Chairman thanked Sir Bernard Lovell for his words and called upon Mrs Margaret Thatcher, Her Majesty's Secretary for Education and Science, formally to open the XIVth General Assembly.

ADDRESS OF HER EXCELLENCY, HER MAJESTY'S SECRETARY OF STATE FOR EDUCATION AND SCIENCE, THE RIGHT HON. MRS. MARGARET THATCHER, M.P.

'Mr Mayor, Vice-Chancellor, Messrs Presidents, Ladies and Gentlemen, there can be few, if any, of the present company who attended the Second General Assembly of the IAU, the only previous one to be held in the UK, because that was as long ago as 1925. It is therefore literally a rare pleasure to welcome you all here today for this, the Fourteenth General Assembly. I do so in my capacity of Secretary of State for Education and Science, the member of Her Majesty's Government responsible for science, of which your subject, astronomy, is such an ancient and yet a continuingly exciting field.

For thousands of years the stars have excited the curiosity and wonder of mankind and the discoveries of astronomers of many lands have led successively to much wider understanding of fundamental physical processes and of the Universe. Contributions by this country have included those of Newton, Herschel and many others and, for a time, we had the largest telescopes in the world. Then, early in this century, the construction of really large telescopes in California gave astronomers there a near monopoly of this particular frontier. It is a healthy development that, within the past few years, the building of some 8 large telescopes, of 3.6 m aperture or more, has been undertaken. The resulting opportunities will undoubtedly enable astronomers of many lands to play a part in going forward from the magnificent achievements centred for half a century in California.

During practically the whole long history of astronomy observations were limited to wavelengths extending only just beyond the visible spectrum. During the last 20 years however, the range of wavelengths observable from the Earth's surface or from space vehicles has been extended by successive steps. It is difficult to realise now that radio astronomy is not yet 25 years old. Within that short span the radio astronomers have dramatically extended Man's basic knowledge of the Universe and there can be little doubt that a great deal remains for them to do. Large radio facilities now exist or are being provided in many countries and we are proud to have in this country two world ranking centres of radio astronomy.

By moving the observing platform beyond the earth's atmosphere, with the use of space vehicles over the past decade, much new knowledge is beginning to be discovered in the ultra violet and X-ray regions. And now initial steps in the infra red and millimetre regions seem at least as promising. Every past entry to new regions of the electromagnetic spectrum, particularly to radio waves, has dramatically extended our knowledge and understanding. There is every reason to foresee comparable dramatic extensions as a result of exploration of the ultra violet, X-ray, infra red and millimetre regions.

I would like to say something of the relevant work in this country for which I am departmentally responsible through the Science Research Council, which was created in 1965 and took over previously dispersed responsibilities for two world famous establishments and an important source of knowledge and expertise. The first of these is the Royal Greenwich Observatory at Herstmonceux, now approaching its tercentenary and the home since 1968 of the 98-inch Isaac Newton Telescope. The Observatory also controls the Observatory at the Cape of Good Hope and, more recently, the Radcliffe Observatory in Pretoria. The second establishment is the Royal Observatory Edinburgh with its outstation near Rome, under happy arrangements for collaboration with the University there.

The third acquisition of the Science Research Council was support of University astronomers including theoreticians, those active in the optical and radio parts of the spectrum and those using space vehicles for investigations in the ultra violet and X-ray regions.

Soon after its creation the Science Research Council declared a policy of giving priority to astronomy. The result has involved commitment to such major projects as participation with our Australian partners in the construction of the 150-inch Anglo-Australian Telescope; the undertaking of a 5 km radio telescope for Sir Martin Ryle's Group at Cambridge University; and a major upgrading of the 250 ft radio telescope of Sir Bernard Lovell's Group at Manchester University.

Then there are a 60-inch flux collector in the infra red now under construction and various rocket and satellite projects in the ultra violet and X-ray regions, some national, some in membership of the European Space Research Organisation and some in collaboration with the National Aeronautics and Space Administrations. And in the theoretical field the Science Research Council has participated, with private Foundations and the Universities of Cambridge and Sussex, in building up theoretical astronomy at those Universities.

I have gone into this detail to demonstrate the extent of the support being accorded to astronomy in this country. We hope in this way to play a full part in the advances which can be confidently expected in the years to come.

Provision of many new large optical telescopes and the growing exploration of the other, and hitherto largely inaccessible, regions of the spectrum have recently combined to make astronomy, despite its long history, one of the most exciting frontiers of knowledge. These are precisely the conditions necessary to ensure lively discussion at a Conference as international as yours and it is therefore with every assurance of your success that I have great pleasure in formally declaring open this the Fourteenth General Assembly of the International Astronomical Union.'

French translation:

'M. le Maire, M. le Vice-Chancelier, MM. les Présidents, Mesdames, Messieurs,

Bien peu d'entre vous ont participé à la Seconde Assemblée Générale de l'UAI, la seule qui ait eu lieu auparavant dans le Royaume-Uni, il y a fort longtemps, en 1925. Aussi est-ce réellement un plaisir rare que de vous souhaiter la bienvenue pour cette Quatorzième Assemblée Générale. Je le fais en ma qualité de Secrétaire d'Etat à l'Education et aux Sciences, en tant que membre du Gouvernement de Sa Majesté chargé de la Recherche Scientifique dont votre domaine, l'astronomie, est l'un des plus anciens et des plus excitants.

Depuis des milliers d'années, les étoiles ont éveillé la curiosité et l'émerveillement des hommes et les découvertes des astronomes, étape par étape, ont conduit à une meilleure compréhension de l'Univers.

La contribution de notre pays est attachée aux noms de Newton, d'Herschel et de beaucoup d'autres et, à une certaine époque, nous possédions les plus grands télescopes du monde. Puis, au début du siècle, la construction en Californie de très grands télescopes a pour ainsi dire donné aux astronomes de là-bas un quasi-monopole. Ce fut un heureux foisonnement, au cours des dernières années, que la construction de quelques huit grands télescopes, d'une ouverture de 3.6 m ou plus. Cela permettra de faire progresser dans le monde entier les magnifiques recherches concentrées pendant un demi-siècle en Californie.

Au long de la longue histoire de l'astronomie, les observations se sont pratiquement limitées aux longueurs d'onde s'étendant tout juste au-delà du spectre visible. La gamme des longueurs d'onde observables s'est élargie par étapes successives. Il est difficile aujourd'hui de se rendre compte que la radioastronomie n'a pas encore 25 ans! Dans ce court laps de temps, les radioastronomies ont très largement étendu notre connaissance fondamentale de l'Univers et nul doute qu'il ne leur reste encore une grande tâche à faire. De puissants moyens sont maintenant mis à la disposition de la radioastronomie dans de nombreux pays et nous sommes fiers de posséder dans notre pays deux centres de radioastronomie pouvant rivaliser avec les plus grands.

L'utilisation des engins spatiaux durant la dernière décade, qui a permis de poursuivre les observations au-delà de l'atmosphère terrestre, a marqué le début de nouvelles découvertes dans le domaine de l'ultraviolet et des rayons X. Et maintenant les premiers pas sont faits dans les régions millimétriques et infrarouges, et laissent apparaître des promesses sans doute aussi belles. Chaque progrès acquis dans le domaine nouveau du spectre électromagnétique, notamment celui des ondes radio, a considérablement élargi nos connaissances et notre compréhension. Aussi nous est-il permis de nous attendre à de semblables progrès dans l'exploration des régions millimétriques, infrarouges, ultraviolettes et des rayons X.

Je voudrais dire un mot des travaux entrepris dans notre pays et vis-à-vis desquels je suis responsable au nom du Gouvernement par l'intermédiaire du Science Research Council. Cet organisme, créé en 1965, a la responsabilité, autrefois, partagée, de deux établissements mondialement connus, et la charge d'une source importante de savoir et d'expérience.

Le premier de ces établissements est l'Observatoire Royal de Greenwich, à Herstmonceux, qui fêtera bientôt son tricentenaire et qui abrite le Télescope Isaac Newton de 98 pouces. L'Observatoire Royal de Greenwich contrôle également l'Observatoire du Cap de Bonne Espérance et, plus récemment, l'Observatoire de Ratcliffe à Prétoria. Le second établissement est l'Observatoire Royal d'Edinbourg, avec sa Station italienne installée près de Rome, grâce à une heureuse collaboration avec l'Université de cette ville.

Le Science Research Council assume une troisième tâche, la responsabilité des astronomes des Universités: théoriciens, chercheurs des domaines optiques et radio, ou ceux utilisant les engins spatiaux pour leurs recherches dans les domaines de l'ultraviolet et des rayons X.

Peu de temps après sa création, le Science Research Council a affirmé une politique donnant priorité à l'astronomie, ce qui eut pour résultat la collaboration à des projets importants tels que la construction, avec nos collègues autraliens, du télescope anglo-australien de 150 pouces; l'entreprise d'un radiotélescope de 5 km à l'Université de Cambridge, pour l'équipe de Sir Martin Ryle; et, à l'Université de Manchester, des améliorations essentielles du radiotélescope de l'équipe de Sir Bernard Lovell.

Nous avons aussi en construction un collecteur de radiation infrarouge de 60 pouces et sont en cours divers programmes de lancements de fusées et de satellites en vue d'observations dans l'ultraviolet et les rayons X, les uns purement nationaux, d'autres en collaboration avec l'ESRO ou avec la NASA.

Dans le domaine théorique, le Science Research Council a participé, avec les Universités du Sussex et de Cambridge et avec des Fondations privées, au développement de l'astronomie théorique dans ces Universités.

J'ai voulu entrer dans ces détails pour vous montrer l'appui que reçoit l'astronomie dans notre pays. Nous espérons ainsi jouer un rôle à part entière dans les progrès que l'on peut attendre avec confiance au cours des années à venir.

Le grand nombre de nouveaux télescopes optiques et les progrès réalisés dans des domaines spectraux jusqu'alors en grande partie inaccessible, font de l'astronomie, en dépit de sa longue histoire, l'un des domaines scientifiques les plus excitants. Ainsi sont réalisées toutes les conditions pour que les discussions durant cette assemblée internationale soient des plus passionnantes, et c'est avec la certitude de votre succès que j'ai le grand plaisir de déclarer officiellement ouverte la Quatorzième Assemblée Générale de l'Union Astronomique Internationale.'

The opening words of Mrs Margaret Thatcher, for which she was thanked by the Chairman, were followed by Lady Jeans' performance on the organ of J. C. Rinck's (1770–1846) Variations for organ on 'Ah vous dirais-je Maman'. The Chairman then invited Professor O. Heckmann, President of the International Astronomical Union, to speak.

ADDRESS BY THE PRESIDENT OF THE INTERNATIONAL ASTRONOMICAL UNION, PROFESSOR DR O. HECKMANN

'Ladies and Gentlemen:

May I propose to address our hosts in what appears to me the natural order, namely that in which they spoke. I ask your indulgence if I transgress protocol.

Alderman Nettleton, Professor Briggs, Lord Blackett, Sir Bernard Lovell, and Mrs Thatcher, I wish to express to each of you in turn the sincere gratitude of the International Astronomical Union for your words of welcome.

All participants in the General Assembly, whether Members of the union, Invited Participants or their Guests, will, I am certain, take full advantage of the magnificent attractions that the town of Brighton so generously provides for us. We are most grateful, Mr Mayor, for having made it possible for us to meet in this historic and delightful seaside town. Our work will benefit greatly from Brighton's excellent facilities set in the magnificent Sussex countryside.

To the Vice-Chancellor of the University of Sussex I can only say that the influx of 2300 astronomers must indeed have placed a considerable strain on the resources of a still very young university. But in your proximity to and close association with Brighton there was a fortunate predestination for us to meet here. Thank you for everything the University has done for us.

Scientists throughout the world have the deepest respect for the Royal Society of London, and we astronomers are honoured that its distinguished President should have come to Brighton to welcome us on its behalf. Over 300 years old, the Society has enjoyed continuous and vigorous life, in spite of the fact that the whole field of science is growing exponentially. May be the IAU could learn something about how to continue its, as I believe, equally vigorous own life by studying carefully the history of the Royal Society.

I do not quite know how to place Sir Bernard Lovell: not only is he one of us, but he is also President of the Royal Astronomical Society and Chairman of the British National Committee for Astronomy, as well as being Chairman of the National Organizing Committee responsible for the organization of this General Assembly. I thank through you the astronomers of the United Kingdom for the extremely careful preparations for our Assembly. May astronomy in this country, in its turn, benefit from the intense contacts with foreign astronomers offered by our Assembly.

We have learnt something of the organization of astronomy in this country and the dominant part that is now played by the Science Research Council for which the Secretary of State for Education and Science has the overall responsibility to Parliament. And we have heard of the marvellous progress astronomy is making and, in particular, of some ambitious projects which have been realized recently or will be realized in the near future. You are to be congratulated, Mrs Thatcher, on the provision that is being made for astronomy. And, I am sure, the astronomers of the United Kingdom feel full of confidence that they will always have your strong support.

At this Inaugural Ceremony there are certainly many participants present who have come into their first contact with the IAU only today. To them I should like to say that the Union was founded in 1919. It is a non-governmental organization whose objects are to promote the study and development of astronomy as well as to further and safeguard its interests by international cooperation. The Union adheres to the International Council of Scientific Unions and has, like other Unions, National Members or adhering countries. But, unlike other Unions, it has also individual members, a fact which intensifies its activity and yet makes our Assemblies somewhat bulky.

I believe it is no unjustified selfpraise when I say that the International Astronomical Union is one of the most alive and active among the Unions. It is full of healthy problems – as it should be. And we all should be optimistic that even the fantastic growth of our science, the increasing influx of new people, new methods, material and results from fields completely unknown or foreign to our astronomical predecessors will be met by the Union with understanding and with an elastic response.

From the earliest days the IAU has had great benefit from the active cooperation of the astronomers of the United Kingdom. Three of our Presidents, Dyson, Eddington, and Spencer Jones, and three General Secretaries, Fowler, Stratton, and Sadler, were British. The United Kingdom was among the nations which founded the Union in 1919. International cooperation in astronomy lies in the very nature of our science. It is, therefore, only natural that a country with such a strong national activity should support as far as possible all serious international plans.

Even today we must confess that modern science cannot understand itself if it does not contemplate its own history. I wish I could, leisurely, go through the rich history of British astronomy; but time will allow us only a few minutes' glance.

In a country which is so much determined by pragmatic ideas it is no wonder that already in 1556 the Almanach of John Field was based on the system of Copernicus because of its practical superiority in calculating planetary ephemerides.

England could not 'rule the waves' without a permanent interest in the stars, in time-control, in clocks, chronometers, and watches. Therefore it was practical reasons again which led to the foundation of the Royal Observatory, nearly 300 years ago. It had the primary task of studying the orbit of the Moon against the background of the fixed stars for navigation purposes, including time control in all its aspects. Thus positional astronomy and the proper motions of stars became a principal part of Greenwich's work through the centuries.

Inevitably pure research made its entrance into the Royal Observatory very early. In the days of Halley and Bradley, as well as under Airy, and later in our present century, the respective authorities knew that he who wants to promote the practical applicability of science has to further science as a whole.

Many of us are looking forward to seeing the Royal Greenwich Observatory in its present setting at Herstmonceux, continuing its old duties, doing much more with modern instruments; and at the same time, equipped with the Isaac Newton Telescope, aiming at the most ambitious modern research problems.

A typical British line of development, the pragmatic use of which is less evident than that of observational astronomy, is astronomical theory, and even speculation. In this country the Copernican theory did not stir up the bewilderment comparable to that on the European continent. It is remarkable that, already a hundred years before Newton, Thomas Digges in his description of the Copernican system suggested an infinite universe, an idea foreign to Copernicus himself and rejected by Kepler. This connection of the Copernican system with the idea of an infinite universe became the basic picture for the scientific and philosophical development of the following centuries. It is undergoing alterations only in our time.

It is a thankless task to talk sketchily about Sir Isaac Newton. His stature surpasses normal measure to such a degree that comparisons are almost useless. His vigorous imagination and his powerful mathematics allowed him to integrate and to crown the endeavours of many forerunners and to point towards great future achievements. His theory of dynamics penetrated the disciplines from mechanical engineering to modern cosmology.

In our century Britain's contribution to speculative theory has become almost as unique as 250 years earlier. It is the theory of stellar structure, possible only as a result of an extremely long deductive chain of theoretical inferences, which in our days received its more or less final shape. The names of Eddington, Jeans, and Milne characterize the epoch and we shall never forget their dramatic fights.

Another phenomenon is characteristic of England: It is the remarkably high number of amateurs who have promoted astronomy. Bradley and Maskelyne, William Herschel and Lord Rosse, Huggins and Lockyer are a few of those who started as amateurs or even remained amateurs throughout their lives.

I am not to forget one last point: It was the British astronomers who started observational work in the southern hemisphere, in South Africa and in Australia. They were the most efficient predecessors of and are still among those astronomers who, in our time, aim at organizing astronomical research of the southern sky with most modern methods.

Thus, coming to a close, and taking a final look at present day astronomy in this country, leaving alone the historical viewpoints, we can congratulate ourselves on having the chance of seeing most modern equipment, both in radio- and optical astronomy, of learning about new projects and of witnessing the permanent internal clearing process which is so typically British. We feel happy to breathe the scientific atmosphere of the country.'

French translation:

Allocution du Professeur Heckmann

'Mesdames, Messieurs,

Puis-je m'adresser à nos hôtes dans l'ordre qui me paraît natural, c'est-à-dire celui dans lequel ils ont parlé. Et si ce faisant, je transgresse le protocole, vous voudrez bien m'en excuser.

Alderman Nettleton, Professor Briggs, Lord Blackett, Sir Bernard Lovell, et Mrs. Thatcher, je tiens à mon tour à exprimer à chacun de vous la sincère gratitude de l'Union Astronomique Internationale pour vos souhaits de bienvenue.

Les participants de l'Assemblée Générale, qu'ils soient Membres de l'Union, Participants Invités ou Hôtes, apprécieront tous, j'en suis persuadé, les merveilleux attraits que la ville de Brighton nous offre si généreusement. Nous vous sommes particulièrement reconnaissants, Monsieur le Maire, de nous avoir permis de nous réunir dans cette ville balnéaire charmante et historique. Nos travaux bénéficieront pleinement des avantages qu'offrent Brighton et la campagne splendide du Sussex.

A M. le Vice-Chancelier de l'Université du Sussex, je ne puis que dire combien nous sommes conscients de la lourde charge que l'afflux de 2300 astronomes doit faire peser sur les ressources d'une unversité encore bien jeune. Mais grâce à votre voisinage et à vos liens étroits avec Brighton, c'est pour nous une très heureuse circonstance que de nous réunir ici. Merci vraiment pour tout ce que l'Université a fait pour nous.

Les scientifiques du monde entier ont le respect le plus profond pour la Royal Society et c'est un honneur pour nous, astronomes, que son Président soit venu à Brighton pour nous accueillir en son nom. La Société a plus de 300 ans, mais elle est toujours aussi jeune et vigoureuse, malgré l'accroissement exponentiel du domaine scientifique. Peut-être l'UAI pourrait-elle étudier soigneusement l'histoire de la Royal Society et apprendre ainsi à perpétuer sa propre vie que j'estime, elle aussi, fort vigoureuse.

Je ne sais pas très bien comment parler de Sir Bernard Lovell, car non seulement il est l'un des nôtres, mais il est aussi Président de la Royal Astronomical Society, Président du British National Council for Astronomy et Président du Comité National d'Organisation responsable de l'organisation de cette Assemblée. Je voudrais que les remerciements que je lui adresse soient étendus à tous les astronomes du Royaume-Uni qui ont si soigneusement préparé notre Assemblée. Puisse l'astronomie de votre pays bénéficier à son tour des nombreux contacts dont cette Assemblée nous offre la possibilité.

Nous connaissons un peu l'organisation de l'astronomie dans votre pays, ainsi que le rôle dominant que joue maintenant le Science Research Council dont le Secrétaire d'Etat à l'Education et aux Sciences a l'entière responsabilité auprès du Parlement. Et nous avons appris les extraordinaires progrès réalisés par votre astronomie, et tout particulièrement les grands projets déjà réalisés ou en cours de réalisation. Permettez-moi de vous adresser toutes mes félicitations, Mrs Thatcher, pour l'aide accordée à l'astronomie. Je suis persuadé que les astronomes britanniques ont une confiance entière en l'appui que vous leur donnerez.

Pour beaucoup, cette Cérémonie Inaugurale représente sans doute leur premier contact avec l'Union Astronomique Internationale. Et c'est pour eux que je voudrais dire quelques mots de l'Union. Cette organisation non-gouvernementale a été créée en 1919 et a pour but de promouvoir l'étude et le développement de l'astronomie, ainsi que de sauvegarder ses intérêts grâce à une coopération internationale. L'Union adhère au Conseil International des Unions Scientifiques et comprend, comme toute autre Union, des Membres Nationaux ou Pays Adhérents. Mais, contrairement aux autres Unions, elle a également des membres individuels, ce qui rend son activité plus intense, mais ce qui, aussi, rend nos Assemblées quelque peu encombrantes.

Je crois que ce n'est pas une excessive prétention que de dire que l'Union Astronomique Internationale est l'une des plus vivantes et des plus actives parmi les Unions. Elle s'occupe d'une foule de problèmes passionnants – comme cela doit être. Et nous devons envisager avec optimisme l'avenir de l'Union qui résoudra toujours avec compréhension et souplesse les problèmes soulevés par le fantastique développement de notre science, l'afflux incessant de nouveaux membres, l'emploi de nouvelles méthodes, de nouveau matériel, et la profusion des résultats obtenus dans des domaines complètement inconnus de nos prédécesseurs.

Dès ses origines, l'UAI a tiré un grand profit d'une active coopération avec les astronomes britanniques. Trois de nos Présidents, Dyson, Eddington et Spencer Jones, et trois Secrétaires Généraux, Fowler, Stratton et Sadler, étaient britanniques. Le Royaume-Uni est un des pays qui a participé à la création de l'Union en 1919. La coopération en astronomie repose sur la nature même de notre science. C'est pourquoi il n'est que naturel qu'un pays ayant une si grande activité nationale, accorde son plus large appui à tout projet international sérieux.

Même aujourd'hui il nous faut reconnaître qu'une science moderne ne peut se comprendre que si elle ne se complaît pas dans la contemplation de sa propre histoire. J'aimerais m'étendre plus à loisir sur la riche histoire de l'astronomie britannique, mais le temps ne me permet que d'y jeter un regard.

Dans un pays où dominent les idées pragmatiques, il n'est pas étonnant que dès 1566 l'Almanach de John Field ait adopté le système de Copernic en raison de sa supériorité *pratique* dans le calcul des éphémérides planétaires.

L'Angleterre ne pourrait pas 'commander aux flots' si elle ne s'intéressait pas constamment aux étoiles, au contrôle du temps, aux horloges, chronomètres et montres. Aussi est-ce à nouveau pour des raisons pratiques que fut fondé, il y a environ 300 ans, l'Observatoire Royal. Sa tâche première était d'étudier, en vue de la navigation, l'orbite de la Lune sur un fond d'étoiles fixes, et de s'occuper du contrôle du temps dans tous ses aspects. Puis l'astronomie de position et les mouvements propres des étoiles devinrent, pendant des siècles, le but principal des travaux de Greenwich.

Il était inévitable que la recherche pure fasse son entrée très tôt à l'Observatoire Royal. A l'époque de Halley et de Bradley, ainsi qu'à celle de Airy, et plus récemment au cours de ce siècle, les autorités savaient que celui qui désire promouvoir l'application pratique de la science doit faire progresser la science toute entière.

Beaucoup d'entre nous se font un plaisir de visiter l'Observatoire Royal de Greenwich, maintenant installé à Herstmonceux, et qui poursuit les mêmes travaux avec des instruments bien plus modernes; à Herstmonceux se trouve également le Télescope Isaac Newton dont le programme couvre les recherches les plus modernes et les plus ambitieuses.

Une direction très britannique de développement, dont l'utilité pragmatique est moins évidente que celle de l'astronomie observationnelle, est le domaine de la théorie, même la plus spéculative. La théorie de Copernic n'a pas produit dans ce pays un bouleversement comparable à celui qui a secoué le continent européen. Il est remarquable de noter que, un siècle déjà avant Newton, Thomas Digges parlait, dans sa description du système de Copernic, d'un univers infini – une idée étrangère à Copernic lui-même, et rejetée par Kepler. Ce rapprochement entre le système de Copernic et l'idée d'un univers infini devint le point de base du développement scientifique et philosophique des siècles suivants. Des changements n'intervinrent qu'à notre époque.

Il ne m'est pas agréable de me limiter à seulement quelques mots sur Sir Isaac Newton. Sa personnalité dépasse tellement les normes communes qu'il est à peu près impossible de faire des comparaisons. Sa vigoureuse imagination et sa prodigieuse connaissance des mathématiques lui permirent de poursuivre et de faire triompher les efforts d'un grand nombre de ses prédécesseurs, et de préparer les grandes réalisations futures. Sa théorie de la dynamique sert de base à toutes les disciplines, depuis les techniques mécaniques jusqu'à la cosmologie moderne.

A notre époque, la contribution de la Grande-Bretagne à la recherche théorique est presque aussi extraordinaire qu'il y a 250 ans. La théorie de la structure stellaire, qui n'est probablement que le résultat d'une très longue suite de déterminations théoriques, reçoit plus ou moins aujourd'hui sa forme définitive. Les noms de Eddington, de Jeans, de Milne marquent cette époque, et jamais nous n'oublierons leurs luttes épiques.

Un autre phénomène est à noter: c'est le nombre remarquablement élevé des amateurs qui, en Angleterre, ont fait progresser l'astronomie. Bradley et Maskelyne, William Herschel et Lord Rosse, Huggins et Lockyer, entre autres, débutèrent comme amateurs et même demeurèrent des amateurs toute leur vie.

Il me faut mentionner un dernier point: ce sont les astronomes britanniques qui ont entrepris les premiers travaux d'observation dans l'hémisphère sud, en Afrique du Sud et en Australie. Ils furent des pionniers et sont encore aujourd'hui de ces astronomes qui s'occupent de l'organisation des recherches astronomiques dans le ciel austral en utilisant les méthodes les plus modernes.

Pour finir et pour en venir aux temps présents et laisser le point de vue historique, nous ne pouvons que nous féliciter de notre chance de pouvoir admirer dans votre pays l'équipement le plus moderne, en radioastronomie comme en astronomie optique, d'être mis au courant de vos nouveaux projets et d'être les témoins des progrès incessants de l'astronomie britannique. Nous nous sentons heureux de respirer l'atmosphère scientifique de votre pays.'

After Professor Heckmann's address of thanks, the Chairman formally declared the Inaugural Ceremony closed.

REPORT OF THE GENERAL ASSEMBLY
(WITH RESOLUTIONS INCLUDED)

AGENDA

First Session

1. Formal opening of the General Assembly, by the President.
2. Appointment of official interpreters.
3. Report of the Executive Committee:
   (a) Discussion of the printed Report.
   (b) Report of decisions taken at meetings in Brighton, in particular of admission of new Adhering Countries and new IAU Members.
4. Report by the General Secretary.
5. Report by the President of the proposals for membership of the Executive Committee.
6. Announcement of:
   (a) The names of representatives of Adhering Countries, empowered to vote on their behalf.
   (b) The names of representatives to serve on the Nominating Committee.
   (c) The names of Acting Presidents of Commissions.
7. The appointment of the Finance Committee.
8. The appointment of the Resolutions Committee.
9. Revisions of the Statutes and By-laws. (The French and English texts of the revised version, as proposed by the Executive Committee, have been distributed to Adhering Organizations and National Committees. Moreover, the English text has been published in *IAU Information Bulletin* No. 23, and the French text in *IAU Information Bulletin* No. 24). Reference is made to the proposal of the U.S.S.R. National Committee of Astronomy under point 10, as regards the interval between two General Assemblies.
10. Consideration of proposals for resolutions submitted by Adhering Organizations or National Committees of Astronomy. The proposals formally submitted before the statutory deadline are shortly as follows:
    (a) *(Australia)* to reconsider the decision to allocate the relative length of the Reports on Astronomy of the various Commissions on the apparently arbitrary basis of the length in the 1967 report.
    (b) *(Belgium)* to recommend the introduction of a fundamental course in Astronomy and Geodesy in the curricula of Technical Universities.
    (c) *(Poland)* to hold under the auspices of the IUHPS and IAU a Symposium commemorative of the 500th anniversary of the birth of Copernicus in Toruń, Poland, and a Symposium on Cosmology in Warsaw, Poland, in 1973.
    (d) *(U.A.R.)* to create a special Commission for Exchange of Equipment within the IAU.
    (e) *(U.S.S.R.):*
    (i) To proclaim the year 1973 the year of Copernicus and to bring this proposal before the UNESCO.
    (ii) To hold the XVth General Assembly of the IAU in Warsaw, Poland, in commemoration of the 500th anniversary of the birth of this great scientist (see also point 19).
    (iii) To investigate the possibility of less expensive IAU publications and thus to create conditions for a reduction of the unit of contribution paid by Adhering Countries.

(iv) To extend the interval between two General Assemblies of the IAU from 3 to 4 years as from 1973 onwards, and to pay more attention to scientific meetings with smaller attendance.

(v) To create within one of the IAU Commissions a Sub-commission on the Physics of Unstable Stars.

The proposals will be submitted to the relevant Commissions prior to the General Assembly.

11. Consideration of proposals for resolutions submitted by Commissions or by Inter-Union Commissions. Eight such proposals have been submitted by Commissions:

(a) *(Commission No. 6)* recommends that the IAU subvention to the IAU Telegram Bureau be maintained at its present value of $ 1 600 per triennium.

(b) *(Commission No. 16)* proposes that $ 1 500 be allocated to the Centre de documentation sur les planètes á Meudon for 1971–1973.

(c) *(Commission No. 17)* propose que l'Assemblée Générale accepte les 500 noms proposés par la Commission n° 17 pour désigner les cratères de la face arrière de la Lune et qu'elle attribue à la Commission n° 17 pour les années 1971–1973 une subvention de $ 2 000 pour aider le fonctionnement des groupes de travail spécialisés de cette Commission.

(d) *(Commission No. 23)* proposes that a resolution be discussed by the General Assembly concerning the future of the Helsingfors Observatory which is threatened to be closed down by the authorities.

(e) *(Commissions No. 23 and 24)* propose the merger of Commission No. 23 'Carte du Ciel' and Commission No. 24 on Stellar Parallaxes and Proper Motions.

(f) *(Commission No. 33)* proposes that the General Assembly resolve to omit the superscript II from the new galactic coordinates that should be simply 1, b; the old coordinates should retain the superscript I.

(g) *(Commission No. 27)* proposes to allocate to Commission No. 27 an annual subvention of $ 1 000 for the continued publication of the *General Catalogue of Variable Stars*.

(h) *(Commission No. 46)* recommends the introduction of a fundamental course in astronomy in University curricula.

Other proposals for resolutions have only been presented in draft form and will first be discussed by the Commissions. No proposals for resolutions have arrived from Inter-Union Commissions.

Second Session

12. Report of the Finance Committee:
(a) Accounts for 1967–1969.
(b) The additional budget for 1970.
(c) The budget for the ensuing period.

13. The unit of subscription for the ensuing period. The Executive Committee will propose to the Finance Committee that the unit of subscription be maintained at 900 gold francs.

14. The appointment of the Special Nominating Committee.

15. Proposals for resolutions submitted by the Executive Committee.

16. Proposals for resolutions submitted by Commissions, subject to the recommendation by the Resolutions Committee.

17. The announcement of new Members of the Union.

18. *Commissions:*
(a) Dissolution and creation of Commissions:
(i) Merger of Commissions Nos. 23 and 24.
(ii) Creation of Commissions Nos. 47 on Cosmology, and 48 on High Energy Astrophysics.
(b) The election of Presidents and Vice-Presidents.
(c) The membership of Organizing Committees.

19. *The Place and date of the XVth General Assembly:*
(a) Consideration of the Australian invitation to hold the XVth General Assembly in Sydney, Australia, in 1973.

(b) Consideration of the U.S.S.R. proposal to hold the XVth General Assembly in Warsaw, Poland, in 1973.

20. The election of a President, three Vice-Presidents, a General Secretary, and an Assistant General Secretary.

21. Addresses by the retiring and newly-elected Presidents.

22. Closing ceremonies.

At the first session of the General Assembly, to be held on TUESDAY 18 AUGUST, 1970, items 1–11 of the above agenda will be considered. The remaining items, and any which may have been adjourned from the first session, will be considered at the second session, to be held on THURSDAY 27 AUGUST, 1970.

FIRST SESSION

Held in The Dome at Brighton, Sussex, England, on Tuesday, 18 *August* 1970 *at* 15^{h}00^m

Professor Dr Otto Heckmann, President, in the chair

1. *Formal Opening.* The President spoke a few words of welcome to the Members of the Union, invited participants and their guests, representatives of Adhering Countries, representatives of sister Unions and other organizations, and formally declared open the first session of the fourteenth General Assembly of the Union.

He welcomed in particular the official representatives of ICSU and other scientific Unions as follows:

ICSU: Professor V. A. Ambartsumian

IUGG (International Union of Geodesy and Geophysics): Professor P. Tardi

IUGS (International Union of Geological Sciences): Dr G. Fiedler, Dr P. M. Millman

URSI (Union Radio Scientifique Internationale): Professor W. N. Christiansen

COSPAR (Committee on Space Research): Professor C. de Jager, Dr. Z. Švestka

IAB (ICSU Abstracting Board): Dr J. B. Sykes

CST (Committee on Science Teaching): Professor Edith A. Müller

IUCSTP (Inter-Union Commission on Solar-Terrestrial Physics): Dr H. Friedman, Dr Z. Švestka

IUTAM (International Union of Theoretical and Applied Mechanics): Dr M. J. Lighthill

Joint Commission on Spectroscopy: Dr G. Herzberg, Professor B. Edlén

BIMP (Bureau International des Poids & Mesures): Dr J. Terrien

IUAA (International Union of Amateur Astronomers): Messrs L. Baldinelli, F. M. Flinch, H. Miles, O. Oburka

The official representative of UNESCO could not attend for other serious engagements.

The President then suggested that a message of thanks be sent to H.R.H. The Duke of Edinburgh. This proposal was approved by acclamation.

The President continued by asking Members to stand while the General Secretary would read the names of Members who had died since the last meeting, or whose death had not been known at the time of that meeting. The General Secretary then read the following list:

Mrs Z. N. Aksent'eva, H. D. Babcock, A. Birkenmayer, Miss M. Bretz, S. Chapman, L. Cichowicz, L. d'Azambuja, G. Demetrescu, A. J. Deutsch, E. J. Dijksterhuis, J. Dufay, L. Egyed, G. Fayet, J. H. Focas, D. A. Frank-Kamenetskij, C. W. Gartlein, E. Gullón, S. E. Hajkin, K. Heinemann, L. G. Henyey, E. Hertzsprung, C. Hoffmeister, J. Ikaunieks, C. C. Kiess, H. Knox-Shaw, A. Kohlschütter, J. J. Kubikowski, N. I. Kucherov, A. I. Lebedinskij, Mrs S. L. McDonald, S. G. Makover, A. V. Markov, O. F. Mathias, R. W. Michie, A. V. Nielsen, Miss R. J. Northcott, Mrs S. V. Romanskaya, L. I. Semenov, E. C. e Silva, E. Vassy, S. C. Venter, M. H. Wrubel.

2. *Appointment of official interpreters.* The following four official interpreters were appointed by acclamation:

from English to French: L. Houziaux, S. A. J. Malaise

from French to English: P. R. Demarque, H. Reeves

1. (*cont.*) The President took the floor and addressed the Assembly as follows:

'During the past three years the members of the Executive Committee, the Officers of the IAU and – I am sure – many members of the Union have felt with deep concern what our predecessors felt and what the responsible men in the Union will feel in the forseeable future: Namely that the growth and the health of the Union could come into conflict with each other. Changes of one sort or other will be inevitable.

But the present Executive Committee is of the opinion that the development of the Union has been sound until now. The Executive Committee has authorized the President to convey to you, in very short words, a few directives, which are only meant as an advice, but which are supported by careful consideration of many factors and as a resonance of many single voices:

1. It has been proposed that it would be a good thing to split the Union into a small number of larger sections. Well, you do not find in the Statutes, neither in the old nor in the new version, a paragraph stating that it is an aim of the IAU to preserve the unity of astronomy. But we believe that this is the vital and basic principle according to which the Union has lived so far and will have to live for quite a number of years to come. It is true that new methods, techniques, theories and discoveries change the scope of astronomy very rapidly. However, everybody knows quite well what astronomy is and how the new developments fit into it. A radioastronomer may get deeply involved in fundamental astronomy and astrometry; a high frequency astrophysicist may derive deep satisfaction from his contact with celestial mechanics; and so on. There still exists the common language, even though we have to confess that some newcomers will have to learn it. We believe very strongly that splitting the Union would very soon result in disrupting the unity of astronomy as such. We believe that such a splitting is unnecessary because the structure of the Union is very elastic and with the new version of the Statutes will become even more elastic.

It might be found necessary in the future to create new Commissions, or to merge and discontinue others when their tasks so require, more frequently than it has been done before. This will allow rapid adaptations.

2. It has been argued that Individual Membership has unnecessarily added to the growth of the Union. The present Executive Committee believes that Individual Membership is one of the Union's principal assets. Since the Membership of the Union is awarded only for active participation in astronomical research, it is rightly regarded as an international recognition of the individual's qualification. It is coveted by young astronomers all the world over and by scientists from neighbouring fields. This Individual Membership in one and the same organisation maintains very strongly the sense of unity of our science. Very few international organisations have introduced it and one may well ask whether those, which have not, have a life similar in intensity to that of the IAU.

3. Our General Assemblies are, to a certain extent, a consequence of Individual Membership. The Membership would have little sense if Members could not meet at one place and at one time at such intervals as to still remember the previous meeting. We know very well that the difficulties of organizing Assemblies like ours are very great. Yet we know also that other branches of science successfully organize congresses with at least twice the attendance of the largest IAU Assemblies. We see a promising line of development: A cluster of smaller meetings, symposia and colloquia, are organized immediately before and after our General Assemblies. It may be necessary, that the Union should pay more attention to the participation of astronomers from such countries which find it difficult to provide visa for more than one foreign country. Therefore the following development seems to be recommendable:

The General Assemblies should be made shorter, cutting off one or two days of their duration, concentrating more on topics which interest a wide attendance, such as Joint Discussions, Invited Discourses, and Joint Meetings of several commissions. On the other hand, Commission Meetings ought to be limited to a bearable number in order to increase the intensity of personal contacts. Commissions desiring to have 10 meetings or more should organize Colloquia of 2–3 days' duration in the same country.

4. A last point should not be forgotten: All the success of the Union depends on work, and very frequently on heavy work, which requires first class experts, and yet brings never any material compensation. The time spent by an individual on IAU matters is limited and is in addition to his main occupation.

The recollection 'how it is done' is an extremely important factor. We believe therefore very strongly that all transformations of the Union must be made continuously.

These are the ideas which I had to convey to you. They are not new but they are not always at hand at moments where they – according to our opinion – should be of some benefit.'

French translation:

Allocution du Président

Au cours des trois années qui viennent de s'écouler, les membres du Comité Exécutif, le bureau de l'UAI et, j'en suis convaincu, bien des membres de l'Union ont éprouvé profondément le même souci que nos prédécesseurs ont éprouvé, le même qu'éprouveront dans un avenir prévisible ceux qui auront la responsabilité des affaires de l'Union: je veux dire le sentiment que la croissance et la santé de l'Union pourraient bien entrer en conflit l'une avec l'autre. Des changements d'une sorte ou d'une autre seront inévitables.

Mais le Comité Exécutif actuel pense que, jusqu'à présent, le développement de l'Union est resté raisonnable. Le Comité Exécutif a donné au Président l'autorisation de vous exprimer, en peu de mots, quelques idées directrices qui n'ont d'autre prétention que d'être un avis, mais qui sont appuyées sur une étude soigneuse de nombreux facteurs, et sont aussi comme l'écho de nombreuses opinions individuelles.

1. On a proposé comme une bonne chose de diviser notre Union en un petit nombre de grandes sections. En fait, vous ne trouverez dans les Statuts (que ce soit sous l'ancienne ou sous la nouvelle version) aucun paragraphe donnant à l'UAI comme l'un de ses objets de préserver l'unité de l'astronomie. Mais nous pensons que ceci est pourtant le principe vital, et fondamental, selon lequel l'Union a vécu jusqu'à présent, et selon lequel elle devra vivre encore pendant pas mal d'années. Il est vrai que la nouveauté dans les méthodes, les techniques, les théories, et les découvertes, modifie très rapidement les perspectives de l'astronomie. Pourtant, chacun sait bien ce qu'est l'astronomie, et comment s'y insèrent les développements récents. Un radioastronome peut devenir profondément impliqué dans des recherches d'astronomie fondamentale et d'astrométrie; un astrophysicien des hautes fréquences peut tirer de profondes satisfactions de ses contatcts avec la mécanique céleste... les exemples sont nombreux. Il existe aussi un langage commun – encore devons-nous admettre que quelques nouveaux venus doivent l'apprendre. Nous croyons très fermement que le découpage de l'Union aboutirait très vite à la destruction de l'unité de l'astronomie en tant que telle. Nous croyons qu'un tel découpage est inutile, car la structure de l'Union est d'une grande souplesse, et que, grâce aux nouveaux Statuts, elle deviendra plus souple encore.

On pourrait considérer comme necessaire dans l'avenir de créer de nouvelles commissions, d'en fusionner d'autres, voire d'en supprimer, quand l'évolution de leurs tâches l'exige ainsi – et ce, plus fréquemment que cela n'a été fait auparavant. Cette technique permettra des adaptations rapides...

2. On a dit que l'existence de membres de l'Union à titre individuel a contribué inutilement à la croissance de l'Union. Le Comité Exécutif actuel pense que l'appartenance individuelle à l'Union est l'un des atouts essentiels de l'Union. Puisque l'on n'accorde cette appartenance qu'à ceux qui participent activement à la recherche astronomique, elle est justement considérée comme la reconnaissance internationale des qualifications de l'individu concerné. Elle est briguée par les jeunes astronomes du monde entier comme par des chercheurs travaillant dans des disciplines voisines. L'appartenance individuelle à une unique organisation maintient fortement le sentiment de l'unité de notre science. Un très petit nombre d'organisations internationales ont introduit cette disposition

dans leurs statuts, et l'on peut se demander si les organisations qui ne l'ont pas fait ont une vie aussi intense que celle de l'UAI.

3. Nos Assemblées Générales, dans une certaine mesure, sont une conséquence de l'appartenance individuelle. Celle-ci n'aurait aucun sens si les Membres ne pouvaient se réunir en un seul endroit, au même moment, à des intervalles assez courts pour qu'ils n'aient pas le temps d'oublier leur réunion précédente. Nous savons fort bien les énormes difficultés que rencontre l'organisation d'assemblées comme les nôtres. Nous savons aussi que d'autres disciplines scientifiques organisent des congrès auxquels participent deux fois plus de personnes qu'aux assemblées les plus importantes de l'UAI. Une direction très prometteuse de développement est maintenant en vue: une collection de réunions moins vastes, symposiums et colloques, s'organise immédiatement avant ou après nos Assemblées Générales. Il peut être nécessaire pour l'Union de considérer avec une attention toute spéciale la participation d'astronomes originaires de pays où il est difficile d'obtenir des visa plus d'un pays étranger; par suite, le processus que je vais maintenant décrire semble à recommander.

Les Assemblées Générales devraient être plus courtes, leur durée étant réduite d'un ou deux jours, et concentrer leurs activités sur des questions susceptibles d'intéresser le plus grand nombre, grâce aux discussions communes, aux conférences solennelles, aux réunions communes à plusieurs commissions. D'autre part, les réunions de commissions devraient être limitées à un nombre supportable, afin de permettre l'accroissement en intensité des contacts personnels. Les commissions qui souhaiteraient organiser une dizaine de réunions, ou même plus, devraient plutôt mettre sur pied des colloques de deux ou trois jours dans le même pays.

4. Un dernier point ne devrait pas être oublié: toute l'efficacité de l'Union dépend du travail de quelques-uns: c'est très souvent un travail très lourd, qui exige des experts de premier plan, et qui n'entraîne pour eux aucune compensation matérielle. Le temps passé par un individu au service de l'Union est limité, et il s'ajoute à son occupation principale. Il est extrêmement important de se rappeler comment se font les choses, et nous croyons par conséquent très fermement que les transformations de l'Union doivent se faire dans la *continuité*.

Telles sont les idées dont je voulais vous faire part. Elles ne sont pas neuves – mais elles ne sont pas toujours en mémoire aux moments précis où, à notre avis, elles pourraient être de quelque utilité...

3. *Report of the Executive Committee*. The General Secretary invited the Assembly to discuss the Report of the Executive Committee (see pp. 1–28), as printed and distributed before the General Assembly, and said that he was prepared to answer any questions that might be asked. There were no questions, and the Report was approved.

4. *Report by the General Secretary*. The General Secretary then read the following addendum to the Report of the Executive Committee, covering the period January–August 1970:

'My complementary report covers the activity of the Union during the period from 1 January to 31 July 1970.

The Executive Committee decided by correspondence on the following items:

(*a*) It approved the list of candidates for the officers of the IAU during the ensuing period of three years.

(*b*) It prepared the Activities Report, as printed and distributed before the General Assembly (see pp. 4–10), and a preliminary comprehensive budget of the IAU for 1971–73.

(*c*) It authorized the General Secretary to approach the Secretary General of ICSU with the request to have the year 1973 proclaimed the Year of Copernicus.

(*d*) It decided to propose to ICSU the creation of an Inter-Union Commission on the Moon, and it authorized the General Secretary to take this problem up with the International Unions concerned, and with COSPAR.

The Executive Committee proposes to the General Assembly the creation of two new Commissions:

*Commission No. 47, Cosmology, presided over by Professor W. B. Bonnor in the first session.

*Commission No. 48, High Energy Astrophysics, to be presided over by Professor H. Friedman.

The Executive Committee also proposes to merge Commission No. 23 Carte du Ciel and Commission No. 24 'Stellar Parallaxes and Proper Motions' into one Commission No. 24 on Photographic Astrometry.

I trust these proposals will find your approval. (Proposals approved by show of hands.)

The Executive Committee examined in detail the applications of Cuba and Uruguay for adherence to the Union, and was happy to admit these two new Member Countries to the IAU.

The 50th anniversary of the foundation of the Union fell into the period which is just coming to its end. There were several occasions to look back at the past of the IAU and to compare it with other international organizations, similar in character. I could not help noticing that our predecessors, during all those past fifty years, always knew what they wanted. They shaped the IAU into an organization with a clear aim at international cooperation, an organization based on personal friedship and on willing assistance of its individual Members.

The reverse of the coin is finances. We had to restrict somewhat the bilinguality, we had to ask the Presidents of Commissions to shorten their reports on astronomy, we could only symbolically finance our scientific meetings. I wish sincerely to thank you for your understanding support which helped the Executive Committee to bring the finances of the Union back to normal, or almost so. And the Executive Committee is considering ways and means how to remove all restrictions on the length of Commission Reports.

However, the economies we had to achieve did in no way interfere with the scientific activity of the Union. We are looking back upon 9 *Symposia* and 12 *Colloquia* held by the IAU until the end of 1969. *Six more Symposia* are being held in conjunction with the General Assembly. The Summer Schools for Young Astronomers, organized with the financial aid of UNESCO in 1967, 1968 and 1969 were an outspoken success. We maintained over the period under review our traditional projects, and we are especially proud of the way Professor Minnaert knew to handle that of the Exchange of Astronomers. Thanks to an additional grant from ICSU it was possible to increase by $ 4000 the sum at the disposal of Commission 38 and to award 50 grants to young astronomers for attending this General Assembly.

The Reports on Astronomy, prepared by the Presidents of Commissions, are a magnificent piece of work. I wish to thank all those who contributed to this success. I wish to thank UNESCO and ICSU for their financial help, I wish to thank the organizers of our Symposia and Colloquia, I wish to thank the organizers of the Summer Schools and I wish to thank the Presidents of Commissions for their selfless work and devotion to the Union.

My thanks are also due to our publishers. Eleven volumes were published – or will appear soon. The printing time, that is the time between the receipt of the manuscript by the printer and the appearance of the book, was 7 months on the average with a maximum of 10 and minimum of 5 months. The time needed by the Scientific Editors was on the average four and a half months and varied between 2 and 7 months. Our publications were well and widely advertised and their sales increased. They are not cheap. I did not find a solution to this problem.

You have in your hands the proposal by the Executive Committee for *new Statutes of the Union*. All changes and the reasons for them were mentioned in *Information Bulletin Nos.* 23 and 24 and I do not propose to repeat them here. You will find that the new proposal affects very little the current life of the Union. It makes the administration and organization of the growing Union easier, more flexible and more democratic. Permit me to stress one aspect:

There is no unique solution to the Statutes. Even if the substance is agreed upon, there remains an infinite number of possibilities in wording, sequence, degree of freedom and intentional vagueness. The question put before the General Assembly will not be to find all existing solutions to a mathematical game but to decide – with a certain amount of generosity – whether that particular solution is acceptable or not.

After this General Assembly the Union will have 2600 individual Members, 46 Adhering Countries, and 39 Commissions.'

5. *Report of the President of the proposals for membership of the Executive Committee.*

The President reminded the Assembly that the rules adopted at the Moscow General Assembly provide for a Special Nominating Committee, appointed within one year following each General Assembly, to advise the Executive Committee on nominations for the new Executive Committee. The Special Nominating Committee elected at the XIIIth General Assembly in Prague, consisted of the present President, the former President Professor P. Swings, a Vice-President of the present Executive Committee, Professor W. N. Christiansen, and four Union Members:

Prof. M. Golay, Professor L. Goldberg, Professor Dr E. R. Mustel, and Dr D. H. Sadler.

The Special Nominating Committee made proposals to the Executive Committee which approved them unanimously as follows:

as President for the period 1970–73:	Professor B. Strömgren, Denmark
as Vice-Presidents for the period 1971–76:	Professor B. J. Bok, United States of America
	Professor Sir Bernard Lovell, United Kingdom
	Professor Dr E. R. Mustel, Union of Soviet Socialist Republics
as General Secretary:	Professor C. de Jager, The Netherlands
as Assistant General Secretary:	Professor Dr G. Contopoulos, Greece

The full Executive Committee would thus consist of the above six together with:

continuing Vice-Presidents:	Dr M. K. V. Bappu, India
	Professor L. Gratton, Italy
	Dr J. Sahade, Argentina
in an advisory capacity:	Professor Dr O. Heckmann, former President, German Federal Republic
	Dr L. Perek, former General Secretary, Czechoslovakia

In accord with the By-laws the formal election would take place at the final session of the General Assembly.

6. *Announcements.* At the request of the President, the General Secretary read
(*a*) the names of representatives of Adhering Countries, empovered to vote on their behalf;
(*b*) the names of representatives of Adhering Countries on the Nominating Committee.

Country	(a) Official Representative	(b) Representative on Nominating Committee
Argentina	J. Sahade	J. Sérsic
Australia	H. W. Wood	W. N. Christiansen
Austria	K. Ferrari d'Occhieppo	K. Ferrari d'Occhieppo
Belgium	H. L. Vanderlinden	H. L. Vanderlinden
Brazil	P. Kaufmann	P. Kaufmann
Bulgaria	N. Bonev	N. Bonev
Canada	J. L. Locke	K. O. Wright
Chile	C. Anguita	J. Stock
Colombia	J. Arias de Greiff	J. Arias de Greiff
Cuba	L. Larragoiti	L. Larragoiti
Czechoslovakia	V. Guth	L. Kresák
Denmark	M. Rudkjøbing	K. Gyldenkerne
Finland	L. Oterma	L. Oterma
France	A. Lallemand	J. Delhaye
German Democratic Republic	H. Lambrecht	G. Ruben
German Federal Republic	L. Biermann	P. Wellmann

Greece	J. Xanthakis	D. Kotsakis
Hungary	L. Detre	L. Detre
India	G. Swarup	S. D. Sinvhal
Iran	A. Therian	A. Therian
Ireland	P. A. Wayman	T. E. Nevin
Israel	Y. Ne'eman	Y. Ne'eman
Italy	G. Righini	G. Righini
Japan	S. Miyamoto	M. Huruhata
Korea (P.D.R.)	not represented	not represented
Mexico	P. Pishmish	A. Poveda
Netherlands	J. H. Oort	A. Blaauw
New Zealand	W. Davidson	W. Davidson
Norway	E. Jensen	Ö Elgaröy
Poland	W. Iwanowska	W. Zonn
Portugal	M. G. Pereira de Barros	M. G. Pereira de Barros
Roumania	C. Dramba	C. Dramba
South Africa	F. J. Hewitt	F. J. Hewitt
Spain	R. Carrasco	M. L. Arroyo
Sweden	E. Holmberg	G. Larsson-Leander
Switzerland	M. Golay	P. Bouvier
Taiwan	M. Tsao	M. Tsao
Turkey	N. Gökdogan	M. Hotinli
U.A.R.	M. K. Aly	M. K. Aly
United Kingdom	B. Lovell	P. O. Redman
U.S.A.	J. L. Greenstein	B. J. Bok
U.S.S.R.	V. A. Ambartsumian	A. G. Massevich
Uruguay	G. Vergara	G. Vergara
Vatican City State	P. J. Treanor	P. J. Treanor
Venezuela	S. Sofía	S. Sofía
Yugoslavia	P. Djurković	P. Djurković

Note: In some cases National Committees designated substitute delegates. In order to simplify the above list the names of the substitute delegates have been omitted.

(c) *Acting Presidents.* The General Secretary announced that the Executive Committee had invited:

W. Dieckvoss to act for A. A. Nemiro as President of Commission No. 8
R. G. Athay to act for M. N. Gnevyshev as President of Commission No. 12
G. M. R. Winkler to act for F. Zagar as President of Commission No. 31
A. Reiz to act for M. G. J. Minnaert as President of Commission No. 38
W. B. Bonnor to act for Ya. B. Zeldovich as President of Commission No. 47

The Acting Presidents were appointed for the duration of the General Assembly.

7. *Finance Committee.* In accord with the Statutes, the following Finance Committee, consisting of one representative of each Adhering Country, was appointed by the General Assembly:

Argentina	A. Feinstein	Japan	S. Miyamoto
Australia	J. P. Wild	Korea (P.D.R.)	not represented
Austria	K. Ferrari d'Ochieppo	Mexico	M. Mendez
Belgium	A. G. Velghe	Netherlands	C. A. Muller
Brazil	P. Kaufmann	New Zealand	W. Davidson
Bulgaria	M. Popova	Norway	Ö. Elgaröy
Canada	V. Gaizauskas	Poland	S. Piotrowski
Chile	C. Anguita	Portugal	M. G. Pereira de Barros

Colombia	J. Arias de Greiff	Roumania	C. Dramba
Czechoslovakia	B. Onderlička	South Africa	F. J. Hewitt
Denmark	A. Reiz	Spain	J. O. Cardùs
Finland	L. Oterma	Sweden	A. Elvius
France	H. Andrillat	Switzerland	U. Steinlin
G.D.R.	K. H. Schmidt	Taiwan	H. Tsao
G.F.R.	W. Fricke	Turkey	A. Kiral
Greece	L. N. Mavridis	U.A.R.	A. S. Asaad
Hungary	L. Dezsö	U.K.	D. W. N. Stibbs
India	R. V. Karandikar	U.S.A.	K. A. Strand
Iran	A. Therian	U.S.S.R.	E. K. Kharadze
Ireland	P. A. Wayman	Vatican	F. C. Bertiau
Israel	Y. Ne'eman	Venezuela	S. Sofía
Italy	M. Cimino	Yugoslavia	P. Djurković

Note: In some cases National Committees designated substitutes. For the sake of simplicity the names of the substitute delegates have been omitted in the above list.

8. *Resolutions Committee.* At the request of the President, the General Secretary explained that the duties of the Resolutions Committee were to advise the Executive Committee in respect of proposals for resolutions submitted by Commissions, with a view to recommending to the General Assembly which proposals should formally be submitted for consideration by the General Assembly. The General Assembly appointed J. H. Oort and D. H. Sadler to serve on the Resolutions Committee. The General Secretary and the Assistant General Secretary were to attend the meeting of the Resolutions Committee in an advisory capacity.

9. *Statutes and By-laws.* The General Secretary said that the text of the revised Statutes and By-laws had been prepared by a special Statutes Committee and approved by the Executive Committee. The corrections suggested by the French National Committee have been taken into consideration in the French version published in Information Bulletin No. 24. With this in mind, the General Secretary moved that the new Statutes and By-laws be adopted and that they become valid from 1 September 1970 on. This motion would be put to the vote at the final session.

10 and 11. *Consideration of proposals for resolutions.* The proposals for resolutions submitted by Adhering Organizations or National Committees were read by the General Secretary as follows:
- (a) (*Australia*) to reconsider the decision to allocate the relative length of the Reports on Astronomy of the various Commissions on the apparently arbitrary basis of the length in the 1967 report.
- (b) (*Belgium*) to recommend the introduction of a fundamental course in Astronomy and Geodesy in the curricula of Technical Universities.
- (c) (*Poland*) to hold under the auspices of the IUHPS and IAU a Symposium commemorative of the 500th anniversary of the birth of Copernicus in Toruń, Poland, and a Symposium on Cosmology in Warsaw, Poland, in 1973.
- (d) (*U.A.R.*) to create a special Commission for Exchange of Equipment within the IAU.
- (e) (*U.S.S.R.*):
- (i) To proclaim the year 1973 the year of Copernicus and to bring this proposal before the UNESCO.
- (ii) To hold the XVth General Assembly of the IAU in Warsaw, Poland, in commemoration of the 500th anniversary of the birth of this great scientist (see also point 19).
- (iii) To investigate the possibility of less expensive IAU publications and thus to create conditions for a reduction of the unit of contribution paid by Adhering Countries.
- (iv) To extend the interval between two General Assemblies of the IAU from 3 to 4 years as from 1973 onwards, and to pay more attention to scientific meetings with smaller attendance.

(v) To create within one of the IAU Commissions a Sub-commission on the Physics of Un-stable Stars.

The General Secretary then read the proposals for resolutions submitted by Commissions as follows:

(a) (*Commission No.* 6) recommends that the IAU subvention to the IAU Telegram Bureau be maintained at its present value of $ 1 600 per triennium.

(b) (*Commission No.* 16) proposes that $ 1 500 be allocated to the Centre de documentation sur les planètes à Meudon for 1971–73.

(c) (*Commission No.* 17) propose que l'Assemblée Générale accepte les 500 noms proposés par la Commission N°. 17 pour désigner les cratères de la face arrière de la Lune et qu'elle attribue à la Commission N°. 17 pour les années 1971–73 une subvention de $ 2 000 pour aider le fonctionnement des groupes de travail spécialisés de cette Commission.

(d) (*Commission No.* 23) proposes that a resolution be discussed by the General Assembly concerning the future of the Helsingfors Observatory which is threatened to be closed down by the authorities.

(e) (*Commissions No.* 23 *and* 24) propose the merger of Commission No. 23 'Carte du Ciel' and Commission No. 24 on Stellar Parallaxes and Proper Motions.

(f) (*Commission No.* 33) proposes that the General Assembly resolve to omit the superscript II from the new galactic coordinates that should be simply 1, b; the old coordinates should retain the superscript I.

(g) (*Commission No.* 27) proposes to allocate to Commission No. 27 an annual subvention of $ 1 000 for the continued publication of the *General Catalogue of Variable Stars*.

(h) (*Commission No.* 46) recommends the introduction of a fundamental course in astronomy in University curricula.

The General Secretary explained that these proposals would first be considered by the relevant Commissions and then passed on to the Resolutions Committee for further recommendation. A number of other proposals received from Commissions in draft form would first be discussed by the Commissions.

The General Secretary reminded that the formal vote on resolutions would take place at the final session of the General Assembly, according to By-laws.

The General Secretary then announced that all other items of agenda would be deferred until the next session of the General Assembly, pending reports from various Commissions and Committees.

The President thanked the interpreters and formally adjourned the meeting at 16^h15^m.

FINAL SESSION

Held in The Dome at Brighton, Sussex, England, on Thursday 27 *August* 1970 *at* 10^h00^m

Professor Dr O. Heckmann, President, in the chair

The Chairman referred to the General Secretary's motion made at the first session and put to the vote of the General Assembly the new Statutes and By-laws of the International Astronomical Union, as proposed by the Executive Committee. The Statutes and By-laws were adopted by show of hands of the official representatives of Adhering Organizations.

12 and 13. *Report of the Finance Committee and consideration of the unit of subscription.*

At the first session of the General Assembly a Finance Committee had been appointed to examine the accounts of the Union for the preceding three years and to consider the budget of the IAU for the coming three years, as prepared by the Executive Committee. The Chairman now called upon the President of the Finance Committee, Prof. A. G. Velghe, to present his report to the General Assembly, which was read as follows:

REPORT OF THE FINANCE COMMITTEE TO THE GENERAL ASSEMBLY

1. The Finance Committee, as appointed at the first session of the XIV General Assembly, appointed a Sub-Committee composed of Professor A. G. Velghe (Chairman), Professor E. Kharadze, Professor H. Elsässer, Professor L. Dezsö, and Professor P. A. Wayman. Professor D. W. N. Stibbs and Dr K. A. Strand attended meetings of the Sub-Committee by invitation of the Chairman.

2. The Sub-Committee examined the accounts for the years 1967 to 1969 as given in the Activity Report of the Executive Committee, together with reports of the Auditors for each year separately and explanatory notes presented by the General Secretary, Assistant General Secretary and Executive Secretary. Approval of these accounts was agreed and thanks were conveyed to those persons responsible for their preparation.

It was noted that a considerable improvement in the financial state of the Union had occurred between 1 January 1967 and 31 December 1969, as represented by an increase of assets of Dollars 23,498. There were two principal reasons for this improvement. Firstly there was an increase in annual subscription unit together with a larger subvention from UNESCO through ICSU. Secondly the publication contract with a publisher had been revised to take the form of a royalty for the Union, the publisher bearing the risk, and this had proved more satisfactory financially.

It was noted that the contributions of all countries were paid up-to-date by the time of the XIV General Assembly, a very satisfactory position.

It was noted that repayments of the substantial loan payments necessary in 1967 had been made according to the prepared schedule.

The Finance Committee recommends to the General Assembly that the accounts for the years 1967, 1968 and 1969 be approved.

3. The Sub-Committee considered the Supplementary Budget for the year 1970 as proposed by the Executive Committee. The net excess of Dollars 5,245 required for the year was regarded as a normal requirement and was approved.

The Finance Committee recommends that the General Assembly accept the Supplementary Budget for 1970 as proposed by the Executive Committee (see p. 24).

4. The Sub-Committee considered the budget for the three years 1971–73 and prepared, on the advice of the General Secretary, figures for consideration by the Finance Committee. In doing so the following points were noted:

(a) That the unit of contribution remains fixed at 900 gold francs until 1973;

(b) That it was henceforth necessary to allow a budgetary item to cover Officers' expenses in order to facilitate administration between Secretaries and President;

(c) That the revised form of distribution of Draft Reports, Transactions, etc. had resulted in a resolution of the financial difficulties occurring in 1967.

5. The Finance Committee recommends that the unit of contribution from Adhering Countries be maintained at the present level of 900 gold francs for the period 1971–73.

6. The Finance Committee considered the recommendations of the Sub-Committee and proposes the budgetary provisions for adoption at the second session of the XIV General Assembly as given in the Appendix.

If the General Assembly decides to hold the XV General Assembly in Australia Appendix A contains the recommended Budget. If the General Assembly decides to hold the XV General Assembly in Poland Appendix B contains the recommended Budget.

7. In conclusion, the Finance Committee recommends that the thanks of the Union be conveyed to the administrative staff for their unremitting work in effecting an improvement in the financial situation by executing the recommendations of the XIII General Assembly in this respect.

APPENDIX A
(Australia)

to the Report of the Finance Committee to the XIV General Assembly

COMPREHENSIVE BUDGET for 1971–73 (without UNESCO contracts)

Estimated Income

Heading	Origin	U.S. $
1.	Contributions from Adhering Countries	$161 000
2.	Sales of Publications	15 000
3.	Interest on reserves in Savings Banks	7 000
4.	UNESCO subvention through ICSU	42 000
	Total estimated income	$225 000

Estimated Expenditure

Heading	Purpose	U.S. $
1.	Administrative Office	$ 70 000
2.	Contribution to ICSU	3 000
3.	Specific projects of Commissions	32 700
4.	General Assembly expenses	41 000
5.	IAU publications	20 000
6.	Meetings of Executive Committee	14 000
7.	Meetings of Officers	5 000
8.	Symposia and Colloquia	* 25 000
9.	Interunion Commissions	8 000
10.	Projects authorized by the Executive Committee	1 300
11.	Reimbursement of loan from ICSU	5 000
12.	Summer Schools	**
	Total estimated expenditure	225 000

* Provided the XV General Assembly is held in Australia and a major astronomical meeting is arranged in Poland in 1973, the Finance Committee recommends the General Assembly to authorize the General Secretary to incur expenditure of up to Dollars 10 000 in excess of the sum of Dollars 25 000 specified under heading 8, and, if insufficient, this excess expenditure could be increased up to Dollars 15 000 with the consent of the Executive Committee, due regard being paid to the financial situation of the Union.

** If the category of adherence of some countries were sufficiently raised, or if additional unforeseen funds were made available to the Union, the item of Summer Schools should receive preferential consideration by the Executive Committee.

Details of individual items under heading 3 (Specific projects of Commissions) are given below:

Commission No.	Project	Sum requested by Commission (U.S. $)	Sum proposed by Finance Committee (U.S. $)
4	Information Bureau on Astronomical Ephemerides	600	600
6	IAU Telegram Bureau	1 600	1 600
16	Centre de documentation sur les planètes à Meudon	1 500	1 500
17	Special Working Groups of Commission 17	2 000	1 000
20	Minor Planet Center, Cincinnati	2 000	1 500
27	Catalogue of Variable Stars	3 000	3 000
38	Exchange of Astronomers	22 000	22 000
	Information Bulletin for the Southern Hemisphere	1 500	1 500
	Total	34 200	32 700

APPENDIX B
(Poland)

to the Report of the Finance Committee to the XIV General Assembly

COMPREHENSIVE BUDGET for 1971–73 (without UNESCO contracts)

Estimated Income

Heading	Origin	U.S. $
1.	Contributions from Adhering Countries	161 000
2.	Sales of Publications	15 000
3.	Interest on reserves in Savings Banks	7 000
4.	UNESCO subvention through ICSU	42 000
	Total estimated income	225 000

Estimated Expenditure

Heading	Purpose	U.S. $
1.	Administrative Office	70 000
2.	Contribution to ICSU	3 000
3.	Specific projects of Commissions	32 700
4.	General Assembly expenses	20 000
5.	IAU publications	20 000
6.	Meetings of Executive Committee	14 000
7.	Meetings of Officers	5 000
8.	Symposia and Colloquia	30 000
9.	Interunion Commissions	8 000
10.	Projects authorized by the Executive Committee	1 300
11.	Reimbursement of loan from ICSU	5 000
12.	Summer Schools	16 000
	Total estimated expenditure	225 000

Details of individual items under heading 3 (Specific Projects of Commissions) are given below:

Commission No.	Project	Sum requested by Commission (U.S. $)	Sum proposed by Finance Committee (U.S. $)
4	Information Bureau on Astronomical Ephemerides	600	600
6	IAU Telegram Bureau	1 600	1 600
16	Centre de documentation sur les planètes à Meudon	1 500	1 500
17	Special Working Groups of Commission 17	2 000	1 000
20	Minor Planet Center, Cincinnati	2 000	1 500
27	Catalogue of Variable Stars	3 000	3 000
38	Exchange of Astronomers	22 000	22 000
	Information Bulletin for the Southern Hemisphere	1 500	1 500
		34 200	32 700

A. G. VELGHE
Chairman, Finance Committee
August 1970

The President then formally requested the General Assembly to give approval to the report of the Finance Committee, explaining that voting would be by country, each country having a number of votes by one greater than the number of its category of adherence (Article 4 and 7(*b*) of the Statutes). The recommendations of the Finance Committee were approved unanimously.

The President thanked Professor Velghe, the members of the Finance Committee and the General Secretary for their valuable work.

14. *Special Nominating Committee.* The President reminded the General Assembly of the agreed procedure by which a Special Nominating Committee – to consist of the President of the Union, the past President (now in the chair), one retiring member of the Executive Committee (to be appointed by the incoming Executive Committee) and four other members (to be appointed by the General Assembly) – is appointed within one year following each General Assembly to propose to the General Assembly candidates for the new Executive Committee.

The President then moved that the General Assembly elect the following Members into the Special Nominating Committee:

G. A. Chebotarev (U.S.S.R.)
J. Delhaye (France)
D. S. Heeschen (U.S.A.)
A. Poveda (Mexico)

This list was adopted to the unanimity, with one abstention.

15 and 16. *Resolutions.* The General Secretary explained that the proposals for resolutions submitted to the General Assembly by the Executive Committee, the National Committees and by Commissions had eventually been passed on to the Resolutions Committee for consideration. Thus, before passing to the vote on the resolutions, Professor Oort would present to the General Assembly the report of the Resolutions Committee.

The President then called upon Professor Oort to take the floor.

REPORT BY THE PRESIDENT OF THE RESOLUTIONS COMMITTEE

In addition to the proposals for resolutions of Adhering Organizations, National Committees of Astronomy and Commissions formally included in the Agenda, the Resolutions Committee received a great number of proposals for resolutions from Commissions during the General Assembly.

All proposals for resolutions have been carefully examined, in some cases together with the appropriate Commissions, and the Resolutions Committee wishes now to make to the General Assembly the following proposal:

* To refer the proposal under 10a of the Agenda to the incoming Executive Committee
* To refer the proposal under 10b of the Agenda to Commission No. 46 as an internal matter of that Commission
* To deal with the proposal under 10c of the Agenda in conjunction with point 19 of the Agenda
* To adopt the proposal under 10d of the Agenda in the wording to be read later
* To adopt the proposal under 10e i of the Agenda in the wording to be read later
* To deal with the proposal under 10e ii of the Agenda in conjunction with point 19 of the Agenda
* To adopt the proposal under 10e iii of the Agenda in the wording to be read later taking into account the fact that the words '…and thus to create conditions for a reduction of the unit of contributions paid by Adhering Countries' have formally been withdrawn by the appropriate delegation from the original proposal.
* To refer the proposal under 10e iv of the Agenda to the consideration of the incoming Executive Committee
* To adopt the proposal under 10e v of the Agenda in the wording to be read later
* To approve the proposal under 11a of the Agenda as recommended by the Finance Committee
* To approve the proposal under 11b of the Agenda as recommended by the Finance Committee
* To adopt the proposal under 11c of the Agenda in the wording to be read later and to approve the financial part of that proposal as recommended by the Finance Committee
* To acknowledge the withdrawal by Commission No. 23 of the proposal under 11d of the Agenda

* To adopt the proposal under 11f of the Agenda in the wording to be read later
* To approve the proposal under 11g of the Agenda as proposed by the Finance Committee
* To refer the proposal under 11h of the Agenda to Commission No. 46

It should be noted that the proposal under 11e of the Agenda had already been approved at the first session of the General Assembly.

I shall now read the text of the resolutions proposed by the Resolutions Committee for adoption by the General Assembly. This text also covers such proposals of Commissions which had been received during the General Assembly and were found of sufficiently wide interest for consideration by the General Assembly (see next pages).

RÉSOLUTION NO. 1

Proposée par le Comité Exécutif

Sur les Émulsions Photographiques

L'Union Astronomique Internationale recommande aux Organisations Adhérentes et/ou aux Comités Nationaux d'Astronomie d'user de leur influence pour obtenir la suppression de toutes les restrictions qui gênent l'approvisionnement en émulsions photographiques d'intérêt astronomique.

RÉSOLUTION NO. 2

Proposée par le Comité National d'Astronomie de la R.A.U.

Sur la création d'une Commission pour les Échanges d'Équipement

La XIVème Assemblée Générale de l'Union Astronomique Internationale décide la création d'un Groupe de Travail Commun aux Commissions 9 et 46 pour les Échanges d'Equipement.

RÉSOLUTION NO. 3

Proposée par le Comité National d'Astronomie de l'U.R.S.S.

Sur la proclamation de 1973 comme Année Copernic

L'Union Astronomique Internationale recommande de proclamer l'année 1973 Année Copernic.

RÉSOLUTION NO. 4

Proposée par le Comité National d'Astronomie de l'U.R.S.S.

Sur l'abaissement du prix des publications de l'UAI

La XIVème Assemblée Générale de l'Union Astronomique Internationale charge le Comité Exécutif d'étudier la possibilité d'abaisser le prix des publications de l'UAI.

RÉSOLUTION NO. 5

Proposée par le Comité National d'Astronomie de l'U.R.S.S.

Sur la création au sein d'une des Commissions de l'UAI d'une Sous-Commission de la Physique des Étoiles Instables

La XIVème Assemblée Générale de l'Union Astronomique Internationale demande au Comité Exécutif d'inciter les Commissions Nos. 27, 35 et 36 à organiser des réunions spéciales pour discuter de la Physique des Étoiles Instables et pour envisager la création éventuelle d'un Groupe de Travail sur ce sujet.

RÉSOLUTION NO. 6

Proposée par la Commission No. 4 de l'UAI

Sur le Temps Universel

L'Union Astronomique Internationale,

Proposed by the Executive Committee

On Photographic Emulsions

The International Astronomical Union recommends that Adhering Organizations and/or National Committees of Astronomy use their influence to secure removal of all restrictions hindering the supply of photographic emulsions of interest to astronomers.

RESOLUTION NO. 2

Proposed by the U.A.R. National Committee of Astronomy

On the creation of a Commission for Exchange of Equipment

The XIVth General Assembly of the International Astronomical Union resolves to create a Joint Working Group of Commissions 9 and 46 for Exchange of Equipment.

RESOLUTION NO. 3

Proposed by the U.S.S.R. National Committee of Astronomy

On the proclamation of 1973 the Year of Copernicus

The International Astronomical Union recommends to proclaim the year 1973 the Year of Copernicus.

RESOLUTION NO. 4

Proposed by the U.S.S.R. National Committee of Astronomy

On less expensive IAU publications

The XIVth General Assembly of the International Astronomical Union instructs the Executive Committee to investigate the possibility of less expensive IAU publications.

RESOLUTION NO. 5

Proposed by the U.S.S.R. National Committee of Astronomy

On the creation within one of the IAU Commissions of a Sub-Commission on the Physics of Unstable Stars

The XIVth General Assembly of the International Astronomical Union requests the Executive Committee to encourage Commissions Nos. 27, 35 and 36 to organize special meetings for the discussion of the Physics of Unstable Stars, and to consider the desirability of creating a Working Group on this subject.

RESOLUTION NO. 6

Proposed by IAU Commission No. 4

On Universal Time

The International Astronomical Union,

ayant pris note des diverses propositions faites en vue de modifier la base actuelle du Temps Universel Coordonné (TUC) et

désirant insister sur le fait que les observateurs visuels dans les domaines astronomiques et voisins ont besoin de connaître le Temps Universel (TU1) avec la même précision, c'est-à-dire $0\overset{s}{.}1$, que celle définie par les possibilités sensorielles de l'homme,

demande instamment aux autorités compétentes de s'assurer que les moyens adéquats ont été utilisés pour que le temps TU1, ou la différence TU1-TUC, soit accessible à ces observateurs, et avec cette précision, avant de pouvoir permettre au TUC de différer du TU1 de plus de $0\overset{s}{.}1$.

RÉSOLUTION NO. 7

Proposée par le Groupe de Travail sur les Données Numériques

Sur la création d'un Groupe de Travail Permanent sur les Données Numériques

La XIVème Assemblée Générale de l'Union Astronomique Internationale décide de créer un Groupe de Travail Permanent sur les Données Numériques.

RÉSOLUTION NO. 8

Proposée par la Commission No. 17 de l'UAI

Sur la dénomination des cratères de la face de la Lune opposée à la Terre

L'Union Astronomique Internationale accepte les quelques 500 noms proposés par la Commission 17 de l'UAI sur la Lune pour désigner les cratères de la face de la Lune opposée à la Terre.

RÉSOLUTION NO. 9

Proposée par la Commission No. 33 de l'UAI

Sur l'indexation supérieure des nouvelles Coordonnées Galactiques

La XIVème Assemblée Générale de l'Union Astronomique Internationale décide de supprimer l'indexation supérieure II des nouvelles coordonnées galactiques qui, en conséquence, seront simplement désignées par *l, b*; les vieilles coordonnées conservent l'indexation supérieure I. Il serait bon que les auteurs précisent dans leurs articles que *l, b* se rapporte aux nouvelles coordonnées galactiques.

RÉSOLUTION NO. 10

Proposée par la Commission No. 40 de l'UAI

Sur l'utilisation de l'espace dans des buts scientifiques et autres

Tenant compte du fait que l'espace est utilisé d'une façon continue et croissante dans des buts scientifiques et autres, l'Union Astronomique Internationale désire attirer l'attention sur le 'Traité sur les Principes régissant les activités des États en matière d'exploration et d'utilisation de l'espace extra-atmosphérique, y compris la Lune et les autres corps célestes', en date du 27 janvier 1967, en particulier sur les Articles IV et IX de ce Traité, et réitère sa propre Résolution No. 1 adoptée en 1961 par la XIème Assemblée Générale et dont le dernier paragraphe est le suivant:

'L'Union Astronomique Internationale... demande instamment à tous les gouvernements engagés dans des expériences spatiales qui pourraient affecter la recherche astronomique, de prendre l'avis de l'Union Astronomique Internationale avant d'entreprendre de telles expériences, et de ne procéder à aucun lancement sans qu'il soit établi d'une manière irréfutable qu'aucun dommage ne peut en résulter pour la recherche astronomique.'

having noted the various proposals to modify the present basis of Co-ordinated Universal Time (UTC) and

wishing to emphasize that visual observers in astronomical and related fields require a knowledge of Universal Time (UT1) to a precision of the same order, namely $0\overset{s}{.}1$, as that of human time discrimination,

formally requests that the appropriate authorities ensure that adequate means have been provided of making UT1, or the difference UT1-UTC, available to such observers, and to such precision, before they permit UTC to depart from UT1 by more than about $0\overset{s}{.}1$.

RESOLUTION NO. 7

Proposed by the Working Group on Numerical Data

On the establishment of a Permanent Working Group on Numerical Data

The XIVth General Assembly of the International Astronomical Union resolves to establish a Permanent Working Group on Numerical Data.

RESOLUTION NO. 8

Proposed by IAU Commission No. 17

On the names of craters on the far side of the Moon

The International Astronomical Union accepts the approximately 500 names proposed by IAU Commission No. 17 on the Moon for designating the craters on the far side of the Moon.

RESOLUTION NO. 9

Proposed by IAU Commission No. 33

On the superscript of the new Galactic Co-ordinates

The XIVth General Assembly of the International Astronomical Union resolves that the superscript II be omitted from the new galactic co-ordinates: henceforth it will simply be *l, b*; the old co-ordinates retain the superscript I. It is desirable that authors indicate in their papers that *l, b* refer to the new galactic co-ordinates.

RESOLUTION NO. 10

Proposed by IAU Commission No. 40

On the uses of space for scientific and other purposes

Having regard to the continued and increased uses of space for scientific and other purposes the International Astronomical Union wishes to draw attention to the 'Treaty on Principles Governing the Activities of States in the Exploration and Use of Outer Space, including the Moon and other Celestial Bodies' of 27 January 1967, in particular to the Articles IV and IX of that Treaty, and to re-affirm its own Resolution No. 1 of the XI General Assembly, 1961, the final paragraph of which reads:

'The International Astronomical Union... appeals to all Governments concerned with launching space experiments which could possibly affect astronomical research to consult with the International Astronomical Union before undertaking such experiments and to refrain from launching until it is established beyond doubt that no damage will be done to astronomical research.'

RÉSOLUTION NO. 11

Proposée par la Commission No. 40 de l'UAI

Sur les Emissions Radio

L'Union Astronomique Internationale
considérant

(a) que les observations radioastronomiques sont faites au moyen de récepteurs d'une grande sensibilité,

(b) que, dans des conditions normales à la surface de la Terre, une certaine protection contre les émetteurs voisins est réalisée grâce à la courbure de la Terre et aux inégalités de terrain,

(c) que, en revanche, les engins spatiaux émettent vers les antennes de radio-astronomie par une transmission en ligne droite,

(d) qu'en raison de la haute sensibilité des récepteurs radioastronomiques, de l'étalement de modulation et des harmoniques de la fréquence porteuse, des interférences proviennent d'émetteurs opérant en dehors des bandes allouées à la radioastronomie,

(e) qu'une Conférence Mondiale Administrative Radio (WARC: World Administrative Radio Conference) doit avoir lieu en 1971 sous les auspices de l'Union Internationale des Télécommunications (UIT) et que son ordre du jour comprend l'étude des problèmes relevant de la Radioastronomie et de l'Astronomie Spatiale,

(f) que le Conseil International des Unions Scientifiques (CIUS) a créé le Comité Inter-Union pour les Allocations de Fréquence (IUCAF) afin de traiter en son nom les problèmes concernant l'utilisation du spectre des fréquences radio,

recommande que

(1) l'IUCAF fasse une démarche urgente auprès de la WARC pour lui demander d'imposer des règles strictes à l'emploi des émetteurs situés à bord des engins spatiaux, afin de réduire les émissions faites à partir de ces engins en dehors des bandes de fréquence allouées à la recherche spatiale, et ce d'une façon dépassant largement ce qui est exigé actuellement par les règlements de l'UIT pour les émetteurs au sol;

(2) l'IUCAF fixe les limites des émissions issues des émetteurs spatiaux en dehors des bandes allouées, en vue d'en faire la recommandation à la WARC;

(3) l'IUCAF insiste pour qu'un programme suivi soit entrepris par l'UIT afin de réduire encore plus les émissions de tous les émetteurs spatiaux et au sol en dehors des bandes allouées;

(4) l'IUCAF insiste pour que l'on décourage les émissions provenant des satellites dans celles des bandes allouées à la Radioastronomie sur la base d'un partage.

RÉSOLUTION NO. 12

Proposée par le Comité des Résolutions

Sur les résolutions adoptées par les Commissions

Considérant qu'il lui est pratiquement impossible d'accorder une attention particulière à chaque résolution adoptée par chacune de ses 39 Commissions et ayant une confiance entière en ses Commissions, cette Assemblée Générale désire exprimer son approbation des Résolutions adoptées par ses différentes Commissions et recommande aux astronomes de les appliquer dans toute la mesure du possible.

Quant aux autres propositions de résolutions, il est recommandé de les soumettre aux Commissions compétentes. Si elles sont adoptées en tant que Résolutions de Commissions, elles seront appuyées par la résolution 'générale' proposée ci-dessus.

RESOLUTION NO. 11

Proposed by IAU Commission No. 40

On Radio Transmission

The International Astronomical Union
considering that
(a) Radio Astronomy observations are made with receivers having great sensitivity,
(b) under normal earth surface conditions some protection from neighbouring transmitters is effected by earth curvature and terrain effects
(c) transmitters in space vehicles are exposed to radio astronomy antennae under line of sight conditions
(d) due to the high sensitivity of radio astronomy receivers interfering signals are received from transmitters operating outside the assigned radio astronomy bands due to modulation spillover and to harmonics of the carrier frequency
(e) there is to be a World Administrative Radio Conference (WARC), under the International Telecommunication Union (ITU), in the year 1971 whose agenda includes matters dealing with 'Space and Radio Astronomy'
(f) the International Council of Scientific Unions (ICSU) established the Inter-Union Committee on the Allocation of Frequencies (IUCAF) to speak for it in matters pertaining to the use of the radio frequency spectrum,
recommends that
(1) The IUCAF take urgent steps to ask the WARC to impose stringent regulations on transmitters located in space vehicles to insure that emissions from the space vehicles outside the assigned space frequency bands will be reduced to an extent greatly in excess of that presently required by ITU regulations for ground based transmitters;
(2) The IUCAF determine the limits on out of band transmissions by space transmitters to be recommended to the WARC;
(3) The IUCAF urge that a continuing program be undertaken by the ITU to further reduce out of band transmissions of all transmitters both earth and space bound;
(4) The IUCAF urge that transmissions from satellites in those bands allocated to Radio-Astronomy on shared basis be discouraged.

RESOLUTION NO. 12

Proposed by the Resolutions Committee

On the resolutions adopted by Commissions

Considering the impracticability of giving individual attention to every resolution adopted by each of its 39 Commissions, and having full confidence in its Commissions, this General Assembly wishes to give its endorsement to the Resolutions adopted by its individual Commissions, and recommends that astronomers give effect to these Resolutions in so far as they are able.

As to the remaining proposals for resolutions, it is recommended to refer them to the appropriate Commissions. In the form of Resolutions of Commissions they will be supported by the 'blanket' resolution, as proposed above.

J. H. OORT
Chairman, Resolutions Committee

The President then moved that the General Assembly approve the report of the Resolutions Committee and adopt resolutions 1 through 12 as recommended. This motion was carried unanimously.

17. *New Members of the Union.* The General Secretary announced that the Executive Committee had, on the proposal of Adhering Organizations and on the advice of the Nominating Committee, admitted 596 new members to the Union, and deleted 27 names from the membership list of the IAU. The names of new Members had been prominently displayed and the General Secretary did not propose to read them. He informed the General Assembly that these names would be incorporated in the alphabetical list of Members to appear in print.

18. *Commissions.*

(*a*) The General Secretary reminded that the General Assembly had, at the first session, approved the creation of two new Commissions, that is Commission No. 47 'Cosmology' and Commission No. 48 'High Energy Astrophysics', and the merger of Commissions Nos. 23 and 24 into Commission No. 24 on Photographic Astrometry.

(*b*) The General Secretary submitted, on behalf of the Executive Committee, the following list of Presidents and Vice-Presidents of Commissions for election by the General Assembly, subject to their willingness to serve:

Commission No.	President	Vice-President
4	J. Kovalevsky	R. L. Duncombe
5	J. B. Sykes	K. F. Ogorodnikov
6	P. Simon	J. Hers
7	G. N. Duboshin	P. J. Message
8	W. Fricke	G. van Herk
9	V. B. Nikonov	A. B. Meinel
10	J. T. Jefferies	K. O. Kiepenheuer
12	R. G. Athay	R. G. Giovanelli
14	A. H. Cook	R. H. Garstang
15	V. Vanýsek	A. H. Delsemme
16	G. H. Pettengil	S. K. Runcorn
17	M. G. J. Minnaert	G. P. Kuiper, B. J. Levin
19	H. M. Smith	C. Sugawa
20	F. K. Edmodson	L. Kresák
21	H. Elsässer	J. L. Weinberg
22	R. E. McCrosky	B. A. Lindblad
24	S. Vasilevskis	P. Lacroute
25	D. L. Crawford	M. Golay
26	J. Dommanget	Miss S. L. Lippincot
27	O. J. Eggen	M. W. Feast
28	Mrs E. M. Burbidge	E. B. Holmberg
29	Y. Fujita	O. C. Wilson
30	R. Bouigue	R. F. Griffin
31	G. M. R. Winkler	H. Enslin
33	S. W. McCuskey	L. Perek
34	F. D. Kahn	H. van Woerden
35	E. Schatzman	L. Mestel
36	R. N. Thomas	R. Cayrel
37	G. Larsson-Leander	I. King
38	A. Reiz	P. M. Routly
40	D. S. Heeschen	Yu. N. Parijskij

41	O. Gingerich	W. Hartner
42	M. Plavec	T. Herczeg
43	R. Lüst	R. J. Tayler
44	V. K. Prokof'ev	A. D. Code
45	B. Westerlund	C. O. R. Jaschek
46	Miss E. A. Müller	D. McNally
47	Ya. B. Zeldovich	P. J. E. Peebles
48	H. Friedman	M. J. Rees

This list was approved by the General Assembly

(c) The General Secretary explained that, according to the By-laws, the members of the Organizing Committees of Commissions are appointed by the Commissions themselves, subject to the approval of the Executive Committee and the General Assembly. He accordingly submitted for general approval the list of Organizing Committees which had been approved by the Executive Committee on the advice of the Nominating Committee. The composition of the Organizing Committees of Commissions as approved by the General Assembly are given in the list of Commissions on pages 289–302.

19. *The place and date of the XVth General Assembly*. The President explained that two proposals had been received as to the place of the XVth General Assembly: one, from the Australian Academy of Sciences, to hold the XVth General Assembly in Sydney in 1973, the other, from the USSR Academy of Sciences, to hold the XVth General Assembly in Poland in 1973. The Australian and Polish delegations had agreed on a compromise to be submitted to the General Assembly. This compromise had been worded as follows by the Executive Committee:

'The Executive Committee, in its meeting of 25 August 1970, has resolved to recommend to the XIVth General Assembly:

(I) To hold the XVth General Assembly in Sydney, Australia in the second half of August 1973 with the normal Agenda. A number of Symposia and Colloquia in conjunction with this General Assembly may be foreseen in Australia.

(II) To hold an Extraordinary General Assembly in commemoration of the 500th anniversary of the birth of N. Copernicus in the first half of September 1973 in Poland. This General Assembly will consist of the following three groups of Symposia:

A 1. Historical Symposium 'Colloquia Copernicana' in Toruń (Commission 41 of the IAU and the IUHPS Comité Copernic)

2. Planetary and Galactic Dynamics (IAU Commissions 7 and 33)

B 1. Symposium on Cosmology and Related Problems (IAU Commissions 28 and 47)

2. A topic from the field of Relativistic Astrophysics (IAU Commissions 47, 48 and the IUPAP)

C 1. Explorations of the Planetary System (IAU Commissions 15, 16 and 44)

2. A topic from the field of Stellar Evolution (IAU Commission 35).

of 3 Invited Discourses on the three above fields and of excursions 'On the Footsteps of Copernicus'.

The President then moved that this proposal be put on the Agenda for the General Assembly, and if approved, voted on by ballot. The General Assembly carried the motion to put the compromise on the Agenda with 35 against 7 votes, with no abstention. The General Assembly then approved the compromise with a majority of 28 against 14 votes, with no abstention. Scrutineers were J. B. Bok, T. Elvius and E. Rybka.

After having announced the result of the ballot, the Chairman called upon Dr H. W. Wood to address the Assembly. Dr Wood spoke as follows:

'The First suggestion that Australia should offer an invitation to the International Astronomical Union to hold a General Assembly in Australia came from Professor Otto Struve in 1954. He was then President of the Union and was at a meeting of the Australian National Committee for Astronomy. Soon after this invitations began to be offered and our President has outlined the actions taken by the Executive Committee and the correspondence with the Australian representatives. I am pleased now that I shall be able to report to the Australian Academy of Science the acceptance by the Union of the invitation.

Some preliminary organization has been done. The accommodation which can be offered for commission meetings at the University of Sydney, where the meeting would be held, is of very satisfactory standard and the Campus of the University would provide a pleasant setting. Preliminary arrangements have been made for residential accommodation in some of the University Colleges. Enquiry has also been made for halls in which the large meetings such as General Assemblies and invited discourses could be held and it appears that satisfactory accommodation would be available.

Although it is a hazardous thing to try to predict weather so far ahead, a table of weather information I have compiled shows that there would be a good chance of pleasant sunny days in Sydney towards the end of August. For those who are acquainted with the Mediterranean Coast, it seems that August weather conditions in Sydney are generally similar to those in Marseille in May.

I look forward to seeing many of you in Sydney at the next General Assembly of the Union and hope that you will enjoy your stay in Australia.'

The President thanked Dr Wood and invited Professor W. Iwanowská to speak. Professor Iwanowská presented the Polish invitation as follows:

'Mr. President, Ladies and Gentlemen,

In the name of the Polish Academy of Sciences I have the honour to invite the International Astronomical Union to hold an Extraordinary General Assembly in commemoration of the 500th anniversary of the birth of Nicolaus Copernicus in 1973 in Poland.

The Polish Government warrants a visa and the access to Poland to each legal participant of this Assembly.

The Polish Academy of Sciences will provide grants for about 100 young astronomers, mostly from developing countries, in order to help them to participate in this Assembly.

There are sufficient and appropriate rooms in Warsaw and in other cities for the meetings of this Assembly, as well as hotel accommodations for more than 4000 participants. There are also available student houses in Warsaw and new University cities in Toruń and Cracow, not so splendid as the University of Sussex, yet we hope that our guests will find their stay comfortable and pleasant in our country.

As we have heard, the programme of this Extraordinary General Assembly will contain not only a commemorative session and a historical symposium, but also five other symposia covering those branches of astronomy in which considerable progress has been made in last years. We consider that this is the most appropriate way for the Astronomical Union to give its tribute to Copernicus. His importance for science consists not so much in achievement as in starting: he started modern astronomy and modern science. Therefore we believe that he would be pleased not only by our appreciation of his own work, but also by hearing what has been done in astronomy after him, in the span of less than five hundred years.

To this official invitation I should like to add an informal one. Young Polish Astronomers which were privileged to stay and work at many excellent Observatories and Institutes abroad, asked me to say, how much they would be honoured and glad to see their patrons and many colleagues from abroad in their country as their guests.'

The President thanked Professor Iwanowská for her words.

20. *Election of the new Executive Committee.* The Chairman formally proposed that Professor Bengt Strömgren be elected the new President of the IAU. This was approved by acclamation.

The President reminded the Assembly of the names put forward for consideration at the first session, and formally proposed that Professor B. J. Bok, Professor Sir Bernard Lovell and Professor

E. R. Mustel be elected Vice-Presidents in place of Professors W. N. Christiansen, M. Schwarzschild and A. B. Severny. This was unanimously approved. He then formally proposed that Professor Dr C. de Jager be elected General Secretary in place of Dr L. Perek, and Professor Dr G. Contopoulos Assistant General Secretary in place of Professor de Jager. This was unanimously approved.

The Chairman then invited Professor Strömgren, Professor B. J. Bok, Sir Bernard Lovell and Professor G. Contopoulos to take seats on the platform.

21. *Addresses by the retiring and newly elected Presidents.* The retiring President, Professor O. Heckmann, addressed the Assembly as follows:

'Ladies and Gentlemen, dear Colleagues:

We are approaching the end of our Assembly, when the President takes the opportunity to thank all those with whom he was in close cooperation during the three past years.

It was a great satisfaction indeed to work with the *Executive Committee:* To the whole group and to every single person in it I feel obliged for the atmosphere of frankness and friendship which dominated all our sessions. Our deliberations, sometimes after tough and serious discussions, ended with an astonishing degree of unanimity; this was due to the fortunate talent throughout the group of carefully listening to arguments and weighing them according to their merits for the welfare of the Union, a principle foremost in all our efforts. A cordial vote of thanks goes to the three Members who are now leaving the Executive Committee: Christiansen, Schwarzschild, and Severny. Each one of them made, on various occasions, decisive contributions to our work.

The Executive Committee was very fortunate to have had the permanent active advice of the former President, Professor Swings, and the former General Secretary, Professor Pecker. Pol Swings has served in four successive Executive Committees and Jean Claude Pecker in three. Their long experience was always a safeguard for the continuity of our work. I ask you kindly to show your appreciation by your applause.

The quality of the work of the Union largely depends on the *Special Nominating Committee.* They deserve a special vote of thanks. I was impressed by the clarity of their judgement and I am certain that you will find their proposals, now accepted, satisfactory in the future.

I also should like to express my indebtedness to the *Working Group for the Revision of the Statutes and By-Laws.* I do hope that the life of the Union will be reasonably guided by the new system of rules, which contain the whole substance of the former ones, yet are easier to deal with.

When, in 1967, I took over the Presidency of the Union I declared that I felt happy and even blessed to have the support of two outstanding *General Secretaries.* Today I have to state that my confidence has been fully justified. The permanent contact and frequent meetings I had with Luboš Perek and Kees de Jager gave me the reassurance, in far-reaching matters as well as in smallest details, that with such Secretaries the Union would have a good chance to solve all problems inherent in the explosive growth of astronomy: Their deep love for our science, their clear and careful analysis of complex and delicate problems, their courageous and realistic proposals in preparing decisions – to see all this at work with the most industrious concentration and passionate precision was an experience which filled me with deep admiration and gratitude.

Luboš and Kees, your personal friendship is among the most precious gifts I ever received. Thank you for everything you did for the Union during my three years' term.

Yet there is a third man involved, Dr Arnost Jappel. As Executive Secretary he has been connected with the Union over five years; and each time with a new General Secretary he had and has to change his place of residence. During the time he has been in office the Union has had the full-time service of a man who not only speaks at least six living languages but has, as a professional jurist and experienced administrator, just those qualities the Union has to rely on now and in the future. In particular, I should like to say that without Dr Jappel's juristic conscience the new Statutes would never have gained the quality which I hope the next Executive Committees will esteem. Again, I ask you kindly to show your appreciation in which you also should include the work of our Secretary Mrs Daňková by your applause.

Now I have to turn to our British friends:

First of all our thanks go to our principal host, the Royal Society, acting through the British National Committee of Astronomy. Under the able Chairmanship of Sir Bernard Lovell it approved the appointment of the National Organizing Committee, of which again Sir Bernard was the Chairman, of the Finance and Policy Committee with its Chairman Professor Tayler, and of the Local Organizing Committee chaired by our old friend Donald Sadler. All these Committees, according to their respective level and task, showed their hospitality to our big Assembly and deserve our gratitude. The associated host-organizations were the Corporation of Brighton under Alderman Nettleton and the University of Sussex with its Vice-Chancellor Professor Briggs. We know very well that all these bodies are no abstract entities, but living beings, who had to display much personal initiative and enthusiasm and to transfer them to numerous helpers. We feel deeply obliged to everyone involved. As for the County of Sussex with its beautiful coastline, many astronomers will wish to return to it; and with respect to the University, I am certain I am not the only one who would wish to be young again and to study here.

Our indebtedness is also due to the Astronomer Royal, Sir Richard Woolley, who made the doors of the Royal Greenwich Observatory at Herstmonceux wide open to all of us. Many of us will have gained a deep impression of the enviable possibilities this great institution offers for vast fields of modern research.

Returning to the Local Organizing Committee I feel obliged to say a few words about Donald Sadler.

Having been General Secretary and Adviser to the Executive Committee for 9 years, he became a member of the Special Nominating Committee and of the Working Group for the Revision of the Statutes and By-Laws. Finally, 3 years ago, he took over the heavy burden of the Chairmanship of the Local Organizing Committee. In all of these functions his efforts to achieve a maximum of clarity in all branches of the Union's activities will not be forgotten. When, in later years, someone will write the history of the Union he will easily show to what an astonishing extent the Union's structure is a monument of Donald Sadler's work – The Union may have quite a number of weaknesses. At this moment I feel it is a very weak point indeed that we cannot give Sadler a high rank in an order of merit.

Finally I thank all Members of the Union for their support and every participant for his or her contribution to this Assembly.'

French translation:

Allocution finale du Prof. Heckmann

'Mesdames, Messieurs, Mes chers Collègues,

La fin de notre Assemblée Générale approche, et c'est pour le Président le moment opportun de remercier tous ceux avec lesquels il a eu l'occasion de coopérer étroitement au cours des trois années qui viennent de s'écouler.

Ce fut en vérité une source constante de satisfaction que de travailler avec le *Comité Exécutif*. A ce groupe tout entier et à chacun de ses membres en particulier, je tiens à exprimer ma gratitude pour l'atmosphère franche et cordiale qui a caractérisé nos réunions. Nos débats, souvent après des discussions serrées et sérieuses, se sont toujours conclus par une remarquable unanimité; ceci est dû à l'heureux talent qu'ont eu les membres du groupe pour écouter attentivement les arguments, et pour les apprécier justement dans le souci de l'intérêt de l'Union, qui fut le principe intangible de tous nos efforts. L'expression cordiale de nos remerciements va aux trois membres du Comité Exécutif qui le quittent maintenant: Christiansen, Schwarzschild et Severny. Chacun d'eux a contribué, en des occasions diverses, et de façon décisive, à nos travaux.

Le Comité Exécutif a eu la chance d'avoir l'avis permanent et actif de l'ancien Président le Professeur Swings et de l'ancien Secrétaire Général le Professeur Pecker. Pol Swings a servi dans quatre Comités Exécutifs, J.-C. Pecker dans trois. Leur longue expérience a toujours permis d'assurer la continuité de notre travail. Je vous demande de leur montrer votre appréciation par vos applaudissements.

La qualité du travail de l'Union dépend grandement du *Comité Spécial des Nominations*. Ses membres méritent une expression spéciale de nos remerciements. J'ai été impressionné par la clarté de leur jugement et je suis sûr que vous trouverez que leurs propositions, acceptées maintenant, se montreront dans l'avenir très satisfaisantes.

Je souhaite aussi exprimer ma reconnaissance au *Groupe de Travail pour la Révision des Statuts et des Règlements*. Je veux espérer que la vie de l'Union sera raisonnablement guidée par le nouveau système de règles qui, sans s'éloigner sur le fond des règles anciennes, seront d'un emploi plus pratique.

Quand en 1967 je suis devenu Président de l'UAI, j'ai déclaré que j'étais heureux, et même que c'était pour moi une bénédiction, d'avoir l'aide de deux remarquables Secrétaires Généraux. Aujourd'hui je dois dire que ma confiance a été pleinement justifiée. Le contact permanent, les réunions fréquentes que j'ai eues avec Luboš Perek et Kees de Jager m'ont renforcé dans mon sentiment que, pour des problèmes de grande importance aussi bien que dans les plus petits détails, l'Union, avec de tels Secrétaires, aurait les plus grandes chances de pouvoir résoudre tous les problèmes liés à la croissance explosive de l'astronomie: leur amour profond de notre science, leur analyse claire et soigneuse de problèmes délicats et complexes, leurs propositions courageuses et réalistes dans la préparation des décisions – de voir tout cela aller de pair avec la concentration la plus active et une précision passionnée, fut pour moi une expérience qui m'a rempli d'une admiration et d'une gratitude profondes.

Luboš et Kees, votre amitié personnelle est l'un des dons les plus précieux que j'ai jamais reçus. Merci de tout ce que vous avez fait pour l'Union au cours de ces trois années.

Un troisième homme encore est concerné, le Dr Arnošt Jappel. Comme Secrétaire Exécutif, cela fait cinq ans qu'il est associé avec l'Union; chaque fois qu'un nouveau Secrétaire Général fut désigné, il a dû, et doit encore, changer de résidence. Pendant son activité au sein de l'Union, il a eu les responsabilités, à plein temps, d'un homme qui non seulement parle six langues, mais encore possède, en tant que juriste de profession et administrateur expérimenté, précisément ces qualités dont l'Union a besoin maintenant comme demain. En particulier, je voudrais dire que, sans la conscience des problèmes juridiques qu'a le Dr Jappel, les nouveaux Statuts n'auraient jamais atteint la qualité que, j'espère, le nouveau Comité Exécutif sera à même d'estimer. A lui aussi je vous demande de montrer notre appréciation et de lui associer, dans vos applaudissements, notre Secrétaire, Madame Daňková.

Je me tourne maintenant vers nos amis britanniques. Tout d'abord, nos remerciements vont à notre hôte principal, la Royal Society, qui a mené son action par l'intermédiaire du Comité National Britannique d'Astronomie. Sous la présidence efficace de Sir Bernard Lovell, la Royal Society a approuvé la désignation du Comité National d'Organisation dont Sir Bernard Lovell fut aussi le Président, celle du Comité des Finances et de l'Orientation, sous la présidence du Professeur Tayler, et le Comité Local d'Organisation, présidé par notre vieil ami Donald Sadler. Tous ces comités, dans le cadre de leurs responsabilités respectives, ont montré leur hospitalité à notre grande Assemblée et ont droit à notre gratitude. Les organisations associées dans cette hospitalité furent la Corporation de Brighton sous la présidence de Alderman Nettleton et l'Université du Sussex avec son Vice-Chancelier, le Professeur Briggs. Nous savons très bien que tous ces organismes ne sont pas des abstractions, mais des êtres vivants qui ont dû déployer un grand enthousiasme, susciter beaucoup d'initiatives individuelles, s'associer des aides nombreuses. Nous nous sentons profondément reconnaissants vis-à-vis de tous ceux qui nous ont ainsi aidés et accueillis. Quant à la Province du Sussex, riche de ses admirables rivages, nombreux sont les astronomes qui voudront y revenir; et pour ce qui concerne l'Université, je suis sûr de ne pas être le seul à souhaiter redevenir jeune pour y étudier.

Notre dette est grande aussi vis-à-vis de l'Astronome Royal, Sir Richard Woolley, qui nous a grand ouvert les portes de l'Observatoire Royal de Greenwich à Herstmonceux. Beaucoup d'entre nous auront conservé une profonde impression des possibilités enviables que cette grande institution offre à de vastes domaines de la recherche moderne.

Je reviendrai maintenant au Comité Local d'Organisation, car je me sens obligé de dire quelques mots de Donald Sadler.

Il a été Secrétaire Général, puis Conseiller du Comité Exécutif pendant neuf ans; il est devenu membre du Comité Spécial des Nominations et du Groupe de Travail pour la Révision des Statuts et Règlements. Enfin, depuis trois ans, il assume la lourde charge qu'est la présidence du Comité Local d'Organisation. Dans chacune de ces fonctions, ses efforts pour arriver à un maximum de clarté dans toutes les branches des activités de l'Union ne pourront pas être oubliés. Plus tard, quand quelqu'un écrira l'histoire de l'Union, il montrera sans mal à quel point la structure de l'Union est en vérité un monument étonnant à Donald Sadler. L'Union peut en vérité avoir ses faiblesses. Aujourd'hui, il me semble qu'elles ne peuvent nous empêcher de donner à Sadler un rang très élevé dans la hiérarchie du mérite.

Enfin, je remercie tous les membres de l'Union de leur appui, et chaque participant, ou participante, de sa contribution à cette Assemblée.

The newly elected President, Professor Strömgren, then addressed the Assembly as follows:

'Dear Colleagues, Ladies and Gentlemen,

In addressing you I wish, first, to express my deep gratitude to the Assembly for having elected me President of the IAU. It is indeed a great honor to serve in an office that has been held by so many outstanding Astronomers; and it is both a challenge and a source of great satisfaction to be given the opportunity to work for the Union that means so much to the astronomical community.

To you, Otto Heckmann, I want to say that for me it is not only a great honor, but also a heartwarming experience to have been elected *your* successor. We have been friends for forty years, and to me, as to others, you have been a central and admired figure in our scientific lives. Only a few days ago, when the University of Sussex bestowed upon you a great honor, we listened to an admirable account of your achievements. On this subject I must be brief. However, three great contributions come to mind on this occasion. Your pioneer work on the color-magnitude diagram of galactic clusters is an example of great scientific achievement resulting from *continuity* in effort and *innovation* in the formulation of the goal. In your work for ESO you and your collaborators have succeeded in building a splendid foundation for future developments that will be of the greatest importance for European Astronomy. And as President of the IAU you have headed an Executive Committee with an admirable record of achievements.

To the newly-elected Executive Committee it means very much indeed that Otto Heckmann and Luboš Perek will council us and support our efforts through the next three-year period.

At the Berkeley General Assembly and at every General Assembly since then both the opportunities and the difficulties stemming from the growth of the IAU have been emphasized. In addressing the Thirteenth and the Fourteenth General Assemblies Otto Heckmann has spoken strongly and convincingly against fragmentation of the IAU as a solution.

I share his view on this, but the problems remain, and the newly-elected Executive Committee is accepting a challenge of great magnitude. I trust that in its work the Executive Committee will, as in the past, have the invaluable cooperation of the National Committees and the Commissions of the Union, also between Assemblies.

Among Scientific Unions the IAU has an admirable record in that almost every scientific activity and scientific center of the astronomical community is strongly represented in the life of the Union. I want to express the hope that this will continue, and that we shall see further growth and strengthening in this respect.

May optical-, radio-, and space-astronomers as well as physicists working on astronomical problems regard the IAU as an excellent forum for the presentation of their results and for strong scientific interactions with colleagues working in other fields. And may the IAU be able to contribute to the solution of science-administrative problems arising out of the diversity of the research efforts in Astronomy.

When the Fourteenth General Assembly of the IAU opened the Union had received two fine and gratefully acknowledged invitations for the next General Assembly. The present General Assembly in a spirit of collaboration has made its decision, and when we look ahead we see great opportunities. But it may take some effort, by all of us, to take full advantage of the opportunities without losing out, temporarily, in the struggle for integration and unity in the scientific life of the IAU.

I am convinced that the IAU, in good or difficult times, has enormous strength because its members are in considerable measure united: in striving for addition to knowledge in a great field of science, and through their conviction that the goals are best attained in co-operation.

May the International Astronomical Union thrive in the coming years.'

22. *Closing Ceremonies*. The Chairman called upon Professor A. Blaauw to propose a comprehensive vote of thanks. Professor Blaauw then spoke as follows:

'Dear Colleagues, Ladies and Gentlemen,

It is my privilege to have the opportunity of proposing, on behalf of the participants, a vote of thanks to all those, through whose efforts this General Assembly has become so successful, and of conveying to them our feelings of gratitude and indebtedness.

We were with more than 2000 participants, and if we would take a poll, there would be undoubtedly as many opinions expressed as to what the Union and this General Assembly have meant to us, – but I am sure we are *unanimous* in our deep appreciation of what the organizers of this Assembly have done for us, and for what it implies for the IAU.

As one who has attended a fair number of assemblies – since I had, as a student, my first cautious look at the one in Stockholm in 1938 – one notices the changing nature of the Assemblies. Some things have changed enormously. The audience has become ten times larger, the number of commissions has grown considerably; the participants, I believe, are on the average much younger than they were then; many new discoveries entered the picture, but much also remains the *same:* The feeling of anticipation with which we travel to the meetings to hear about recent developments and discoveries, and the anxiousness to meet colleagues and friends for private discussion of our work. One wonders whether not, indeed, this part of the role of the Union's General Assembly has become of greater importance than ever: providing a meeting place for individuals and informal gatherings *as well* as for commission sessions.

A strong recollection of this assembly will be that of the spacious rooms at Falmer House with their colourful chairs, occupied by participants in conversation in all degrees of seriousness, a continually changing, lively population. I guess that for many of us *this*, in the midst of the green lawns and the modern lecture facilities, will be the image we carry home of one of our excellent hosts, the University of Sussex. Before we leave, we want to express our gratitude to all those who made this event possible.'

'Le Président a exprimé sa gratitude à nos Secrétaires Généraux pour leur travail exceptionnel pour l'Union. Je suis sûr que vous me permettrez, Monsieur le Président, d'ajouter quelques mots au nom des participants sans grade.

Louis XIV disait: 'l'État, c'est moi...'; pour nous, les astronomes ordinaires, quelque soit notre pays ou notre spécialité, nous pensons: l'Union, ce sont nos Secrétaires Généraux. Nous attendons d'eux les directives et les sages conseils dans de si nombreuses questions concernant l'Union. Le fardeau des problèmes à résoudre par l'Union repose normalement sur eux. Une grande partie de leur travail est faite en silence, sans que nous ne nous en apercevions jamais. Nous ne prenons vraiment conscience de l'énorme travail qu'ils font pour nous, et de leur dévouement désintéressé à la cause de l'Union, qu'à l'occasion des Assemblées Générales.

Nous savons que Kees de Jager continuera trois ans encore à travailler pour nous. Mais Luboš Perek va nous quitter, et je suis certain d'exprimer un sentiment partagé par tous en vous remerciant profondément, Luboš, pour tout ce que vous avez fait pour nous durant ces six années. Nous vous souhaitons un retour heureux et bénéfique à vos travaux de recherche auxquels vous êtes si attaché.'

Our hosts were manifold: the Royal Society through the National Committee for Astronomy, the University of Sussex, the Corporation of Brighton and the Royal Greenwich Observatory. The President has already expressed our deep obligation with respect to all of these. We, participants, are filled with the greatest admiration for what the Local Organizing Committee has achieved under the chairmanship of Dr Sadler. As with so many great enterprises, success or failure depends to a large extent on one man in whose hands lies the direction of the programme. To you, Dr. Sadler, go our warmest thanks for the outstanding way in which you guided the organization for these meetings. The President, who has witnessed the preparatory work of your Organizing Committee from so much closer than we could do, has just expressed his feelings of deep appreciation, and I can do no better than tell you how much these are shared by all of us. Thank you very much indeed.

Busy as we have been, I guess that many of us have had only a quick glance, if any, on those pages of our programme booklet which list the *names* of the Local Organizing Committee and their aids. We were *too* busy running from room I to room X, to room IV, to room XI etc. guided by the *specially erected* IAU street signs, but at this moment we think of the admirable way in which these and countless other provisions were made under the responsibility of the registrar of the University of Sussex, Mr. Shields. We are reminded at *this moment* again of the many arrangements that were made here in *Brighton* and deeply appreciate the work done by Mr. Bedford of the Corporation of Brighton and his staff. If, during one of the excursions we had a chance to withdraw for a while from the busy dealings on the Campus, and to experience the serene atmosphere in one of the beautiful old churches in the lovely countryside; or if we enjoyed one of the many cultural evening events, this was, as I understand, made possible through the excellent arrangements made under the responsibility of our astronomical colleagues Dr Smith of the Royal Greenwich Observatory and Dr Tayler of the University of Sussex. We thank them for all they did, and include in it their excellent administrative and secretarial staff Mr Pepperel, Miss Hanning, Miss Adams, Mrs Gillingham and Mrs Norris.

A special word of thanks we owe also to the local participants – mostly junior members of the scientific staff and research students, who assisted us and the Organizing Committee so magnificently – here and at Herstmonceux – and we are equally grateful for the helpful and cheerful assistance of the many staff members of the University of Sussex.

May I conclude by assuring all of those who in any way contributed to the success of this assembly that what they did goes beyond providing us with the smoothly running facilities of a very large and complicated assembly. It is one of the very good things of the IAU that it has been able to keep the various branches of astronomy together. We have noticed how desirable this remains, seeing how domains that may at one time have appeared to be unrelated, unexpectedly became fields of common interest. But to continue, having them all dealt with at one assembly will only be possible if such large meetings as this one can really be handled. By showing us that it could be done indeed, the Local Organizing Committee has rendered us a great service, for which we feel deeply obliged. We thank you most cordially.'

On behalf of the ladies, Mrs Christiansen addressed the Assembly as follows:

'Mr President, Ladies and Gentlemen,
    It is with the warmest pleasure that, today on behalf of the women guests, I thank our hostesses and hosts for all that they have done for us.
    Every woman present realizes and appreciates the enormous amount of time, energy and thought which has gone into the preparation and the carrying out of the programme. Every possible arrangement has been made for our comfort and convenience. The hospitality of the common room, the music, the excursions, the play, the receptions and informal gatherings have been a constant delight.

I suppose that the primary object of the organizers of the women's activities has been to give pleasure to their guests. This they have done abundantly.

But we should like them to realize that we do appreciate the deeper purpose that has been served. This has been yet another opportunity for new friendships to be made among women who come from many lands and who speak many different languages; an opportunity for old friendships to be strengthened.

In this way also the Assembly has contributed to that all important task of mankind, the furtherance of international understanding.

Will you please accept our sincere thanks.'

The Chairman then called upon the retiring General Secretary, Dr Perek, who spoke as follows:

'Three years ago I expressed from this platform – in another country and in another city – the confidence that pleasant aspects of the post of the General Secretary of the IAU would prevail. Today I can say that they did and overwhelmingly so.

The administration of the Union requires considerable work and in this I was helped by an understanding Executive Committee, by a deeply interested President and by a friendly and most dependable Assistant General Secretary. In the Secretariat itself I have always had the benefit of legal advice and opinion of Arnošt Jappel, of the tallying accounts kept by Mrs. Daňková, and of the perfect typing of Mrs. Hermanová and later of Miss Tůmová. Here, at the General Assembly, all matters concerning membership were expertly handled by Mrs. Kohoutek-Sachs and Miss Drouin spared no effort in collecting documents to be published in the proceedings. The youngest member of our staff, Miss Reijnen, was prevented by a sudden illness, which confined her to a hospital here in Brighton, from continuing in her enthousiastic help.

My thanks are due to many Members of the Union for their patience in reading the 3713 letters I wrote to them from Prague. May I express this time my hope that all those letters were understood in the same friendly and cooperative spirit in which they had been written.

The administration of the Union becomes more and more complex. It is not only the growth in membership which requires more organization but also the increasing involvement in interdisciplinary programmes and, evidently most important, the spectacular expansion of many branches of astronomy. My successor will have a pleasant but difficult job. He will need all the support you have given to me, and perhaps more. I wish him full success.'

The Chairman then invited the new-elect General Secretary, Professor de Jager, to take the floor. Professor de Jager had the following to say:

'Mr. President, dear Colleagues,

Being asked to serve the Union for another three years, in a function held in the past by so many eminent astronomers, I would, first of all, thank the General Assembly for their confidence in me. I realise the weight of my new task, yet I accept it with a feeling of happiness. Our Union, though growing rapidly, has remained a large family, closely held together by our beautiful science. In this family I feel my future function somewhat like that of the kitchen master. I shall carefully watch that the food is never burnt or served uncooked!

The past three years were, for me of course, a period of some hard working, but above all, one of most cordial contacts and friendships with so many astronomers; too many to mention. Yet I hope that you will allow me, once again, to thank our retiring President, Otto Heckmann, for the energetic, but at the same time so human and wise way he has guided the Union. To serve under his Presidentship was a great experience; I thank him for his continuous friendship.

Most admirable was the way Luboš Perek dealt with so many intricate problems of the Union's secretariat. Gradually, cordially and friendly in the literal sense of the word he prepared me for my future function, and thus guaranteed the continuity in the chain of Union Secretaries.

I am very pleased that the IAU staff, my good friend, the executive secretary Arnošt Jappel, and

Mrs Daňková have accepted my invitation to move with the Secretariat to Utrecht. The Union owes very much to these devoted staff members; the continuation of their services is the best guarantee for a sound future of the IAU.

Thank you all again!'

The Chairman then closed the meeting around noon with the following words:

'I thank you for your attendance, and I wish you all a happy return home. I hereby formally declare closed the fourteenth General Assembly of the International Astronomical Union.'

PART 3

REPORTS OF MEETINGS OF COMMISSIONS

——

COMPTES RENDUS DES SÉANCES DES COMMISSIONS

COMMISSION 4: EPHEMERIDES (EPHÉMÉRIDES)

Report of Meetings 19, 20 and 25 August 1970

PRESIDENT: G. A. Wilkins.
SECRETARY: A. T. Sinclair.

The Commission held five meetings, of which the first and fourth were concerned with general matters. The second and third meetings on 20 August, were devoted solely to an open discussion of the resolutions adopted at IAU Colloquium No. 9 on the IAU system of astronomical constants; many members of other Commissions were present. The fifth meeting was held jointly with Commission 31 on time scales and its report appears with those for the meetings of Commission 31.

ORGANIZATIONAL MATTERS

It was agreed that the Officers and Organizing Committee for the period 1970-3 should be as follows: President, J. Kovalevsky; Vice-President, R. L. Duncombe; Organizing Committee, V. K. Abalakin, W. Fricke, A. M. Sinzi and G. A. Wilkins. Proposals for 17 new members and Consultants were accepted. It was agreed to remove from the list of members those persons who had not indicated in any way during the past three years their continued interest in the activities of the Commission. The final list of members is given on page 289.

The reports of the President and of the Directors of the national ephemeris offices, which had been previously circulated to all members of the Commission, were accepted; they are published *in Trans. IAU* XIVA, 4-9, 1970. The President's suggestion that the Commission should keep under review the comparison between ephemerides and observations was unanimously agreed, although it was noted that Commission 17 had appointed a working group on the figure and motion of the Moon, and wished to resume its interest in the results obtained from observations of occultations by the Moon.

INFORMATION BUREAU ON ASTRONOMICAL EPHEMERIDES

The President reported that the proposal of the Working Group on Ephemerides for Space Research for the establishment of an International Information Bureau on Astronomical Ephemerides (see pp. 84, 85 for details) had been provisionally accepted by the Executive Committee of the Union. Further *J. Kovalevsky* had generously agreed to provide appropriate facilities at the Bureau des Longitudes in Paris, and to appoint B. Morando to manage its activities. On the proposal of *R. L. Duncombe* and *F. M. Sadler*, the Commission formally recommended that the Information Bureau should be established. It was then agreed that the Commission's representatives on the joint committee should be R. L. Duncombe and T. Lederle; the other representatives will be appointed by the Bureau des Longitudes and COSPAR Working Group I.

IAU WORKING GROUP ON NUMERICAL DATA

The President drew attention to the formation by the Executive Committee of the Union of a Working Group on Numerical Data in Astronomy and Astrophysics. At the open meeting held on 21 August he and T. Lederle had reported briefly on the experiences of Commission 4 in attempting to coordinate the exchange of astronomical ephemerides and positional catalogues. Others had discussed the difficulties of compiling catalogues of physical data and the possibilities of establishing comprehensive data centres which would provide facilities for information retrieval using computer techniques. There seemed to be general agreement to the view that priority should be given to the collection and dissemination of information about the availability of relevant data. The experience gained in the establishment of the Information Bureau on Astronomical Ephemerides should provide useful guidance for similar activities in other fields. (See report on p. 79).

M. Davis reported on the proposals of Commission 7 to set up data and program banks in the field of celestial mechanics; at first the emphasis would be on the collection of general algorithms rather than of machine-dependent programs.

LUNAR-RANGING EXPERIMENTS AND DATA

J. Kovalevsky reported that, at its meeting in Leningrad in May 1970, COSPAR had set up a working party to encourage international cooperation in all aspects of lunar ranging experiments, including the prompt exchange of observational data. Commission 17 had adopted a resolution (see pp. 141, 142) suggesting that all interested Commissions of the IAU should be represented on the Working Party; its Chairman, C. O. Alley, said that such representation would be welcomed. It was agreed that Commission 4 would support the resolution and that J. Kovalevsky should represent the Commission.

BASIS OF STANDARD TIME SIGNALS

The President introduced a discussion on the proposals of CCIR for changing the basis of the standard-frequency and time-signal emissions by stating that correspondence with members of the Commission had indicated that the majority were agreeable to the emissions being based on the SI-second with step adjustments of exactly one second to maintain approximate agreement with universal time. *H. M. Smith* stated that a CCIR Working Party, of which he was Chairman, had the responsibility for recommending a suitable method for indicating the difference between the transmitted time-scale and universal time; he would welcome the views of Commission 4. *D. H. Sadler* stated that he was strongly opposed to the proposed system, and he considered that insufficient effort had been made to develop techniques appropriate to step adjustments of 0.1 s, which corresponded to the limit of human time-discrimination; he considered that the change would give rise to unnecessary difficulties for navigators and others who required immediately a knowledge of universal time to a precision of 0.1 s. After further general discussion it was agreed that the following resolution, which was proposed by *D. H. Sadler* and *R. L. Duncombe*, should be submitted to the General Assembly:

"The International Astronomical Union, *having noted* the various proposals to modify the present basis of Coordinated Universal Time (UTC) and *wishing to emphasize* that visual observers in astronomical and related fields require a knowledge of Universal Time (UT1) to a precision of the same order, namely 0.1 s, as that of human time discrimination, *formally requests* that the appropriate authorities ensure that adequate means have been provided for making UT1, or the difference UT1-UTC, available to such precision, *before* they permit UTC to depart from UT1 by more than about 0.1 s."

It was also agreed that the attention of the General Secretary should be drawn to the fact that the Union has not yet been officially informed of the relevant recommendations adopted by CCIR at its twelfth Plenary Assembly in New Delhi in February 1970.

The further discussions on the CCIR proposals, on the supplementary recommendations of Commission 31, and on the methods for the dissemination of the corrections UT1-UTC are included in the report of the Joint Meeting of Commissions 4 and 31 on p. 198.

DEFINITION OF THE ASTRONOMICAL UNIT

The President stated that, at the request of the General Secretary of the Union, he had drawn up the following definition of the astronomical unit for transmission to the Bureau International des Poids et Mésures for inclusion in a booklet on the International System (SI) of Units;

"The astronomical unit of distance (UA) is the length of the radius of the unperturbed circular orbit of a body of negligible mass moving around the Sun with a sidereal angular velocity of 0.017202098950 radians per day of 86400 ephemeris seconds"

After discussion it was agreed that there should be a formally recognized definition of the astronomical unit but that alternative definitions should also be considered. There was no general agreement to a common symbol in all languages. *J. Kovalevsky* suggested that the unit should be named the 'cassini', while *V. A. Abalakin* suggested the 'kepler'. *I. I. Shapiro* questioned the need for an independent system of astronomical units of mass, length and time. Finally, all these questions were referred to the Working Group on units and time-scales.

IAU SYSTEM OF ASTRONOMICAL CONSTANTS AND RELATED MATTERS

The President gave a brief account of the background to the resolutions that had been adopted at *IAU Colloquium* No. 9, which had been held at Heidelberg during the previous week. (It is hoped that the proceedings and papers will be published in a separate issue of *Celestial Mechanics* early in 1971). The resolutions were then discussed in turn.

"1. Considers that any changes in the precessional constants and in the system of planetary masses be introduced into the national and international almanacs together, at a time that is closely linked with the introduction of the next fundamental star catalogue."

This is to avoid having two separate discontinuities in the basis of the planetary ephemerides and to ensure that the apparent places of the planets and of the fundamental stars are on the same system.

"2. Recommends that a Working Group be set up to report, in time for consideration in 1973, on the consequences of changes in the precessional constants and on the procedure for the introduction of new values at a later date. The Group should feel free to discuss actual values, if it wishes."

Although many participants in the Colloquium had been doubtful about the desirability of attempting to adopt new values of the precessional constants in 1973, the final consensus of opinion had been that the Working Group should not be inhibited from recommending new values if the evidence were considered to be sufficiently strong. *J. Kovalevsky* emphasized the importance of investigating and publicizing the full consequences of any changes in the precessional constants before such changes were adopted. *K. C. Blackwell* considered that for those engaged on the determination of proper motions it would be preferable to leave the precessional constant unchanged.

"3. Considers that no changes be made in the series for nutation until a decision is made about the precessional constants, but considers that a new theory of nutation be developed, based upon a more realistic model of the Earth and consistent with recent developments of the tidal potential."

P. J. Melchior drew attention to the possible derivation by a simple arithmetical process of a series for the nutation from a harmonic development (such as that by A. T. Doodson) of the tidal potential. Recent work on the tides took account of the nonrigid structure of the Earth and led to results that are in closer accord with observation than the adopted series for the nutation. A new and extensive development of the tidal potential by D. E. Cartwright would provide a suitable basis for the nutation.

"4. Recommends that no changes in the basis of the ephemerides published in the national almanacs be made before 1980."

This delay before the possible introduction of new constants is required to allow adequate time for the preparation of new fundamental catalogues and ephemerides, for the subsequent preparation of derived data and explanatory material for publication, and for their printing, proofreading and distribution well in advance of the year to which the data refer. It seems unlikely that the preparation, at the Astronomisches Rechen-Institut, of a new fundamental catalogue will require the use of final constants until 1976, and so publication will not take place until 1978. In normal circumstances the almanacs for 1980 would be scheduled for publication not later than the end of 1978, although the distribution of the data for the first part of the *Astronomical Ephemeris* would take place in 1975 or 1976. *R. L. Duncombe*, in a reply to a question by *D. H. Sadler*, confirmed that the resolution implied no change in the theories to be used and, in particular, that a new theory for Mars should not be introduced before 1980.

"5. Recognizes the need for ephemerides of higher precision for use in many applications and recommends that Commission 4 seeks ways by which a standard set of such ephemerides may be made available in machine-readable form as often as is practicable, together with adequate documentation."

D. H. Sadler drew attention to the likelihood that the ephemerides printed in the almanacs would not be used for the comparison with observations once ephemerides of higher precision were available and that this would lead to confusion. (This can be largely avoided if the new ephemerides are only used where the additional precision is *necessary* for the proper interpretation of the observations as, for example, when looking for variations in the periods of pulsars.) *R. L. Duncombe* considered that the Directors of the national ephemeris offices should propose, after consultation with other interested organizations, how best to implement this resolution.

"6. Recommends that a Working Group be set up to specify, in time for consideration in 1973, the basis for the planetary ephemerides to be published in the almanacs for 1980 onwards, and suggests that ephemerides on this basis be made available in machine-readable form at the earliest opportunity."

D. H. Sadler considered that the resolution would permit the introduction of the new ephemerides before they had been considered by the Commission in 1973. The President stated that this was certainly not the intention; it was merely hoped to shorten the period for which the special ephemerides referred to in the previous resolutions would be required.

"7. Considers that further changes in the values of the primary constants adopted in 1964 should not be made at the Brighton meeting, and endorses the values of the secondary constants compiled by IAG in 'Geodetic Reference System 1967' (special publication of *Bulletin Geodesique*, 1970)."

The new values of the constants adopted in the 1964 appeared to be of adequate accuracy for the purposes for which the system is used, although there had been some criticism of the choice of primary constants. *J. Kovalevsky* explained that the second part of the resolution had been included to meet the wishes of the International Association of Geodesy that only one set of extra figures for the secondary geodetic constants be recognized.

"8. Recommends that a Working Group be set up to review the definition of ephemeris time, its relation to other time scales, and the possible effects of changes in the primary constants on its definition and determination."

In addition to investigating the full effects on astronomical time-scales of the earlier change in the constant of aberration and of any future changes in the precessional constants and planetary masses, it is desirable that the Group consider carefully the extent to which the ephemeris time-scale should be more closely related to the atomic time-scale. *T. C. Van Flandern* drew attention to the results of an analysis of occultation observations which suggested that the currently adopted values of the secular accelerations of the Sun and Moon would lead to a large acceleration of ephemeris time, as determined from the Moon, with respect to atomic time. (See also report of Joint Meeting of Commission 4 and 31 on p. 198.)

"9. Recommends that the next standard equinox be that of 2000.0 and that it be introduced in the next fundamental catalogue."

Although *A. M. Sinzi* argued that the next standard epoch should be that of 2050.0, there appeared to be general agreement with the recommendation, which leaves unspecified the frequency of subsequent changes.

"10. Recommends that the Working Group on Precessional Constants should also consider the desirability of changing the mean places of stars by the effects of the *E*-terms of aberration when the next new fundamental catalogue is produced."

The wording of the resolution was criticized on the grounds that the nature of the change was not precisely specified; the President stated that this had been done deliberately as this was a matter for the Group to consider very carefully. Any change from the current practice would lead to a change in the mean place of each star, but should greatly simplify the theory and practice of aberration corrections.

"11. Urges that the observational data of the International Latitude Service be made available on a uniform system in machine-readable form as soon as is practicable, for use in the determination of nutations."

"12. Urges that the promising techniques of radio interferometry and laser ranging be developed for astrometric purposes, that regular observations by radar of the positions of planets be made to help resolve the uncertainties in the orbits and masses of the planets, and that the observational data be made available to the scientific community as soon as possible."

"13. Urges that all significant observational data be preserved in machine-readable form, in as raw a state as is practicable."

The President pointed out that these resolutions went well beyond the field of responsibility of Commission 4, but they had arisen naturally from the discussions at the Colloquium. *D. H. Sadler* considered that such resolutions were not helpful to committees responsible for the allocation of funds to different projects as they did not take account of relative values and costs. *P. J. Melchior* stated that Commission 19 had set up a working group to implement resolution 11.

At the final meeting the President suggested that the Commission should endorse the resolutions of Colloquium No. 9 as a whole, on the understanding that this did not imply complete acceptance of all details. This was agreed without objection. In order to avoid administrative complications it was agreed that the Commission should be responsible for the three working groups suggested in resolutions 2, 6 and 8, but that the members of the groups need not be members of Commission 4 since it is hoped to cover the interests of other commissions. The terms of reference of each group would be drawn up so as to include the related problems that had arisen during the meetings of the Commission. It was agreed that the convenors of the three groups should be as follows:

Working Group on Precessional Constants: W. Fricke

Working Group on Planetary Ephemerides: R. L. Duncombe

Working Group on Units and Time-Scales: G. A. Wilkins

Further, the membership of the groups should be decided by the new President (J. Kovalevsky), who will be an ex-officio member of the three groups, and the respective convenors. It is intended that all three groups should endeavour to publish reports by the end of 1972 in ample time for consideration before the IAU General Assembly in 1973.

PREPARATION AND DISTRIBUTION OF EPHEMERIDES

G. A. Wilkins reported that H. M. Nautical Almanac Office had recently distributed copies of the advanced data for *A.E.* 1974 and 1975 in the form of reduced-size Xerox copies of computer listings; it appeared that the quality of reproduction was considered to be adequate.

R. L. Duncombe reported that the determination of the constants of Clemence's theory of Mars was almost complete, but that full publication would probably require another year.

T. Lederle stated that some minor changes were to be made in the Introduction to *Apparent*

Places of Fundamental Stars, and he would be grateful for assistance in translating them into other languages. Further the tables for the components of double stars will be improved in the volumes for 1975 onwards. A continuation from 1975 to 2000 of the tables of double stars given in FK4 is available on request.

It was agreed that the new President should consult with the Directors of the national ephemeris offices to determine how best to prepare and make available a set of ephemerides of higher precision than those published in the almanacs (see resolution 5 above).

The President drew attention to the resolutions of Commission 16 concerning new systems of planetographic coordinates for Mercury and Venus (see p. 128); they were approved, subject to detailed examination. *R. L. Duncombe* offered to prepare ephemerides on this basis at the U.S. Naval Observatory and to consider publishing them in *USNO Circulars* until such time as they could be introduced into the *A.E.*

PRINTING TECHNIQUES

R. Haupt described the facilities of the Linotron 1010 filmsetter that was now being used by the U.S. Naval Observatory for the production of navigational almanacs and tables; the system gives results of high quality, and it is very flexible and reliable. (Later, after the meeting had been formally closed, a film about the Linotron 1010 filmsetter was shown.) *G. A. Wilkins* mentioned that the use of Monophoto filmsetter for the preparation of the first part of *A.E.* 1972 has not been as successful as the early trials had indicated, owing to errors in tape conversion and difficulties in the preparation of overlays, and the final results would be below the usual standards; the use of a Linotron 505 filmsetter was now under trial.

The President closed the meeting by thanking all members and others who had helped to further the activities of the Commission during his period of office.

ANNEX

PROPOSAL (OF JULY 1969) FOR THE ESTABLISHMENT OF AN 'INTERNATIONAL INFORMATION BUREAU ON ASTRONOMICAL EPHEMERIDES'

Introduction

This proposal for the establishment of an 'International Information Bureau on Astronomical Ephemerides' is submitted to the IAU Executive Committee by the President of IAU Commission 4 (Ephemerides), after consultation with the Chairman of COSPAR Working Group I (Tracking and Telemetry of Satellites). The proposal originated from a meeting of the IAU Working Group on Space Ephemerides that was held during the period of the twelfth Plenary Meeting of COSPAR in Prague in May 1969; members of both IAU Commission 4 and COSPAR Working Group I attended this meeting.

Throughout this proposal the term astronomical ephemerides shall be deemed to refer only to: ephemerides of the coordinates of the Moon, the major planets, the four principal minor planets, and the natural satellites; star catalogues of a positional character; collected observations of the positions of these bodies. The proposal is primarily concerned with such ephemerides that exist in machine-readable form.

Although IAU Commission 4 has done much to reduce unnecessary duplication of effort in the production of fundamental ephemerides, there is no doubt that the increased demands for improved ephemerides for radio astronomy and for space research have resulted in much duplication of effort. This has been due partly to a lack of knowledge about the availability of ephemerides and partly to an apparent reluctance (sometimes stemming from administrative difficulties) on the part of some institutions to make new ephemerides available to others. It is believed that the establishment of an internationally sponsored organization of the type that is now proposed will help to overcome both of these obstacles to the exchange of astronomical ephemerides.

Purpose

The principal aim is to provide information to the international scientific community on the availability of astronomical ephemerides for use in astronomical and space research in order to facilitate cooperation between institutions and individuals and to avoid unnecessary duplication of effort.

Activities

The Bureau shall:

(a) receive details of astronomical ephemerides that are currently available or in course of preparation;

(b) maintain indexed lists of such ephemerides and of cooperating institutions and individuals;

(c) publish from time to time information bulletins giving summary lists, details of significant new ephemerides and other relevant information; and

(d) answer requests for information about the availability of ephemerides in printed or machine-readable form.

In addition the Bureau may, in appropriate circumstances:

(a) make recommendations to cooperating institutions as to convenient standards for the specification and supply of ephemerides; and

(b) receive details of, and answer questions about, new determinations of astronomical constants, but the Bureau is not expected to carry out original work in this field.

Establishment

The establishment of the Bureau shall be dependent on the offer by a suitable European institution (preferably one already engaged on the production of astronomical ephemerides) to act as sponsor and provide the services of suitable staff and appropriate office facilities. It is expected that the demands on the Service would require only part-time activity by one astronomer with appropriate secretarial assistance. It is suggested that IAU should contribute towards the costs of postage and of the duplication or printing of the information bulletins; an annual sum of 200 dollars is expected to be sufficient for such purposes.

The day-to-day activities of the Bureau shall be the responsibility of a Manager, who shall act on the advice of the Chairman for the time being of a small joint committee containing 2 representatives of IAU Commission 4, and one each of COSPAR Working Group I and of the sponsoring institution. The Manager shall report to IAU Commission 4 at every General Assembly.

It is hoped that the Bureau can begin its operations soon after the IAU General Assembly in August 1970.

COMMISSION 5: DOCUMENTATION

Report of Meeting, 19 August 1970

PRESIDENT: J. B. Sykes.
SECRETARY: A. J. Meadows.

1. *President's Report*. Pecker moved that the report be accepted; seconded by Heintz.

2. *Composition of the Commission*, 1970–1973. It was proposed that the present officers of the Commission should continue in office for another three years, with no organizing committee. These proposals were approved by Commission members. Five new members and four new consulting members were named.

3. *Primary publications*. Heintz mentioned that Commission 26 had discussed the increasing difficulty of publishing observational work. In particular, the publication of observational results in certain journals was being delayed, and the editorial requirements were felt to be too stringent. There was a tendency on the part of editors to think that observations should be published in catalogue form. A new journal might eventually be necessary. The President pointed out that associations of editors are under consideration in a variety of subjects, and enquired whether this might be done in astronomy. There was general agreement that such an association was not needed at present, but that closer association of editors with Commission 5 would be an advantage. It was pointed out that this idea had been suggested at Prague, but had not yet progressed very far. Feuillebois remarked that the Paris Observatory kept a file on names of editors, prices, page charges, etc., and agreed to circulate a list to Commission members.

4. *Secondary publications*. The Commission carried unanimously the following resolution:

> Commission 5 recommends that the three principal abstracting journals in astronomy:
> *Astronomy and Astrophysics Abstracts*
> *Bulletin Signalétique: Astronomie*
> *Referativnyj Zhurnal: Astronomiya*

be authorized to state that they are prepared under the auspices of the IAU. Further discussion of secondary publications was postponed to the meeting on Desiderata for Abstracting Services (see below).

5. *Tertiary publications*. Ogorodnikov described the new series of *Itogi nauki*: periodic reviews of restricted areas of astronomy, containing detailed bibliographies of some 150 items. These are now to be published at a rate of one per year. The series will repeat at intervals so that there will be a continuous coverage of subjects.

A discussion of review papers was mainly concerned with the reduction in coverage of the IAU Transactions. This led to the formulation of the following resolution, which was carried unanimously.

> Commission 5 considers that the thorough and unbiassed reports on the progress of astronomy as formerly published triennially in Vol. A of the IAU Transactions are of great value. It urges the Executive Committee to maintain this series of reviews in some form, despite the financial difficulties.

> Commission 5 would welcome the opportunity of being consulted by the Executive Committee before further decisions are taken about the nature of the Reports of Commissions.

It was agreed that the Commission should not approve a proposal by the Smithsonian Astrophysical Observatory for a subvention towards the distribution costs of *Observatories of the World*.

6. *ICSU Abstracting Board*. The President gave a review of recent activities of the Board relevant to astronomy. The following resolution was carried unanimously:

> Commission 5 records its support for the recommendation of the ICSU Abstracting Board that

abstracts prepared by authors or editors in accordance with the UNESCO *Guide for the Preparation of Authors' Abstracts for Publication* should be published with all scientific papers.

7. *Subject Classification.* An extensive discussion on the correlation of subject classifications used by abstracting journals, the revision of the Universal Decimal Classification, and thesaurus construction led to the setting up of an *ad hoc* sub-committee, whose deliberations are recorded below.

8. *Current-awareness publications.* Ogorodnikov reported that a current-awareness journal for astronomy (the contents of which would later appear in the *Referativnyj Zhurnal*) was under consideration. He would appreciate comments on the possible style of publication and the likely demand for it. The project cannot start before 1972, but sample copies will appear in autumn, 1971.

Second Meeting, 22 August 1970

9. *Information use.* Meadows presented a description of the work which had been carried out at Leicester University on the usage and characteristics of astronomy and space science literature. The first report on this work – an analysis of a questionnaire on information usage by astronomers and space scientists in the U.K. – was published in 1969 (*Information Sources and Information Retrieval in Astronomy and Space Science* A. J. Meadows and J. G. O'Connor, Leicester University, 1969). The second report – a study of selected astronomical libraries in the U.K. – has just been issued (*Library Provision in Astronomy and Space Science* J. G. O'Connor and A. J. Meadows, Leicester University, 1970); and the third – a computer investigation of the characteristics of the main astronomical journals – will appear in 1971.

10. *Bibliography of Astronomy, 1881–1898.* The President announced that this bibliography was now available on microfilm, and that it was accompanied by an explanatory booklet written by himself.

11. *Ancient Books.* Grassi Conti described briefly the card index for astronomical books of the fifteenth and sixteenth century in European observatory libraries, which she was compiling. Cards were currently being prepared for relevant titles at Edinburgh. One card will be prepared for each title; this card will refer to each observatory library where a copy of the book is held. The results of this survey will eventually be published as a printed catalogue. Further details are given in *Journal for the History of Astronomy* 1, 165–6, 1970; *International Journal of Special Libraries* 5, 36–7, 1970.

12. *Transliteration.* In view especially of likely problems with machine readability in the future, the President proposed the following resolution, which was passed unanimously:

Commission 5 recommends to the Executive Committee that the scheme for the transliteration of the Russian alphabet shown on page 13 of *Transactions* XIVA be adopted by the IAU.

Wilkins enquired whether there should not be a common computer code for different alphabets, so that tapes might be exchanged. The President suggested that Wilkins might like to circulate a specific proposal to members of Commission 5 for comment.

13. *List of non-commercial publications and catalogues from observatories.* Rosenberg reported on the current state of this project. It is intended that the compilation should be loose-leaf, and arranged to show both current and ceased publications from observatories. Any future changes in the status of observatory publications will be circulated as amendments. Rosenberg suggested that many observatory publications were becoming simply reprint series, and that it would be better, therefore, if they ceased publication. There was general agreement that this was true in some cases, but that, in others, the publication of an observatory series was still valuable: no general rule concerning the need to discontinue such publications could be laid down.

14. *Library holdings.* The President noted that, at a recent meeting of U.S. astronomical librarians, it had been decided to draw up a list of the serial holdings of their respective institutions.

15. *Title abbreviations.* It was agreed that, until new proposals regarding such abbreviations had been formulated, the proposals contained in the *Astronomer's Handbook* should be adhered to.

16. *Notation.* It was agreed that this was a matter which should be kept under surveillance by Commission 5.

17. *Data centres*. The President commented on their current status, as discussed in Brighton by the Working Group on Numerical Data.

18. *Working Group of Astronomical Librarians*. It was agreed that this Working Group should be allowed to lapse, since the Commission now included five observatory librarians as consulting members, and it was undesirable to duplicate the work of the corresponding sub-section of the International Federation of Librarians' Associations.

Third Meeting, 24 August 1970
Desiderata for Abstracting Services in Astronomy

The President explained that the purpose of the meeting was to discuss whether the current abstracting services sufficiently covered the needs of astronomers, and whether review reports, such as had been previously produced by the IAU, would still be valuable (a resolution to this effect having already been passed at an earlier meeting of Commission 5).

Dewhirst enquired to what extent *Astronomy and Astrophysics Abstracts* covered report literature such as NASA special publications. Henn said that some reports were included, but not all; there were no holdings of STAR at AAA in Heidelberg. Merrill (Commission 42) suggested that other Commissions should follow the example of his own, which published a twice-yearly list of relevant papers. These were collected by astronomers from different countries who were working in the field. The lists were circulated to all members of the commission, and to anyone else who was interested. Such lists were very useful in drawing up draft reports. Dewhirst suggested that IAU members should again be reminded of the need to send reprints of publications appearing in uncommon journals to *Astronomy and Astrophysics Abstracts*. The President replied that he would be sending a note on abstracting services to the *IAU Information Bulletin*, and that he could add an appropriate statement to that. Wray described a personalized computer file which he had prepared for himself. This included all papers on extra-galactic research listed in the *Astronomischer Jahresbericht* since 1950 (it was being updated at intervals). Authors and titles are included (not abstracts), and items can be recalled by author, source, word-in-title, or added keyword (there is no standard thesaurus, however). Some 2000 items are currently on file, and each search costs $3–4. Wray suggested that members at every institution might find such personal files useful. Henn noted that *Astronomy and Astrophysics Abstracts* would be publishing a cumulative index in terms of authors and keywords every 3–4 years. There followed some discussion of the problem of abstracting fringe fields. The President remarked that there would hopefully be an exchange of abstracts of mutual interest to different subject areas through the ICSU Abstracting Board. The problem was more important for *Astronomy and Astrophysics Abstracts* than for the other two main abstracting services, since each of these formed part of a larger system. It was noted that *Astronomy and Astrophysics Abstracts* was on punched cards, and therefore not available for distribution on tape. Feuillebois emphasized that there was a current international system for putting bibliographical information on tapes (MARC) and that astronomy should follow this if possible. She and Kemp agreed that the system was at present unsuitable for information searches, but might well be improved, from this point of view, in the immediate future. Wilkins suggested that, since INSPEC tapes had been designed with information retrieval in mind, it might be better to make an astronomical system compatible with these. Dewhirst remarked, amidst general agreement, that, whatever search methods were devised, they might be found eventually to be too expensive for astronomers to use. Finally, Heintz proposed that the habit of referring only to abstracts in review articles should not be pressed until such time as the various abstracting services' classifications were unified.

Meeting of Sub-Committee on Classification (20 August 1970)

The President summed up the discussion at the meeting of 19 August. He pointed out that classification was required for: (1) Abstracting systems; (2) UDC; (3) Thesaurus development for machine retrieval.

Kemp remarked on the need to keep the current research on classification – as by the Classification Research Group in the U.K. – in mind. Feuillebois described the keyword system being developed in Paris. The NASA thesaurus is used as a basis, and the keywords are gradually compiled by machine. Both broad terms and narrow terms are used; the former can be used for shelf placement. There was a general feeling that this scheme might be used for all three purposes outlined by the President. It was agreed, however, that the UDC and keyword systems would need to be fully inter-related; it was therefore necessary to hasten the current UDC revision. To this end, it was agreed to establish a Working Group on Universal Decimal Classification 52 (Astronomy), to work with the FID. The members suggested were Kemp, Wilkins, Lavrova and an astrophysicist.

JOINT MEETING OF COMMISSIONS 5, 9, 38 AND 46

The meeting took place on 25 August under the chairmanship of J. Rösch. Its purpose was to consider the construction and publication of a listing of observatories, astronomers, instruments and programmes in a form suitable, in particular, for the needs of Commissions 9 and 46.

Existing documents of this kind serving as a basis for discussion were Rigaux' *Les observatoires astronomiques et les astronomes*, last published in 1959, and Page's *Observatories of the World*, revised in 1970, of which the compiler had sent 100 copies for distribution at the General Assembly.

The need for up-to-dateness was emphasized and it was agreed that this would have to be given precedence over completeness.

The following information was agreed to be desirable (list compiled largely by Heintz):

1. Name and address of observatory or institution, followed by its stations elsewhere (these being noted also under their country of location, with cross-reference to the parent foundation).
2. Geographical coordinates and altitude.
3. Names of permanent research and teaching staff, with their astronomical interests.
4. Number of temporary staff, students, etc.
5. Brief listing of instruments (including computers), fuller details being available on request from some specified location.
6. Research and teaching programmes.
7. Titles of publications and reports issued.
8. Names of those responsible for providing information about facilities for observation and study.
9. Indexes.

Radio observatories and data centres should be included, but not private and amateur observatories. Müller showed a specimen listing of Swiss observatories. Velghe showed a computer print-out prepared at Brussels and described a plan whereby each country would be responsible for issuing updating sheets in respect of its own observatories, using an agreed computer programme, and distributing these updatings at appropriate intervals. The need to computerize the system was questioned but was thought to be defensible having regard to the advantages of uniformity, ease of index preparation, and selective print-out of data.

Velghe said that the Brussels Observatory was willing to compile the complete world list for the initial distribution, but thereafter could accept responsibility only for the Belgian lists. The responsibility for the completeness and correctness of the information given for each other country would lie with the respective National Committee for Astronomy. It was agreed that Velghe should in the first place send to Sykes copies of papers describing the proposed system, for forwarding to the Chairmen of National Committees together with a copy of the present minutes and a covering letter explaining the matter under discussion and requesting comments. The plan will be carried through as far as possible and the situation will be reviewed at the next General Assembly.

COMMISSION 6: ASTRONOMICAL TELEGRAMS
(TÉLÉGRAMMES ASTRONOMIQUES)

Report of Meeting, 21 August 1970

PRESIDENT: F. L. Whipple.
SECRETARY: B. G. Marsden.

The Commission approved the election of P. Simon as incoming President and J. Hers as incoming Vice-President. A. Mrkos and Y. Kozai were elected as new members, the former to replace E. Buchar, who was retiring from the Commission because of ill-health.

Noting that the request for an IAU subvention to the Central Telegram Bureau of $ 1600 for the triennium 1971–3 had already been submitted to the Finance Committee*, the President remarked that the Smithsonian Astrophysical Observatory continued to subsidize the Bureau's operation, although now almost entirely in the form of personnel. He was pleased to see that Miss Virginia Lincoln was present at the meeting and expressed the appreciation of the Commission for the great service performed by the International Ursigram and World Day Service in relaying the Bureau's telegrams.

The President also mentioned that, in the United States at any rate, the telegram was fast becoming a dying institution, and the Secretary circulated a recent newspaper item on this subject. While the Commission should not at present be too alarmed by this development, it was suggested that some initial thought be given to the possible future impact on the work of the Central Bureau. The trend is toward the increased use of the telex system and possibly the introduction of 'mailgrams', where a telegraph company telephones the contents of a telegram to the postmaster in the recipient's town, the postmaster then forwarding the message by mail.

Since his presence was simultaneously required at another meeting the President at this point invited the new President, Dr Simon, to take the Chair.

Dr Marsden reported that, contrary to the hope expressed in *Trans. IAU* **XIVA**, 16, the billing for the *Circulars* had not yet been changed to a yearly basis. Part of the reason for this was that there had been frequent changes in the secretarial assistance available to the Central Bureau; the matter was still very much under consideration, however, and he was confident that the desired change would be made before too long.

The seemingly trivial problem was brought up as to a suitable character to indicate that certain information was being suppressed from an astronomical telegram. Some telegraph companies were taking it upon themselves to suppress this character itself or even the whole group containing it. After some discussion by O. J. Gingerich, J. V. Lincoln, B. G. Marsden, E. Roemer and P. Simon and consideration also of a suggestion sent by J. Hers it was agreed that, while the choice was not perfect, and it would thus be inadvisable to make an official change in the code, the Central Bureau should for the time being experiment by indicating a suppressed digit by a slash (/).

Dr Marsden remarked that he felt that, since the Bureau had issued more than 100 *Circulars* during the previous twelve months, it should be more strict concerning the items accepted for publication. A decision had already been made concerning the Bureau's intention to limit the number of precise positions of comets that were published, but it was also true that many other items, while often referring to topics of great astronomical interest (e.g. pulsars), were better suited for publication in other media. Several others attending the meeting, however, expressed their pleasure at acquiring from the *Circulars* immediate, reliable information on such topics.

---

* and was approved; see the report of the Finance Committee.

COMMISSION 7: CELESTIAL MECHANICS (MÉCANIQUE CÉLÈSTE)

Report of Meetings, 19, 20, 22, 24 and 25 August 1970

PRESIDENT: W. J. Eckert.
VICE-PRESIDENT: G. N. Duboshin.
SECRETARY: P. J. Message.

Session (a): Business Meeting

The President referred to the Report on progress in the field in the previous three years, which was now presented in its final form, and not as a 'Draft Report', as formerly. He thanked those who had reviewed the various sections, and expressed satisfaction with the mechanized method which he had used to deal with the references. There had been no attempt to make the report exhaustive, but rather a cross-section of the types of work going on, and of the relation of Celestial Mechanics to other fields of study.

The Commission endorsed the President's strong support for the tradition that workers in Celestial Mechanics be autonomous, and work independently, without formal organization of projects by the Commission.

Dr A. Deprit raised the question of the difficulties in dissemination of numerical and computer information arising in large quantities, as in large scale numerical experiments, literal theories taken to a large number of terms, and raw observational data, where the publication of such material in the usual journals was not feasible. This appeared to be a matter of some widespread concern. Knowledge of the existence of such material is often only accidentally come by. A committee of the Commission, under the chairmanship of Dr M. S. Davis, was charged to discuss the question, and report to a later session.

Dr Déprit acted as interpreter in this and all later sessions.

Sessions (b), (c), and (d): *Colloquium on: 'Analytical Methods for Orbits of Artificial Celestial Bodies'*, organized by Professor G. N. Duboshin, with the assistance of Drs. E. P. Aksenov, B. Garfinkel, A. H. Cook, B. Morando, and D. G. King-Hele.

Session (b)

Chairman: Professor G. N. Duboshin.
Speakers: Yu. V. Batrakov, on 'Studies of the Motion of Artificial Earth Satellites at the Institute for Theoretical Astronomy at Leningrad'.

A. Deprit, 'On the Theory of an Artificial Satellite'.

P. Sconzo, on 'Time Series Solution in Explicit Form of the Asteroidal Three-Body Problem Using Generalized Lagrangian Functions f and g'.

R. R. Allan, on 'Resonance Effects for Close Earth Satellites'.

Session (c)

Chairman: Dr B. Garfinkel.
Speakers: V. Szebehely and P. Nacozy, 'On the Use of Chebeshov Polynomials in a General Perturbation Method of Earth-to-Moon trajectories'.

V. A. Brumberg, L. S. Evdokimova, and N. I. Kochina, on 'Analytical Methods for the Orbits of Artificial Lunar Satellites' (presented by Dr Abalakin).

K. Aksnes, on 'A Complete Second-Order Theory, Based on an Intermediate Orbit'.

I. Stelmacher, on 'Influence du champ magnétique sur le mouvement autour de son centre de gravité d'un satellite artificiel de la Terre' (presented by Dr J. Kovalevsky).

P. Bretagnon, on 'Expression analytique des termes en J_2 au carré dans la théorie des satellites artificiels' (presented by Dr B. Morando).

Session (d)

Chairman: Professor A. H. Cook.

Speakers: S. Herrick, on 'A Universal Singularity-Free Determination of an Orbit from Two Positions and Time-Interval'.

G. Hori, on 'A Second-Order Theory for an Artificial Satellite, Based on a Kepler Ellipse' (presented by Dr B. Garfinkel).

P. Sconzo, on 'Mechanized Algebraic Manipulation of a Second-Order Theory for Artificial Satellites'.

S. Ferraz-Mello, on 'Earth's Shadowing Effects in the Long-Periodic Perturbations of Satellites' Orbits'.

Sessions (e), (h), and (f): *Colloquium on 'The Impact of Precise Measurements of Distances on Celestial Mechanics'*, organized by Dr G. M. Clemence, with the assistance of Drs R. L. Duncombe, J. Kovalevsky, and J. D. Mulholland.

Session (e)

Chairman: Dr. Clemence.

Speakers: C. A. Lundquist, on 'Laser Ranging to Artificial Satellites'.

P. Bender, on 'Lunar Ranging'.

A. Orszag, on 'Laser Ranging to the Moon'.

S. J. Peale, on 'Evolution of the Earth-Moon System'.

J. Kovalevsky, on 'Lunar Theory'.

Session (h)

Chairman: Dr J. Kovalevsky.

Speakers: I. I. Shapiro, on 'Solar System Tests of General Relativity by Radar Ranging'.

J. G. Davies, on 'Radio Tracking at Jodrell Bank'.

W. H. Michael, on 'Gravity Fields from Radio Tracking Data'.

G. C. McVittie, on 'Questions of Interpretation in Relativistic Celestial Mechanics'.

Session (f)

Chairman: Dr R. L. Duncombe.

Speakers: C. Oesterwinter, on 'Numerical Integration and Analytical Theory'.

J. Chapront, 'Literal Planetary Theory'.

A. Deprit, on 'Literal Developments in Celestial Mechanics'.

W. J. Klepczynski, on 'Planetary Masses'.

D. A. O'Handley, on 'Ephemeris Improvement and the Topography of Mars'.

A short *Business Meeting* was held at the end of Session (e), at which the President's recommendations for the officers of the Commission for the following three years were approved.

The following list of Consultants of the Commission was approved:

D. G. King-Hele, J. Marchal, J. Moser, I. D. Zhongolovich, and E. A. Grebenikov.

Session (g): Business Meeting

The President referred to the working party on lunar laser ranging data being set up by COSPAR and Commission 17. The President of that Commission had invited representatives from Commissions 4, 7, 17, 30, and 31 to join this working party. The Organizing Committee of Commission 7 asked Dr. Eckert to accept this invitation on behalf of the Commission.

The report of the committee on Program and Data Banks was presented by Dr Davis. The members of the committee were: M. S. Davis (chairman), Yu. V. Batrakov, J. Chapront, H. Debehogne, A. Deprit, and A. Ollongren. The report, which is given as the Appendix, was accepted in principle by the Commission. Dr Davis has accepted the Chairmanship of the Standing Working Group on Program and Data Banks, whose establishment was recommended by the committee, and membership of this Group has been accepted by Yu. V. Batrakov, H. Debehogne, A. Deprit, and A. Ollongren.

APPENDIX

REPORT OF COMMITTEE ON PROGRAM AND DATA BANKS OF COMMISSION 7

The committee recommends the establishment of a Standing Working Group of Commission 7 on 'Program and Data Banks', whose functions would be:

1. To collect and disseminate information on data and programs relating to celestial mechanics.

2. To recommend international standards for those interested in participating in this exchange of data and programs.

3. To serve as a clearing house for questions related to (1) and (2) whenever it is feasible and practicable to do so.

The committee distinguishes several kinds of data.

1. *Observational data*. This type of data is regarded as the chief concern of Working Group 1 and the several Commissions working with it to ensure the preservation of observations in the rawest form possible and with all relevant references. Commission 7 should be represented at all of the sessions of the IAU dealing with numerical data to be certain that its interests are represented.

2. *Numerical Data derived from Theories*. Examples of this class of data are (a) Clemence's theory of Mars, (b) the Improved Lunar Ephemeris, and (c) numerical experiments with periodic solutions in the restricted problem of three bodies. The purpose of having this kind of information in machine-readable form is, among other things, for comparison of theories, or extension of theories, or calculation of new numerical values by substitution of new numerical values as arguments.

3. *Analytical Data*. The lunar theories of Delaunay and Eckert in machine-readable form (for comparison with other theories, or for comparison with other derivations of the theory) is one example of analytical data.

The numerical data derived from theories (item 2 above) and analytical data (item 3) would be the proper concern of the subcommittee.

The committee also distinguishes two kinds of program for exchange.

1. *Numerical Programs*. This has been the most common use of computers. Examples would be programs for orbit determination and correction, calculation of Hansen perturbations, Schubart and Stumpff's *n*-body solar system numerical integrations.

2. *Analytical Programs*. Literal programs for symbol manipulation such as D. Barton's programs, Deprit and Rom's MAO, and the symbolic compiler at the Smithsonian Astrophysical Observatory.

The committee does not at this time recommend the establishment of a physical repository of data and programs in machine-readable form, but does recommend that the Standing Working Group be charged with the responsibility of collecting and disseminating information. In this capacity its function is that of a broker who establishes connections between requestors of programs and suppliers thereof.

To ensure the international character of this brokerage service, this function should be carried out under the auspices of the IAU. The sub-committee should investigate what role supporting national committees can play in implementing its purposes. The advice and assistance of well-established groups which have dealt with similar problems, such as the U.S. Naval Observatory, the Royal Greenwich Observatory, and the Astronomisches Recheninstitut should be solicited.

The committee fully recognizes the serious difficulties involved in program exchange and therefore stresses that the principal goals can be realized by adequate documentation and the establishment of international standards. What is primarily intended here is the algorithmic presentation of theories so that any person or group can write programs *ab initio*, if necessary. It would be regarded as fortuitous, in general, if one institution were able to use another institution's programs intact.

It was stated that users of this service, following standard scientific practice, would be expected to acknowledge authorship of programs and data used.

Other suggestions made were:

1. Editors of astronomical journals should be alerted to the publication of information about material which would prove very useful in machine-readable form.

2. An inventory of computers in observatories, departments of Astronomy, Astronomical Institutes and Laboratories would prove very useful, in particular to visiting astronomers.

3. Guide lines should be set up for the mechanics and financing of interchanged material.

M. S. DAVIS (Chairman)

COMMISSION 8: POSITIONAL ASTRONOMY
(ASTRONOMIE DE POSITION)

Report of Meetings, 19 and 26 August 1970

ACTING PRESIDENT: W. Dieckvoss.
SECRETARIES: G. van Herk and R. H. Tucker.

President Nemiro, who is absent, proposes G. van Herk for Vice-President. Symms proposes R. H. Tucker for the same position. Tucker refuses to accept a nomination. The meeting accepts G. van Herk unanimously.

The Commission then nominated the members of the Organizing Committee.

The Sub-committee on photographic catalogues up to the 9th magnitude is reported to die out.

The following resolutions are unamimously adopted:

1. Commission 8 reaffirms its recommendation that meridian circle observers should include time and latitude stars in their observing programmes (A. A. Nemiro).

2. Commission 8 recommends that as many observatories as possible which possess suitable meridian instruments should organize absolute observations of bright and faint stars (A. A. Nemiro).

3. Commission 8 considers as very important the inclusion in the fundamental catalogues of a number of faint stars up to the 9th magnitude (especially the FKSZ stars) (A. A. Nemiro).

4. Commission 8 approves the initiative of the Nikolaiev observatory in respect of the meridian observations of zodiacal stars, and recommends that the observation of these stars be organized at other observatories on the basis of international co-operation (A. A. Nemiro).

5. La Commission 8 souligne le grand intérêt qu'il y aurait à préparer et effectuer des observations spatiales en astrométrie pour obtenir rapidement dans le domaine qui l'intéresse des progrès très importants difficiles, ou impossible, à obtenir autrement (P. Lacroute).

6. Commission 8 encourages the developments of new instrumentation and techniques which endeavor to improve the fundamental inertial reference system (B. L. Klock).

7. Considering the necessity and the actuality of the problem of refraction for the further advance of astrometry, Commission 8 proposes a symposium before the next General Assembly of the IAU, be dealing with the astronomical refraction (G. Teleki).

With respect to the name of the Commission, no change is considered at this moment in view of the oncoming merging of Commissions 23 and 24 (A. Reiz).

The following papers were presented at the meeting:

B. L. Klock: The Automatic Transit Circle of the U.S. Naval Observatory.

P. Lacroute: Intérêt de mesures spatiales en astrométrie.

E. Høg: A Theory of a Photoelectric Multislit Micrometer.

J. E. B. von der Heide: Mounting Errors of a Photoelectric Micrometer.

G. Teleki: Need for an International Agreement on Astronomical Refraction.

H. J. Fogh Olsen: Results obtained with the Copenhagen Transit Circle at Brorfelde.

D. Saletić, S. Sadzakov: Quelques résultats du travail sur le catalogue des étoiles de latitude.

D. Saletić, S. Sadzakov: Une méthode différentielle pour la détermination des erreurs du cercle.

C. Anguita, G. Carrasco, P. Loyola, D. D. Polojentsev, K. N. Tavastsherna, M. S. Zverev: The SPF1 Catalogue of Right Ascensions.

J. Petit, E. Portugal, R. Tapia, D. Polojentsev, G. Silva, G. Timashkova, T. Polojentseva, R. Taibo, D. Viveros: The Pulkovo Large Transit Instrument in Chile.

B. Guinot: A Full-Aperture Astrolabe.

J. E. B. von der Heide: SRS-Program. Preliminary Results of the Observations made at Perth Observatory, Bickley, Western Australia by the Hamburg Observatory Expedition.

K. N. Tavastsherna: The Catalogues of Declinations Compiled at the Pulkovo Observatory on the Base of Observations Made at the Melbourne Observatory.

D. A. Pierce: Star Catalogue Corrections Determined from Photographic Observations of Selected Minor Planets (Summary and Results).

I. Pakvor: The Meridian Marks of the Large Transit Instrument in Belgrade.

F. Noël: Corrections for Some FK4 Stars Deduced from Astrolabe Observations at Santiago, Chile.

J. L. Schombert: Status of the SRS Program, 1 June 1970.

Observatory	Zone	Commitment	Date started	Completed Obs'n	Completed Redn's	Final results expected
Abbadia	$+5°$ to $-15°$	1560×4	62.3	100%	100%	1968.8
Bordeaux	$+5$ to -15	1560×4	62.5	100	100	1970.5
Bucharest	$+5$ to -10	1176×4	62.6	100	50	–
Nicolaiev	0 to -20	5984×2	64.3	100	95	1971.0
San Fernando	-10 to -30	3709×4	63.3	93	88	1972.0
Tokyo	-10 to -30	3560×4	63.3	100	95	1971.0
USNO 6-in	$+5$ to -30	8706×2 1233×4	66.5	85	70	1972.0
Bergedorf (Bickley)	$+5$ to -90	20495×4	68.4	69	95	–
Cape	-30 to -40	10082×4	61.3	95	95	–
	-40 to -50			100	100	1966
	-50 to -90			8	0	–
San Juan	-40 to -90	7190×2	69.5	40	60	–
Santiago ⎰	-47 to -90	11496×4	63.1	95	35	–
Pulkovo ⎱	-47 to -90			100	60	–
USNO 7-in (El Leoncito)	$+5$ to -20	7683×2				
	-20 to -75	12121×4	68.7	51	57	1974.0
	-75 to -75SP	1382×4				

Summary:

1. As of 1 June 1970, 72% of the observational work of the SRS-Program was completed.

2. It is expected that the final results from all observatories should be on hand about 1975.5. The commission expressed its thanks to Mr. F. P. Scott for his valuable work on the SRS-Program.

Mme E. Marcus: A Catalogue of Faint Stars in the FK4-System.

H. Yasuda: A Proposal for the Meridian Observations of PZT Stars.

R. d'E. Atkinson: On the Advantages of a Mirror Transit Circle.

Mlle S. Débarbat: Weights of Observations Made with Astrolabes in France.

W. Fricke and A. A. Nemiro were appointed delegates to represent Commission 8 on the working group dealing with Precessional Constants now being set up by Commission 4.

Names of new members should be proposed to any member of the Organizing Committee.

Besides the many papers relating to progress made in meridian circle or astrolabe work, W. Gliese pointed out that the FK4-system of the southern hemisphere at least in right ascension happens to be very poor because it is based mainly on the work of one observatory only, the Cape Observatory.

He asked which programmes are under way for the determination of absolute right ascensions in the southern hemisphere, or which have been finished already since the compilation of the FK4. Special attention should be paid to the determination of absolute azimuths which are really independent of the system of any basic catalogue! The answer to the question is that there is only one programme, fulfilling Gliese's requirements, going on in the southern hemisphere, namely that with the Large Transit Instrument of Pulkovo, now at Cerro Calan.

Other series of observations are only quasi-absolute; there remains always some doubt as to the validity of the system.

E. Høg wanted to stress the importance of the inclusion of observations of the four bright minor planets in meridian circle programmes performed at southern observatories. (Incidentally, the U.S. Naval Observatory includes these observations in its work in South America.) The experience of the USNO has been that it is impossible to obtain enough observations in both quadratures of each planet. This problem could be solved by combining observations from the northern and the southern hemisphere since a quadrature during the northern summer could be more easily observed from a station in the south.

COMMISSION 9: INSTRUMENTS ET TECHNIQUES
(INSTRUMENTS AND TECHNIQUES)

Comptes Rendus de séances, 20 et 24 Août 1970

PRÉSIDENT: J. Rösch.
SECRÉTAIRE: A. H. Delsemme.

Affaires administratives

Ordre du jour:

1. Election du nouveau bureau.
2. Proposition du Comité National Belge relatif à un lexique des termes d'optique.
3. Proposition de la République Arabe Unie relative à l'échange d'équipement et de documents, notamment pour les pays en voie de développement.
4. Rôle et fonctionnement de la Commission 9.

1. *Elections*

Les propositions suivantes du Comité d'Organisation sont approuvées à l'unanimité des membres présents:

Président: Nikonov; Vice-Président: Meinel

Membres du Comité d'Organisation: Baum, Livingston, McGee, Mikhelson, Ring, Rösch, Valníček.

2. *Proposition de la Belgique*

Le Dr Dommanget explique les raisons de la proposition Belge. L'Union Internationale de Physique possède une commission internationale d'optique qui a demandé à ses membres d'étudier la possibilité de constituer un glossaire des termes d'optique. La délégation belge y a intéressé le Comité Belge d'Astronomie, qui juge utile de reporter la question devant des instances internationales, en l'occurence la Commission 9.

Il résulte de la discussion qu'il convient de ne pas limiter le glossaire aux 2 langues de l'UAI (anglais et français) mais aussi de l'étendre au moins au russe et à l'allemand.

Un group de travail de 3 ou 4 personnes sera constitué, sous la responsabilité de J. Rösch.

3. *Proposition de la R.A.U.*

En l'absence d'un représentant de la République Arabe Unie, la proposition de ce pays est lue et commentée par le président de la Commission.

Il résulte de la discussion qui suit que la proposition de la R.A.U. est importante et mérite un examen approfondi, mais que la création d'une nouvelle commission *ad hoc* ne se justifie pas.

Il est donc proposé qu'un groupe de travail soit établi pour la mise en oeuvre de la proposition. A la suite de l'intérêt officieusement exprimé par la Commission 46, ce groupe de travail pourrait être établi en commun avec cette commission. Une proposition sera transmise en ce sens au Comité Exécutif par le président de la Commission 9.

Le Comité d'Organisation de la Commission 9 assurera provisoirement la liaison avec la Commission 46 et désignera ultérieurement les représentants de la Commission 9 au sein du groupe de travail commun.

4. *Rôle et fonctionnement de la Commission*

La constitution d'un certain nombre de groupes de travail au sein de la Commission est mise en discussion. Le groupe de travail sur les Tubes à Images a été particulièrement actif et son exemple est encourageant. Par ailleurs, la grande variété d'intérêts des membres de la Commission et leur nombre croissant suggère d'adopter la même procédure dans d'autres domaines.

Une discussion générale révèle la difficulté de partager *a priori* le travail de la commission en une série de groupes d'étude bien délimités. Différents découpages sont proposés en groupes et sous-groupes. Par exemple, Dr Hooghout suggère le découpage suivant:

A. *Basic Instruments*

(1) Optical. (2) Radiotelescopes. (3) Infrared telescopes. (4) Solar. (5) Astrometric. (6) Space.

B. *Auxiliary instruments*

(1) Photometers. (2) Spectrometers. (3) IR detectors. (4) Image Tubes.

C. *Data handling and automation*

(1) Computers on line. (2) Programming and formats. (3) Storages and tapes. (4) Photographic plates.

Plusieurs membres se prononcent contre toute classification, et surtout contre une classification à deux dimensions, comme trop rigide et trop artificielle. En particulier:

Lasker souligne le danger de cloisonner la Commission en un trop grand nombre de groupes de travail.

Ulrich mentionne qu'une classification des sujets en fonction des longueurs d'onde ne serait nullement moins commode que la division proposée.

Atkinson pense qu'aucun groupe ne devrait être constitué si un intérêt spontané ne s'est pas préalablement manifesté pour sa constitution.

Wlérick souligne que la tendance qui se dégage est d'être moins cartésien et plus pragmatique.

Plusieurs autres interventions confirment qu'il convient de bâtir quelques groupes de travail autour de tâches spécifiques qui ont une signification concrète, plutôt que de suivre une véritable classification exhaustive.

A l'occasion de cette discussion générale, la question du mandat exact de la Commission est abordée, en particulier par Sisson. Il ressort de la discussion qu'il y a un accord général sur les termes de référence de la Commission 9.

En particulier, le Président appuyé par Strand rappelle que l'objectif majeur de la Commission est de rester un centre d'information et d'échange d'idées sur les problèmes d'instrumentation; Wlérick souligne que les personnes qui bénéficient du travail de la Commission se divisent en deux catégories:

(a) les constructeurs d'instruments, pour lesquels la Commission constitue le seul forum, pour échanger leurs idées,

(b) les astronomes intéressés qui viennent s'y instruire sur les nouvelles techniques.

Le Président insiste sur le fait que la conjonction des *deux* catégories au sein de la Commission est essentielle.

Conclusions

Il ressort clairement de cette discussion:

(a) que la Commission 9 a un mandat clair qu'il n'y a pas lieu de changer, et que son utilité est incontestable,

(b) qu'il serait utile de constituer plusieurs nouveaux groupes de travail, d'une manière pragmatique, en s'inspirant de l'exemple du groupe sur les Tubes à Images, et autour de tâches spécifiques pour lesquelles un intérêt spontané se manifeste.

Séance plénière du 24 Août 1970

Avant la présentation de communications scientifiques, le Président apporte les informations ci-après:

1. *Proposition de résolution du Comité National de la R.A.U.*

La Commission 46, qui a créé un Groupe pour l'échange de documents, estime que ce Groupe répond à une partie de la proposition; elle n'a pas cru devoir aller au-delà, et laisse à la Commission 9 le soin de s'occuper des échanges d'instruments si elle juge bon de le faire. Il n'y aura donc pas lieu de créer un Groupe commun aux deux Commissions.

2. *Nouveaux Groupes à l'intérieur de la Commission 9*

A la suite de la discussion qui a eu lieu lors de la séance précédente, un certain nombre de membres ont décidé de former un Groupe sur l'Automation et les calculateurs associés aux télescopes ('Data systems') et invitent les astronomes intéressés à assister à une réunion constitutive.

(N.B. – Cette réunion a effectivement eu lieu le 24 août et a abouti à la désignation de P. Boyce, Lowell Observatory, Flagstaff, Arizona 86001, U.S.A., comme responsable du groupe, assisté de B. Lasker, I. G. Van Breda et Ed. Dennison; P. Boyce diffusera une Lettre d'Information). Il est suggéré, en séance, de former aussi un groupe sur 'Les Grands Télescopes'. (N.B. – Un appel aux astronomes intéressés a fourni, avant la fin de l'Assemblée Générale, une première liste de noms; J. Rösch assurera provisoirement la centralisation et la diffusion des informations en ce qui concerne ce Groupe). Il y aura lieu d'envisager également la formation d'un Groupe couvrant l'ensemble des techniques de l'Infra-Rouge (N.B. – Le Comité d'Organisation, dans sa dernière réunion, a décidé de proposer au Comité Exécutif d'inscrire au programme des années à venir un Symposium sur ce sujet).

La suite de la séance est consacrée aux communications suivantes:

R. N. Wilson, L. Müller: Astrographic Objectives with Reduced Secondary Spectrum.

D. Rudolph, G. Schmall: Gratings Produced by Holography.

O. G. Franz: Instrumental Profiles for Double-Star Scanning Methods.

A. A. Wyller: Interferometric Echelle Scanner.

E. H. Richardson: An Efficient Coudé System (48-Inch Telescope, Dominion Astrophysical Observatory).

K. Serkowski: Ten-Channel Stellar Polarimeter.

G. M. Lasker, J. E. Hesser: A Data-Acquisition System (Cerro-Tololo).

K. L. Hallam: Photo-Image Sensors for Space Telescopes.

H. D. Greyber: Space Station and Large Space Telescope (Résumé).

GROUPE DE TRAVAIL SUR LES RÉCEPTEURS PHOTOÉLECTRIQUES D'IMAGES (20 ET 24 AOÛT, 1970)

PRÉSIDENT: G. Wlérick.

La première séance est consacrée à l'électronographie. Elle commence par un exposé de B. Morgan sur 'Les propriétés photométriques des émulsions pour électrons.' Il s'agit d'un *travail collectif*, effectué à l'initiative de J. D. McGee à l'Imperial College, avec la collaboration de six observatoires. Le principal investigateur a été E. Kahan. Un résultat intéressant a été obtenu: on trouve que la caractéristique 'Eclairement-Densité optique' d'émulsions Ilford G5 neuves et sensibles, est toujours linéaire jusqu'à une densité $D=2$ et quelquefois jusqu'à $D=4$. Cette propriété donne une base solide à la photométrie par électronographie; celle-ci fait l'objet des trois communications suivantes:

G. E. Kron, H. Ables et A. Hewitt: 'Astronomical Photometry with the USNO Electron Camera: Very Faint Objects, Planets, Satellites, Double Stars...' (Dans la discussion, J. Rösch rappelle la méthode originale utilisée au Pic du Midi depuis 1960 pour l'étude des étoiles doubles).

M. Walker: 'The Application of the Spectracon to Electronographic Photometry of Globular Clusters in the Magellanic Clouds'.

G. Wlérick: 'Identification et photométrie des radiosources optiquement faibles du catalogue 3C R de Cambridge' (travail effectué avec G. Lelièvre et P. Véron).

Ces trois exposés montrent:

(a) La photométrie par électronographie est maintenant opérationnelle jusqu'à la magnitude 23,5 avec des télescopes de 1,5 à 2 m de diamètre.

(b) Il est possible de mesurer l'éclat d'un astre très faible situé au voisinage immédiat d'un astre brillant.

(c) Avec des émulsions Ilford L4 de bonne qualité, on obtient une caractéristique 'Eclairement-Densité optique' qui peut être linéaire jusqu'à $D=6$.

Les huit exposés suivants sont consacrés à des développements instrumentaux dans le domaine des tubes électronographiques. Les trois premiers décrivent l'amélioration de tubes déjà en service; les cinq derniers traitent des réalisations nouvelles, dont deux sont déjà opérationnelles tandis que les trois autres devraient l'être bientôt. Voici les titres de ces communications:

G. Kron: 'Optical System Correcting the Distortion of the USNO Electron Camera.'

J. D. McGee: 'Recent Developments of the Spectracon Image Tube for Astronomy.'

M. Duchesne: (lu par G. Wlérick)

(a) 'Utilisation de la Caméra électronique à des niveaux lumineux très faibles; linéarité.' – (b) 'Etude d'une optique électrostatique à grand champ.'

A. Lallemand: 'Présentation d'un tube électronographique à grand champ destiné à la photométrie.'

P. Felenbok: 'Construction d'une caméra électrostatique Lallemand de type spécial et utilisation de celle-ci pendant une éclipse totale de Soleil.'

M. Combes: 'Étude théorique et résultats expérimentaux concernant une caméra électronique à grand champ et à haute résolution, utilisant un champ magnétique intense.'

McMullan: 'Development of an Electronographic Image Tube at the Royal Greenwich Observatory.'

P. Griboval: 'The High Resolution Electronographic Camera of the Department of Astronomy of the University of Texas.'

Le nombre important de ces exposés et leur qualité témoignent de l'effort intense fourni actuellement dans divers établissements astronomiques pour doter les astronomes de récepteurs électronographiques ayant un champ étendu et donnant des images très fines. Le principal stimulant de ces entreprises est sûrement la qualité photométrique exceptionnelle que permet l'électronographie; c'est un plaisir de noter que A. Lallemand aura été, à deux reprises, un pionnier dans ce domaine, c'est un autre plaisir de constater qu'un certain nombre de jeunes astronomes n'hésitent pas à s'engager dans cette voie.

2ème Session

Cette deuxième session est consacrée à l'organisation du groupe de travail, à deux exposés complémentaires sur l'électronographie et à une série de communications sur les convertisseurs d'images et la télévision.

I. ORGANISATION DU GROUPE DE TRAVAIL

Pendant les trois dernières années, le Groupe de travail a fonctionné avec un petit nombre de personnes, c'est à dire que son activité a été comparable à celle du comité d'organisation d'une Commission. Il semble préférable de constituer un Groupe de travail élargi, au sein duquel un groupe restreint s'occupe des tâches d'organisation et d'administration. La Commission 9 approuve les propositions suivantes:

1. Le Groupe de travail est ouvert à tous les membres de la Commission intéressés.

2. Le groupe restreint est composé comme suit:

(a) Un 'Comité des Sages': W. A. Baum, J. D. McGee, G. Kron, A. Lallemand, V. B. Nikonov.

(b) Un noyau d'astronomes qui se sont engagés plus récemment dans ce domaine: K. Ford, R. Lynds, D. McMullan, P. V. Shcheglov, M. Walker, G. Wlérick (Président).

(c) Un certain nombre de jeunes astronomes dont les noms seront proposés pendant la période 1970–73.

L'importance et la qualité des communications présentées au cours des deux sessions montrent que l'intérêt pour les récepteurs photoélectriques d'images est manifeste, à l'intérieur de la communauté astronomique; le comité restreint espère donc que l'activité du Groupe de travail, dans sa version élargie, sera grande; les suggestions constructives seront les bienvenues.

II. EXPOSÉS SUR L'ÉLECTRONOGRAPHIE

Il y a d'abord une communication de M. Walker sur le sujet: 'Optimization of Instrumentation for Astronomical Photometry.' Il s'agit d'un important exposé de Synthèse dont le résumé sera adressé aux membres du Groupe de travail.

Vient ensuite un exposé de D. Palmer (Herstmonceux) 'A Low and Intermediate Spectrograph of Modular Construction for Use with Image Intensifiers. Report of First Trials.'

III. CONVERTISSEURS D'IMAGES

Sept exposés leur sont consacrés dont cinq pour les utilisations astronomiques et deux pour les développements. Voici les auteurs et les titres:

V. B. Nikonov: 'Research with the Aid of Image Intensifiers at the Astronomical Observatories of the U.S.S.R.' (communication tirée d'un texte préparé par P. V. Shcheglov).

W. A. Baum, T. Pettauer, and D. Busby: 'A Planetary Image Tranquilizer Utilizing an Image Converter with Deflection Coils.'

B. Oke: 'Photography with Image-Tube at the Hale 200-in. Telescope.'

M. Schmidt: 'Cassegrain Image-Tube Spectrograph for the 200-in. Telescope.'

R. Lynds: 'Spectroscopy with Various Image-Tubes at the Kitt Peak 84-in. Telescope.'

J. S. Hall: 'Activity of the Carnegie Image-Tube Committee (experiments with new tubes; optical systems associated with tubes,...).'

J. D. McGee: 'Recent Developments of Cascades Image Intensifiers at Imperial College.'

Ces sept communications conduisent aux remarques suivantes:

– L'emploi des convertisseurs s'est largement développé au cours des dernières années.

– Ils sont utilisés même avec le plus grand télescope: dans ce cas, avec une dispersion de 200 Å/ mm, on obtient le spectre d'une étoile de magnitude $B = 18,0$ en 3 min; pour une étoile présentant des raies d'absorption, on s'arrête en général à la magnitude $B = 19,0$ mais, s'il s'agit d'un quasar à raies d'absorption fortes, on peut aller jusqu'à une magnitude $B \approx 21$ à 22.

En photographie directe, au premier foyer, ces tubes permettent d'atteindre la magnitude $B = 23$ en quelques minutes.

– Avec un télescope moyen, comme le 84-in. de Kitt Peak, et des dispositifs spéciaux, on peut encore faire de la Spectroscopie d'objets très faibles. Ainsi avec une dispersion de 200 Å/mm, une résolution de 8 à 10 Å et un temps de pose de 100 min, on obtient le spectre d'une étoile de type G ou K, de magnitude 21, montrant les raies H et K, la bande g et éventuellement Hγ.

IV. TÉLÉVISION

L'emploi de la télévision a fait l'objet de trois communications:

Boksenberg: 'Photon Counting Image Device for Astronomical Photometry.'

Hynek: 'New Photometer Using an Image Orthicon Tube.'

Nikonov: 'Astronomical Research with the Aid of Television Tubes in the U.S.S.R.'

On peut rappeler enfin que la télévision a apporté beaucoup de résultats nouveaux concernant la Lune et Mars et que ceux-ci ont été présentés dans les réunions des commissions spécialisées correspondantes.

V. REMARQUE FINALE

Ces deux sessions ont témoigné de la vitalité du Groupe de travail. Avec la collaboration de tous ses membres, le Groupe va s'efforcer de faire largement circuler les informations utiles à la communauté astronomique pendant les trois années qui vont s'écouler jusqu'à la prochaine Assemblée de l'UAI.

COMMISSION 10: SOLAR ACTIVITY (ACTIVITÉ SOLAIRE)

Report of Meetings, 19, 20 and 25 August 1970

PRESIDENT: Z. Švestka.
SECRETARY: A. D. Fokker.

Business Meeting

The President opened the meeting by referring to the death of the Commission members L. d'Azambuja and O. F. W. Mathias. Dr d'Azambuja was a former president of Commission 10. It was decided to write a condoleance, on behalf of Commission 10, to Mme d'Azambuja.

I. ORGANIZING COMMITTEE

The President announces the proposed composition of the Organizing Committee of Commission 10: Jefferies (President), Kiepenheuer (Vice-President), Hyder (Secretary), de Feiter, Giovanelli, Krat, Michard, Nagasawa, Newkirk, Švestka.

The proposal is approved.

II. REPORT OF THE WORKING GROUP ON THE PRESENTATION OF SOLAR DATA

M. Waldmeier reports on the present status of the Quarterly Bulletin on Solar Activity. The system of reporting on solar flares and on radio emission remained the same. The reportings of coronal data from High Altitude Observatory and Sacramento Peak Observatory came to an end. Although the data from different coronal stations continue to show up discrepancies, it is possible to derive from them heliographic maps which picture coronal intensities in a qualitative way. An important and valuable addition to the Quarterly Bulletin's content has been the introduction of synoptic charts of solar magnetic fields, supplied by the Mount Wilson Observatory. Reports on X-ray observations are still not contained in the Bulletin; so far no centre has been found for collecting the data and preparing them from publication. Suggestions as to a possible X-ray data centre will be welcome.

III. DISCUSSION ON FLARE DATA

(a) P. Simon reads a report prepared by R. Michard. Notwithstanding the prescriptions given for rating the importance of a flare, notable discrepancies continue to exist between different patrol stations. Although the Hα patrol is continuous, there is a marked diurnal variation in numbers of reported flares. A new method of processing the reported data is now being developed by which known systematic discrepancies are accounted for. By this procedure the diurnal variation will be largely eliminated.

(b) C. Sawyer reports on efforts made to define indices of reliability of flare reportings and flare patrols. For a given observatory one can derive an index $u = $ number of unconfirmed reports/total number of reports, and an index $m = $ number of confirmed flares that were missed/number of confirmed flares during patrol.

From these two indices a 'confidence index' C between 0 and 1 can be derived for different observatories.

(c) H. W. Dodson-Prince reports on attempts to define a comprehensive major flare index. Such an index was defined as the sum of the five indices that characterize different aspects of the flare event:

    1. importance of Hα flare

2. importance of SWF
3. magnitude (logarithmic) of $\lambda \approx 10$ cm flux
4. dynamic spectrum (type II, continuum or type IV)
5. magnitude (logarithmic) of ≈ 200 MHz flux

IV. PROPOSED SPRAY PATROL

Upon Y. Öhman's proposal, the following recommendation has been adopted: "*Considering* the fact that many Doppler-shifted sprays are not recorded with ordinary Hα-filters, it is *recommended* that a small Working Group be formed in order to improve the optical methods of observation".

Institutes participating in the Working Group should be prepared to cooperate in a test patrol during a few months in the course of next year. Öhman will settle the composition of the Working Group by himself after consulting the interested observatories and individuals.

V. SPECTRAL CLASSIFICATION OF MICROWAVE BURSTS

J. P. Castelli reports on attempts of finding a simple classification of microwave outbursts by which the overall spectral characteristics and the intensity level of a burst are summarized in a compact way. A few letters serve to symbolize the spectral shape and a quantitative index is given for the spectral index or the relative bandwidth. The suggested classification is related to physical mechanisms. It might serve to sort out the observed bursts in a physically relevant way.

VI. RECOMMENDATION ON THE SOLAR FLUX MEASUREMENTS

A. E. Covington draws the attention to the importance of the solar radio flux at centimetre and decimetre wavelengths as an index of solar activity. Efforts to maintain homogeneous and standardized series of daily observations should therefore be encouraged. The following recommendation has been proposed and adopted:

"Commission 10 of the IAU *continues* to be aware of the value of solar flux observations in the decimetre portion of the radio spectrum as an index of the slowly varying component of solar activity in research studies of the Sun and in solar-terrestrial relationships, as well as for predictions of radio propagation characteristics of the ionosphere for practical telecommunications; and *notes* that the generally adopted international standard endorsed by CCIR and other groups is the series of measurements made on a frequency of 2800 MHz by the National Research Council of Canada for more than two solar cycles; and further *notes* that a Working Group on Absolute Solar Calibration in URSI has removed major discrepancies at specific frequencies in the band 536 MHz to 9400 MHz to derive a smooth solar spectrum for the epoch 1968 and *expresses* the strong hope that this series of measurements be carried on indefinitely as a contribution to international and technical knowledge and that the reliability of the measurements be monitored and improved as needed. Commission 10 of the IAU therefore *urges* the appropriate national agencies to undertake the operation of facilities which will ensure the provision of these services".

VII. PROPOSAL TO ESTABLISH A WORKING GROUP ON MAGNETOGRAPH CALIBRATION

D. Rust, speaking also on behalf of J. M. Beckers, explains why it is desirable to make a coordinated effort to improve the calibration procedures of magnetic field measurements. He introduces the following proposal.

"*Because* of the fundamental importance of solar magnetic field measurements for the understanding of diverse phenomena in all observable levels of the solar atmosphere and

because of the difficulties and controversies encountered in calibrating the signals of solar magnetographs in terms of magnetic field intensities, Commission 10 of IAU *establishes* a Working Group to make a comparative study of the solar magnetic field measurements obtained at the many observatories where a magnetograph is available. Commission 10 *recommends* that all these observatories map the magnetic field in an active region to be designated by the Working Group. The Working Group should consult with the researchers operating magneto-graphs to determine the most convenient and appropriate operating parameters (i.e. entrance aperture, spectral line etc.) to be used with all the magnetographs during the study period. The resulting observations and the detailed descriptions of the techniques used to deduce the values of both the longitudinal and transverse components (when available) should be collected, compared and compiled for distribution to all interested researchers. In this way it should be possible to estimate the reliability of the magnetic field measurements and to recommend improved calibration procedures."

A. Severny comments upon this proposal that work of this kind has been done already in the past. For instance, comparisons have been made between the measurements made at the Crimea and Mount Wilson observatories. He recommends that the Working Group takes advantage of the exist-ing correspondence on this subject.

The president proposes the following persons as members of the Working Group: Beckers (chair-man), Bumba, Cacciani, Deubner, Giovanelli, Howard, Ioshpa, Krat, Livingston, Orrall, Rayrole, Rust, Schröter, Severny, Stepanov, Tandberg-Hanssen, von Klüber, Vrabec. Of course others may join the Working Group if desired.

The proposal to establish the Working Group and its proposed composition is adopted.

VIII. THE INTERNATIONAL PROGRAMME FOR SUN-EARTH MONITORING

A. H. Shapley reports on the programme for international monitoring of the Sun-Earth environ-ment (MONSEE) and related matters. At the Leningrad meeting, May 1970, the IUCSTP Bureau has taken a resolution in which the IUCSTP Group 1 on monitoring of the Solar-Terrestrial Environment was requested to act as a steering committee to organize the international MONSEE, in consultation with all other IUCSTP working groups and other appropriate bodies and to make periodic reports and recommendations to IUCSTP concerning the current and future operation of MONSEE. IUCSTP Working Group 1 is contemplating an organizing or planning meeting at the time of the 1971 IUGG general assembly. Any comments from IAU Commission 10 in the area of solar activity monitoring will be welcome. The IUCSTP Working Group 1 naturally considers Commission 10 the authorative group as solar activity is concerned.

IX. THE JOSO PROJECT

K. O. Kiepenheuer reports on the activities of the Joint Organization for Solar Observations. Site testing has been conducted at various places in southern Europe: Sicily, the isle of Lampedusa, Isola della Corrente, Faro and Peseguera on the southern and west coasts of Portugal. Participating countries in the project are France, Germany (D.B.R.), Italy, The Netherlands, Norway, Sweden and Switzerland.

Joint Meeting with Commissions 40 and 44 on Solar Mapping

Prior to starting with the scientific programme the President called attention to a proposal of V. Bumba to have an IAU Symposium organized on Solar Activity to be held in the course of 1972 in Czechoslovakia. The question arises whether there might be an overlap in content of this pro-posed symposium with the one which is planned to take place in Australia in 1973. R. G. Giovanelli gives some information: this symposium will be devoted to the Sun's outer layers. Provided care is taken to avoid overlap between the two symposia, the proposal of Dr Bumba is adopted. The following Organizing Committee for the Symposium on Solar Activity has been approved: Bumba (chairman), Dodson-Prince, Gnevyshev, Kiepenheuer, Kopecký, Leighton, Martres, Nakagawa, Vitinsky and Wilson.

The meeting on solar mapping was organized by C. W. Allen and A. D. Fokker who acted as chairman for respectively the XUV aspect and the radio aspect of solar mapping. The following papers were given:

C. W. Allen: Solar Mapping and Its Use.

J. P. Wild: Review of Culgoora Radio Heliographic Results. Dr Wild presented a movie of dynamic solar radio phenomena observed with the Culgoora heliograph at the frequency 80 MHz. This movie, the first in its type, demonstrates impressively the potentialities of the Culgoora instrument.

J. L. Bougeret: Results of Nancay Multilobe Interferometer Observations.

M. Simon (also on behalf on D. Buhl and A. Tlamicha): Mapping of the Sun at Millimetre Wavelengths.

W. N. Christiansen: New Facilities at Fleurs.

R. W. Noyes: X-Ray Studies of Active Regions by the Harvard Observatory.

L. W. Acton: A Mapping X-Ray Heliometer for OSO.

T. Takakura (also on behalf of several others): A Balloon Observation of the Position and Size of a Hard X-Ray Burst.

A. Title: New Observational Techniques in Harvard for Solar Mapping.

An additional paper, not fitting in the subject of solar mapping, was presented by Z. Suemoto on behalf of

H. Yoshimura: Differential Rotation and Solid Body Rotation of the Active Regions.

Joint Meeting with Commission 40 on Radio Studies of the Sun and the Interplanetary Medium

The meeting was organized by M. R. Kundu; its report is to be found in the proceedings of Commission 40.

Meeting of the Organizing Committees of Commissions 10 and 12

The Organizing Committees, together with some invited guests, discussed on a possible reorganization of the solar commissions of IAU. The president of Commission 10 had distributed in advance a number of opinions on this matter received from members of Commission 10. He pointed out that there are three reasons for reconsidering the organization of Commission 10:

1. the great number of members;
2. the overlap between Commissions 10, 12, 40, 44;
3. the fact that the agenda contains too many non-scientific topics. As to the third point the President suggested that a sub-commission of Commission 10 might be formed which would deal with all matters concerning the reporting of solar data and international cooperation in the data acquisition.

The president asked J. T. Jefferies, as President of Commission 10 for the coming three years, to take the chair.

Apart from the three issues mentioned by Švestka, two more were raised: the content of meetings during the General Assemblies (Jefferies) and the question whether there should be only one commission for the Sun (Newkirk). As to the number of members, nobody saw a possibility of reducing it. Sub-commissions might be created for such subjects as sunspots, flares etc., but this would tend towards increasing the overall number of meetings during a General Assembly. As to the overlap with other commissions, Jefferies suggested that in the draft report Commission 10 deals only with ground based optical research. Space observations and radio work on the Sun then should exclusively be covered by resp. Commissions 44 and 40. There is also the question of the overlap of membership. Membership of solar radio astronomers eventually may be restricted to Commission 10 (Wild). As to the programme of meetings, the system of joint meetings is perhaps liable to still more perfection; there should be much more consultation between Presidents of Commissions.

The suggestion to create a sub-commission for all matters relating to solar data found a favourable response. Such a sub-commission could naturally act as advising body for the IUCSTP and also for the Quarterly Bulletin. In the meetings of Commission 10 more time will thus become available for scientific matters. Jefferies proposed to have only a few review papers on respectively ground-based optical research, radio work (with Commission 40) and solar observations from space (with Commission 44). Moreover, one or two topics might be dealt with more specifically, largely by invited papers followed by discussions on short communications.

As to the question of one or more solar commissions, no clear opinion emerged from the discussions.

Meeting of the Full Commission (the Vice-President in the chair)

I. Highlights from the Symposium on Solar-Terrestrial Physics held in May 1970 in Leningrad.

C. de Jager gave an account of some of the main topics dealt with on this Symposium.

II. Report on the previous meeting.

Before reporting on the outcome of the discussions in the previous (restricted) meeting, J. T. Jefferies, as the new President, requests to send titles and abstracts of forthcoming papers to him. He intends to prepare each year a list of titles and to send this to members of Commission 10. The question of unifying the solar commissions will have to be considered by the Organizing Committee which may be expected to prepare a report on this topic.

The proposal to establish a Sub-Commission 10a for reporting of solar data and for international cooperation in data acquisition is adopted and P. Simon is approved as its chairman for the coming three years. Within a few months he will submit his proposal for the sub-commission's membership to the President for approval.

L. W. Acton suggests that the work of the Working Groups and the evaluation of homogeneity and reliability of data be supervised by the sub-commission.

The other proposals, regarding the organization of meetings during General Assemblies, do not meet any opposition. The proposals imply that the Presidents of Commissions 10, 12, 36, 40, 43, 44 act as a programme committee for joint meetings.

III. Report of the Working Group on the normalization of intensity measurements of the solar corona.

J. Rösch reports on the activities of the Working Group. Although no second meeting was held after the one in 1967, much work was done which concentrated mainly on the definition of a suitable altitude above the limb and on the establishment of an empirically derived standard scale which, though being not an absolute scale, brings a reasonable uniformity in the various series of measurements. The work will be continued by correspondence.

IV. Report on IUCSTP activities.

H. Friedman, reporting on IUCSTP, refers to the International Active Sun Years and to the Leningrad Symposium. A Symposium on results of the 1970 eclipse will be organized jointly with COSPAR in Seattle, U.S.A. Furthermore, IUCSTP stimulates the coordination of efforts during the next minimum of solar activity. The emphasis should from now on be in the gathering of simultaneous information from satellites at a variety of suitable positions in space. Data on the existence of gradients in interplanetary space may thus be acquired. Moreover, IUCSTP recommends to place probes in heliocentric orbits and to send probes out of the plane of the ecliptic.

V. Report on activities of the IUCSTP Working Group 2.

Z. Švestka reports on IUCSTP Working Group 2 which is concerned with proton flares. The Working Group made preparations for a second Proton Flare Project after the one successfully carried through in 1966. Unfortunately no proton event showed up in the designated period May–July 1969. A third PFP period is not envisaged, but the study of selected retrospective intervals is fairly encouraging. Extensive compilations for a few proton flare events have been made by J. Virginia Lincoln, by ESLAB in Noordwijk and two more of such studies are in preparation. The Working Group has decided to prepare a detailed catalogue of all particle events before the end of 1969 which will include the particle data, X-ray, radio and optical data on the responsible flares and on the active regions concerned. Comments on the problems of a future Proton Flare Project, on retrospective intervals and on possible forms of international cooperation will be welcome.

COMMISSION 12: RADIATION AND STRUCTURE OF THE SOLAR ATMOSPHERE
(RADIATION ET STRUCTURE DE L'ATMOSPHERE SOLAIRE)

Report of Meetings, 22, 24 and 25 August 1970

CHAIRMAN and ACTING PRESIDENT: R. Grant Athay.
SECRETARY: L. Delbouille.

Business Session

The chairman gave first a welcoming address and best wishes for the work of the Commission in the name of Dr Gnevyshev, who was unable to attend the General Assembly.

He then transmitted an inquiry from the General Secretary of the IAU asking that the commission inform the General Secretary of the manner in which the commission arrives at official decisions.

The new officers of the Commission were then proposed to the members, and approved by a vote.

The chairman then read a proposed set of rules for the operation of Commission 12. After some discussion it was decided that a formal set of rules should not be adopted until those members of Commission 12 not in attendance at Brighton had had an opportunity to study them. The chairman agreed to circulate the proposed rules to all members of the commission for their critical comment.

The chairman then asked the members if they had questions to raise.

de Jager: Is it worthwhile to plan a new 'Bilderberg Conference?'

Answers:

Bonnet: Yes, it gives good opportunities for contacts between experimenters and theoreticians.

Giovanelli: Yes, but some thought should be given as to how far in the chromosphere to push the model.

Namba: Is it not time now to include some non-homogeneous structure?

Wilson: Perhaps we should have a special meeting on inhomogeneities.

Jefferies: The first Bilderberg conference was very stimulating. I am personally very favorable to another meeting in 2 or 3 yr.

Athay: It is important to have, at all times, a reference model to which a large number of people have contributed. One of the most useful features of the Bilderberg model was that enough people contributed to the model that it became a genuine reference model.

Athay: Have you any suggestions about colloquia or symposia to be organized by Commission 12? Plans are already in progress for a joint discussion, in Sydney, on solar fine structure. This will involve several commissions besides Commission 12.

Athay: Another question we need to discuss is that relative to the reports, both the past one as published and the next one, to be prepared for 1973. The general guidelines for the past report were that it should be very short, should not include references to work already published, and should not include a bibliography. It seems that a bit more freedom will be given for the next one. Three points were then discussed:

(a) Bibliography in the future report.
The strong consensus was that all future reports should contain complete bibliographies.

Houtgast introduced an idea given by Thomas at Commission 36.

Thomas gave more particulars suggesting a distribution, through the president, at the end of each year of a neswletter including the titles only (or the titles plus abstracts) of all the papers brought to the attention of the commission president. The newsletter can probably be distributed even to interested people not in the Commission.

Newkirk asked about the possibility to distribute, as in the past, the report first in a draft form. This was not possible for the present report.

(b) In the past, the report was supposed to mention the work made between two IAU Assemblies. What is the opinion of the members on this question?

Pasachoff favors the inclusion of all the work done during each three year period.

Newkirk – Inclusion of only the work in progress attenuates the usefulness of the report, which was, in the past, an easy source of references.

Other members of the Commission also strongly favored this point of view.

(c) Should the report include evaluated comments by its author?

Thomas: Giving a personal opinion is always dangerous. In any case, nothing should be suppressed by a personal choice.

Orrall insists once more on the usefulness, if practical, to include a complete bibliography.

Thomas: Perhaps in the form of a separate report?

Athay: A separate report can more easily be lost. What about the use of 'Astronomy Abstracts,' giving only their number and title?

Müller agrees with Thomas about the difficulty to be objective, and asks if the report should not be just a listing of work done, without any comment.

Newkirk agrees also. It seems to him that the IAU reports are not the most appropriate place for comments.

Houtgast stresses the point that excluding comments will shorten the report, and take less pages in the printed book.

Athay concludes the discussion in accepting the idea that comments should be reduced to a minimum.

Other problems raised:

Orrall: What can be done about common interests of Commission 12 with Commission 10, perhaps also with Commissions 40, 44 and 36?

Athay announces his intention to propose that Commissions 10 and 12 work jointly in organizing their scientific sessions. This will be proposed and discussed in a joint meeting of the Organizing Committees of the two Commissions a few days later.

Pierce suggested that the Commission should discontinue the work of the group on central line intensities, but should create two new Working Groups:

(a) On sunspots spectra (excluding magnetic fields problems).

(b) On the high resolution atlas of the photospheric spectrum.

He suggests Delbouille as chairman for this second new group.

Jefferies expressed his interest in such a group, insisting on the usefulness for its members to include theoreticians, as advisers about what has to be done.

He suggests Pierce as chairman of a new working group on sunspots spectra, and Pierce accepts, on the condition that the 'central intensities' group be discontinued.

A vote of approval was cast for the formation of the two new groups:

(a) On sunspots spectra (Chairman A. K. Pierce),

(b) On a high resolution atlas of the photospheric spectrum (Chairman L. Delbouille) and for the dissolution of the previous group on central intensities.

SCIENTIFIC COMMUNICATIONS

A. Wyller: 'Fabry-Pérot Observations of Sunspots'.

A scanning type Fabry-Pérot filter with a band pass of 0.04 Å in the Na D line region has been built for observations of sunspots.

C. E. Moore-Sitterly: 'The Revision of the Multiplet Tables'.

This was mainly an announcement that the Section III (C I to C VI) is now in press, and the Section IV (N) is in preparation.

P. Lemaire (presented by R. M. Bonnet): 'High Resolution Spectra of the Mg II Resonance Lines Observed from a Balloon-Borne Instrument.'

To observe the Mg II lines, a balloon-borne instrument was built, comprising a cassegrain telescope ($F/20$, focal length 300 cm) and a spectrograph with a theoretical resolution of about 10 mÅ. In flight, 25 mÅ spectral and 3 s spatial resolutions have been achieved. The spectra show strong local variations on the solar disk. In general the shape of the lines is asymmetric and seems single peaked across the spots.

Second Meeting

Chairman: E. A. Müller.

A. Dupree and L. Goldberg: 'New Information on the Structure of the Quiet Chromosphere and Corona from OSO IV and OSO VI Data.'

A report was given based on the two OSO satellites (observations between 300 and 1400 Å with about 3 Å resolution) and on a 1969 rocket experiment with a photoelectric scanning spectrometer for the region 1350 to 1825 Å. A model of the chromosphere, the transition region and the corona has been built to fit these observations.

R. J. Bray: 'Preliminary Observations of the Chromosphere with a Computer-Controlled Tunable 1/8 Å Filter.'

This filter is installed in the 30 cm chromospheric telescope of the CSIRO Solar Observatory (Australia).

V. N. Karpinsky: 'Some Results of Solar Granulation Photography in Pulkovo Observatory.'

A resolution of 0.3 to 0.5 arc sec has been achieved in various monochromatic pictures of the solar granulation. Very good correlation is found between images taken simultaneously at 4800 and 6000 Å.

V. A. Krat: 'Granulation Studies with Balloon-Borne Instruments in Pulkovo Observatory.'

Two successful flights took place, one at the end of 1966 and one in 1970. For the 1970 flight, observations lasted six hours from an altitude of 20.5 km. About 6 % of the 900 pictures obtained show a resolution corresponding to the theoretical diffraction limit of of the 50 cm objective. An attempt to measure the Dα deuterium line in the blue wing of Hα in the absence of most of the telluric H_2O absorption leads to a value of $D/H < 2 \times 10^{-5}$.

J. Rösch: 'Images recentes de taches solaires.'

Various exposure times give the possibility to see many detailed features at different brightness levels between the chromosphere and the central region of the spots.

P. R. Wilson and C. J. Cannon: 'Velocity and Brightness Fluctuation Correlations in Chromospheric Lines.'

Report of line-center intensities and velocity fluctuations measured for Mg b, Na D and Ca II K lines. Near the limb, the strucuture is consistent with the supergranulation cells. At disk center, regions of significant correlation may extend over distances of the order of 25000 km.

There are points on the disk at which the K_2 emission feature appears only in either the red or the blue wing of the line, but also regions in which the doubly peaked profile is real.

Z. Suemoto: 'Emission Line Profiles and the Inhomogeneous Model of the Chromosphere.'

High spatial resolution spectrograms of the Ca II K line suggests that the averaged line profile may result from a more or less random distribution of many narrow local emission and absorption features.

Third Meeting

Chairman: Z. Suemoto.

V. N. Karpinsky: 'The High Precision Investigation of Fraunhofer Lines in Pulkovo Observatory.'

The Pulkovo Observatory is equipped with a photoelectric high resolution solar spectrometer used since 1966 to record line profiles and central intensities in the visible.

L. Delbouille, G. Roland, and R. Zander: 'Recent High Resolution Solar Spectroscopy by the Liège Group."

Tables of wavelengths and identifications of the 10450 lines measured in the *Atlas of the Solar Spectrum between* λ 7498 to λ 12016 published in 1963 have been prepared by W. S. Benedict and J. W. Swensson and are under print. With the collaboration of L. Neven, the Jungfraujoch photoelectric spectrometer is currently used to remap the photospheric spectrum between 3000 and 12000 Å. The region 5000 to 6300 Å is ready to be printed. A balloon-borne infrared spectrometer was flown in April 1970. The solar spectrum between 1.8255 and 1.8570 μ has been recorded with a resolution limit better than 0.04 cm^{-1}. 72 water lines and 67 solar lines have been measured in this domain.

N. Grevesse: 'New Solar Abundances and Identifications Obtained at Liège.'

The following results were recently obtained:

(a) Solar abundances of Ni from the photospheric [Ni II] lines by N. Grevesse and J. P. Swings ($\log N_{Ni} = 6.30$) – *Solar Phys.* **13**, 19, 1970.

(b) [Fe I] lines: their transition probabilities and occurrence in sunspots, by N. Grevesse, H. Nussbaumer and J. P. Swings ($\log N_{Fe} = 7.50$).

(c) Identification of SiH$^+$ in the solar photospheric spectrum, by N. Grevesse and A. J. Sauval.

(d) Oscillator strengths for SiH and SiH$^+$ deduced from the solar spectrum, by N. Grevesse and A. J. Sauval.

(e) Solar and meteoritic abundances of mercury, by N. Grevesse.

(f) Deuterium in the solar photospheric spectrum, by N. Grevesse. To appear in the proceedings of the 16th Liège Colloquium.

E. A. Müller: 'On the Solar Lithium and Beryllium Abundances.'

E. A. Müller reported on a detailed investigation of the solar lithium feature carried out in collaboration with J. W. Brault and E. Peytremann. They obtain $\log \varepsilon_{Li} = 1.0 \pm 0.10$ and Li^7/Li$^6 \geqslant 10$. Center-limb observations secured by Brault and Müller of the solar absorption features at $\lambda 3321$ and $\lambda 3130$–3131, and of the lines of Be I and Be II recorded with a laboratory source were presented. The solar feature at $\lambda 3321$ proved not due to Be I. Both Be II resonance lines are present, but severely blended, in the solar spectrum. Both the equivalent width and the abundance of beryllium in the Sun have up to now been much overestimated. This work is still in progress.

G. Brückner: 'Line Profiles in the Extreme UV.'

Recent results obtained at the Naval Research Laboratory with high resolution far UV spectrometers were reported.

H. L. Demastus and R. B. Dunn: 'Coronal Movies.'

Two movies were shown containing the best coronal scenes taken from the Sacramento Peak Observatory between October 1966 and November 1969.

The scenes contained a number of flare-associated fast-accelerated expansions and also several fine loops.

Fourth Meeting

Chairman: A. K. Pierce.

D. Hall: 'Observations of the Infrared Sunspot Spectrum between 1 and 2.5 μ.'

The infrared umbral spectrum has been observed, at Kitt Peak National Observatory, in the atmospheric windows λ 11340–λ 13528, λ 14872–λ 18037 and λ 19614–λ 24778, with a spectral resolution of 100000.

Preliminary analysis of the spectra has established the presence of umbral lines due to the molecules OH, CO, CN, HF and H$_2$O.

J. W. Harvey: 'A Photographic Sunspot Spectrum Atlas (λ 3700–λ 9200).'

A photographic atlas of the spectrum of a large sunspot is in preparation at Kitt Peak National Observatory.

G. Gonczi and F. Roddier: 'A Model of the Solar Photospheric Velocity Field.'

The model presented includes two layers of different microturbulence and, for convection, an ascending and a descending columns.

H. Wöhl: 'Photoelectrically Scanned Sunspot Spectra λ 4000 to λ 8000.'

Newly identified molecules in the sunspot spectra are NiH and CoH. Another similar sunspot spectrum was obtained by H. Wöhl, A. Wittman and E. H. Schröter (Gottingen Univ. Observatory). A section of the uncorrected spectrum (λ 4500–λ7000) is available on magnetic tape.

O. Hauge: 'The Presence of CuH Lines in the Sunspot Spectrum and the Solar Isotope Ratio of Copper.'

CuH lines from the R and P branches of the $^1\Sigma - X^1\Sigma$ (0, 0) band are shown to exist in the sunspot spectra obtained at the Kitt Peak National Observatory. From two CuI lines of relatively high isotope splitting, an isotope ratio $Cu^{63}/Cu^{65} = 2 \pm 1$ is obtained.

P. Turon: 'High Resolution Sunspot Images at 10 μ.'

A slow raster scanning technique with a germanium bolometer has been used to build TV images of the Sun at 10 μ. The Kitt Peak National Observatory solar telescope allows a diffraction limited resolution of 2 arc sec.

O. Namba and G. Hafkenscheid: 'Profiles of the CI 5052 Line from Individual Elements of the Granulation.'

This line is appreciably stronger in the granules than in the intergranular regions, but the outer parts of the profiles are broader in the intergranular regions than in the granules. The profiles from the intergranular regions are red-shifted relative to those from the granules by 0 to 0.6 km/sec.

WORKING GROUP ON SOLAR ECLIPSES

Chairman: M. Rigutti.

D. H. Menzel and J. M. Pasachoff reported on prospective sites for the 30 June 1973 solar eclipse in Africa. The sites fall in the Sahara desert and the Aïr mountains in Niger and Mali. They suggest, because of the limited facilities available there, that substantial IAU-coordination on an international basis would be advisable, and that local preparations must be commenced long in advance. Preliminary weather forecasts are most favorable in the visited sites.

ECLIPSE OF MARCH 7, 1970

J. Houtgast reported on observations made by the Dutch party at Miahuatlan, Mexico. A Littrow type slit spectrograph with a resolution of 150000 equipped with slit jaw camera for the orientation of the solar image, a photoelectric exposure time device, and a photoelectric multiplexing device in the Mg b region were used.

B. Fort reported on an experiment with a specially built new Lallemand type electronic camera made by the 'electronic camera group' of the Meudon Observatory. The observations were made by direct photography through a very narrow band pass filter. Brightness distribution of both intensity and polarization of the green line up to a height of about 1.25 solar radii with a good spatial resolution was obtained.

Z. Suemoto reported on observations made by the Japanese party at Puerto Escondido, Mexico. A quadruple camera (focal length of 5 m) with a Newkirk filter and four polaroids just in front of the film was used by K. Saito in order to study three-dimensional structures even in the innermost corona. M. Makita took direct photographs (focal length of 5 m) with a continuous spectrum of about 100 Å centered around 6900 Å excluding the Hα-lines of hydrogen and the 6678 Å-line of HeI. M. Kanno got a number of flash spectra in the range from 4686 to 6678 Å with an oblique incidence slitless grating spectrograph and a solar image of 46 mm in diameter.

T. C. Van Flandern reported on observations from near the path edge made by himself, D. E. Dunham and J. E. Bixby.

M. Waldmeier presented a documentary film on the instrumentation of a number of observing parties and on the partial and totality phase of the eclipse.

V. Gaizauskas provided the audience with copies of a map prepared by the Canadian N.R.C.

Astrophysics Branch of the totality path through Canada of the total eclipse of 10 July 1972. He announced also the preparation of tables for weather forecasts.

Z. Suemoto announced the preparation by the National Meteorological Bureau of Japan of weather forecasts for the 30 June 1973 eclipse in Africa.

The Chairman gave a short report on the activity of the Working Group on Solar Eclipses.

First of all he remarked on the impossibility of preparing a report on the past 1968 and 1970 eclipses with a summary of the main subjects covered by successful experiments because of the very limited information provided to him by the party leaders. For this reason the last issue of the beautiful 1970 eclipse bulletin, edited by the U.S. Coordinator Dr E. Belon, will be the only comprehensive source of information of successful observations during the 1970 eclipse.

After this, the Chairman remarked on the rather poor activity of the Working Group in the past three years and asked the members and the audience to reconsider the usefulness of the existence of the Working Group itself. A long discussion took place and the following points were unanimously reached.

1. The Working group on Solar Eclipses must continue its work which can be very useful to all solar astronomers. Accordingly, the members will actively participate in the needs of the group. New members willing to join the group should be accepted and old ones not really active should be cancelled.

2. The Group should get a representation in the COSPAR.

As far as the activity of the working group in the near future is concerned, it has considered the possibility of organizing a symposium on 'Theoretical and experimental problems related to solar eclipse observations.' This symposium should be an occasion to discuss problems both from the theoretical and experimental point of view and would have the aim of pointing out the most interesting problems which can be considered for experiments during solar eclipses, and to suggest to solar astronomers what kind of observations could be worthwhile to acquire. It came out from the discussion that such a symposium should interest a great number of people and that, for this reason, it should be necessary to find a supporting organization and a host country. G. Athay, President of Commission 12, will then contact the COSPAR and the IUCSTP for trying to get the necessary support.

The usefulness of preparing a comprehensive bibliography of as many as possible papers appearing as a consequence of solar eclipse experiments and studies has been examined. The Chairman will test the actual feasibility of such a bibliography by inquiring personally and through the IAU Bulletin among solar astronomers.

The need for information about the next eclipses has been recognized. Contacts with governments of countries through which the total eclipse path will pass will be necessary but it appeared clear that it will be wise to establish them through important organizations. G. Athay will ask the U.S. National Science Foundation whether they will continue to act as an information centre. This problem is, of course, of the greatest importance for the 1973 eclipse.

Concluding the meeting, the Chairman on behalf of the whole group expressed thanks to Dr E. Belon and the people who worked with him for the publication of the 1970 eclipse bulletin, Dr E. R. Dyer Jr., secretary of the IUCSTP, Dr G. Haro, Dr J. L. Locke, Dr A. de Tuddo, co-director of the 'Estudio de los recursos del Estado de Oaxaca' and all the persons who in one way or another made easier for the observers the work connected with the eclipse experiments in 1968 and 1970.

WORKING GROUP ON SUNSPOT SPECTRA

Chairman: A. K. Pierce.

An organizational meeting of the working group on the spectra of sunspots was held at Brighton, England. All interested persons are invited to participate in the activities of the group. The work of the group will primarily be in the exchange of information with respect to: atlases (photographic, photoelectric), identifications, wavelengths, models, and abundances.

WORKING GROUP ON CENTRAL LINE INTENSITIES

Chairman: A. K. Pierce.

The preliminary results of a number of workers have been brought together. The following tabulation is a partial listing of corrected central intensities together with their continuum points. Further details are to be found in the literature or from: Goldberg-Rogozinskaya, Karpinsky, Krat, Pravdjuk, Babiy, Gurtovenko, White, Delbouille, Brault, or Pierce.

Ni	3411.500	100.0	Fe	5574.796	100.0
	3414.799	2.3		5576.099	22.2
	3419.932	100.0		5577.960	100.0
Y II	3600.525	98.4	Ni	5846.882	99.7
	3600.739	9.4		5847.006	76.9
	3601.113	99.8		5849.395	100.0
Fe	4069.424	100.0	Na	5888.220	96.4
	4071.749	3.2		5889.973	4.1
	4073.401	98.1		5894.000	98.0
Ca	4220.909	100.0	Na	5894.000	98.0
	4226.740	2.3		5895.940	4.8
	4230.803	100.0		5898.620	98.6
Mg	4571.104	14.5	Fe	6085.818	99.9
				6089.574	64.1
Mg	5167.330	8.0		6089.881	99.9
Mg	5172.700	5.8	Ni	6128.751	100.0
				6128.984	72.1
Mg	5183.621	5.2		6130.482	100.0
Fe	5294.932	99.9	Ni	6128.751	100.0
	5295.321	66.7		6130.141	77.6
	5295.556	99.8		6130.482	100.0
Fe	5304.708	100.0	Fe	6677.357	100.0
	5307.369	21.7		6677.997	27.5
	5308.163	100.0		6679.088	99.9
Ni	5432.170	100.0	Ni	6767.120	100.0
	5435.866	46.0		6767.784	32.9
	5440.334	100.0		6768.205	100.0

O₂ 6870–6877 zero central intensity at 5 air masses.

WORKING GROUP ON THE HIGH RESOLUTION ATLAS OF THE PHOTOSPHERIC SPECTRUM

Chairman: L. Delbouille.

A first informal meeting to organize the working group was held in Brighton. A first list of persons interested in its activity has been made, but can, of course, always be expanded in the future. After discussion, a few guidelines have been established for the activities of the group, whose main purpose is to coordinate observational work made by several groups and to try to organize the centralization of all the assignment work in a single location. As a first scheme, it has been recognized that the Liège group, from the Jungfraujoch high altitude station, is progressing in the record of the region λ 3000–λ 12000 Å and will publish it in sections, with wavelength calibration done in collaboration with Kitt Peak National Observatory. Coordination of future work further in the infrared will be attempted as soon as possible. Contacts will be taken by Dr Delbouille about the possibility to concentrate all the work on the assignments of all the measured lines in Washington, D.C., U.S.A. Practical exchange of information will be made in the form of newsletters by the chairman and private communications to him.

COMMISSION 14: FUNDAMENTAL SPECTROSCOPIC DATA
(DONNÉES SPECTROSCOPIQUES FONDAMENTALES)

Report of Meetings, 19 and 25 August, 1970

PRESIDENT: M. V. Migeotte.
SECRETARY: J. G. Phillips.

Silent tribute was paid to one member lost by death since the 1967 meeting: C. C. Kiess, who died on 16 October 1967, just two days before his eightieth birthday. Dr Kiess' 40 years service at the National Bureau of Standards was highlighted by his contributions to the understanding of the complex spectra of heavy elements, notably the rare-earths, and by his work in the red and infrared, leading to the detection of phosphorus in the solar spectrum, and to the demonstration that certain strong solar lines were due to silicon atoms.

The draft report compiled from contributions by the five committees has been distributed to all members of the Commission.

The Organizing Committee has approved a proposal by Vice-President A. H. Cook that Commission 14 sponsor a colloquium on 'Experimental Techniques for the Determination of Fundamental Spectroscopic Data'. This has been organized as IAU Colloquium No. 8, and will be held at Imperial College, London, from September 1 to 4, 1970.

Dr R. Wilson, the President of Commission 44, has proposed a joint discussion on the determination of fundamental atomic data relative to space astronomy. Dr M. J. Seaton has agreed to organize this discussion under the title 'Atomic Data of Importance for Ultraviolet and X-Ray Astronomy'.

Mrs Sitterly has played a very important role in organizing a meeting for those interested in the general problem of handling numerical data in various fields of research. Notice was made of the fact that she had been asked by the Union to act as chairman of a 'Working Group on Numerical Data in Astronomy and Astrophysics'.

Of interest to molecular spectroscopists and astrophysicists is the publication (1970) of an up-dated 'Données spectroscopiques relatives aux molécules diatomiques', established under the direction of B. Rosen. In addition, R. F. Barrow is drawing up plans for the annual publication of critical abstracts; the first volume should appear in 1971.

G. Herzberg described the purpose and function of the 'Inter-Union Commission on Spectroscopy' of which he is the President. One important function is to seek uniformity of notation used by various disciplines; for instance, Mrs Sitterly is urging the universal adoption of two-letter abbreviations for each element. The role of the Inter-Union Commission is purely advisory; it has no power to require compliance by member Unions.

Proposals have been made to the Executive Committee regarding the composition of Commission 14 for the period 1970–73.

New Consultants are: G. H. C. Freeman, L. J. Kieffer, and W. R. G. Rowley.

M. J. Seaton has expressed the wish to be relieved of the Chairmanship of Committee 3 on 'Collisional Cross-Sections and Line Broadening'. H. Van Regemorter has been appointed to replace him.

First Scientific Meeting

A. COMMITTEE 1 – STANDARD OF WAVELENGTH

Chairman: B. Edlén.

E. Engelhard has measured the pressure-dependent wavelength shifts in the Lamb dip of the 6328 Å helium-neon laser radiation. As the neon pressure was varied from 0.1 to 1.2 Torr at constant helium pressure of 2.45 Torr the following variation was found for λ_{vac} as function of pressure:

$$\lambda_{vac} = 0.632991\,433\ \mu m - 19.4 \times 10^{-9}\ p$$

If, on the other hand, the neon pressure was held constant while the helium pressure was varied from 1 to 4 Torr, the following relation was found:

$$\lambda_{vac} = 0.632991\,499\ \mu m - 30.64 \times 10^{-9}\ p.$$

Applying these pressure corrections, it is estimated that the wavelength is now known with an accuracy of 10^{-8}.

Interferometric wavelength determinations in ^{136}Xe I and ^{136}Xe II by *C. J. Humphreys* and *E. Paul, Jr.* have been compiled, together with calculated wavenumbers and wavelengths. Covering manuscripts have been accepted for publication.

V. Kaufman has compiled a comprehensive list of reference wavelengths accurately determined by interferometry, grating measurements, the Ritz combination principle and by a combination of these methods. They should be useful as standard wavelengths, although the great majority have not been formally adopted as standards. Below 2000 Å he lists 1481 lines from 29 atoms or ions, while above 2000 Å he lists 1656 lines from 9 atoms or ions. The present listing is not in its final form, new measures by various investigators of Ne I, Ar I, Ar II and Al II will be included before publication, as will a table of wavelengths of Lyman lines of H through Si xiv.

A. H. Batten called attention to the possibility of introducing systematic errors in radial velocities of stars as a consequence of increased popularity of Fe-Ne discharge tubes as source of comparison spectra rather than the iron arc in air. At a dispersion of 2.5 Å per mm, the difference between λ_{arc} and the low pressure wavelengths from the discharge tube averages 0.002 Å or the possibility of 0.1 km/sec error in radial velocity. The use of the Revised Rowland Table to check errors led to the following residuals:

$$(O - C)_{discharge} = +0.0035 \pm 0.0006\ \text{Å}$$
$$(O - C)_{arc} = +0.0005 \pm 0.0007\ \text{Å}$$

B. COMMITTEE 4 – STRUCTURE OF ATOMIC SPECTRA

Chairman: Charlotte Moore-Sitterly.

Bengt Edlén presented a critical compilation, as of June 1970, of references to analyses of atomic spectra. The selected papers supplement the information contained in Charlotte E. Moore's 'Atomic Energy Levels' Volume 1 (1949). Only the first ten spectra of each of 22 elements are included, and hydrogen-like spectra are omitted. The list represents a condensation, with recent material added, of Moore-Sitterly's 'Bibliography on the Analyses of Optical Atomic Spectra' Section 1, Nat. Bur. Stand. Special Publ. 306, Washington, D.C. (1968).

W. R. S. Garton, *E. M. Reeves* and *F. S. Tomkins* reported on their recent observations of absorption line series and autoionization in Sc I, Yt I and La I.

C. J. Humphreys and *E. Paul, Jr.* in collaboration with L. Minnhagen have extended observations of the spectra of Cl I, Br I, and II in the 4 μ region. They hope to improve the intensity estimates. One new 5g level has been found in Br I by Minnhagen.

W. C. Martin reported that observations and analyses had been obtained for the following spectra at NBS: Rb II, *Sr III, Y III, *Y IV, †Y v, Cs II, †Cs III, †Ba IV, *La V, Lu III, *Lu IV, and *Lu V. An (*) indicates that no energy levels were known previously, and the (†) indicates that extension of the analysis has corrected important errors in earlier work. All the new observations are wholly or partly in the vacuum-ultraviolet region ($\lambda < 2000$ Å).

In the Atomic Energy Levels Data Center, compilation of levels of the lanthanide (^{57}La $- ^{71}$Lu) and actinide (^{89}Ac $-$?) atoms and free ions is underway for eventual publication in Vol. IV of *Atomic Energy Levels*. A great increase in the data (especially for first and second spectra) during the past ten years has come from outstanding work in several laboratories. Revisions of AEL Vols. I, II, and III are also planned. Revised or extended analyses are available for more than half the 483 spectra previously included in these three volumes, and levels have been found for more than 40 new spectra of various ions of the included elements. Compilation of tables for Vols. II and III has begun. Pending a revision of Vol. I, C. E. Moore's *Selected Tables of Atomic Spectra* are being published for a number of the elements ^{1}H $-$ ^{23}V.

The present status of identifications in the ultraviolet solar spectrum was described by *R. Tousey*. Recent observations included material gathered during the solar eclipse, and a rocket firing giving the spectral region from 1400 Å to 1900 Å with 0.1 Å resolution. Monochromatic ultraviolet spectro-heliograms included one showing a solar flare. They are also using OSO IV and OSO VI data. Somewhat more than half of the 700 lines measured in the solar spectrum longward of 2000 Å have been identified. Extensions of laboratory measures and identifications to shorter wavelengths are needed; for example, the spectra of S VIII and S IX are confused near 200 Å. Laboratory and theoretical work is needed on lines shortward of 30 Å, as well as more highly resolved spectra of flares. A feature found near 1.9 Å in a flare may be identified with Fe XXV, but better laboratory and solar measures are needed.

W. M. Burton reported on researches carried out at the Astrophysics Research Unit, Culham Laboratory, subsequent to work reported in an appendix to the *Draft Report*. Using the technique of laser heating of solid targets (1), classification in the 150–250 Å region has been extended to inner-shell transitions in S x, K XI–XIII, Ca XII–XIV, Sc XIII, and Ti XIV (2). In the wavelength range 20–300 Å, more than 50 lines of S X–XIV, 18 lines of P X–XIII, and several lines of Si X–XI have been newly classified and a further 150 previously identified lines of P VI–XIII have been confirmed (3). More recently in the range 240–750 Å, Fawcett (4) has classified 180 spectral lines associated with second-shell transitions in Na VII–IX, Mg VI–X, Al VII–XI, Si VIII–XII, P X–XIII, S XI–XIV, Cl XII–XV, K XIII and Ca XIV, In the wavelength range 17–40 Å, 59 lines in Cl XII–XV have been classified and a further 10 lines have been listed as unclassified transitions in Cl X–XII. This work has led to improved predicted wavelengths for several strong lines in Fe XIX–XXIV near 12 Å which confirm earlier classification of these lines in solar flare spectra (5). Finally, transitions have been studied between low term levels in highly ionized elements between chlorine and iron (6); more than 230 lines in the range 250–900 Å have been classified as low-level transitions in ions of the Na I, Mg I, Al I, Si I, P I, S I and Cl I isoelectronic sequences. These measures will aid in the identification of forbidden transitions; for instance, the classification of a multiplet in Fe XIII near 300 Å has provided the first direct confirmation of infrared coronal Fe XIII forbidden lines at 10 747 Å and 10 798 Å.

The Culham group have found many new spectral line identifications and classifications from emission spectra from the solar limb using rocket-borne instrumentation (7, 8). The low density chromosphere and corona enhance intersystem and forbidden transitions. In the 300 Å to 2803 Å range, identifications are proposed for new intersystem lines in C III, N IV, O III–V, S IV–V and Si XI which provide term values for several previously uncertain metastable levels. Between 1200 Å and 1500 Å several coronal lines have been observed and some have been tentatively classified as forbidden transitions in Fe XI–XII. In addition, accurate wavelengths have resulted for emission lines in C V, N VI and O VII near 41 Å, 29 Å and 22 Å. Three lines near 22 Å led to the identification of the forbidden transition $[1s^2\ ^1S_0-2s\ ^3S_1]$ in O VII at 22.09 Å (9). In each of the ions C V, N VI and O VII, the resonance line $(^1S-^1P)$, the intersystem line $[^1S-^3P]$ and the forbidden line $[^1S-^3S]$ have all been observed in the new solar spectra.

Charlotte Moore-Sitterly reported that, at the National Bureau of Standards, in the Office of Standard Reference Data, special effort is being made to publish additional Sections of the 'Tables of Atomic Spectra', i.e. Atomic Energy Levels and Multiplet Tables for spectra having revised and extended analyses. Two Sections have been published, Si II, Si III, Si IV in 1965 and Si I in 1967. Similar tables for the six carbon spectra are now in press (10) and work is in progress on the nitrogen spectra. The demand for these tables increases with the rapidly accumulating space observations of ultraviolet solar and flash spectra and other stellar spectra.

In 1956, the late A. S. King generously sent to W. F. Meggers a hand-written manuscript containing his Temperature Classifications of more than 4500 Dy I and Dy II lines. This table was unpublished at the time of his death, but is needed for the analyses of these spectra, as King fully realized. He used wavelengths taken from the early literature, since no complete line list was available. These are now superseded by recent observations made at the Lawrence Radiation Laboratory in Berkeley. The present authors have edited and published King's material to fit the current line list (11).

A Monograph on Ybɪ is in course of preparation. It is based on the unpublished analysis by the late W. F. Meggers, left unfinished at the time of his death, in November 1966.

The table of revised ionization potentials mentioned in the Draft Report is now in print (12).

Second Scientific Meeting

A. COMMITTEE 2 – TRANSITION PROBABILITIES

Chairman: R. H. Garstang.

The chairman reviewed recent work on transition probabilities. He drew attention to the increased use of coincidence techniques in experimental measurements in both allowed and forbidden processes. Important measurements have been made on two-photon emission from the Heɪ 2^1S state by Pearl in Ottawa. Two-photon emission has also been observed by R. Marrus (Berkeley) in Sixɪv, Sxvɪ and Arxvɪɪɪ, from the $2\,^2S_{1/2}$ states, and hopefully lifetime measurements will soon be possible. Several new determinations of the transition probability of the auroral green line of Oɪ have been made, the best value (1.18 sec^{-1}) being that of Sinanoglu.

W. L. Wiese discussed the new oscillator strength scale for Feɪ. Several recent measurements of the oscillator strengths of prominent iron lines (13, 14, 15, 16) have been in serious disagreement with the widely accepted comprehensive data of Corliss and Bozman (17) and King and King (18). Various deviations ranging from factors of 0.6 to more than 20 have been found, from which for example, a drastic revision of the solar photospheric iron abundance would follow. Unfortunately, these measurements do not overlap to any significant extent. Therefore, a comprehensive measurement of prominent iron lines has been undertaken at NBS (19) with the objective to measure all the recently determined oscillator strengths and thus tie together the new results on one common scale. Photoelectric emission intensity measurements have been performed with a wall-stabilized arc. The principal result of the NBS experiment is that it is fully consistent with all other recent experiments. The various deviations against the older work are due to a strong dependence of the Corliss and Bozman (17) and King and King (18) data with excitation energy. Thus all the new f-value material on Feɪ indicates much smaller oscillator strengths and therefore a much higher photospheric solar iron abundance than assumed up to now.

G. W. Wares described his shock-tube measurements of absolute gf-values for Feɪ. The attempt was made to vary the shock-tube conditions over as wide a range as possible, resulting in a 250-fold change in concentration of the emitting atoms. He presented experimental data on line intensities versus excitation which made possible measurements of temperature. Comparisons of the resultant gf-values led to good overall agreement with other recent results, confirming the conclusion that the Corliss and Bozman monograph contained temperature and wavelength errors.

J. P. Swings described his calculations of transition probabilities of forbidden lines of [Feɪ], [Feɪɪ] and [Crɪɪ]. In the case of Feɪ, he has carried out magnetic dipole and electric quadrupole calculations and made comparisons with sunspot spectra. Magnetic dipole calculations for Feɪɪ have been compared with earlier results by Garstang and J. P. Swings, showing basic agreement for stronger lines but discrepancies for the weaker transitions. Finally, he has found that a puzzling line at 4581 Å in the spectrum of the star η Car should be attributed to [Crɪɪ], which produces a doublet at 4580.9 Å at an excitation χ of 2.7 eV. He has found good agreement between calculated and observed stellar lines in the infrared, though in other cases many misidentifications have been made.

R. J. Takens discussed possible temperature errors in modern oscillator strengths. From the comparison of two sets of oscillator strengths, one obtained in absorption and one in emission, it is possible to separate the temperature errors in the two sets. One obtains two equations, connecting the temperature errors with the difference between the two sets in a 'continuous' wavelength-dependent error. So one additional argument is always needed to solve the equations. At present there is no suitable set of absorption measurements available.

The method has been used to compare the Corliss and Warner data with the Sun for sufficiently weak lines between 4000 and 7000 Å and upper level below 47 kK. The equations are solved by the single correction $= -0.15 \pm 0.02$. However, by putting into the equations $= -0.27$ as required by the new measurements from Kiel and NBS, the solar excitation temperature increases to above the effective temperature, which is very unlikely from the astrophysical point of view.

R. Viotti described his work on emission lines in some peculiar stars (η Car, XX Oph, AG Peg, V 380 Ori), and suggested that [Tiɪɪɪ] may be important.

B. COMMITTEE 3 – COLLISION CROSS SECTIONS AND LINE BROADENING

Chairman: M. J. Seaton.

D. Richards presented the results of calculations of probabilities of collisional transitions between highly excited states. Use is made of classical theory and a generalized correspondence principle (I. C. Percival and D. Richards, *J. Phys. B.* 3, 1035, 1970). Results are valid for incident energies E in the range $(2M/m)(ZZ'/n) < (E/\mathrm{Ry}) < (M/m)Z'^2$ where n is the initial quantum number, Z the nuclear charge, M the mass of the incident particle of charge Z', m the electron mass. Results are obtained in analytic form and should be correct to better than 20 % for n, $n' \gg 1$ and $ZZ' < 15$. Preliminary results are published in *Astrophys. Lett.* 4, 235, 1969.

K. L. Bell has been considering Penning ionization of metastable helium, He $(2\,^3S) + \mathrm{H}(1\,^2S) \to$ He$(1\,^1S) + \mathrm{H}^+ + e$. At a temperature of 300 K the rate coefficient is 7.5×10^{-10} cm^3 sec^{-1}, to an accuracy of 20 %, and in the range 300 K to 10^4 K the rate coefficient differs from that at 300 K by a factor less than 2.

R. W. Hindmarsh described investigations of line broadening by impact of neutral particles. Laboratory observations show a broadening and shift significantly different from that calculated assuming van der Waals forces only. Improved agreement can be obtained on taking account of additional forces of shorter range.

S. Brechot reported that broadening of Naɪ $\lambda 5891$ by helium impact had been studied experimentally at Meudon. The assumption of van der Waals forces gave reasonable agreement with experiment at 450 K, but the observed broadening was much larger at higher temperatures.

D. D. Burgess reported on work done at Imperial College, in collaboration with C. J. Cairns, on Stark broadening of Heɪ $\lambda 4471$, and its forbidden companion at $\lambda 4470$ (*J. Phys. B* 3, L67 and L70, 1970). The laboratory results agree with B star observations but do not agree with previous theoretical predictions.

C. COMMITTEE 5 – MOLECULAR SPECTRA

Chairman: J. G. Phillips.

A. J. Sauval described some results that he and *N. Grevesse* had obtained on the oscillator strengths for SiH and SiH$^+$. Recent identifications of SiH (A $^2\Delta - $ X $^2\Pi$) and SiH$^+$ (A $^1\Pi - $ X $^1\Sigma^+$) in the solar photospheric spectrum have enabled them to derive absolute oscillator strengths for the (0, 0) bands of these transitions: f_{00} (SiH) $= 0.0033$ and f_{00} (SiH$^+$) $= 0.0005$. Their result for SiH is compared with other values. The full paper will be published in *J. Quant. Spectr. Radiative Transfer*.

G. Herzberg outlined briefly the work presently going on at the National Research Council of Canada.

Spectroscopic work at Ottawa is going on in the optical region, the microwave region, and the radio-frequency region. In addition, theoretical studies on topics in molecular spectroscopy are being carried out.

In the optical region much attention is still paid to the study of diatomic molecules. Among the astronomically interesting unpublished observation special mention was made of Douglas's observation of the first Rydberg transition in OH. The observations of J. W. C. Johns on the spectrum of ArH, while intrinsically of great interest, are less important for astronomy. Another observation of astronomical interest is that of the 0-0 band of CS by Horani at 3622 Å. It may well be that the CS molecule could be detected in astronomical objects by means of this transition.

Among the studies of polyatomic molecules and radicals may be mentioned the work of Woodman on HNF, of Billingsley on two new spectra of HSiX, where it is not yet certain what X is, and of the spectrum of C_2O by Devillers and Ramsay.

The work on spectra of molecular ions has been actively pursued. Spectra of SiH^+ have been observed in emission by Douglas and Lutz, spectra of CS^+ and H_2S^+ by Horani. The study of absorption spectra of molecular ions is proceeding more slowly but the flash radiolysis technique has been developed and appears to be promising for the future.

In the microwave region the work of Oka may be of considerable astronomical interest. He is studying the exchange of rotational energy by collisions, a process that is of vital importance in the understanding of the microwave lines in astronomical objects.

R. W. Nicholls described work going on at York University on oscillator strength measurements on astrophysically important diatomic oxide spectra. A computer-oriented synthetic spectrum method has been developed, and used parametrically to reproduce in great detail optically thick emission spectra of diatomic molecules. It has been applied to the comparison of synthetic spectra of the following oxides: BeO ($B^1\Sigma - X^1\Sigma$); AlO ($A^2\Sigma - X^2\Sigma$); TiO-α ($C^3\Delta - X^3\Delta$); TiO-β ($C^1\Phi - C^1\Delta$); VO ($B^4\Sigma^- - X^4\Sigma^-$) with the same spectra produced experimentally in emission by thermal excitation in shock tubes. It has also been used on the CN violet ($B^2\Sigma - X^2\Sigma^-$) system. The comparison between experimental and synthetic spectra has allowed the determination of band strengths (some absolute and some relative) for many bands of each system. Much of this work was published recently or will be published shortly (**20–25**).

P. Swings presented an outline of programmes in molecular spectroscopy at present in progress in Liège. *B. Rosen* is continuing work on C_3 and C_2, with emphasis on the formation of these molecules in discharges and in photolysis, starting from parent molecules of carbon. He is examining the possible photochemical reactions, as well as the radiative and non-radiative mechanism of energy transfer. *B. Rosen* and *F. Remy* plan to publish soon their work on the excitation of C_3 in a hollow cathode discharge. *I. Dubois*, *H. Bredohl* and *H. Leclercq* are paying special attention to the long-neglected molecules containing silicon, in particular the oxides and their ions. The observed spectrum of the $A^1\Pi - X^1\Sigma^+$ system of SiO has been extended longward to $\lambda 3800$. The new extension corresponds to bands of high vibrational quantum numbers, e.g. up to $v'' = 25$ and $v' = 16$. In addition, 3 new violet-degraded bands have been found of SiO at 2948.8 Å, 2955.6 Å and 2962.3 Å implying the $^3\Pi$ level of SiO. Three redward-degraded bands at 3763.8, 3831.4, and 3902.1 Å will soon be studied with high dispersion, while 4 bands with complex structure between 3922 and 4072 Å might be produced by a polyatomic species. The spectrum of Si_2 has been obtained in absorption with high resolution in a flash discharge in SiH_4. In addition to $H^3\Sigma^-_u - X^3\Sigma^-_g$ and $K^3\Sigma^-_u - X^3\Sigma^-_g$, the transition $L^3\Pi_u - D^3\Pi_g$ is also shown in absorption. Attempts are being made to get information on the triatomic radicals: SiCC, CSiC, SiCSi, CSiSi, Si_3 and the diatomic radical SiC, using flash discharges or hollow cathode discharges. They are continuing the analysis of the visible bands of the radical SiH_2 ($A^1B_1 - X^1A_1$), and extending the search for other electronic transitions of SiH_2. Finally, experiments on the flash discharge in simple hydrocarbons (allene, propylene, acetylene) have revealed unknown absorptions. *F. Remy* has been conducting experimental determinations of the lifetimes of excited states in molecules. Results for $N_2(C^3\Pi_u)$ and $N^+_2(B^2\Sigma^+_u)$ are as follows:

$$B^2\Sigma^+ \text{ of } N^+_2; \quad \text{lifetime} = (660 \pm 4.8) \times 10^{-9} \text{ sec.}$$
$$C^3\Pi_u \text{ of } N_2; \quad \text{lifetime} = (42.3 \pm 3.5) \times 10^{-9} \text{ sec.}$$

For the excited state ($A^2\Delta$) of CH one obtains $(568 \pm 60) \times 10^{-9}$ sec and a similar result for $A^2\Delta$ in the course of the study of the 4050 group of C_3. The lifetime of $A^3\Pi_g$ of C_2 is strongly affected by pressure. The extrapolation to zero pressure is in agreement with the determinations by others, e.g. 780×10^{-9} sec derived by Jeunehomme and Schwenker. The estimated lifetime of Π_u of C_3

is about 300×10^{-9} sec. The work by Dubois and others on radicals containing Si is being extended by Remy to include lifetimes. These experiments are being made with a high temperature furnace combined with electron collisions or electric discharges. Finally, a high dispersion study is underway on an emission near 1850 Å which has been very tentatively assigned to OH.

J. G. Phillips described progress being made at Berkeley on the high dispersion spectrum of TiO. All band systems from 4000 Å to the infrared are being analyzed. To date, analyses include 20 bands of the α-system, 4 bands of the β-system, and 19 bands of the γ-system. A new system with its strongest band at 4113.7 Å has been shown to be produced by a $^1\Sigma - {}^1\Sigma$ transition. Perturbations in the $C^3\Delta$ state with $v' = 4$ to 7 are probably produced by interactions with a singlet electronic state, but it has not yet been identified.

H. Wöhl mentioned that he identified CoH in sunspot spectra and that it is perhaps possible to find the unknown dissociation energy of CoH by using a sunspot model and the equivalent widths of the CoH lines identified in the photospheric and the umbral spectra. Unfortunately, at present there is no definitive sunspot model to do the needed calculations.

REFERENCES

1. Fawcett, B.C., Gabriel, A.H., Saunders, P.A.H.   1967, *Proc. Phys. Soc.* **90**, 863–7.
2. Fawcett, B.C., Burgess, D.D., Peacock, N.J.   1967, *Proc. Phys. Soc.* **91**, 970–2.
3. Fawcett, B.C., Hardcastle, R.A., Tondello, G.   1970, *J. Phys. B: Atom. Molec. Phys.* **3**, 564–71.
4. Fawcett, B.C.   1970, *J. Phys. B: Atom. Molec. Phys.* **3**, in press.
5. Neupert, W.M., Gates, W., Schwartz, M., Young, R.   1967, *Astrophys. J.* **149**, 79–83.
6. Fawcett, B.C.   1970, *J. Phys. B: Atom. Molec. Phys.*, in press.
7. Burton, W.M., Ridgeley, A., Wilson, R.   1967, *Mon. Not. R. astr. Soc.* **135**, 207–23.
8. Burton, W.M., Ridgeley, A.   1970, *Solar Phys.* **14**, 3.
9. Gabriel, A.H., Jordan, C.   1969, *Mon. Not. R. astr. Soc.* **145**, 241–8.
10. Moore, C.E.   Nat. Std. Ref. Data System, Nat. Bur. Std., NSRDS-NBS 3, Section 1 (1965); Section 2 (1967); Section 3, in press, 1971.
11. King, A.S., Conway, J.G., Worden, E.F., Moore, C.E.   *J. Research Nat. Bur. Std.* **74A**, 355–94.
12. Moore, C.E.   Nat. Std. Ref. Data System, Nat. Bur. Std., NSRDS-NBS 34 (1970).
13. Huber, M., Tobey, Jr., F.L.   1968, *Astrophys. J.* **152**, 609.
14. Grasdalen, G.L., Huber, M., Parkinson, W.H.   1969, *Astrophys. J.* **156**, 1153.
15. Garz, T., Kock, M.   1969, *Astron. Astrophys.* **2**, 274.
16. Whaling, W., King, R.B., Martinez-Garcia, M.   1969, *Astrophys. J.* **158**, 389.
17. Corliss, C.H., Bozman, W.R.   1962, *N.B.S. Monog.*, No. 53.
18. King, R.B., King, A.S.   1938, *Astrophys. J.* **87**, 24.
19. Bridges, J.M., Wiese, W.L.   1970, *Astrophys. J.* **161**, L71.
20. Drake, G.W.F., Tyte, D.C., Nicholls, R.W.   1967, *J. Quant. Spectr. Radiative Transfer* **7**, 639.
21. Linton, C., Nicholls, R.W.   1969, *J. Phys. B. Ser. 2* **2**, 490.
22. Linton, C., Nicholls, R.W.   1969, *J. Quant. Spectr. Radiative Transfer* **9**, 1.
23. Linton, C., Nicholls, R.W.   1970, *J. Quant. Spectr. Radiative Transfer* **10**, 311.
24. Harrington, J., Nicholls, R.W.   1970, unpublished.
25. Myer, J.A., Nicholls, R.W.   1970, unpublished.

COMMISSION 15: PHYSICAL STUDY OF COMETS
(L'ÉTUDE PHYSIQUE DES COMÈTES)

Report of Meetings, 19 and 22 August 1970

PRESIDENT: L. Biermann.
SECRETARY: J. Rahe.

First Meeting

I. ADMINISTRATIVE MATTERS

During the meeting the following administrative matters were discussed:

As new officers of the Commission were proposed and unanimously nominated by the Commission for approval by the Executive Committee of the Union: President: V. Vanýsek; Vice-President: A. H. Delsemme.

As new members of the Commission were approved: J. C. Brandt, B. Donn, M. Harwit, D. Malaise, A. Mrkos, J. Rahe, Z. Sekanina.

W. F. Huebner was suggested as Consultant.

II. SCIENTIFIC PRESENTATIONS

The following communications were presented during the meeting of the Commission, in order of their presentation:

Comments on OAO Observations of Comets 1969g and 1969i (A. D. Code)

An extensive H-envelope was discovered in Comets 1969g and 1969i. Comet 1969g was optically thick in $L\alpha$ up to about $500\,000\,000$ km. The intensity of the nucleus was approximately 70 kR for 1969g and 72 kR for 1969i and measurable out to about 2, respectively 3°. Variation of $L\alpha$ with heliocentric distance was not linear but showed a decrease in brightness near perigee. The H-atom temperature was found to be 1600 K. The (0,0) band of OH at 3090 Å was observed. OH was optically thick to about 3′ from the nucleus. The number density of OH is of the order $N = 2 \times 10^{23}/r^2$ cm^{-3}. Variations of OH intensity with solar distance yield a dependence on distance to -5.8 power.

Internal Motions in the Head of Comet Ikeya-Seki 1965 VIII from High Resolution Spectra
(Z. Sekanina)

Doppler displacements of the emission and absorption lines of Comet 1965 VIII have been compared with predicted radial velocities from the comet's orbit. The emissions suggest a blue shift corresponding to a residual radial velocity of, on the average, 0.8 km/sec; the absorptions indicate a red shift of about 2 km/sec. The deviations should be interpreted in terms of internal motions of ejected material in the comet's head.

Solar Wind and Comets
(J. C. Brandt)

The analysis of a new theoretical treatment of the orientation of ionic comet tails indicates that the effects of coplanarity on the determination of the average solar wind velocities are small.

A report is given on a new telescope that has been designed and built specifically for observations of the large scale structure of comets (field-flattened $f/2$ Schmidt with 14″ corrector plate, 23″ mirror, 4″ × 5″ film or plate size, 8 × 10° field). First results are presented that indicate the great potential of colour photography in cometary research.

Physico-Chemical Phenomena in Comets
(A. H. Delsemme)

The photometric profile of the continuum and the molecular bands as predicted by Delsemme's new model of the cometary head (*Planetary Space Sci.* **18**, 709–30, 1970) is compared with the observed continuum and C_2 emission in Comet Burnham.

A Search for Radio Emission from Formaldehyde in Comet 1969i
(L. E. Snyder, W. F. Huebner)

L. E. Snyder reported on the unsuccessful search for the $1_{10} - 1_{11}$ transition of formaldehyde in emission at 4830 MHz from Comet 1969i using the 140 ft NRO radio telescope at Green Bank, W. Va. An upper limit of the H_2CO projected density is placed at approximately 10^{14} molecules/cm^2.

Search for Microwave H_2O Emission in Comet 1969i
(T. Clarke, B. Donn, W. Sullivan)

W. Sullivan reported on the unsuccessful search for the $6_{16} - 5_{23}$ rotational line of H_2O at 22235 MHz (1.347 cm) in the nucleus of Comet 1969i on April 1, 1970. Assuming a kinetic temperature of 100 K and a molecular cloud of 1′ in diameter, an upper limit is placed at 10^{18} H_2O molecules/cm^2 in the nucleus.

Second Meeting

I. ADMINISTRATIVE MATTERS

During the meeting the following administrative matters were discussed:
The Draft Report of the Commission was approved.
As members of the Organizing Committee were elected by the Commission: Biermann, Herzberg, Levin, Richter, Roemer, Whipple, Wurm.

II. PROPOSALS AND SUGGESTIONS

The following propositions were presented and discussed:
Dr J. Virginia Lincoln suggested that Institutions making observations on the geometry or physical characteristics of comet tails suited to solar-terrestrial studies notify 'World Data Center A, Upper Atmosphere Geophysics, ESSA, Boulder, Colorado 80302, USA' annually (e.g., in March) or oftener concerning their observing programme since their last report. This should include the identity of the comets observed, the periods of observation and the general nature of the observations. The Commission recommended to support the proposal and established the following working group that will be concerned with the matter: Babadjanov, Dossin, Rh. Lüst, Marsden, Rahe, Vanýsek.
Dr F. D. Miller suggested the exchange of information on filters, and standardization. At the present time it is difficult to intercompare comet observations made with different sets of filters, and the UBV system is not suitable for this work. It is therefore suggested that there be a discussion of the characteristics of filters suitable for isolating particular components of comet radiation (C_2 bands, scattered sunlight, etc.). Interference filters are not very expensive, and it would be possible for an observer to have a 'standard' set, as well as filters designed according to his own ideas.

The President proposed to submit the suggestion to the working group for international cooperation in the physical observation of comets that was established at the XIII General Assembly in
Prague 1967. Dr Delsemme agreed to succeed Dr Biermann as chairman of the working group.

Following a suggestion by Dr F. D. Miller, consideration was given to the desirability of alerting
observers by telegram when a comet shows some sudden and particularly interesting physical
development. The Central Bureau for Astronomical Telegrams agreed to cooperate in this matter.
This type of message does not conform with the existing categories of telegrams, however, and
subscribers are therefore requested to contact the Bureau if they would like to receive such messages;
suggestions concerning the type of information that might be transmitted are also solicited.

III. PROGRESS OF COMET ATLASES

Dr Rahe outlined the plans for the second part of the *Atlas of Cometary Forms* (J. Rahe, B. Donn,
K. Wurm, 1969) which will deal with cometary ion tails as discussed with Dr Donn and Dr Wurm.
A letter from Dr Wurm was read suggesting to limit the second part of the Atlas to a few selected
items thus avoiding long delays in the publication.

The Commission was informed of Dr Richter's plan to publish a second part of the *Isophotometrischer Atlas der Kometen* (W. Högner, N. Richter, 1969). Dr Levin suggested that in the second
part a scale in km is included for each picture.

Dr Haser summarized the plans for an extension of the *Atlas of Representative Cometary Spectra*
(P. Swings, L. Haser, 1955) including in particular the high dispersion spectra obtained since the
publication of the Atlas.

It was suggested by the President that the Commission recommend the publication of additional
parts of the *Atlas of Cometary Forms*, the *Isophotometrischer Atlas der Kometen*, and the *Atlas of
Representative Cometary Spectra*.

IV. SCIENTIFIC PRESENTATIONS

The following communications were presented during the meeting of the Commission, in order of
their presentation:

Observations of Comets 1969g and 1969i in Lα
(L. Biermann)

From discussions with Code it appears that Code's value of the OH density $(2 \times 10^{23}/r^2)$ refers to
regions in which the larger part of the OH is likely to be already dissociated. A report is given on the
work of Bertaux and Blamont on Comet 1969i. The total number of H-atoms seen around the inner
part that is optically thick in $L\alpha$, is 1.2×10^{36} each of which is excited every 350 sec.

Lα Observations of Comet 1969g
(E. B. Jenkins, D. W. Wingert)

E. B. Jenkins reported that Comet 1969g had been photographed by an objective grating camera
launched in a rocket on January 25, 1970. A preliminary interpretation of the picture suggests that
the hydrogen cloud appears as a nearly circularly symmetric peak whose width (FWHM) equals
$0°.5$ (5×10^5 km). There is some suggestion that in addition to the broad peak a small nucleus may
be visible, and if further analysis confirms its existence, one could infer that the main cloud is not
optically deep to $L\alpha$ radiation.

Photographs of Comet 1969g in Lα
(F. L. Whipple)

Professor Whipple presented several photographs in $L\alpha$ of Comet Bennett. The image of the comet shows a large elliptical coma of about 1 million km in diameter which is in agreement with the $L\alpha$ measurements made with OAO 2 and OGO 5.

The 'Dust' Effect in $L\alpha$ Cometary Radiation
(V. Vanýsek)

Possible effects of the dust content in comets on the $L\alpha$ radiation of comets are discussed.

The scattering of the solar $L\alpha$ line on small particles depends strongly on the actual albedo in the UV region and most probably is generally weak.

More important, however, seems to be possible additional net and continuous production of neutral hydrogen atoms from the dust particles in the cometary tail. The production of 10^6 atoms per second per particle in a 'dust-rich' comet can lead to significant prolongation of isophotes and therefore also to the misinterpretation of the observed distribution of the UBV cometary radiation.

Interpretation of $L\alpha$ Emission in Comets
(L. Biermann)

An interpretation of the $L\alpha$ observations is given following JILA Report No. 93, 1968. The H-atom temperature found by Code was 1600 K as compared to 2000 K assumed then which means a reduction of the probable velocity of the outflow by $\geqq 10\%$ to ≈ 5 km/sec. With this the production of H-atoms was estimated to be $\geqq 10^{29}$/sec ster. The light pressure in $L\alpha$ reduces the flow velocity in the solar direction by 40% over a distance from the nucleus of $\approx 10^6$ km.

UV-Spectroscopy from Stabilized Rockets
(L. Haser)

3 payloads for the observation of the UV spectrum of comets with a resolution of 30 Å are being prepared for a launching period beginning 1972.

Collisional Effects in Cometary Atmospheres
(J. Malaise)

see *Astron. Astrophys.* **5**, 209–7, 1970.

The Spike of Comet 1969g
(F. D. Miller)

In Comet 1969g a faint spike or 'antitail' (projected length 230000 km, at most a few 1000 km thick) was visible on December 26 and January 2, but not on January 8. Miller has no plates between January 2 and January 8.

Particulate Matter in Comets as Studied in the Infrared
(C. R. O'Dell)

Infrared observations are reported for Comets 1969g and 1969i. The optical albedo of the particles is about 0.3 and the emissivity drops from about 0.7 at optical wavelengths to about 0.2 at 10 μ producing temperatures significantly higher than for black bodies. An average particle radius of 0.1 μ could be derived.

The 0.9–1.6 μ Spectrum of Comet 1969i
(R. P. Kovar)

Emissions due to the CN red system and the C_2 Ballik-Ramsay system are tentatively identified, emission due to the C_2 Phillips system appears to be absent. Preliminary analysis seems to indicate that the C_2 molecules are formed in a triplet rather than a singlet state.

COMMISSION 16: PHYSICAL STUDY OF PLANETS AND SATELLITES (L'ÉTUDE PHYSIQUE DES PLANÈTES ET DES SATELLITES)

Report of Meetings: 20, 24, 25 and 26 August 1970

PRESIDENT: John S. Hall.
SECRETARIES: C. Sagan, B. Middlehurst, and G. H. Pettengill.

The Commission sponsored the Joint Discussion with Commissions 15, 17, 21, 22 and 35 on the Origin of the Earth and Planets, which took place on 21 August. A joint meeting, sponsored by Commission 40, and including also Comissions 16 and 17, was held on 24 August.

Presentations of the Mariner 6 and 7 results, given on August 26, included visual imaging, infrared and ultraviolet spectroscopy. Descriptions of projected NASA missions to Mercury, Venus, Mars and Jupiter were presented on August 25.

The abstracts of the individual papers, presented by members of Commission 16 only, were submitted by Sagan, the reports on the Joint Discussion were compiled by Middlehurst, and those of the joint meeting with Commissions 17 and 40 by Pettengill.

RESOLUTION ON NEW PLANETOGRAPHIC COORDINATE SYSTEMS

A. *Guiding Principles*

1. The rotational pole of a planet or satellite which lies on the north side of the invariable plane shall be called north, and northern latitudes shall be designated as positive.

2. The planetographic longitude of the central meridian, as observed from a direction fixed with respect to an inertial coordinate system, shall increase with time. The range of longitudes shall extend from 0° to 360°.

B. *Definitions and Numerical Values for Mercury and Venus*

1. For Mercury (having a direct rotation) the origin of planetographic longitudes is defined by the meridian containing the subsolar point at the first perihelion passage of 1950 (J.D. 2433292.63). The rotational axis shall be provisionally defined as perpendicular to the orbital plane of Mercury (1950). For purposes of obtaining longitude at earlier or later time, a provisional value for the sidereal rotational period of $58^d.6462$ is adopted.

2. For Venus (having a retrograde rotation), the origin of planetographic longitudes is defined such that the central meridian of Venus as observed from the center of the Earth is 320°.0 at 0^h on 20 June, 1964 (J.D. 2438566.5). The rotational axis shall be provisionally defined as having a north pole direction of $\alpha = 273°.0$, $\delta = +66°.0$ (1950.0). For the purposes of obtaining longitude at earlier or later time, a provisional value for the sidereal rotational period of $243^d.0$ is adopted.

RESOLUTION ON MAKING JUPITER INFORMATION MORE READILY AVAILABLE

In order to place on a systematic basis the collection, reduction and reporting of the rotation periods of Jupiter's radio sources and visible features (e.g. Great Red Spot), together with related information such as the appearance, position, and dimensions of the feature observed, Commission 16 recommends:

1. that a comprehensive bibliography be compiled without delay;

2. that copies of reports in preprint or reprint form be sent by authors to a 'central office' for inclusion in supplements to the bibliography;

3. that the bibliography supplements and summaries of the rotation periods, presented with error estimates, be published annually in one of the regular international journals (e.g. *Icarus*);

4. that the President and Organizing Committee be invited to implement these proposals.

ACTION ON MARTIAN NOMENCLATURE

The Martian Nomenclature Committee, consisting of Gerard P. Kuiper (Chairman), A. Dollfus, John S. Hall and Robert Leighton held two meetings between General Assemblies XIII and XIV. There was also an exchange of information by mail between other members of Commission 16 and the Subcommittee. All members of Commission 16 were invited to participate in the discussions and were kept fully informed by News Letters.

The Subcommittee met at Brighton and drew up the following proposals which were subsequently approved by the members of Commission 16:

(A) The hundred or so largest and most prominent craters on Mars are to receive names of deceased persons whose works were related to the Planet Mars.

(B) For smaller features deserving names, about 110 of the 125 names adopted by the IAU in 1958 (*Transactions* X, 262, Plates I and II) that cover *regions* (*bright* or *dark*) rather than 'narrow features', be used as *provinces*. The names are to be *abbreviated by* 4-*letter* symbols (not unlike the 3-letter abbreviations of constellations), in such a manner as to retain the alphabetical arrangement of the 1958 table. These are to be followed by Roman letters *A*, *B*, ... to designate the approximately 25 largest topographic features within a province. *Two-letter designations*, Aa, Bb, etc., are to be used to designate several hundred *smaller* features requiring symbols, with the *first* letters in each case referring to a *sub*-province (or district). This scheme can be continued indefinitely, as far as may be required.

(C) A Working Group is to be appointed, charged with *three tasks* to be completed before the next General Assembly of the IAU to:

1. substantially define the province boundaries and, in the process, possibly add or delete a few provinces, as may appear desirable; and circulate among members a map showing coordinates and proposed boundaries;

2. apply the principles of topographic nomenclature to the regions adequately covered by the Mariner 1969 data, and by such 1971 data as may become available;

3. propose appropriate names (cf. A above) for some prominent Martian topographic features (with reference to the lists of names already submitted by Commission 16 members).

The President of the Commission, with the advice of the Subcommittee, appointed the following members to the Working Group on Martian Nomenclature: de Vaucouleurs (Chairman), Miyamoto, Sagan, Dollfus, B. A. Smith and M. D. Davies (consultant).

BETTER COMMUNICATION BETWEEN ASTRONOMERS AND GEOSCIENTISTS

Commission 16 recommended that:

(A) The IAU *records* its view that recent developments in the observation and theory of the planets and of the earth make it timely to improve communication between geologists, geochemists, geophysicists and astronomers.

(B) The IAU *invites* the members of the IUGG, the IUGS, and other interested Unions to participate more fully with the Members of the IAU in exchanging knowledge of fundamental concepts and data of their disciplines and to facilitate the collaboration of their members in new terrestrial and planetary investigations.

(C) The IAU *notes* the promotion of, and *would welcome* liaison with, the inter-union geodynamics commission, the work of which has relevance to planetary interiors.

MERIDIAN LONGITUDES FOR SATURN

Members of Commission 16 expressed the urgent need for a continuous table of central meridian longitudes for Saturn and recommends that a Working Group be established to undertake a study of this matter with a view to proposing a definitive system for adoption as soon as practicable.

JOINT MEETING OF COMMISSIONS 16, 17 AND 40 ON 'RADIO AND RADAR STUDIES OF THE MOON AND PLANETS'

First Morning Session

J. E. B. Ponsonby: 'Moon Mapping by CW Radar at 162 MHz'.

The mapping of the Moon has been carried out at Jodrell Bank using a CW-radar technique similar to aperture synthesis. By observing the echoes in the same sense of circular polarization as the transmitted waveform, sensitivity only to diffuse scattering by small-scale surface roughness is assured. It is possible that a significant fraction of the energy is reflected from a depth of some 20 m. Observations were reported for several days in early 1970, when the lunar-terrestrial geometric configuration was favorable for the application of this technique. Angular resolution as viewed from earth was about 2′ arc, with the crater Tycho forming a particularly prominent scattering feature.

I. I. Shapiro, M. E. Ash, R. P. Ingalls, G. H. Pettengill, A. E. E. Rogers, M. Slade, W. B. Smith, and S. H. Zisk: 'Topography of the Moon, Mercury and Venus'.

Four radar methods of gaining topographic information on the surfaces of the Moon and planets were discussed. The first method depends on the accumulation of measurements of the radar time-of-flight to the sub-earth point. As the target rotates these may be used to explore the topography along the sub-earth track. Resolution is determined by the pulse-width used. The second method uses both delay and Doppler resolution to derive a profile along the equator of apparent rotation which may be compared with that expected from a spherical target. The third method involves stereoscopic comparison of delay-Doppler maps taken at different rotations. While applicable to large areas of the target, this method does not possess high accuracy. The fourth, and most sophisticated, method uses interferometry to fill in the coordinate otherwise missing in delay-Doppler mapping.

The second and fourth methods have been applied at M.I.T. to derive contours of the lunar surface good to an absolute accuracy of better than 500 m, while the first two methods have been applied to Venus and Mercury. The preliminary lunar results so far obtained show good agreement with optically-derived lunar maps over short distances, but seem to indicate 'drifts' in the absolute control of the optical method over longer distances. Very little topographic variation, at least at the resolution so far available, has been found on Venus in the equatorial belt, with only one region of about 2 km height in evidence. No departure from sphericity can yet be reliably reported for Mercury.

C. C. Counselman, M. E. Ash, G. H. Pettengill, A. E. E. Rogers, I. I. Shapiro and W. B. Smith: 'Radar Map of Martian Topography'.

Observations of Mars with the Haystack Radar ($\lambda = 3.8$ cm) during the oppositions of 1967 and 1969 have been combined to map variations of surface elevation, reflectivity, and roughness for nearly all longitudes and for latitudes from 2° to 22° North. Variations in height of approximately 15 km have been found. The radar maps show that the locations of dark visible markings are not systematically high or low, but certain prominent dark features occur on slopes. The dark visible features Syrtis Major, Trivium Charontis, and Cerberus are relatively smooth and have high radar reflectivity.

Following this paper, *R. A. Wells* presented his interpretation of Martian topography based on optical maps and an assumed internal structure.

G. C. Pimentel called attention to topographic results for Mars obtained from spectrographic observation of CO_2 pressures during the Mariner 6 and 7 flyby's. These consisted of a number of spot measurements between 0° and 40°S.

G. H. Pettengill, D. B. Campbell, R. B. Dyce, R. P. Ingalls, R. F. Jurgens, and A. E. E. Rogers: 'Venus Radar Maps at 3.8 and 70 cm'.

During a four-week period surrounding inferior conjunction in April 1969, Venus was mapped using radars at a wavelength of 3.8 cm (M.I.T. Haystack Observatory) and 70 cm (Cornell University, Arecibo Observatory). The technique used was delay-Doppler mapping augmented with sufficient interferometry to remove the hemispheric ambiguity. The linear surface resolution was about 300 km in both maps, with coverage confined to a region between $+40°$ and $-50°$ in latitude and extending over about 80° (3.8 cm) and 120° (70 cm) in longitude. Only the sense of received polarization corresponding to quasi-specular scattering was analyzed. The maps agreed in respect to the location of most outstanding anomalously scattering features, including several dark, crater-like ring structures. The relative intensities differed between the maps, however, suggesting variability in the small-scale roughness from feature to feature.

D. O. Muhleman and G. L. Berge, 'Interferometric Observations of Venus and Mercury'.

Venus has been observed interferometrically at the Calif. Inst. of Technology's Owens Valley Observatory using baselines up to 1 km at wavelengths of 6 and 21 cm. Contrary to expectation a uniform disk brightness was found, both along the equator and towards the poles. The polarizations of the emission were in good agreement with those computed from a model assuming a surface dielectric constant of 4.8 and CO_2 atmosphere. A value of 6063.1 ± 3.7 km for the apparent radius of Venus was calculated from the zeroes of the 6-cm visiblity function. This may be compared with a value of approximately 6050 km for the surface radius as determined by radar. The 6-cm brightness temperature was found to be 660 ± 33 K.

For Mercury, a mean brightness temperature of 365 ± 15 K was observed at 3.1 cm. An observation at 6 cm near a 'hot equatorial pole' yielded a brightness temperature of 417 ± 43 K.

Second Morning Session

J. R. Dickel: 'The Microwave Spectrum of Venus'.

Observations of Venus at a number of microwave frequencies show that the brightness temperature of the planet increases markedly toward longer wavelengths as the emission arises from deeper within the atmosphere. Although calibration is difficult because there are no good 'absolute' standard sources for comparison, differential observations with respect to the single source Hydra A (assumed to have a straight spectrum) have given a good indication of the shape of the spectrum. The maximum brightness temperature of about 700 K occurs at a wavelength of about 6 cm where the surface is presumably reached but the evidence suggests that the temperature then decreases again toward still longer wavelengths. This apparent decrease has not been accounted for by any present models of the lower atmosphere and surface of Venus.

C. H. Mayer and T. P. McCullough: 'New Observations of Venus, Mars, Jupiter, Uranus and Neptune'.

Observations of Venus at 2.7 cm wavelength over a complete cycle of phase have yielded a brightness temperature of 614 ± 9 K (rms), independent of phase.

Observations of Mars and Jupiter at 1.65 and 2.7 cm are consistent with most other recent measurements. Observations of Uranus and Neptune at 1.65, 2.7 and 6 cm yielded brightness temperatures near 200 K for both planets at all three wavelengths.

E. E. Epstein: 'Microwave Spectrum of Mars and 3-mm Data on Mars, Jupiter, Saturn and Uranus'.

The microwave and millimeter-wave, illuminated-hemisphere brightness temperature of Mars appears to show no significant change with wavelength or phase. At 3 mm no secular variation in the brightness temperature of Jupiter and Saturn was observed between April, 1966 and November, 1969. The brightness temperature of Uranus at 3 mm was measured to be 105 ± 13 K.

A. G. Kislyakov, V. A. Efranov, I. G. Moiseev and A. I. Nauprov: 'Observations of Mars at 2.3 and 8 mm'.

The brightness temperature of Mars has been measured at 2.3-mm and 8-mm wavelengths as 240 ± 30 K and 210 ± 30 K, respectively. From the observed increase at short wavelengths, and using Troitsky's lunar radiation theory, we find for Mars:

$$\tan \delta/\rho = 0.015 \, ^{+0.01}_{-0.008} \text{ and } m = \delta/\lambda = 1.2 \, ^{+0.6}_{-0.4}.$$

These are approximately the same values as observed for the Moon.

A. D. Kuzmin, B. Losovsky and Yu. Vetukhnovskaya: 'Radiometric Observations of Mars at 8.22 mm'.

The radio emission of Mars at 8.22 mm wavelength has been measured using the 22-m Lebedev Radio Telescope and a maser receiver. The ratio of the Mars-to-Jupiter brightness temperatures was found to be 1.22 ± 0.04, which corresponds to $T = 176 \pm 5$ K for Mars. The best fit of these data to theory yields a ratio for loss tangent-to-density of: $\tan \delta/\rho = 5 \times 10^{-3}$. Thus the Martian surface material is a good dielectric.

A. D. Kuzmin and B. Losovsky: 'Radiometric Observations of Uranus at 8.22 mm'.

The radio emission of Uranus at 8.22 mm wavelength has been measured using the 22-m Lebedev Radio Telescope. The ratio of the Uranus-to-Jupiter brightness temperatures was 0.91 ± 0.01, corresponding to $T = 131 \pm 15$ K for Uranus. This value agrees well with a model for the atmosphere of that planet which requires saturated ammonia.

T. D. Carr: 'Jupiter's Magnetospheric Rotation Period'.

Five independent decametric determinations of the mean value of Jupiter's rotation period averaged over approximately one orbital period yield a value of $09^h55^m29^s.76 \pm 0.04$ sec. An analysis of all the available decametric data meeting certain criteria confirms the existence of a cyclic effect due to the variation in the Jovicentric declination of the Earth. The most probable value of the Jovian magnetospheric rotation period, based on both decametric and decimetric observations, is 09 hr 55 min 29.75 ± 0.04 sec.

S. Gulkis and B. Gary: 'Circular-Polarization and Total-Flux Measurements of Jupiter at 13.1 cm.'

Circular-polarization and total-flux measurements of Jupiter at a wavelength of 13.1 cm were made during April and May, 1969, with the 210-foot JPL radio telescope at Goldstone, California. An upper limit of 1 percent to the degree of circular polarization can be placed over the longitude range, 10°–100° and 160°–250° System III (1957.0). This contrasts with an earlier determination deduced from interferometric fringe data which suggested the presence of 2 percent circular polarization at this wavelength and within the same longitude coverage. Total flux data have been used to derive a magnetosphere rotation period of $09^h55^m29^s.72 \pm 0.11$. This period is longer than the standard IAU System III (1957.0) by $0^s.35$.

R. G. Conway: 'Polarization of Jupiter's Decimetric Emission'.

Observations made at 49 cm using Jodrell Bank radio telescopes have deduced a rotational period for Jupiter in close agreement with the values given in the preceding two papers. The degree of circular polarization noted is consistent with a magnetic field of a few tenths of a gauss at a distance of 3 Jupiter radii.

G. L. Berge: 'The Position of Jupiter's Radio Emission Centroid at 21 cm'.

The interferometer at the Owens Valley Radio Observatory has been used to determine the position of Jupiter's 21-cm emission centroid in a rotating coordinate system fixed to the planet. There appears to be a small displacement between the emission centroid and the planet's center amounting to 0.13 ± 0.06 polar semidiameters. The difficulties associated with measuring the position of the source to an accuracy of 2 or 3% of its size were overcome by the excellent phase stability of the instrument which allowed high angular accuracy despite the use of baselines short enough to avoid resolution of the source.

Summaries of contributed papers to IAU Commission 16

G. de Vaucouleurs: 'Areographic Coordinates and New Maps of Mars.'

A twelve-year Mars map project, now 90 % complete, has as its principal objectives the derivation in a uniform coordinate system of the areographic positions of several hundred individual markings on Mars from all available groundbased observations (since 1877); and the production of albedo maps from this coordinate net. The principal results include a new value of the rotation period $[P_t = 24^\text{h}37^\text{m}22^\text{s}. 665 \pm 0.003$ (m.e.)]; a new estimate of the phase effect in longitudes, improved areographic longitudes, areographic coordinate maps of Mars and the development of planning charts for the Mariner Mars 1971 mission.

T. Gehrels: 'Recent Polarimetric Observations of Planets.'

Gehrels reported on the polarimetric work of himself and his collaborators. The wavelength dependence of polarization between 0.3 and 1.0 μ is flatter for Mercury than for the moon and may indicate an appreciable difference in surface composition between the two bodies. The integrated disc of Venus was observed at 97.9° phase angle with the Polariscope balloon system; at 2150 Å the polarization was 23%. Groundbased scans of Venus show that the cloud pattern contrast in the ultraviolet has a rather sudden onset near 4000 Å attaining a constant value over the range of 3600–3100 Å. Ultraviolet polarization scans appear to correlate inversely with the intensity scans. Coffeens's computations of single Mie scattering and multiple Rayleigh scattering give the following conclusions: the atmospheric pressure at the cloud tops is less than 50 mb; the bulk of the highest cloud particles (responsible for single scattering) have diameters of about 2.5 μ; and a real part of the refractive index between 1.43 and 1.55 with some uncertainty should the particles be non-spherical. The opposition effect is found to be the same for the asteroids (20) Massalia, (4) Vesta, and (110) Lydia. The opposition effect appears connected with surface texture only and not with composition.

Neil B. Hopkins, William M. Irvine, and Adair P. Lane: 'Monochromatic Albedos for Saturn.'

Multicolor photoelectric photometry at wavelengths $0.315 \leqslant \lambda \leqslant 1.06$ μ for Saturn during 1963–65 has provided albedos for the disk of Saturn. The disk spectrum appears generally similar to that of Jupiter, but Saturn has a steeper slope for $\lambda \leqslant 6250$ Å and so appears redder. A marked opposition effect, apparently due to the rings, is noted.

Brian O'Leary: 'The Halo Effect of Venus and the Opposition Effect of Mars.'

Photometric observations of Venus near the 1969 inferior conjunction indicate an anomalous brightening of about 0.07 magnitude at 158° phase angle. The width of the brightness maximum is about 3° and its peak is between 1.1 and 4.4 standard deviations, depending on color, from the expected background phase curve. This is precisely the behavior expected if the Venus cloud tops were to contain at least a few percent of hexagonal water-ice crystals, producing a halo effect analogous to the common terrestrial 22° halo phenomenon. The results may be considered provocative but not unambiguous.

Photometry of the full disk of Mars near opposition 1969 confirms the existence of a moderate opposition effect, i.e. a non-linear surge in brightness toward zero phase. Observations of small areas on the disk suggest a very strong opposition effect for Syrtis Major; the extent is somewhat similar to the lunar case. We also confirm a 13% absorption feature near 1.0 μ in the ratio spectrum Syrtis Major/Arabia. Contrary to earlier analyses, the color of Syrtis Major/Arabia appears to have changed from previous observations such that the brightness contrast is greater in the red and near-infrared than at shorter wavelengths.

Peter B. Boyce: 'Specific Martian Studies.'

Direct photoelectric area scans and photometric analysis of the Lowell planetary patrol films have yielded the following results: (1) Contrast in the red, 6200 Å is a linear function of contrast in the blue, 4340 Å; red contrast is high at times of blue clearing. (2) For Syrtis Major the contrast in all spectral regions shows a definite dependence upon phase angle, being highest at small phase angles and decreasing symmetrically with increasing phase angle. (3) For Syrtis Major the contrast in red, green, and blue light shows a diurnal variation. Maximum contrast apparently occurs at the angle of specular reflection. (4) In red light Syrtis Major shows no significant limb darkening out to viewing angles of 45° from the center of the disk. These conclusions have been shown to be unaffected by seeing variations in the Earth's atmosphere. These data were obtained for the noted features at the 1969 opposition and, in view of the variability of the Martian features, extrapolation to other features and times is probably not valid.

J. Rösch, H. Camichel, F. Chauveau, M. Hugon, and G. Ratier: 'An Upper Limit to the Diameter of Mercury.'

During the 7 November 1960 transit of Mercury across the Sun, we previously obtained by the Hertzsprung photoelectric method a value for the diameter of Mercury at unit distance of 6.84 arc sec. An attempt to confirm these results by the transit of 9 May 1970 was thwarted by cloudy skies both at Pic du Midi and at Athens. Nevertheless, despite the dispersion in the data points, significant results seem to have been obtained with the 40 cm Doridis refractor, giving a value consistent with our earlier results (corresponding to a diameter of 4920 km) and implying an upper limit to the density of the planet at 5.10 gm cm^{-3}.

T. Encrenaz, D. Gautier, L. Vapillon, and J. P. Verdet: 'Far Infrared Spectrum of Jupiter.'

The far infrared spectrum of Jupiter, between 20 and 250 cm^{-1}, is computed with high resolution from three atmospheric models. The results of the spectroscopic analysis of H_2 and NH_3 are used for the radiative transfer calculations. The detailed study of the derived spectra shows how the H_2/He ratio and the thermal Jovian structure can be deduced for far infrared measurements of the Jovian flux. In particular, such measurements provide an obvious test of the existence of an inversion in the Jovian temperature profile; the experimental accuracy needed for this information is shown to be obtained with the existing equipment.

G. E. Thomas: 'Results of the Mariner Mars 1969 Ultraviolet Spectrometer Experiment.'

The results for the atmospheric composition of Mars are summarized. The dominant constituent is carbon dioxide with less than 5% nitrogen by volume. Laboratory experiments on pure CO_2 have now simulated every emission feature observed on the bright atmospheric limb. Carbon monoxide, atomic hydrogen, carbon and oxygen are minor constituents. Three topics which were discussed in some detail are: (1) the observed reflected spectrum in terms of atmospheric scattering, ground reflection, aerosols, and large dust particles; (2) the use of the spectrometer as a means of determining local relief, 'UV cartography'; and the data which show that the Hellas region is a very extensive low region on Mars; and (3) the discovery of a strong absorption feature at 2500 Å over the polar cap. Various explanations are put forward to account for the presence of ozone on the polar caps, but not over the land areas. Either adsorbed ozone in the solid CO_2 itself could account for the observations or ozone distributed uniformly over the planet, but in a thin boundary layer near the surface.

Harlan J. Smith: 'Seasonal Changes in Atmospheric Water Vapor on Mars.'

McDonald Observatory coudé spectra and analysis, primarily by E. Barker, R. Tull, and S. Little of the University of Texas, A. Woszczyk of Torun University, and R. Schorn of the Jet Propulsion

Laboratory, establish that water vapor appears at detectable levels (about 10 μ) in the Martian atmosphere during the late spring of each successive hemisphere. By midsummer it has risen to about 40 μ of precipitable H_2O in a vertical column. The water vapor then falls below detectable levels over the entire planet, effectively disappearing for the several months of early autumn, before reappearing with the new late spring season of the opposite hemisphere. The data suggest that the appearance may be delayed a few weeks in the southern hemisphere spring, but that the amount of water may be slightly greater during the southern hemisphere summer.

F. Link: 'On the Optical Radius of Venus.'

The aim of this paper is to connect the provisional Venus atmospheric model of Marov, based on Venera data, with observed optical phenomena. Horizontal refraction is calculated and compared with the refraction curve deduced from the Regulus occultation. A positive correction of about 0.5 in logarithms is required to the model at 6165 km. The aureole (the refracted image of the Sun produced by the atmosphere of Venus) observed during the past 4 transits indicates a refraction of about 1 minute, but at an indefinite altitude. The refraction curve gives the transit level at between 6128 and 6132 km. This should be the level of horizontal transparency. For a pure atmosphere the extinction is too small. Another source of extinction may be indicated in the Venera 5 and 6 results which indicate a layer of high water content whose upper limit is just at the transit transparency level. The upper water level in the Venera results should therefore be considered as the cloud level of the Venus atmosphere. These results provide an indirect confirmation of Marov's model in this region.

Stephen E. Dwornik: 'Mariner Venus-Mercury 1973 Mission.'

The National Aeronautics and Space Administration will conduct a mission to the planets Venus and Mercury in the fall of 1973. The spacecraft will be a Mariner Class spacecraft and will weigh about 455 kg. The launch period will be 12 October to 20 November 1973, with Venus encounter between 19 March and 3 April 1974. The primary mission objective is to conduct exploratory investigations of the planet Mercury's environment, atmosphere, surface and body characteristics and to obtain environmental and atmospheric data from Venus. Secondary objectives include interplanetary experiments. The following is the science payload, with principal investigator and his organization: Imaging – B. Murray, California Institute of Technology; Radio Science – H. T. Howard, Stanford University; Plasma Science – H. Bridge, Massachusetts Institute of Technology; Magnetometer – N. Ness, Goddard Space Flight Center; Ultraviolet Spectrometer – A. L. Broadfoot, Kitt Peak National Observatory; Infrared Radiometer – S. Chase, Santa Barbara Research Center, and Energetic Particles – J. Simpson, University of Chicago.

In discussion A. D. Kuzmin stressed the utility of passive microwave observations from Venus flybys or orbiters.

William E. Brunk: 'Mariner '71, Viking '75 and Pioneer F and G Missions.'

Brief summaries of three future planetary missions planned by the National Aeronautics and Space Administration were presented. In 1971 two Mariner spacecraft will be placed into highly eccentric orbits about Mars with nominal 90-day lifetimes. One space vehicle is intended for mapping a large fraction of the planet at 1 km resolution; the other for looking for time-variable features. Experiments include two television cameras, an ultraviolet spectrometer, infrared interferometric spectrometer, two infrared radiometers, an S-band occultation experiment and a celestial mechanics experiment. A wide range of information on the atmosphere, surface and interior of Mars are anticipated. It is planned to follow up with the Viking 1975 mission, currently envisioned as involving two orbiters and two soft landers, each to be ejected from its own orbiter. Rendezvous with Mars occurs in 1976. Among a variety of Lander experiments now planned are some intended to search for life. The nominal life time of the landers and orbiters is 90 days. The Jupiter 1972–73 mission involves two space vehicles which fly by Jupiter and perform a detailed investigation of its particle and field environment as well as single line imagery and photopolarimetry of the surface, and examination of Jupiter in the vicinity of the 584 Å line of helium. In addition there are experiments devoted to the passage through the asteroid belt which precedes the Jupiter rendezvous. Emphasis was placed on the anticipated contributions of these three programs to our understanding of the solar system.

In discussion Carl Sagan stressed that mission B of the Mariner Mars 1971 program is designed to have an orbital period four-thirds the Martian rotational period so that every four days the spacecraft observes the same area under the same lighting conditions. In this way intrinsic Martian albedo changes can be distinguished from effects due to the scattering phase function of surface material. He also mentioned the possibility that photographic mapping of Phobos and Deimos by the Mariner Mars 1971 mission would provide cartography of these moons superior to the best groundbased cartography of Mars.

PLANETARY PATROL – AN INTERNATIONAL EFFORT

W. A. Baum

Lowell Observatory

Abstract. An international photographic planetary patrol network, consisting of the Mauna Kea Observatory in Hawaii, the Mount Stromlo Observatory in eastern Australia, the Republic Observatory in South Africa, the Cerro Tololo Inter-American Observatory in northern Chile, and the Lowell Observatory, has been in operation since April 1969. The Magdalena Peak Station of the Mexico State University also participated temporarily. New stations are now being added at the Perth Observatory in western Australia and at the Kavalur Station of the Kodaikanal Observatory in southern India. During 1969 Mars and Jupiter were photographed through blue, green, and red filters; and the network produced more than 11 000 fourteen-exposure filmstrips with images of a quality suitable for analysis. Observations of Jupiter and Venus in 1970 are expected to add another 15 000.

All telescopes of the network have apertures in the 60-to-70 centimeter range, have been designed or modified to produce identical image scales, and are equipped with identical planet cameras that provide for the automatic recording of basic data associated with each exposure. All of the patrol observations are being calibrated, processed, edited, copied, and catalogued by the staff of the Planetary Research Center at the Lowell Observatory. The support of NASA Headquarters is gratefully acknowledged.

ACTIVITIES OF THE
PLANETARY RESEARCH CENTER OF THE LOWELL OBSERVATORY

W. A. Baum

Lowell Observatory

Abstract. The research program of the Planetary Research Center at Lowell Observatory, established through the IAU, includes photoelectric measurements at the telescopes, the development of new instruments for planet observation, and the analysis of photographic images. In addition, the Center is managing the International Planetary Patrol Program (described separately).

Photoelectric observations have particularly utilized pulse counting and multichannel storage in the scanning of planetary spectra, planetary brightness profiles, planetary polarization distribution, and satellite brightness changes. The spectrum scanning and area scanning methods have been applied by Boyce to Mars and Jupiter. Hall and Riley have made photoelectric scans of Mars, Jupiter, and Saturn. Millis and Franz have used area scanning to show that the Jovian satellite Io does not brighten anomalously on its emergence from eclipse.

Present instrument development includes work by Baum, Pettauer, and Busby on a planet image stabilizer utilizing deflection coils on a cascaded image converter tube, in combination with a servo-driven tiltingplate guider.

Current or recent analyses of photographic images with our planetary image projectors include extensive measurements of the boundaries of the Martian polar caps by Martin and Fischbacher, the investigation of transient brightenings and their progressive displacements on Mars by Martin and Baum, measurements of the vertical shear in the equatorial region of the Jovian atmosphere by Layton, and positional wanderings of the Red Spot by Millis and O'Dell. With the help of micro-photometry in addition to visual estimates, studies of the Martian blue haze phenomenon and of the diurnal brightness changes of Martian clouds have been carried out by Thompson, Faure, and Boyce.

These image analyses have utilized the IAU planetary plate and film collection of the Center, which is the most extensive in the world. A typical plate or film includes a sequence of planetary images taken in rapid succession in a particular color. As of late 1970, the number of different cata-logued sequences of usable quality from various sources will be as follows:

Source	Image Sequences
Patrol, 1969 + 1970	26 000 +
Lowell, 1903 → 1968	10 086
Lick	1 902
New Mexico	1 663
Meudon	1 177
Table Mountain	763
Miscellaneous	800
	42 000 +

This includes a high percentage of all such material that now exists. The complete facilities of the Center, including access to this collection and the associated machine-searchable IBM card catalogue, are available to all qualified investigators. As in the case of the Planetary Patrol, the support of NASA Headquarters is gratefully acknowledged.

COMMISSION 17: THE MOON (LA LUNE)

Reports of Meetings, 20, 24, 25 and 26 August 1970

PRESIDENT: A. Dollfus.
SECRETARY: J. O'Keefe.

First Session: Business Meeting

The first meeting was held in the large Chemistry Theater of Sussex University, Falmer. It was called to order at 14:15 with the President, A. Dollfus in the chair.

Dollfus announced the death of A. I. Lebedinsky, Vice-President of the Commission, and of J. H. Focas. He appointed a secretary for the session: Dr J. O'Keefe.

Professor Millman was invited to inform about the special session scheduled about the Moon at the Montreal Colloquium (Canada) on 22–29 August 1972. IAU is co-organizer.

Then, Dollfus reported on the enormous increase of complexity in the work of the Commission over the past 3 years as the programmes of lunar study and exploration have been realized. To cope with this problem, it was decided, as a result of extensive correspondence and discussions with members, to organize 3 working groups:

(a) Lunar nomenclature: D. H. Menzel (Chairman), M. Minnaert, B. H. Levin and A. Dollfus (ex officio).

(b) Figure and Motion of the Moon: Chairman Th. Weimer. This group has been working since 1968; it has recently had new members added, mostly interested in lasers.

(c) Geology and Geophysics of the Moon: Chairman G. Fielder. This group is under organization.

The President then asked for comments and advices about this organization of the Commission in three Working Groups. Discussion dealt with the question of putting a laser group in with Figure and Motion. B. Levin and J. O'Keefe commented that the mathematical skills of the senior men interested in Figure and Motion were essential to getting scientific information out of the laser data.

Dollfus then noted that in our times the lunar problems have come to interest unions other than IAU, notably IGGU, URSI, IUGS, IAVCEI, IUCSTP and COSPAR. President Heckmann, visiting the meeting, introduced the General Secretary Perek. Perek noted that consideration is being given to forming an International Council on the Moon, under ICSU. For Dollfus, such a commission would have one member from each Union; its goal should be coordination; the work in the several fields should be done by the Unions themselves. The International Committee on the Moon could presumably meet at the time of the meetings of the Unions involved.

Dollfus raised the question whether NASA or Akademiya Nauk of USSR would be represented, since they are significant contributors to the new technologies. Perek replied that only International Unions would be represented on the lunar commissions.

Returning to the organization of Commission 17, Dollfus then put forward names of a new president, 2 vice-presidents and the organizing committee.

The proposal was unanimously adopted.

Several members were of the opinion that a petrologist should be added. The question of adding Anders to the list was put forward and carefully considered. Levin pointed out that we must ask Anders. Dollfus remarked that it might be possible to give the composition of the Committee to the General Secretary only on the 25th after the 3rd session of the Committee; Perek remarked, in reply to a question, that new members can be added to the Organizing Committee between Assemblies if this is found suitable.

Second Session: Working Group 1: 'Lunar Nomenclature'

The session opened at 16:07 in the same room with the President, A. Dollfus, in the Chair. A telegram was prepared to Professor Minnaert as follows: 'Professeur Minnaert: Commission 17 vous a confié sa présidence, vous exprime ses chaleureuses félicitations, et souhaite votre prompt rétablissement'. The telegram being approved, Dollfus turned the meeting over to Menzel, chairman of Working Group I, who presented the proposed scheme of nomenclature for the far side of the Moon. The proposals are following the principles of guidelines previously adopted by IAU. Names, biographies and coordinates have been established. Six living Soviet cosmonauts and six living U.S. astronauts are exceptionally included. Lists of names are provided, and ACIC prepared maps at the 1:10000000 scale allocating the proposed names. These documents are distributed. They will be complemented by a guide to the pronunciation of the names. Minnaert is working on this problem; it must necessarily be approximate because of the many sounds involved in all the languages. There are to be phonetic transcriptions into French and English.

Moore moved acceptance for the general spirit of the report; motion seconded. A 'resolution' will be proposed for adoption by the full IAU General Assembly.

Improvements on details were then proposed. Arthur objected that the name Sven Hedin had been assigned to a crater on the back side, but for some years, work of the British Astronomical Association had referred to a crater Hedin on the front; there will be confusion in the literature. Menzel pointed out that the designation Hedin was not in IAU Official Blagg and Mueller list.

Koziel remarked that the name Banachiewicz had been moved, also Lamark, Riemann, Rayleigh and some others. There were names given from photographs taken of the Moon's limb; when more nearly vertical photos became available, it appeared that they represented rather poorly-defined craters.

Arthur then objected to naming one crater inside another. In particular, where letters are added, it will be difficult to decide whether these should be after the inner or the outer crater. In particular, he pointed out Krylov and Ingalls inside Korolev. A similar case with Apollo and with Das which is inside Galois.

O'Keefe pointed out that most of these problems involve the large farside craters of dimensions as large as front side maria, for which the Soviets had proposed the name thalassoids. He pointed out that front-side maria do not control the assignment of letters to craters.

Arthur then proposed that the names of these large craters be designated by parentheses in the lists, as a guide to cartographers, but not to letter smaller craters after them.

Whitford asked that the biographical list include another Russell beside H. N. Russell; Menzel accepted.

The report of the Committee on nomenclature as amended was then unanimously accepted.

Menzel then remarked that the Soviets have suggested that crater chains be named after Institutes, these being designated by their initials; for example, the rocket research institute, RNEE. The Working Group remarked that there was not enough data to make a decision; and that it should be referred to a new committee for the interim period.

Menzel then asked whether there should be a pronunciation guide. Markowitz felt that it would be useless; the sense of the meeting was, however, that the pronunciation guide should be prepared.

The session of Working Group I on 'Nomenclature' is adjourned.

Then, the President. A. Dollfus, announced that an additional working session is scheduled at 19:45, in the same room, about the problem of coordination in Laser studies. Attendance is for IAU Members, or others, directly interested in the Laser problems.

Special Working Session on Laser Study of the Moon

A. Dollfus, Chairman, described the widespread interest in the use of lasers with retroreflectors on the Moon; he read a COSPAR resolution appointing a working party to work on the problem:

"*Decision No.* 7 proposed by the COSPAR Executive Council on the proposal of Working Group 1 – (Leningrad, May 1970).

COSPAR,

noting with satisfaction that the lunar laser experiment has been successful in its initial phase, but *recognizing* that the scientific value of the experiment will be fully realized only under the following conditions:

(1) when several retroreflectors are widely distributed on the lunar surface;

(2) when there are as many terrestrial observing stations as practicable, well distributed geographically, with at least one station in the southern hemisphere.

(3) when the data are made available to the scientific community;

urges space agencies launching lunar landing spacecrafts to place on the lunar surface more optical retroreflectors, some of which should be large enough to be used easily with one-meter telescopes,

encourages all countries to develop and build lunar laser ranging systems and to participate in the observations, and

establishes a working party of Working Group 1 on lunar laser ranging under the chairmanship of Dr C. C. Alley (U.S.A.) in order to initiate international coordination and data exchange in this field".

J. Kovalevsky then read a proposed IAU resolution supporting this working party, to be submitted to IAU Commission 4:

"Proposed IAU resolution (draft to be submitted to IAU Commission involved):

IAU,

realizing that the potential scientific value of lunar laser experiment would be highly enhanced if international cooperation in the observations and prompt data exchange is achieved by all groups active in the field,

and *noting* that a working party was appointed by COSPAR (Decision No. 7, 1970) under the chairmanship of C. O. Alley in order to initiate such coordination,

endorses the formation of this working party and requests that it reports on its activities to the relevant IAU Commissions 4, 17 and 19.

The Kovalevsky resolution was then extensively discussed. Dollfus pointed out that although COSPAR is well-adapted to promote the measurements, the IAU is deeply involved for scientific utilization. In the discussion it became clear that a major role is likely to be played by the Commission 17 astronomers interested in the librations of the Moon; D. Y. Martynov emphasized the names of A. A. Nefediev, K. Koziel and Th. Weimer.

J. Rösch argued for COSPAR-IAU working groups at the same level. On the other hand, A. H. Cook pointed out that AIG, AISPEI, and especially IGGU are interested in the problem. P. J. Melchior urged that we avoid a multiplicity of committees on the same subject. G. P. Kuiper felt that IAU should speak with one voice.

The possibility of a separate IAU working group with one or two members from each of the interested Commissions was discussed, with perhaps more representatives from Commission 17 because of their stronger interest.

It was decided that there was not enough information to permit a decision; Dollfus was to contact Perek; Rösch to see some of the other Commission members. An additional meeting for delegates of IAU Commissions involved is scheduled on August 24, at 20:45 at the Dome Amphi-theater. Meeting adjourned near 19:30.

Additional Meeting on Laser

Present: A. Dollfus, C. O. Alley, J. A. O'Keefe, B. Guinot, K. Koziel, J. Kovalevsky, J. Rösch. Dollfus pointed out that there were two possibilities:

(a) At minimum, an IAU Group with representatives from each of the Commissions involved.

(b) At maximum, a mixed COSPAR-IAU group working on the whole thing.

Kovalevsky remarked that de Jager feels that only one group should coordinate efforts. De Jager suggests that IAU representatives be added to the COSPAR party. IGGU may do likewise. The Chairman of this joint working group would be in a position to ask for meeting either from COSPAR or from IAU during an Assembly of either organization.

It was the sense of the meeting that IAU should send representatives to make a joint working party. Names suggested:

Commission 4	To be decided
Commission 7	W. J. Eckert
Commission 9	J. Rösch
Commission 17	K. Koziel, W. H. Michael, J. Rösch
Commission 19	P. J. Melchior
Commission 31	B. Guinot

Kovalevsky's droposed resolution to support COSPAR was amended to indicate support for coordination of analytical studies as well as observations and exchange of data. Final wording of the resolution was left to Dollfus and Kovalevsky. Adjourned 22:00.

Kovalevsky and Dollfus then worked on the wording and drafted the following proposal, submitted to the General Secretary on August 25, morning:

RESOLUTION DE LA COMMISSION 17
APPUYEE PAR LA COMMISSION 4
PROPOSEE POUR LE COMITE EXECUTIF

"L'Union Astronomique Internationale,

consciente que l'importance scientifique des déterminations de distance de la Lune par laser se trouverait grandement valorisée par une collaboration internationale concernant les observations et les développements théoriques ainsi que par des échanges rapides des données à réaliser entre les différentes équipes engagées dans ce domaine,

et *considérant* qu'un groupe a été constitué par le COSPAR (décision No. 7, 1970) dans le but d'assurer cette coordination,

estime appropriée la formation de ce groupe.

Afin de renforcer la coopération internationale et la liaison entre les unions scientifique l'UAI *propose* que des représentants des commissions suivantes:

4 (Ephémérides)
7 (Mécanique Céleste)
9 (Instruments Astronomiques)
17 (La Lune)
19 (Rotation de la Terre)
31 (Temps)

soient nommés dans ce groupe et qu'ils rendent compte de ses travaux à l'UAI."

RESOLUTION FROM COMMISSION 17
SUPPORTED BY COMMISSION 4
PREPARED FOR THE EXECUTIVE COMMITTEE

"The International Astronomical Union,

Realizing that the potential scientific value of lunar laser ranging experiments would be highly enhanced if international cooperation in the observations and theoretical developments, as well as prompt data exchange are achieved by all groups active in the field,

and *noting* that a working party was appointed by COSPAR (decision No. 7, 1970) in order to initiate such coordination, welcomes the formation of this working party.

In order to strengthen international and interunion cooperation, IAU *suggests* that representatives of the following commissions:

    4 (Ephemerides)
    7 (Celestial Mechanics)
    9 (Astronomical Instruments)
    17 (The Moon)
    19 (Rotation of the Earth) and
    31 (Time)

be appointed to the working party and report about its activity to IAU."

Third Session: Working Group 2 'Figure et mouvements de la Lune'

A. Dollfus announced that the papers on laser results would be postponed to the 26th. He then turned the chair over to Menzel, for additional comments about the lunar nomenclature.

Menzel announced that Working Group 1, Lunar Nomenclature, had now completed some minor revisions resulting from the discussion on August 20. The craters Pingré, Krylov, Ingalls, Bobone, Tereshkova, Hagen, Chappel, and Das had been relocated; most of these changes removed craters from the interior of other craters. It is believed that the latter problem is now solved, so that there is no need for the proposed use of parentheses to designate very large craters to prevent their use to sub-letter the small nearby craters.

Menzel stated that three small maria near Mare Orientale had been renamed as 'Lacus', namely Lacus Veris, Lacus Aestatis and Lacus Autumnae.

A number of mountain ranges are found not to be clearly identifiable and are removed: Montes Doerfel, D'Alembert, Leibnitz, Hercynii, Sovieticii.

C. de Jager has agreed to print the revised list in Space Science Reviews.

The revised list was unanimously adopted by voice vote.

Dollfus then presented for the full Commission a report on the laser working session discussions reported above.

He then proposed for approval the above-reported resolution. Resolution adopted.

Dollfus then announced that those members of Commission 17 or others interested in the problem of coordination for geological, geophysical and lunar samples analyses would meet for a working session at 18:00.

He then presented a list of proposed new members of Commission 17. It is not always possible, at this stage, to decide if the names on the list will correspond to 'IAU members' or to 'consultant members'. Such attribution depends on IAU by-laws and will be decided by the Executive Committee. The list of names proposed is: Alley, C. O. – Anders, E. – Baldwin, R. – Brunk, W. E. – Carder, R. – Cook, A. H. – Ewing, M. – Gast, P. – Gavrilov, Z. A. – Hagfors, T. – Hunt, M. – Lipsky, U. N. – Michael, W. H. – Moutsoulas, M. – Orszag, A. G. – Runcorn, S. – Shapiro, I. I. – Sonett, C. B. – Van Flandern, T. L., and the chairman of COSPAR W.G. 7 'Moon and Planets' (presently A. D. Kuzmin).

Such a number of new appointed members is realistic on account of the existence of three working groups, the status of Consultant, and the fact that previous Commission Members are no longer active in the field.

Five additional names were proposed by the Commission, during the discussion:

    D. H. Eckhardt
    J. Guest
    L. P. Morrison
    W. Sjogren
    R. Wildley

The list was accepted.

To the names of the Organizing Committee were then added E. Anders and P. Millman, the latter because of a conference soon to be held on the subject of the Moon at Ottawa.

Then the session of Working Group 2: 'Figure et mouvements de la Lune" was open. Th. Weimer, Chairman of W.G. 2, then spoke on the purpose and goal of Working Group 2. He proposed to report scientifically each year on work in this field, after the manner of the old IAU Draft Report. Koziel felt that an annual report was too often; but the sense of the meeting was that the volume of new work would require an annual report.

Furthermore, bibliographical lists will continue to be prepared. Three lists were already circulated in 1969–70.

The definition of the field covered by the Working Group was discussed, in view of avoiding overlaps with other IAU Commissions, namely Celestial Mechanics. A new title was proposed for the Working Group, adequately indicating its scientific field:

'Figure, rotation et position observée de la Lune'.

Then comes the *scientific session of the Working Group* 2:

First Part – Chariman: K. Koziel.

R. F. Hall: Preliminary Results of the Moon Camera.

S. Chapront: Progrès dans la construction des éphémérides de la Lune.

T. C. Van Flandern: The Motion of the Moon from Occultations.

F. M. McBain-Sadler: Results of Occultations.

I. I. Shapiro: Comparison of Radar Data and Meridian-Circle Observations.

Second Part – Chairman: A. Mikhailov.

A. A. Nefediev: Constants of Libration from Heliometer Observations at Kasan.

K. Koziel: Cracow Selenodetical Works.

I. V. Gavrilov: Comparison of some Selenodetic Coordinates.

Z. Kopal and M. Moutsoulas: Results from the Manchester Team on Star-Calibrated Lunar Photographs.

D. W. G. Arthur: Selenographic and Selenodetic Results with Apollo Photographs.

P. Gottlieb: Recent Results in Lunar Gravity Fine Structure.

Special Working Session on Geology, Geophysics and lunar samples analysis

A. Dollfus opened the meeting by remarking that the NASA Meeting on lunar samples at Houston was in many respects admirable. However, there were very few astronomers at the meeting. He considered that astronomers must play a role in the analysis at the next step, for interpretations of the data, and be asked for ideas.

B. Levin remarked that IGGU is organizing a geological congress in Moscow, which will include a symposium on the figure and motion of the Moon. He remarked that it is the function of Working Group 3 to use these data for studies of e.g. the thermal history of the Moon.

B. Levin and J. A. O'Keefe both agree on the point that interpretation of the data, including lunar sample results, is what is now needed.

B. Levin, R. J. Fryer, W. E. Elston then discussed the difficulty of translating geological nomenclature so as to make it comprehensible to astronomers.

W. E. Elston brought forward the point that there should be some expression of opinion from the IAU concerning the value of the lunar landings and the need for future work, especially to explore the highlands.

A long discussion followed on the kind of action an IAU Commission can undertake to stimulate the lunar space programmes, to orientate its goal in scientifically useful directions and to coordinate the scientific interpretation of the results.

The argument often put forward is that direct exploration of the Moon represents a major step towards the basic problem of the origin and evolution of the Solar System. But it was clearly stressed by the group that such an argument was never clearly explicited in details. Of major importance should be a report, prepared by a group of specialists of the Commission, detailing the scientific approaches involved, previews of the successive scientific steps and the results expected, and the overall limitations. Members of the audience willing to cooperate were: Baldwin, Dollfus, Elston, Levin, O'Keefe, Runcorn, Strom.

Furthermore, as immediate action, a resolution expressing the needs of lunar scientists for further exploration should be drafted.

A small group (B. Levin, R. B. Baldwin, R. G. Strom and K. Runcorn) was appointed to prepare a resolution for IAU, expressing these ideas.

The resolution prepared reads as follows:

Commission 17: *RESOLUTION*

"The IAU expresses its conviction of the great scientific importance of the preliminary physical measurements made in the environment of the Moon and on its surface, and of the collection and analysis of samples returned from the surface, especially in relation to the advance of our knowledge of the origin and evolution of the Earth, the Moon and the whole Solar System.

The IAU strongly hopes that continued programmes will be developed in which a sufficient minimum of sampling sites are included so that conclusions can be drawn about the Moon as a whole. In particular, the IAU stresses the importance of sampling the highlands, and of some special areas which are critical for the understanding of the processes which have shaped the Moon and even the Solar System as a whole.

The IAU recognizes the value to member states of contributions to the scientific exploration of the Moon made from the international scientific community, and expresses the hope that international cooperation in this field will continue to expand."

Fourth Session: Working Group 3 'Geology and Geophysics of the Moon'

This session was opened at 14:15. Gold's paper was called, out of order, on 'Present Information concerning the Lunar Surface'.

Next, information was given about the working session of previous night on geology, geophysics and lunar samples analysis.

The resolution prepared at the issue of this session was brought forward and was passed.

Dollfus reported that the group decided to prepare a report indicating the conclusions from the astronomic standpoint to be obtained from the direct exploration and returned lunar samples. Volunteers being asked for, the committee was composed of R. Baldwin, A. Dollfus, W. E. Elston, B. Levin, J. O'Keefe, K. Runcorn and R. G. Strom. Furthermore, E. Anders could be approached to join the group, G. Fielder, chairman of W.G. 3, has agreed to chair the group.

Runcorn pointed out that a conference on the Moon will be held in Newcastle-upon-Tyne by April 1971, with the support of IAU; he proposed to take this opportunity to discuss the contents of the report. Accordingly, the schedules will be to have the draft ready for March 31, 1971.

Fielder then presented a report on the functions of Working Group 3. He was concerned over the problem of coping with the enormous and very varied information coming in; obviously the Commission must be concerned with the analysis of this mass of information. He proposed a basic plan for an overall committee on the Moon, with representatives of NASA, AGU, IUGS and other international unions. The programme was criticized as sinking in a mass of Union formalities and was tabled.

Fryer then brought forward a request that Commission 17 supports the *Catalog of Terrestrial Crateriform Structures* to be prepared for the sake of astronomical needs. Specimens were already prepared as Part I: *Canada* and Part II: *Indonesia*. This could be a Commission 17 report.

W. Elston, however, objected that the report would duplicate other catalogs such as N. v. Padang's *Catalog of Active Volcanoes*. He jointed out that there are hundreds of thousands of such structures; the job would be monstrous and would strain the available resources. Fryer clarified that the purpose is not at all to be exhaustive, but only to make available to the discipline of Astronomy pre-existing documents of the geological literature. Money would not be requested from IAU.

Elston indicated that the Geological Survey of Canada has a similar project in work.

Levin feels we should not get involved too much with this kind of problem.

Dollfus was of the opinion that although other catalogs exist, they are not in the form in which astronomers want it. The adaptation will be helpful and still a manageable task.

On Levin's motion, the committee expressed its gratitude to Titulaer and Fryer for the work which they have accomplished.

Miss Middlehurst then presented her paper: The International Group of Lunar Transient Phenomena Observers.

Next came the scientific session, chaired in the first part by B. Levin, and in the second part by G. Fielder, as follows:

Scientific Papers

C. O. Alley: Laser Reflections on the Moon.

P. L. Bender: Lunar Ephemeris and Laser Experiments.

M. Hunt: A Lunar Laser Optical Ranging Experiment.

A. G. Orszag and O. Calame: Télémétrie laser en France.

S. K. Runcorn: Ancient Magnetic Field and Convection inside the Moon.

P. Gottlieb: Mass Anomalies.

V. S. Safronov: Mascons and Isostasy.

R. Baldwin: The Flux of Meteoritic Particles throughout Geologic History.

J. E. Geake: Optical Properties of Lunar Samples.

A. G. Kisliakov: Note, due to V.S. Troitskii, on the Lunar Thermal Behaviour.

J. O'Keefe: Implications of the Chemical Composition of the Moon for the Question of Its Origin.

SPECIAL MEETING OF WORKING GROUP 2:
'FIGURE, ROTATION ET POSITIONS OBSERVÉES DE LA LUNE'

Before the end of the General Assembly, some members of W.G.2 met again, to study with Th. Weimer a proposal from D. Bender on the need for more laser-cataphotes on the Moon. They concluded the following statement:

"Le groupe de travail 'Figure, rotation et positions observées de la Lune' de la Commission 17 de l'UAI,

– considérant qu'il serait utile de disposer de deux à trois grands cataphotes et de six à dix petits pour étudier avec soin la libration de la Lune et pour faire une triangulation de sa surface destinée à fournir un canevas précis pour les futurs travaux sélénodésiques et cartographiques,

– formule le vœu que les expéditions futures sur la Lune déposent de tels cataphotes en des endroits choisis en accord avec les astronomes intéressés."

"The working group 'Figure, rotation et positions observées de la Lune' of Commission 17 of the IAU,

– considering that, for a careful study of the Moon's libration and the establishment of a triangulation on its surface to have a precise net of fundamental points for the coming selenodesic and cartographic work, it would be useful to have two or three great cube corner panels and six to ten smaller ones on the Moon's surface,

– expresses the wish that the forthcoming Lunar expeditions shall put such panels on places chosen with the agreement of the interested astronomers."

COMMISSION 19: ROTATION DE LA TERRE
(ROTATION OF THE EARTH)

Report of Meetings, 17, 20, 24 and 26 August 1970

PRÉSIDENT: P. Melchior.
SECRÉTAIRE: G. Billaud.

Comptes-rendus des séances

Le Président rend hommage aux membres disparus: Z. N. Aksent'eva, G. Demetrescu, F. Koeb-cke, S. V. Romanskaya et poursuit:

"First I should like to give you some personal comments on the present situation of the study of the Earth's rotation, the evolution of concepts and methods used, before analysing the future of our research. Without any doubt, investigations in fundamental astronomy are no longer in favor by the greatest part of astronomers and many of them prefer to devote themselves to astrophysics and radioastronomy. It seems to many people that all the problems of positional astronomy are solved and that routine work is the only thing to be done now.

This is a very dangerous situation about which I wish to give you two recent examples. It has been suggested in the IAU to reduce the low number of commissions of fundamental astronomy by fusion of our Commission 19 with Commission 31. The reasons of this proposition were neither very clear nor had a scientific basis. The Presidents and Vice-Presidents of both Commissions opposed, of course, and the idea was abandoned.

An other fact, much more dangerous, is the idea, raised in one of the important observatories, that a time department nowadays has no longer its place in an observatory. These facts and also some others of minor importance have convinced me that many members of the IAU, even those among them having official responsibilities in their countries, are not correctly informed about the modern aspects of the problems of the irregularities of the Earth's rotation.

They remember the old techniques learned many years ago but they probably ignore that the determination of the secular retardation of Earth's rotation is of great importance for the history of the Earth-Moon system. They probably do not know the implications of the Earth's liquid core effects in both phenomena of tides and nutations.

I can, I think, summarize the situation in saying that the astronomers seem to have forgotten that the Earth also is a planet in the Solar system, that when studying the planetary physics and dynamics we must begin with Geodynamics and that the best laboratory to do it is the Earth itself.

But our Commission has some responsibility in this situation. My feeling is that in the past we did not call the attention of the Union on these facts, on the necessity of developing investigations on the problems of the planet Earth in astronomical observatories.

The activity of our Commission, excellent although from the technical point of view, remains extremely passive concerning the actions to be undertaken to promote and to develop the research activity. Till now, we mainly have been concerned with observation and reduction of data, what is extremely important indeed. But now we must be more concerned with a correct theory of the polar motion as well as of changes of speed of the rotation. The geophysicists for many reasons pay more and more attention to these phenomena. Among these reasons the principal ones seem to be:

1. *The Problem of excitation and damping of Chandler Wobble*

Very few data are available on the Earth's viscosity and its internal distribution. The long period of measurements of the polar motion – 70 yr – should provide an exceptional material for such analysis. But to detect relaxation times, we must be sure that our data are reduced in an absolutely homogeneous way and you perfectly know that this is not the case. The first task of our Commission is to decide on this problem and to produce a homogeneous table of x, y. The results of a comparison of a new catalogue, derived from meridian observations, with group corrections of SIL are extremely encouraging to undertake such a task.

2. *The Secular retardation of the speed of the Earth's rotation* yet mentioned is a problem of the same nature. It also depends on the internal viscosity if we consider the phase lag of the sectorial Earth tides. But it also involves the problem of the friction process due to the oceanic tides which is of a very complex nature. Our first duty will be to produce a table of variations of ω. The history of the Earth-Moon system depends indeed upon the frictional processes and several techniques provide information for the past. I mean of course ancient eclipses and the study of coral growth in paleontology.

3. *The Earth's Liquid Core effects* have a fundamental importance in the particularities of the Earth's rotation. A great number of papers have been published on this subject since ten years or so and astronomers should read carefully these papers. Should we remember that the famous paper written by Poincaré in *Bulletin Astronomique* of 1910 indeed remained practically unknown among astronomers? Now it is clear that diurnal Earth tides and precession-nutation are two aspects of the same phenomena and that resonance effects produced by the liquid core can be observed. Therefore the IAU Colloquium on astronomical constants held last week at Heidelberg took two recommendations related to this problem:

(a) IAU Colloquium considers that no changes should be made in the series for the nutation until a decision is made about the precessional constants, but considers that a new theory of nutation, based upon a more realistic model of the Earth and consistent with recent developments of Tidal Potential should be developed.

(b) IAU Colloquium urges that the observational data of the International Latitude Service be made available on a uniform system and in machine-readable form as soon as is practicable, for use in the determination of the nutations.

Many problems are interconnected with the existence of the liquid core: precession nutation, geomagnetism, internal convection, exchange of angular momentum between the several parts of the Earth, tides... all having a great influence on the rotation of the mantle which is what we are measuring.

4. *Correlations between irregularities in the rotation (polar or speed irregularities) and earthquakes mechanisms*

This is of extreme importance for humanity and we should not underestimate it even when it seems at first sight astonishing. An important symposium was held last year in Canada where this matter was discussed and it is hoped that the publication will be soon available. But you already now may know some of the conclusions based on the elasticity theory of dislocations and changes of the Inertia Tensor.

According to Smylie and Mansinha "There now appears to be no difficulty in accounting theoretically for both the secular pole shift and Chandler wobble excitation as being due to earthquakes. Improved pole path measurements should provide a better understanding of earthquake mechanism".

5. I must also mention the secular pole shift which was the subject of our Stresa Symposium.

6. Finally we know that many Earth tides effects are present in our observations
(a) periodic deflections of the vertical
(b) tidal alteration of the moments of inertia of the Earth
(c) tidal effects on nutations as already mentioned.
As our ephemerides can not yet take into account all these effects the constants of which depend on the internal constitution of the Earth (density, elasticity and viscosity) it is clear that any spectral

analysis of observations will reveal many tidal frequencies. A precise determination of their amplitudes will provide important information not only for geophysics but also for the dynamics of the Earth-Moon system.

Now there is hope that new techniques will greatly improve the measurements in the near future. These techniques are

(a) Doppler satellite observations developed by Anderle and Beuglass
(b) Laser measurements on Moon's corner reflector
(c) Radio interferometry.

But it is to be remembered that we have yet in hand this large amount of data accumulated during 70 yr by the observers of the ILS spending so great efforts. I am convinced that a new general reduction is to be made now with improved fundamental constants, improved declinations and proper motions.

I wish to read you what just has been written this month in the last issue of an important geophysical publication:

"The ILS started over half a century ago and has provided data on the motion of the pole, the value of which has, I think, not been fully appreciated by geophysicists... The calculation of the damping period of the Chandler Wobble is quite basic to the understanding of the non-elastic properties of the Earth's mantle but is an involved statistical problem, and the excitation of this motion is one of the most intriguing unsolved problems of geophysics... A careful exposition of the methods of obtaining data and reducing it will be of very great value to geophysicists concerned with these important questions..."

This last sentence is indeed most important and emphasizes the need of a new reduction in order to diminish the noise present in the data proposed for the analysis.

There are the main things I desired to point out before starting our discussions. It is of course quite impossible to develop fully all the subjects during the present General Assembly but you should keep them in mind when discussing the several practical questions of IPMS and BIH activities.

I think that our Commission has to take a resolution to claim the attention of the Union and of all the Astronomical Observatories upon the new developments of geodynamics and another developing the resolution of IAU Colloquium at Heidelberg, on the necessity of a complete revision of ILS data. Next year we shall have now opportunities to discuss fully these problems. Our Japanese colleagues, convinced of the importance of our activity, have invited our Commission to hold an International Symposium on the Rotation of the Earth at Marioka, near Mizusawa. Many distinguished geophysicists will join this meeting which is to be held from May 9th to May 15th with the official support of IAU and IUGG".

I. ADMINISTRATION

Programme de travail

Le Président indique les grands thèmes dont la commission aura à débattre, ainsi que les résolutions sur lesquelles elle aura à se prononcer au cours de la séance du 24 août 1970. E. P. Fedorov est chargé de constituer un groupe de travail pour l'étude des résolutions.

II. RAPPORTS DU SIL, SIMP ET BIH

(a) P. Melchior annonce la sortie du volume X du SIL contenant les résultats de la période1941,0–1949,0 (T. Nicolini et E. Fichera). Les résultats de la période 1949,0–1962,0 sont prêts à être publiés, mais il est regrettable que cet important travail n'ait pas tenu compte de la résolution adoptée à Prague sur le pôle origine (OCI) et ait modifié la latitude de Kitab. P. Melchior suggère que la commission vote une recommandation de subvention pour aider à la publication dans la mesure ou les résultats sont rapportés à l'OCI. P. Melchior signale également l'existence de paires d'étoiles appartenant au programme SIL et qui présentent des résidus importants. Il souhaite que de nouvelles observations méridiennes soient entreprises pour ces étoiles.

(b) S. Yumi présente le rapport sur les travaux du SIMP. Depuis 1962 les résultats préliminaires sont publiés dans les Monthly Notes avec un délai d'environ 2 mois. Les résultats définitifs de 1962–67 ont été publiés dans les Annual Reports; ceux de 1968 sont actuellement sous presse.

(c) B. Guinot présente une communication sur la précision du TU et des coordonnées du pôle déterminées par le BIH. Pour tenir compte de la précision propre de chaque instrument et de la stabilité de ses résultats, les calculs sont exécutés en deux étapes qui conduisent à la publication des résultats bruts de 5 jours en 5 jours. Ce sont ces résultats bruts qui doivent être utilisés chaque fois que la plus grande précision est requise. En ce qui concerne les erreurs accidentelles, les écarts-types calculés par rapport à une courbe moyenne sont de $0''{,}015$ sur x, $0''{,}015$ sur y et $0''{,}0017$ sur TU1. Quant aux erreurs systématiques, on estime à moins de $0''{.}001$ la dérive fictive du pôle introduite par les mouvements propres des catalogues, mais on ne peut rien dire de la dérive du TU1 qui est liée à la précision de l'équinoxe de ces catalogues; les erreurs annuelles peuvent avoir une amplitude totale de 2 à $3\ 10^{-2''}$ en x et y, et de $0''{,}001$ en TU1.

Une meilleure répartition des instruments sur la Terre et des programmes plus denses sur certains des instruments existants permettraient de réduire les erreurs de l'order de 30 %. Cependant, non seulement l'effort nécessaire pour réaliser cette amélioration n'est pas entrepris dans l'ensemble, mais il apparait une désaffection pour les mesures classiques des caractéristiques de la rotation terrestre. B. Guinot désire qu'une résolution affirme l'intérêt de telles mesures jusqu'à ce que des techniques nouvelles et meilleures soient régulièrement mises en oeuvre.

(d) Discussion générale

A. Mikhailov félicite S. Yumi pour la qualité du travail du SIMP.

E. P. Fedorov souhaite avoir les erreurs sur x et y, sous forme d'ellipse de confiance.

B. Guinot précise que la probabilité pour que l'ellipse soit un cercle est d'environ $\frac{2}{3}$.

S. Yumi donne communication d'une lettre de H. Yasuda indiquant que l'observation de 1000 étoiles des programmes PZT débutera au cercle méridien de Tokyo à la fin de 1971. L'Observatoire de Tokyo souhaite la collaboration du plus grand nombre de services méridiens.

P. Melchior demande que cette information soit transmise à la Commission 8.

P. Melchior répond ensuite à une question de A. Mikhailov sur les chaînes d'instruments et fournit des informations récentes sur le transfert de la station SIL de Carloforte à Cagliari. Après intervention de W. Markowitz et de J. Fleckenstein, le dossier concernant ce transfert est transmis au Comité des Résolutions.

R. O. Vicente demande la création d'un groupe de travail qui serait chargé de la publication des coordonnées du pôle dans un système homogène.

P. Melchior lit une note de E. Fichera concernant la mise des données sous forme utilisable dans les ordinateurs. Suit une discussion ou interviennent E. Fedorov, W. Markowitz, P. Melchior, S. Yumi et B. Guinot.

III. PRESENTATION DE TRAVAUX SCIENTIFIQUES

– Communication est donnée des travaux de Y. Wako sur le terme z.

– A. Mikhailov a trouvé, en s'appuyant sur les travaux d'Eötvös, que le pôle Nord devrait se déplacer vers la longitude 83 °W.

– C. Dramba signale qu'il a entrepris des recherches sur la trajectoire du pôle d'inertie.

– W. Markowitz présente une étude sur la rotation de la Terre déduite des observations des PZT de Washington et Richmond. Il n'y a relevé aucune discontinuité de la vitesse de rotation, malgré des changements brusques de l'accélération; les grands séismes de la période considérée n'ont entraîné aucune modification de ces deux paramètres. W. Markowitz a aussi étudié le mouvement séculaire du pôle d'après les données du SIL: la corrélation des variations de latitude des diverses stations le porte à croire que c'est le pôle moyen qui oscille plutôt que les stations elles-mêmes.

S. K. Runcorn demande s'il serait possible de trouver des variations à courte période de la vitesse (un à deux mois). W. Markowitz répond affirmativement, citant comme exemples les écarts à la courbe quasi uniforme de décélération en 1963,1 et 1964,7.

D. V. Thomas veut connaître la précision des points représentatifs du mouvement du pôle. W. Markowitx répond: "quelques centièmes de seconde de degré". D. V. Thomas remarque alors que les fluctuations autour de la courbe moyenne ne paraissent pas significatives.

– Travaillant sur des données paléontologiques, A. et N. Stoyko ont trouvé pour la valeur du ralentissement séculaire de la vitesse de rotation de la Terre: $\Delta\omega = 2{,}15 \times 10^{-8} \, \omega$.

– B. Guinot a étudié les valeurs brutes de TU2 du BIH calculées de 5 en 5 jours sur la période 1967–69. Il y retrouve les termes lunaires de période 13,7, 14,2, 14,8 et 27,6 jours d'ou l'on peut déduire la valeur du nombre de Love $k: k = 0{,}302 \pm 0{,}045$. Une correction des données avant réduction par le BIH réduirait la dispersion et améliorerait le lissage.

– P. Melchior présente ensuite une analyse de Pilnik portant sur 5463 j et répartie en 9 séries se recouvrant mutuellement. On trouve $k = 0{,}300 \pm 0{,}005$.

Répondant à K. Lambeck sur les mélanges de données, S. Débarbat indique qu'un travail exécuté à Paris sur une seule station conduit à un résultat analogue à celui obtenu par B. Guinot.

P. Melchior pense que la valeur $k = 0{,}3$ qui est en excellent accord avec les mesures récentes de marées terrestres suffirait pour introduire les corrections suggérées par B. Guinot.

E. P. Fedorov attire l'attention sur les problèmes que soulèvent le lissage des courbes.

– S. K. Runcorn expose diverses questions géophysiques relatives à la variation de la durée du jour: champ magnétique, tremblements de terre, déplacement de masse....

Suit une discussion à laquelle participe notamment W. Markowitz, E. P. Fedorov, N. Stoyko, P. L. Bender, R. Hide.

– H. Enslin rapporte que les latitudes observées au PZT de Hamburg montrent une fluctuation saisonnière importante de la variation diurne: en tenir compte réduirait sensiblement le terme z.

D. V. Thomas demande à quel moment la latitude apparente, affectée de la variation saisonnière de la variation diurne, est égale à la latitude moyenne de l'observatoire.

H. Enslin pense qu'il est possible que z varie différemment avant et après minuit; mais cela n'affecte pas la valeur moyenne de la latitude.

D. V. Thomas trouve qu'il serait très difficile de distinguer l'effet que peuvent avoir différentes formes de variations diurnes sur les observations (variations linéaires, sinusoïdales...).

N. P. J. O'Hara partage l'opinion de D. V. Thomas sur ce point. A. R. Robbins demande s'il y a une variation liée à la température. D'après H. Enslin la chose est possible.

N. P. J. O'Hara attire l'attention sur la non-linéarité des phénomènes thermiques et sur leur importance.

B. Guinot confirme l'existence de termes annuels constants et, sur une question de P. Melchior, que les termes annuels sont différents d'une station à l'autre.

– E. P. Fedorov présente un travail de Y. Yatskiv sur la nutation presque diurne. Dans la zone de fréquence correspondante on pourrait distinguer plusieurs termes.

P. Melchior considère que c'est une question très importante. On pourrait en déduire les fréquences de résonance et les utiliser pour tester les modèles de l'intérieur de la Terre.

N. Stoyko remarque toutefois que l'étude à partir d'étoiles brillantes fournit des amplitudes différentes de celles obtenues à partir de l'étude des groupes.

– La séance débute par un exposé de R. J. Anderle sur la détermination des coordonnées du pôle par observation de satellites artificiels. La précision obtenue sur la moyenne de 6 jours est de 0,5 m et les écarts avec les résultats du BIH et du SIMP n'ont pas atteint 1 mètre durant les deux dernières années.

– P. Melchior demande si on a recherché une erreur systématique annuelle avec ces résultats. La réponse est négative.

– D. V. Thomas demande s'il serait possible de déterminer la variation diurne d'une station d'observation. R. J. Anderle indique que la précision sur 24 h serait de 1,6 m.

– B. Guinot demande si les résultats sont affectés par le changement du nombre de stations. R. J. Anderle répond qu'il n'y a eu, jusqu'à présent, aucun effet; toutefois il pense que l'abandon d'une station peut perturber les résultats.

– E. P. Fedorov demande qu'elle est l'origine choisie. R. J. Anderle précise qu'il s'agit de l'OCI et qu'il existe une différence systématique en y avec le BIH et le SIMP. Enfin répondant à F. Barlier, il ajoute que l'on ne connaît pas l'erreur sur la position du pôle d'inertie.

C. O. Alley fait le point des mesures laser Terre-Lune à l'aide du réflecteur déposé sur la Lune par les cosmonautes d'Apollo XI. L'erreur sur les premières mesures, qui était de l'ordre de 2,5 m a été ramenée à 0,15 m; on pense atteindre bientôt une précision de l'ordre de quelques centimètres. Les différentes expériences ont montré l'excellent comportement du réflecteur lunaire aussi bien pendant la nuit que pendant le jour lunaires. Ces travaux conduisent à de nouveaux résultats dans de nombreux domaines et certains nous intéressent tout particulièrement: rotation de la Terre, termes à courte période, mouvement polaire et oscillation Chandlérienne, déplacement de la croûte terrestre et dérivé des continents. P. L. Bender complète cet exposé en indiquant que l'on espère améliorer les techniques, ce qui aura pour conséquence de rendre possible l'utilisation de télescopes de 1,5 m comme émetteurs-récepteurs. On pourrait ainsi augmenter la participation et obtenir 3 h d'observation par nuit, 24 nuits par mois. Suit une discussion technique à laquelle participent K. Lambeck, C. O. Alley, P. L. Bender, L. V. Morrison, J. Terrien.

N. N. Parijskij présente un travail sur le mouvement séculaire du pôle et compare les résultats qu'il a obtenus avec Kuznezov avec ceux de Groves et Munck.

ANNEXE I. CONSEIL SCIENTIFIQUE DU SIMP

Le Conseil Scientifique du SIMP s'est réuni deux fois, sous la présidence de B. Guinot.

Le 19 août, le Conseil a examiné la situation créée par la proposition du transfert de l'instrument du SIL, de Carloforte à Cagliari, selon le voeu de la Commission géodésique italienne; il a établi une liste de conditions pour que ce transfert soit acceptable. A la même séance, sur proposition de B. Guinot, il a été décidé que P. Melchior serait proposé comme prochain Président du Conseil à FAGS.

Le 26 août, le Conseil a établi la liste des membres du groupe de travail chargé de la révision des résultats anciens du SIL:
Président: S. Yumi, Membres: E. P. Fedorov, E. Fichera, B. Guinot, G. Hall, W. Markowitz, P. Melchior, R. Vicente. Le Président du groupe de travail devra présenter, deux fois par an, au Président du Conseil Scientifique, un rapport sur l'activité du groupe.

Le groupe de travail pour la révision des résultats du SIL s'est réuni le 26 août (sauf E. Fichera, absent) et a examiné les conditions dans lesquelles sont rassemblées les données de base, sous une forme acceptée par les ordinateurs.

ANNEXE II. RESOLUTIONS

1. Considering that the basic observational data concerning the physics of the Earth, namely its rotation and polar motion, needed for space research, geophysics, and geodesy, are provided by astronomical observations:
Commission 19 recommends that:
(a) The precise astronomical determinations of time and latitude be continued, and
(b) New chains on the same parallel of latitude be formed, in accordance with the resolutions adopted at Prague in 1967.

2. The fundamental importance of the ILS northern chain in providing a long, consistent series of data is recognized by astronomers and geophysicists and the continuation of the visual observations at Mizusawa, Kitab, Carloforte (Cagliari), Gaithersburg and Ukiah is considered of paramount importance.

3. The importance is stressed of planning for the installation of PZT's at Gaithersburg and Ukiah, by the U.S. Coast and Geodetic Survey, to form a complete ILS chain.
(Note; Present status: Mizusawa, in operation; Kitab, being installed; Cagliari, planned).

4. Recognizing the urgent need to provide the best possible polar coordinates from all available observational data, commission 19 recommends that a program of investigation be conducted in several stages. The objectives are:

4.1. Determination of the appropriate frame of reference.

4.2. Anew, homogeneous reduction of the ILS visual results.

4.3. Determination of the best possible polar coordinates from all observations.

To initiate the first stage, Commission 19 requests the Scientific Council of the IPMS to establish a small working group to re-reduce the northern ILS visual observations on a homogeneous basis.

5. The importance of the program for the homogeneous determination of the polar motion and UT from both time and latitude observations made since 1955 is recognized and the BIH is urged to complete this work.

6. Noting the increasing difficulties of obtaining observing staff for Carloforte and noting the importance of obtaining continuous observations, Commission 19 concurs in the proposal of the Italian Geodetic Commission to move the VZT of the ILS station at Carloforte to Cagliari, provided that:

6.1. The systematic difference of the two sites shall be determined by conducting concurrent observations at the two sites with the use of auxilliary VZT.

6.2. The auxilliary VZT must be of the same accuracy as the one now existing at Carloforte.

6.3. Skilled observers will be maintained at both sites.

6.4. The observing program at each site will be that of the normal ILS station.

6.5. The duration of the concurrent observations will be 6 years.

6.6. Steps shall be taken so that observations with a VZT can be resumed at the present location at Carloforte, if so desired, after completion of the 6-year program.

6.7. The Scientific Council of IPMS will advise on the carrying out of this program.

7. Commission 19 notes with satisfaction that the definitive ILS results for 1949,0–1962,0 have been completed. However, it expresses concern that the origin of the pole used is not the Convential International Origin (CIO) adopted by both the IAU and the IUGG in 1967.

Commission 19 recommends that the Executive Committee of the IAU grant a subsidy to aid publication of the 1949,0–1962,0 results, providing that the results are referred solely to the CIO.

(Note: Use of CIO solely will avoid confusion of multiple systems).

8. Commission 19 welcomes the introduction of new techniques which may be used to study the rotation of the Earth and polar motion, such as artificial satellites via Doppler and laser measurements, lunar laser ranging, and radio interferometry, and urges that regular programs of extensive observations be established.

9. Recognizing that modifications in the surrounding areas of observatories affect the precise measurement of time and latitude, Commission 19 recommends that local authorities concerned aid in preserving suitable observing conditions in the vicinity of observatories.

ANNEXE III. ERRATA

Reports on Astronomy **XIVA**

p. 179 ligne 13   au lieu de Billaud lire US Naval Observatory
p. 179 ligne 14   au lieu de Jorge Silva lire Greenwich Observatory
p. 180 ligne 17   .....des nombres de Love k et 1
p. 182 ligne 8   bi-annuelle.

COMMISSION 20: POSITIONS AND MOTIONS OF MINOR PLANETS, COMETS AND SATELLITES (POSITIONS ET MOUVEMENTS DES PETITES PLANÈTES, DES COMÈTES ET DES SATELLITES)

Report of Meetings, 19 and 26 August 1970

PRESIDENT: G. A. Chebotarev.
SECRETARY: L. Kresák.

The meeting was opened by G. A. Chebotarev, who asked F. K. Edmondson to act as the associate chairman of the session. Silent tribute was paid to the memory of the former Vice-President and very active member of the Commission, Professor S. G. Makover.

Since the printed Report of the Commission was not distributed until shortly before the meeting, it was agreed to postpone the discussion on this item until the third session. It was stated, with regret, that due to an unexpected mail delay several members of the Commission received the circular letter of the President so belatedly that their replies could not be included. The following statements are to be added to the printed report:

"I. I. Shapiro has analyzed more than 400 photographic observations of 1566 Icarus. These observations appear to be of little or no value for estimating the mass of Mercury, since the uncertainty of about 5×10^5 (in the reciprocal) is far inferior to that obtained from the analysis of the interplanetary radar data. K. Strand reported that bright minor planets, suitable for a redetermination of the mass of Jupiter, have been observed with the 15-in. astrograph of the U.S. Naval Observatory, Washington. The close approach of 1566 Icarus in 1968 and the opposition of 1620 Geographos in 1969 were observed with the 61-in. astrometric reflector at Flagstaff. Photoelectric observations of the brightness variations of some minor planets were obtained by P. Tempesti with the 40-cm refractor of the Teramo Observatory, Italy."

Mimeographed copies of the Report of the Working Group on Orbits and Ephemerides of Comets, compiled by E. Roemer, were distributed among all members of the Commission present. It was felt advisable to have this report published in full, as a supplement to the Report of Commission 20 which appeared in *IAU Transactions* **XIVA**.

F. K. Edmondson read a letter by P. Herget, who was unfortunately prevented from attending the meeting. In addition to a brief account of the recent work of the Minor Planet Center at the Cincinnati Observatory, this letter presented a number of suggestions for future observing practices in general and a summary of the types of objects the positions of which are most desirable. The proposals by Dr Herget were discussed and incorporated into the resolutions to be approved in their final form at the closing session. Further comments and suggestions were presented by S. Herrick, T. Gehrels, Z. Sekanina, T. Kiang, H. G. Hertz, P. Wild, B. G. Marsden and A. Schmitt.

The secretary read the list of Commission members proposed by the President for the next triennium. In the Working Group on Orbits and Ephemerides of Comets, E. Roemer has been proposed as Chairman, and M. P. Candy, E. I. Kazimirchak-Polonskaya, Y. Kozai, L. Kresák, B. G. Marsden and G. Sitarski as members. These proposals were unanimously approved by the Commission.

The organization and programme of future scientific meetings to be sponsored by the Commission was discussed. It was recommended that the following two IAU Colloquia be held before the next General Assembly:

1. IAU Colloquium 'Physical Studies of Minor Planets'. Date: March 8–10, 1971; place: Tucson, Arizona, U.S.A; chairman of the Organizing Committee: T. Gehrels.

2. IAU Colloquium 'Asteroids, Comets and Interplanetary Matter'. Date: April 4–6, 1972; Nice, France; chairman of the Organizing Committee: B. L. Milet.

It was approved to ask the IAU for a subvention of $ 2000 for the continued work of the Minor

Planet Center, Cincinnati, for the period 1971–73, and subventions of $ 500 each for travel grants towards participation in the two colloquia mentioned above.

Second Meeting

CHAIRMAN: E. Roemer.
SECRETARY: L. Kresák.
INTERPRETER: J. Kovalevsky.

The second meeting was devoted to comets. E. Roemer, Chairman of the Working Group on Orbits and Ephemerides of Comets, presided.

The report of the Working Group, distributed in advance, was unanimously approved. The access of comet observers to large telescopes was discussed as an imperative problem, because there are too few opportunities where such telescopes can be used for the extremely important observations of comets at large distances from the Sun. All efforts should be made to ensure an improvement in the assignments of telescope time at suitable observatories.

B. G. Marsden, the chairman of the Cometography Committee formed at the last IAU meeting, presented the report on the work of the Committee and read letters by G. Sitarski and S. K. Vsehsvjatskij, who were unable to attend. The recommendations of the Committee were approved by the Commission. It was felt that greater emphasis should be laid at present on the improvement of orbits, with nongravitation effects properly accounted for, than on the collection of general descriptions and physical observations.

The situation with regard to the repository of cometary positions in machine-readable form, which is an important preliminary to the definitive correction of orbits, was thoroughly discussed. At present there are four places where extensive, but incomplete, data of this type are available: Cambridge, Mass., Cincinnati, Leningrad, and Warsaw. It appears highly desirable to have a central repository for positional observations of comets, analogous to the Minor Planet Center at Cincinnati, but in the absence of substantial external financial aid there was no proposal for establishing one. On the motion of E. Everhart, seconded by H. G. Hertz, it was approved to transfer the study of the practical aspects of this problem to a special committee, which will prepare its recommendations for the next IAU General Assembly. P. Herget is asked to serve as chairman, with members G. A. Chebotarev, B. G. Marsden, and G. Sitarski.

B. A. Lindblad submitted an informal request from workers interested in different aspects of cometary statistics (such as the relations between meteor streams and comets) for the publication of a new supplement to Porter's catalogue of cometary orbits, or at least a new edition of the material in the appendices. The present state of orbit computations was outlined by B. G. Marsden. He explained that about 250 new or improved orbits have been calculated since the publication of the first supplement to the catalogue, and they could be easily assembled into a second supplement, provided that the necessary financial assistance would be available for its publication. At the same time, however, he pointed out reasons why the compilation of a new fundamental catalogue, or a new supplement, appears premature at the moment. Suggestions for future improvements in the form of the catalogue were made by B. G. Marsden (inclusion of osculating dates), Z. Sekanina (coincidence of these dates with the nearest 40-day standard Julian dates), L. Kresák (information on the number of observations used for the least-square solution, and the length of the arc, in days, covered by these observations) and G. Guigay (information on the brightness and aspect of the comets). It was felt premature to publish a new fundamental catalogue of cometary orbits until further study has been made on some standard way of accounting for nongravitational effects and until the orbits of many more of the older comets have been determined anew. Meanwhile, the orbit lists in the annual comet reports in the *Quarterly Journal of the RAS* can and should be regarded as supplements to the *BAA Comet Catalogue*, and it is suggested that the 1970 report should include a revision of the appendices to the *Catalogue*. For the same reason it is agreed to postpone the compilation of the new *Cometography* (with an emphasis on ephemerides for all comets ever observed) and to disband the Cometography Committee appointed at the Prague meeting. Meanwhile, current cometographical sources (Vsehsvjatskij's catalogue and its supplements, *RAS comet reports*) should be maintained, and a conscious effort should be made to locate and correct errors that have been propagated in such sources in the past.

Since most of the recent results had been reported earlier in the month at *IAU Symposium* No. 45 in Leningrad, in the scientific part of the meeting only two papers were presented. B. G. Marsden gave a brief survey of the manner in which he allows for nongravitational effects on comet motion; in general his procedure predicts the motions of short-period comets rather successfully, but there is evidence for sudden, unpredictable changes in the nongravitational effects in a few cases. Z. Sekanina spoke on the secular variations in the comet deactivation mechanism, particularly with respect to P/Ecke; in addition he gave an interpretation of the coefficients of the nongravitational terms in Marsden's equations and made an application to P/Faye.

Third Meeting

PRESIDENT: G. A. Chebotarev.
SECRETARY: L. Kresák.
INTERPRETER: J. Kovalevsky.

The Report of the Commission, with the amendments passed at the first meeting, was unanimously approved. Then the secretary read the resolutions compiled from the recommendations put forward by G. A Chebotarev, F. K. Edmondson, T. Gehrels, P. Herget, B. G. Marsden, E. Rabe, E. Roemer, Z. Sekanina, and S. K. Vsehsvjatskij. All but one of these resolutions, which are given in full at the end of this report, were carried unanimously. Only the proposal by S. K. Vsehsvjatskij concerning the change in the name of P/Wolf to P/Wolf-Kamieński, was tabled. Although the Commission highly appreciated the splendid work of Professor Kamieński on the evolution of the orbit of this comet, it was felt that the renaming would be at variance with the practice generally adopted and might represent an undesirable precedent for the future.

In the scientific part of the session three papers were presented and discussed. P. Lacroute spoke on the mean errors of star positions in the catalogues AGK2, AGK3, and SAO. C. J. Van Houten reported on the results of the Palomar-Leiden survey of faint asteroids, with an emphasis on the implications for the structure of the Trojan clouds. T. Gehrels gave a brief review of recent physical studies of minor planets.

At the end of the meeting President Chebotarev relinquished the Chair to the incoming President, F. K. Edmondson, who announced his suggestion to assign the name 'Herget' to the minor planet No. 1751 = 1955 OC, discovered at the Goethe Link Observatory. This proposal, acknowledging Dr. Herget's enormous contributions to minor planet research, was greeted with unanimous applause.

RESOLUTIONS

A. *Minor Planets*

1. Observations are encouraged for objects of special interest or unusual circumstances, such as the Earth-approaching asteroids, asteroids with cometary orbits, and librating asteroids. There will always be an enduring interest in fast-moving objects which are found on any plates.

2. Observations are encouraged for those numbered planets whose ephemerides do not yield reasonably small residuals, say less than $1^m.0$ and $15'$.

3. It must be reiterated to the observers that positive identification of a moving object is impossible unless there are confirming observations from two to four weeks later.

4. Observations of selected minor planets for the purpose of establishing the equinox, and equator

and systematic corrections to star catalogues are encouraged until these programmes are completed.

5. Any systematic observing programme, especially with meridian circles, of the bright minor planets for the purpose of supplementing the observations of the Sun, Moon, and major planets should be encouraged.

6. Observers with sufficiently powerful telescopes are urged to further observe the two 'clouds' of Trojans associated with the Lagrangian points L4 and L5.

7. Determination of approximate positions of minor planets has generally lost its importance. It is sufficient to give the approximate position only if there is a reasonable expectation that accurate measures can be provided to satisfy the request of any orbit computer in the years to come.

8. The accurate measures of moving objects should be reduced with AGK3 comparison stars or corrected to the FK4 system whenever possible, and this should be clearly stated.

9. The changes in the form of the *Ephemerides of Minor Planets* published in Leningrad, in particular the extension of the ephemeris intervals around the opposition, are accepted with satisfaction. The recent practice of including extended ephemerides for Earth-approaching asteroids is much to be encouraged. It is hoped that the number of planets given this special treatment will be increased so that the list includes eventually all those with highly eccentric orbits and not necessarily best observable around opposition.

10. As the very limited field of large reflectors imposes strict requirements on the accuracy of the ephemerides of the minor planet, it is desirable to print with each ephemeris a reference to the elements on which the ephemeris is based, or at least the year of the 'Ephemerides' in which the elements were introduced.

11. The present photometric system of asteroids should be revised to conform with the UBV photometric system. The new values are 0.10 magnitude fainter, nearly the same difference for all asteroids; a new list of photoelectric magnitudes referred to the UBV system has been prepared by T. Gehrels (list published in *Surfaces and Interiors of Planets and Satellites*, A. Dollfus, ed., Academic Press, 1970).

B. *Comets*

12. Recognizing that the most important function of large telescopes is the observation of very faint objects, Commission 20 calls attention to the astrophysical importance of observing faint comets. Directors of observatories with large telescopes are urged to include the observation of faint comets in their assignments of telescope time.

13. Further experiments to investigate appropriate methods for allowing for nongravitational effects in the computation of the orbits of comets (particularly short-period comets) are encouraged.

14. It is recommended that, whenever perturbations are taken into account, the published osculating elements of a comet should be referred, in general, to the 40-day standard Julian date nearest the time of perihelion passage.

15. In order to ensure reliable predictions on the observability of comets in different types of telescopes, it is recommended that the notation m_1 be used for the 'total' magnitude of a comet and m_2 for the 'nuclear' magnitude in the publication of both observations and ephemerides. The subscripts are consistent with the code used for the telegraphic reporting of observations.

REPORT OF THE WORKING GROUP ON ORBITS AND
EPHEMERIDES OF COMETS FOR THE TRIENNIUM 1967–70

INTRODUCTION

This report has been compiled from information received in response to an inquiry sent by E. Roemer as chairman of the Working Group to members of the Group (Candy, Kresák, Marsden, Makover, Sitarski) as well as to twelve others known to be working actively in dynamical study of comets whose work might not otherwise be adequately represented in the report. Responses were received from J. L. Brady (Livermore), Candy, Kresák, A. L. Friedlander (Astro Sciences Center, Illinois Institute of Technology Research Institute), Makover, Marsden, Milet, E. Rabe, Sekanina, Schubart, and Van Biesbroeck.

Highlights of the triennium include a reasonably satisfactory level of observational activity, although further application of large reflectors would be highly desirable; attainment of a new high level of predictional accuracy through the more rigorous application of dynamical theory permitted by the availability of large-scale computing machinery; significant progress in the investigation of nongravitational effects; and important studies of the dynamical evolution of comet orbits, including diffusion of comets from near-parabolic to short-period orbits and resonance effects within the Jupiter family.

Several specialized symposia and meetings during the triennium have been devoted to dynamical problems of comets. All-Union Conferences on Comets have been held annually in Kiev, and meetings particularly connected with space missions to comets have been held in the U.S.A. Most significant, however, is the IAU Symposium No. 45, 'The Motion, Evolution of Orbits, and Origin of Comets,' being held in Leningrad just preceding this General Assembly.

I. GENERAL

Most discoveries of new comets during the triennium bear witness to the enthusiasm of amateur astronomers, particularly the Japanese. Amateurs have also been of considerable help in confirming relatively bright new objects in instances wherein the discoverer has had difficulty in making adequate observations, or in which considerable time elapsed between date of a photographic observation and detection of a new object on the plates. Note is taken with satisfaction of the prize offered by the Astronomical Society of Japan, and of the newly established Comet Medal of the Astronomical Society of the Pacific.

Early recovery of returning periodic comets and prolonged observation of both periodic and near-parabolic comets are at least as desirable as ever. The outstanding achievement during this triennium of K. Tomita, who recovered no less than ten periodic comets during 1967, including four on the single night of October 5, deserves special mention. Z. M. Pereyra has given valuable help with observation of comets in southern declinations, including two recoveries and extended observation of a number of comets not followed elsewhere. The last observations of the recent sungrazer, 1970 f, White-Ortiz-Bolelli, are an excellent example of Pereyra's important contributions. Roemer has had consistent success with the 154-cm Catalina reflector, and particularly, since October 1969, with the 229-cm reflector of the Steward Observatory on Kitt Peak. In instances in which positional uncertainty has complicated the search for relatively faint objects, important work has been done by G. A. Tammann and C. T. Kowal, who have had the opportunity to apply the 122-cm Schmidt telescope of the Palomar Observatory to observation of comets.

Preliminary orbit computations seem to be in a satisfactory state at the present time, thanks in significant degree, to good computing facilities available to Marsden, who is in the pivotal position of director of the IAU Central Telegram Bureau at the Smithsonian Astrophysical Observatory. Note should be taken, however, of the calculations of orbits and ephemerides at the Institute for Theoretical Astronomy, Leningrad, on behalf of the several Soviet observatories engaged in observation of comets, and by several individuals, including Candy, S. W. Milbourn, and T. Seki, who have much more modest computing facilities at their command.

Calculations of definitive orbits by present standards typically include allowance for perturbations during the observed arc wherever significant, and a least-squares adjustment that takes account of all valid observations. Such work is carried on at the major centres, including the Institute for Theoretical Astronomy, Smithsonian Astrophysical Observatory, Institute of Astronomy of the Polish Academy of Sciences, Astronomisches Rechen-Institut, and the Minor Planet Center, Cincinnati. In some cases important aid is given through a centre to an individual working independently, as, e.g., between Marsden and Van Biesbroeck.

During the triennium, I. Hasegawa has published an interesting Catalogue of Periodic Comets, in which alternate sets of elements for each object are presented. Sekanina has circulated the first supplement to his Catalogue of Original and Future Comet Orbits.

II. ASTROMETRIC OBSERVATIONS

Astrometric observations of comets with astrographs or Schmidt-type cameras are carried on regularly at Abastumani, Burakan, Berne, Crimean Astrophysical Observatory, Flagstaff (Lowell Observatory), Kleť, Nice, Perth, Pulkovo, Skalnaté Pleso, Tartu, Uccle, Washington, and, among other amateurs, by R. L. Waterfield (Woolston), and T. Seki (Kochi). Several of the Soviet observatories receive assistance with reductions from the Institute for Theoretical Astronomy, Leningrad.

The principal contributors having access to long-focus instruments include Z. M. Pereyra (154-cm reflector, Bosque Alegre Station, Córdoba Observatory for the CNEGH), K. Tomita (91-cm reflector, Dodaira; 188-cm reflector, Okayama, about 20 nights per year), E. Roemer (154-cm reflector, Cataline Station, LPL, about 25 nights per year; 229-cm reflector, Steward Observatory, Kitt Peak, 10-20 nights per year?). H. L. Giclas reports that the 183-cm Perkins reflector at the Lowell Observatory has been equipped for photography of moving objects, and some success has been achieved. The need for observations with powerful instruments such as these is, if anything, greater than ever as computational standards have risen to permit critical comparison of various theoretical interpretations of nongravitational effects and orbital evolution. Several of these telescopes can reach objects of 19th and 20th magnitude, and the Steward reflector has played a critical role on several occasions in reaching objects of magnitude 21.

Note is taken with considerable satisfaction of the progress with redetermination of plate constants for the northern zones of the Astrographic Catalogue. New values, on the system FK4, are soon to be published in *Astronomy and Astrophysics Supplement* (cf. A. Günther and H. Kox, *Astron. Astrophys.* **4**, 156, 1970). Availability of the new constants will be a definite factor in increasing the accuracy and consistency of the positions from long-focus instruments.

III. ORBITS OF SHORT-PERIOD COMETS

By present standards of computation, predictions for returning periodic comets are based on starting orbits that represent in satisfactory manner all valid observations at the most recent previous apparition. Allowance for all significant perturbational effects of the major planets is made in arriving at the predicted orbit and ephemeris. In an increasing number of instances, several apparitions have been rigorously linked, resulting in determination of, and some form of empirical allowance for, nongravitational effects. Many recoveries now involve reobservation of the comet within less than one arcminute of the predicted place. At the principal centres, the procedures to carry out the necessary computations have been standardized to a considerable degree and are carried out with the aid of modern digital computers.

For details of specific results for individual objects, including references, it is very gratifying to be able to refer to the annual reports on comets prepared by Marsden and published in the *Quarterly Journal* of the RAS.

The principal new areas of activity (reported in greater detail in Section 5 of this report) include analyses of the nature of the nongravitational effects, resonances and librations arising from interactions with the principal planets and affecting orbital stability, as well as the persistent problem of the dynamical origin and orbital evolution of the short-period comets.

Another new interest attaches to short-period comets in connection with proposed direct exploration of comets by means of space missions. Orbital analyses of a considerable number of such comets have been carried out in connection with feasibility studies for missions. Practicable missions fall into two basic categories, (1) fly-through missions, and (2) rendezvous missions. In the first type the probe intercepts the comet fairly close to perihelion but passes through with a rather high relative velocity (of the order of 10 to 20 km sec^{-1}, typically), spending only a few hours in data-gathering in the near vicinity of the comet. Farther in the future are the rendezvous missions, in which the probe's trajectory is closely adjusted to that of the comet, intercept being achieved near aphelion and the probe traveling with the comet for prolonged periods of data-gathering during the comet's physical development on approach to the Sun. Extremely high standards of accuracy of positional prediction are required by either type of mission. (For a general discussion see A. L. Friedlander, J. C. Niehoff, J. I. Waters: 'An Interim Report on Comet Rendezvous Opportunities,' *Illinois Institute of Technology Technical Memorandum* No. T-21, November 1969.)

On the observational side, the orbits of several comets of one apparition have been closely investigated by Marsden with the finding that in some cases renewed searches at current apparitions can be sufficiently limited in area to give reasonable chance of reobservation of the comet. The recovery of P/du Toit-Neujmin-Delporte by Kowal as a result of this effort is particularly gratifying.

IV. ORBITS OF NEARLY PARABOLIC COMETS

Definitive orbits have been calculated for several near-parabolic comets of recent years by O. N. Barteneva, L. M. Belous, R. Branham, Jr., M. A. Mamedov, L. E. Nikonova, Z. Sekanina, G. Van Biesbroeck, and G. T. Yanovitskaya. More complete details are included in the annual reports published in the *Quarterly Journal* of the RAS, to which it is again very satisfying to be able to refer.

Original and future orbits for a number of objects have been calculated by Barteneva, Marsden in collaboration with Van Biesbroeck, Sekanina, and Yanovitskaya.

Brady reports that he has finished integrating the orbits of all the available near-parabolic comets that fit his criteria – some 143 objects. He mentions in particular that a critical value of the osculating eccentricity stands out prominently. Of those comets with osculating eccentricity above 1.00017, 75% pick up energy through the effects of planetary perturbations and escape; of those with e below this value, about 50% escape and 50% remain bound.

E. Everhart also has studied the changes of energy of near-parabolic comets in passing through the solar system, more particularly from a statistical point of view. His work is referred to in greater detail in Section V.

It is clear, however, that the proved presence of nongravitational effects in the motion of many short-period comets, and in the near-parabolic comets Arend-Roland (1957 III) and Burnham (1960 II), complicates the interpretation of original and future orbits computed by purely gravitational representations.

V. THEORETICAL INVESTIGATIONS

Nongravitational effects have been investigated by Marsden in a series of papers that establish their presence and empirical character. Deviations of the observed motion of 15 out of 18 long-observed comets from the corresponding rigid representation by gravitational theory have been definitely established. The principal force component seems in general to be radial; that along the orbit is roughly an order of magnitude smaller; and no component perpendicular to the orbit plane is required by observations. There is a correlation between the appearance of the comet and the magnitude of the nongravitational effects; the two comets of most nearly stellar appearance seem to move strictly gravitationally. Nongravitational effects generally seem to show a secular decrease in amount, which leads Marsden to speculations concerning the relationship between comets and minor planets. There are, however, significant dynamical differences between comets and minor planets, particularly with regard to resonance effects and the occurrence of close approaches to principal planets. An outstanding question relates to apparent changes in nongravitational effects associated with close approaches to Jupiter. There is some evidence that random and impulsive effects do occur from time to time, but P/Schwassmann-Wachmann 1, which might be expected to be the prime candidate to show such effects, in fact does *not* show them.

Sekanina also has done important work in the area of nongravitational effects, concentrating on physical models of how such effects may arise. Reference is made to the Bibliography for his extended series of papers. He concludes that both mass loss from a rotating nucleus and nuclear splits, either spontaneous or arising from collisions, contribute to observed orbital deviations. Sekanina also has considered nuclear splits of several comets in a more general context. An interesting finding by Sekanina has been the association of meteor streams, using Harvard radio meteor data, with certain Apollo asteroids. If this finding can be substantiated, weight would be added to the idea of evolution of these asteroids from comets, a suggestion put forward long ago by E. Öpik.

P. Stumpff is continuing his studies of differential perturbations of comet orbits, the motion of P/Brooks 2 since 1886 being under particular investigation. Resolution of nongravitational effects is complicated by the two close approaches to Jupiter during the interval included in the perturbation calculations.

D. E. Gonzales has shown that the solar wind-induced drag on comets is much less than the mechanical effect of mass loss.

The origin of the Kreutz group of sungrazers has been the subject of study by both Marsden and Sekanina. Marsden argues rather convincingly that the known members fall into two distinct groups that very likely arose from tidal disruption of a parent comet near perihelion only one or two revolutions ago. Observed disruptions of several of these comets near perihelion lend credence to the suggestion. Sekanina proposes that a parent comet may have been disrupted by collision with a 'cosmic projectile' far from perihelion, or, alternatively, that the group may have arisen through a glancing collision of two parent comets.

E. Everhart in a series of papers has attempted to allow for selection effects in discovery of long-period comets to reconstruct the intrinsic distribution of perihelia and magnitudes of the parent population. Then he has investigated the changes in total energy that arise in passage of members of such a population of comets through the solar system, first through small perturbations, then through single close encounters. The aim has been to investigate the mode of the presumed diffusion of long-period orbits to short-period ones.

G. Sitarski has studied the approaches of the 494 actually observed parabolic comets to the outer planets, finding that 50% of comets can approach outer planets closer than 1 UA. Orbital changes were calculated for 62 close approaches. Capture into an elliptic orbit by Jupiter seems to be possible even for orbits with e as large as 1.5.

E. M. Pittich also has investigated selection effects on discovery, with interesting findings related to the contribution of photographic discoveries. In a second paper he concludes that sudden brightness changes are a factor in some 8% of comet discoveries. Kresák notes that it is intended to extend the work of Pittich, which is broadly concerned with investigation of the changing population of the inner zone of comets, to the central problem of the equilibrium between perturbational capture of comets from outside, on one hand, and ejection and disintegration in the inner part of the solar system, on the other hand.

H. I. Kazimirchak-Polonskaya is continuing her investigations on evolution of orbits of some 45 short-period comets in the interval 1660–2060, with the aim of clarifying the role of outer planets in this evolution. N. A. Beljaev also has contributed substantially to this investigation.

Cometary motion in the outer solar system has been the subject of investigations by Chebotarev and, more recently, by Sekanina. Both have been interested in the sphere of action of the solar system and perturbations of comets of the Oort cloud by nearby stars.

Hamid, Marsden, and Whipple have investigated the influence of a hypothetical comet belt beyond Neptune, both in its possible influence on motions of periodic comets and on the motions of Uranus and Neptune. A not very restrictive upper limit to the mass of such a belt of one Earth-mass within 50 UA from the Sun has been found.

Hamid and Whipple have searched for evidence of P/Encke in ancient records, especially the Chinese. They have computed the gravitational position 2500 yr into the past as the basis for identification of possible observations, but uncertainties remain as a consequence of unknown nongravitational effects.

VI. COMPUTING TECHNIQUES

Various developments in computing techniques have been implicit in the programming necessary to carry out the orbit determinations, perturbation calculations, and adjustments of elements already referred to as a part of the rigorous application of gravitational theory to the motions of comets. Many of these are described incidentally in the principal papers indexed in Section 5 of the Bibliography.

Makover reports that V. F. Myachin and O. A. Sizova have worked out and programmed a method of simultaneous integration of equations of celestial mechanics, based on ideas by Teylor and Steffensen. In this method each comet coordinate is expanded into a series containing, instead of high-order differences of the function f, the derivatives to any order. The method is particularly suitable for integrations carried out with a variable step. Makover also reports the work of G. T. Yanovitskaya, who has derived a set of formulae for the method of variation of arbitrary constants for the case of nearly parabolic motion.

E. ROEMER

BIBLIOGRAPHY

An effort has been made to include articles from the general literature which have come to attention through communications from authors, supplemented by a non-comprehensive search of the literature. With respect to determination of orbital elements of individual comets, the annual report on comets presently prepared by B. G. Marsden and published in the *Quarterly Journal* of the Royal Astronomical Society includes far more comprehensive references than the present report.

1. *General*

Everhart, E., Raghavan, N.   1970, Changes in Total Energy for 392 Long-Period Comets, 1800–1970, *Astr. J.* **75**, 258.
Hasegawa, I.   1968, Catalogue of Periodic Comets (1967), *Mem. Coll. Sci., Kyoto Univ., Ser. Phys., Astrophys., Geophys., Chem.* **32**, 37. (Alternate sets of orbital elements.)
Marsden, B.G.   1967, One Hundred Periodic Comets, *Science* **155**, 1207. (General discussion of current problems.)
Sekanina, Z.   1968, Supplementary Catalogue 1 of Original and Future Comet Orbits, *Astr. Inst. Charles Univ., Prague, Publ. Ser. II*, no. 56. Related publication by the same author: An Interference Effect in the Calculation of Original and Future Comet Orbits, *Bull. astr. Inst. Csl.* **18**, 369.

2. *Astrometric Observations*

Bejtrishvili, I.   1967, Observations of Comets Kopff (1963 *i*) and Everhart (1964 *h*) at Abastumani Observatory, *Bull. Inst. teor. Astr.* **11**, 286.
Bronnikova, N.   1967, Photographic Observations of the Comet Alcock (1963 *b*) at Pulkovo, *Bull. Inst. teor. Astr.* **11**, 288.
Haupt, H., Schroll, A.   1967, Photographische Positionsbestimmungen des Kometen Kilston (1966 *b*), *Ann. Univ. Sternw. Wien* **27**, 123.
Kurpińska, M., Michalec, A.   1968, Positions of Comet Kilston 1966 *b*, *Acta astr.* **18**, 321.
Milet, B., Soulié, G.   1967, Positions de petites planètes et de comètes, *J. Observateurs* **50**, 219.
Mintz, B.   1968, Observations of Bright Minor Planets and Comets, *Astr. J.* **73**, 49.
Morkowska, B.   1969, Observations of Comets Made at the Poznan University Observatory, *Acta astr.* **19**, 85.
Polyakova, T., Romashin, G.   1967, Observations of Comet Everhart (1964 *h*) at Burakan, *Bull. Inst. teor. Astr.* **11**, 287.

Velthuyse, F.H.M., Wisse, P.N.J.   1969, Photographic Positions of Minor Planets and of Comet Barbon (1966 c), *Bull. astr. Inst. Netherlds., Suppl.* **3**, 117.

3. *Orbits of Short-Period Comets*

Belyaev, N.A.   1967, The Orbital Evolution of Comets Neujmin 2 (1916 I), Comas Solá (1927 III), Schwassmann-Wachmann 2 (1929 I) over the 400 Years from 1660 to 2060, *Astr. Zu.* **44**, 461.
Brady, J.L., Carpenter, E.   1967, The Orbit of Halley's Comet, *Astr. J.* **72**, 365.
Brady, J.L.   1967, Note Regarding Nongravitational Forces on Halley's Comet, *Astr. J.* **72**, 1184.
Herget, P.   1968, Revised Orbit of Comet Schwassmann-Wachmann 1, *Astr. J.* **73**, 729.
Herget, P.   1968, Ephemerides of Comet Schwassmann-Wachmann 1 and the Outer Satellites of Jupiter, *Cincinnati Obs. Publ.*, no. 23.
Kamieński, M., Sitarski, G.   1967, Comet P/Wolf 1 in 1958–1968, *Acta astr.* **17**, 73.
Kazimirchak-Polonskaya, H.I.   1967, Quelques problèmes actuels de l'astronomie cométaire du point de vue de la mécanique céleste contemporaire, *Trudy Inst. teor. Astr.* **XII**, 3.
Kazimirchak-Polonskaya, H.I.   1967, Perfectionnement de la théorie de la comète Wolf 1 pendant la période 1918–1925 contenant son rapprochement de Jupiter en 1922, *Trudy Inst. teor. Astr.* **XII**, 24.
Kazimirchak-Polonskaya, H.I.   1967, Evolution de l'orbite de la comète Wolf 1 durant 400 ans, 1660–2060. Investigations préliminaires, *Trudy Inst. teor. Astr.* **XII**, 64.
Kazimirchak-Polonskaya, H.I.   1967, Sur la possibilité de l'origine commune des comètes Wolf 1 (1884 III) et Barnard (1892 V). Investigations préliminaires, *Trudy Inst. teor. Astr.* **XII**, 86.
Khanina, F.B.   1968, (Investigation of the Motion of Faye comet), *Problems of Cosmic Physics* (Kiev) **4**, 152.
Marsden, B.G., Aksnes, K.   1967, The Orbit of Periodic Comet Kearns-Kwee (1963 VIII), *Astr. J.* **72**, 952.
Rasmusen, H.   1967, The Definitive Orbit of Comet Olbers for the Periods 1815–1887–1956, *Copenhagen Obs. Publ.*, no. 194.
Schrutka-Rechtenstamm, G.   1968, Definitive Bahnbestimmung des ersten periodischen Tempelschen Kometen, *Ann. Univ. Sternw. Wien* **27**, 188.
Sitarski, G.   1967, The Orbit of the Periodic Comet Harrington 1953 VI, *Acta astr.* **17**, 321.
Sitarski, G.   1968, The Improved Orbit and the Ephemeris of the Periodic Comet Kopff (1906 IV) for its Reappearance in 1970/71, *Acta astr.* **18**, 155.
Sitarski, G.   1968, The Orbits and the Ephemerides of the Periodic Comets Tsuchinshan 1 (1965 I) and Tsuchinshan 2 (1965 II), *Acta astr.* **18**, 163.
Sitarski, G.   1968, The Orbit and the Ephemeris of the Periodic Comet Slaughter-Burnham (1958 VI) for 1969/70, *Acta astr.* **18**, 419.
Sitarski, G.   1968, An Attempt to Link the Appearances of Comet P/Perrine-Mrkos in 1955 and 1961/62. *Acta astr.* **18**, 423.
Sitarski, G.   1968, The Orbit and the Ephemeris of the Periodic Comet Kearns-Kwee (1963 VIII) for its Reappearance in 1972/73, *Acta astr.* **18**, 429.
Sitarski, G.   1969, Ephemeris of Comet Grigg-Skjellerup for its Reappearance in 1971/72, *Acta astr.* **19**, 175.

4. *Orbits of Nearly Parabolic Comets*

Barteneva, O.N.   1969, Definitive Orbit of the Comet Whipple-Fedtke-Tevsadse 1943 I, *Bull. Inst. teor. Astr.* **11**, 585.
Branham, R., Jr.   1968, Orbit of Comet 1961 V (Wilson-Hubbard), *Astr. J.* **73**, 97.
Mamedov, M.A.   1969, Definitive Orbit of the Comet 1957 III Arend-Roland, *Bull, Inst. teor. Astr.* **11**, 598.
Sekanina, Z.   1967, Future Orbits for Ten Comets of the General Catalogue of Original and Future Comet Orbits, *Bull. astr. Inst. Csl.* **18**, 1.
Sekanina, Z.   1967, Definitive Orbit of Comet Pereyra (1963 V), *Bull. astr. Inst. Csl.* **18**, 229.
Yanovitskaya, G.   1968, Definitive Orbit of the Comet 1959 IV Alcock, *Bull. Inst. teor. Astr.* **11**, 544.
Yanovitskaya, G.   1969, Original and Future Orbits of the Comet 1959 IV Alcock, *Bull. Inst. teor. Astr.* **11**, 705.

5. *Theoretical Investigations*

Aver'yanova, T.V., Stanyukovich, K.P.   1967, Application of General-Relativity Methods to the Study of Long-Period Comet Trajectories, *Astr. Zu.* **43**, 1301.

Chebotarev, G.   1966, Cometary Motion in the Outer Solar System, *Astr. Zu.* **43**, 435.

Everhart, E.   1967, Comet Discoveries and Observational Selection, *Astr. J.* **72**, 716.

Everhart, E.   1967, Intrinsic Distributions of Cometary Perihelia and Magnitudes, *Astr. J.* **72**, 1002.

Everhart, E.   1968, Change in Total Energy of Comets Passing Through the Solar System, *Astr. J.* **73**, 1039.

Everhart, E.   1969, Close Encounters of Comets and Planets, *Astr. J.* **74**, 735.

Hamid, S., Marsden, B.G., Whipple, F.   1968, Influence of a Comet Belt beyond Neptune on the Motions of Periodic Comets, *Astr. J.* **73**, 727.

Hamid, S.E.   1969, Influence of a Cometary Belt on Uranus and Neptune, *Smithsonian Astrophys. Obs., Special Report*, no. 299.

Kazimirchak-Polonskaya, H.I.   1967, Evolution of the Short-Period Comet Orbits from 1660 to 2060, and the Role of the Outer Planets, *Astr. Zu.* **44**, 439.

Kazimirchak-Polonskaya, H.I.   1968, Rôle des Planètes Extérieures dans l'Évolution des Orbites des Comètes, *L'Astronomie* **82**, pp. 217–27, 323–39, 432–8.

LePoole, R.S., Katgert, P.   1968, The Major-Axis Distribution of Long-Period Comets, *Observatory* **88**, 164.

Lyttleton, R.A.   1968, On the Distribution of Major-Axes of Long-Period Comets, *Mon. Not. R. astr. Soc.* **139**, 225.

Marsden, B.G.   1967, The Sungrazing Comet Group, *Astr. J.* **72**, 1170.

Marsden, B.G.   1968, Comets and Nongravitational Forces, *Astr. J.* **73**, 367.

Marsden, B.G.   1969, Comets and Nongravitational Forces. II, *Astr. J.* **74**, 720.

Marsden, B.G.   1970, Comets and Nongravitational Forces. III, *Astr. J.* **75**, 75.

Marsden, B.G.   1970, On the Relationship between Comets and Minor Planets, *Astr. J.* **75**, 206.

Pittich, E.M.   1969, The Selection Effects on the Discoveries of New Comets, *Bull. astr. Inst. Csl.* **20**, 85.

Pittich, E.M.   1969, Sudden Changes in the Brightness of Comets Before Their Discovery, *Bull. astr. Inst. Csl.* **20**, 251.

Sekanina, Z.   1967, On the Origin of the Kreutz Family of Sungrazing Comets, *Bull. astr. Inst. Csl.* **18**, 198. Also see: Problems of Origin and Evolution of the Kreutz Family of Sungrazing Comets, *Acta Univ. Carolinae, Math, Phys.*, No. 2, 33 = *Publ. astr. Inst. Charles Univ.*, no. 51.

Sekanina, Z.   1968, Splitting of the Primary Nucleus of Comet Ikeya-Seki, *Problems of Cosmic Physics* (Kiev) **3**, 66.

Sekanina, Z.   1967, Nongravitational Effects in Comet Motions and a Model of an Arbitrarily Rotating Comet Nucleus I. Hypothesis, *Bull. astr. Inst. Csl.* **18**, 15; II. Push-effect, *Bull. astr. Inst. Csl.* **18**, 19; III. Comet Halley, *Bull. astr. Inst. Csl.* **18**, 286; IV. Comet Splits, *Bull. astr. Inst. Csl.* **18**, 296; V. General Rotation of Comet Nuclei, *Bull. astr. Inst. Csl.* **18**, 347; VI. Short-Period Comets. Empirical Data, *Bull. astr. Inst. Csl.* **19**, 47; VII. Short-Period Comets. Analysis, *Bull. astr. Inst. Csl.* **19**, 54. See also Nongravitational Effects in Comet Motions and a Model of an Arbitrarily Rotating Comet Nucleus, *Problems of Cosmic Physics* (Kiev) **3**, 82, 1968.

Sekanina, Z.   1968, Non-Gravitational Forces and Comet Nuclei, *Sky and Telescope* **35**, 282.

Sekanina, Z.   1968, Dynamical Effects of Explosive Phenomena in Comet Halley and its Nuclear Rotation, *Problems of Cosmic Physics* (Kiev) **3**, 75.

Sekanina, Z.   1968, Disruption of Comet P/Biela and Explosive Mechanisms of Cometary Splits, *Bull. astr. Inst. Csl.* **19**, 63.

Sekanina, Z.   1968, Motion, Splitting and Photometry of Comet Wirtanen 1957 VI, *Bull. astr. Inst. Csl.* **19**, 153.

Sekanina, Z.   1968, Dynamics of Comet Alcock 1963 III and its Enhanced Activity, *Bull. astr. Inst. Csl.* **19**, 170.

Sekanina, Z.   1968, Anomalous Comet Burnham 1960 II, *Bull. astr. Inst. Csl.* **19**, 210.

Sekanina, Z.   1968, A Dynamic Investigation of Comet Arend-Roland 1957 III, *Bull. astr. Inst. Csl.* **19**, 343.

Sekanina, Z.   1968, Non-Gravitational Impulses on 20 Short-Period Comets, *Bull. astr. Inst. Csl.* **19**, 351.

Sekanina, Z.   1968, Existence of Meteor Showers Associated with Short-Period Comets Anomalous in Motion, in *Physics and Dynamics of Meteors*, L. Kresák and P. Millman (eds.), D. Reidel Publ. Co., Dordrecht (IAU Symposium No. 33).

Sekanina, Z.   1968, On the Perturbations of Comets by Nearby Stars I. Sphere of Action of the Solar System, *Bull. astr. Inst. Csl.* **19**, 223; II. Encounters of Comets with Fast-Moving Stars, *Bull. astr. Inst. Csl.*, **19**, 291.

Sekanina, Z.   1969, Total Gas Concentration in Atmospheres of the Short-Period Comets and Impulsive Forces Upon Their Nuclei, *Astr. J.* **74**, 944.

Sekanina, Z.   1969, Dynamical and Evolutionary Aspects of Gradual Deactivation and Disintegration of Short-Period Comets, *Astr. J.* **74**, 1223.

Sekanina, Z., Vanysek, V.   1967, Irregularities in the Motion of Comet Halley in 1910 and its Physical Behavior, *Icarus* **7**, 168.

Sitarski, G.   1968, Approaches of the Parabolic Comets to the Outer Planets, *Acta astr.* **18**, 171.

Vsehsvjatskij, S. K.   1968, On the question of existence of Oort's comet cloud, *Problems of Cosmic Physics* (Kiev) **3**, 98.

6. *Computing Techniques*

Benima, B., Cherniack, J. R., Marsden, B. G., Porter, J. G.   1969, The Gauss Method for Solving Kepler's Equation in Nearly Parabolic Orbits, *Publ. astr. Soc. Pacific* **81**, 121.

Illyinsky, I.   1968, The Comparison of Computations of the Comet Orbit by the Gauss' Method and the Method Offered by the Author, *Vestnik Kiev Univ., Ser. Astr.*, No. 10.

Mamedov, M. A.   1966, On the Computation of a Nearly Parabolic Orbit from Three Observations of a Comet, *Bull. Inst. teor. Astr.* **10**, 549.

Sitarski, G.   1968, On the Barycentric Method of Computing the Perturbations, *Acta astr.* **18**, 149.

Sitarski, G.   1968, Digital Computer Solution of the Equation of Position of a Comet on a Keplerian Orbit, *Acta astr.* **18**, 197.

COMMISSION 21: LIGHT OF THE NIGHT-SKY
(LUMINESCENCE DU CIEL)

Report of Meetings, 20 and 24 August 1970

PRESIDENT: M. Huruhata.
SECRETARY: J. L. Weinberg.

Business Meeting

The President asked Commission members and guests to stand in memory of Commission members who passed away since the Prague meeting:

S. Chapman
J. Dufay
C. T. Elvey
C. Hoffmeister
A. Lebedinskij.

1. THE DRAFT REPORT

The Draft Report of Commission 21 was approved without change. The President thanked Commission members for forwarding information which enabled him to prepare the Report. F. E. Roach suggested that a full report be prepared and circulated to members in advance of future meetings. The abbreviated version required for publication in the Transactions would be condensed from this full Report.

2. OFFICERS AND MEMBERSHIP

The officers for the next triennium were nominated and approved by Commission members.

3. MISCELLANEOUS

After some discussion, it was agreed that the Commission should sponsor two colloquia: on integrated starlight and diffuse galactic light, and on zodiacal light. Heidelberg was suggested as a possible site for the starlight meeting in the summer of 1972, and Tenerife as the site for a zodiacal light meeting in 1973. Details of the locations and times of these colloquia will be handled by the Commission officers after contacting the membership.

J. L. Weinberg was asked to form a committee of Commission 21 to evaluate the use of modern techniques of star counting in a renewed effort to map the Milky Way. The committee will prepare a report and will submit its recommendations to the Commission at its 1973 meeting.

Commission 21 emphasis in recent years on geophysical problems was questioned, and it was agreed that greater emphasis should be placed on astronomical aspects of the light of the night sky. Present and anticipated research programs discussed in the sessions for scientific papers suggest that this change has already taken place.

Scientific Meeting

One session was held for presentation of scientific papers. At the request of the President, the meeting was chaired by F. E. Roach. The papers, in outline:

1. H. Elsässer: On the Problem of the Absolute Night-Sky Brightness.

Comparison of the results of Elsässer and Haug (Z. Astrophys. **50**, 1960) and of Weinberg (Ann.

Astrophys. **27**, 1964) suggests that differences heretofore ascribed to calibration errors (Roach and Smith, NBS Tech. Note 214, 1964) instead arise from different corrections for atmospheric scattering (elongations, ε, less than 60°) and for star background (ε greater than 60°).

2. R. Dumont and F. Sanchez: Some Results and Present Research Subjects at Observatorio del Teide (Tenerife) about Zodiacal Light and Green Airglow Photometry.

During years of high solar activity, local fluctuations of up to 50% were observed in the brightness of zodiacal light. Evolution of the isophotes suggest that solar streams affect the spatial distribution of zodiacal dust. The zodiacal light at the north ecliptic pole (at 5000 Å) was found to have a total brightness of 70 S_{10} (vis) units and a brightness in polarized light between 9.7 and 12.6 S_{10} (vis). Results were also presented on the symmetry of zodiacal light polarization with respect to the Sun/anti-Sun axis and on the 5577/5200 line/continuum covariance in the nightglow.

3. J. L. Weinberg: A Coordinated Program of Satellite and Ground-Based Observations of the Zodiacal Light.

Description was given of two satellite experiments scheduled for 1972/73 (see *Trans. IAU* **XIVA** (Reports 1970), page 204). Also described was a planned ground-support program which will relate the satellite observations to concurrent observations from the ground and to earlier observations obtained in Hawaii. Interest was expressed by other observers in participating in a network of ground-support stations.

4. R. H. Giese: Current Model Calculations Concerning Zodiacal Light as seen from Space Probes.

Results were presented of theoretical models of zodiacal light as seen from space probes. Mixed-component and off-ecliptic models are also being calculated.

5. N. Kovar: Proposed Manned Spacecraft Observations of the Zodiacal Light and F-Corona.

A description was given of photographic experiments to be carried out from lunar orbit and from the Earth-orbiting Skylab. Programs at the University of Houston involve extensive calculations of Mie scattering functions and of zodiacal light and F-Corona models.

6. R. Robley: Remarques sur la couleur du Gegenschein.

Four-color (B, V, J, R) observations of the Gegenschein at Pic-du-Midi between 1965 and 1968 suggest that the Gegenschein is a little bluer than the Sun.

7. R. G. Roosen: The Gegenschein and Interplanetary Dust Outside the Earth's Orbit (read by J. L. Weinberg).

Photoelectric observations from McDonald Observatory in March 1969 and February 1970 are used in a search for the Earth's shadow in the Gegenschein. The absence of a shadow in these observations (to an accuracy of about one per cent) requires that the spatial density of material must increase outside the Earth's orbit. Model calculations suggest that the brightness distribution of the Gegenschein can be explained by dust produced by collisions between known asteroids. A cometary origin of the material is unlikely. One observation in February 1969 indicated that the Earth's shadow was visible in the center of the Gegenschein, and the result is attributed tentatively to increased dust near the Earth associated with the passage of the Earth through the orbital plane of P/Encke (see, also, IAU Circ. 2266).

8. N. Tanabe and K. Mori: A New Star Counting Device in Tokyo Observatory (read by M. Huruhata).

A new photoelectric, star-counting instrument has been developed to measure the diameters of star images on the *Palomar Sky Survey Atlas* to its limiting magnitude in blue and red. Operation of the instrument is described as are results of measurements in regions near the north celestial and ecliptic poles.

9. K. Mattila: Surface Brightness of Dark Nebulae and the Scattering Properties of Interstellar Grains (Diffuse Galactic Light).

UBV Observations of dark nebulae were made with a 50-cm Cassegrain at Boyden Station. Significant amounts of diffuse galactic light (25 to 80 S_{10} (vis) units) were found, and the results were used to infer the albedo of the interstellar grains.

10. N. K. Reay: $H\beta$ Emission in the Night-Sky.

Observations of $H\beta$ emission were made in northern Italy with a Fabry-Pérot interferometer. Three sources of $H\beta$ emission were detected: galactic (to 25° latitude), night-sky, and nebulosity associated with the diffuse Cetus Arc radio region.

11. R. D. Wolstencroft: Annual Variation of the Zodiacal Light.

Variations in nightglow polarization toward the north ecliptic pole (between December 1968 and October 1969) are attributed to an annual variation in the polarized component of zodiacal light. The maximum brightness in polarized light is 32 S_{10} (vis) units (June 15 ± 30 days) and the minimum is 7.5 S_{10} (vis) units (December 15 ± 30 days).

COMMISSION 22: METEORS AND METEORITES
(MÉTÉORS ET MÉTÉORITES)

Report of Meetings: 19 and 24 August 1970

PRESIDENT: Z. Ceplecha.
SECRETARY: I. Halliday.

First Session

The President called the business meeting to order with about 60 members and guests present. He explained the procedure for the appointment of officers for the 1970–73 term and the Commission approved the slate of officers including the Committee on Meteorites with E. Anders as Chairman and B. J. Levin as Vice-Chairman.

The chairman outlined the limitations on the length of the current Progress Report and invited corrections or additions to the Report. Members discussed the value of periodic review papers in meteor astronomy and also the complete list of references which had formerly been published in the reports.

The President opened discussion on the following proposals which he had received:

1. It was proposed that ICSU be asked to co-ordinate meteorite research directly. Members felt that there were already enough committees dealing with meteorite research and because of the importance of meteorites in the evolution of the solar system the logical place for such co-ordination was the Committee on Meteorites of Commission 22. With the rapid growth of meteorite research the Organizing Committee of the Commission was asked to consider by 1973 whether meteor and meteorite research should be split into separate commissions.

2. It was proposed that a second international symposium on meteorites be organized for 1971 or 1972. It was agreed that the Chairman of the Committee on Meteorites would consider the need for this, keeping in mind the possibility of the IAU acting as a co-sponsor of a symposium.

3. The Commission approved a proposal that an effort be made to identify 'unclassified' meteorites that have never been scientifically described. Members are urged to locate any such meteorites in their countries and to arrange for their investigation by modern methods. The Committee on Meteorites shall be responsible for the implementation of this resolution.

4. The Commission approved a recommendation that programs of visual meteor observations be continued, especially if these observations are co-ordinated with radio or photographic programs. The need for a more vigorous program of visual observations in the southern hemisphere is stressed.

The President stated that he intends to distribute one more up-to-date version of the mailing list for reprints. The value of this list was recognized and the incoming President will continue to maintain it.

C. L. Hemenway offered to act as host for a symposium on meteor physics if there was sufficient interest. Many members indicated an interest in such a meeting which might be held in Albany, New York, to fit with the 1971 meeting of COSPAR in Seattle, Washington, in June. The possibility of IAU sponsorship for the meeting will be investigated.

C. L. Hemenway and B. A. Lindblad stated that the Cosmic Dust Panel, now organized within COSPAR, should strengthen its relations with meteor astronomers of the IAU. The Commission approved the formation of a 'Committee on Interplanetary Dust' with C. L. Hemenway as Chairman and a Vice-Chairman to be named at the next session of the Commission.

R. K. Soberman outlined the probable need for a system of naming meteor streams discovered from space experiments, in particular streams which do not intersect the Earth's orbit.

T. R. Kaiser extended a warm welcome to members of Commission 22 to visit the University of Sheffield at the close of the General Assembly.

Second Session

About 50 members and guests were present at the second session of Commission 22. The President requested that any changes or corrections to the meteor mailing list should be forwarded to him promptly.

The President reported that the IAU might be able to support the proposed symposium on meteor physics in Albany to a maximum amount of $1600. The Commission appointed an organizing committee for the symposium consisting of C. L. Hemenway (Chairman), P. M. Millman, and A. F. Cook.

Millman described plans for a five-day symposium on Planetology to be held as part of the 24th International Geological Congress in Montreal, August 21–29, 1972. The IAU and Meteoritical Society are among the co-sponsors of the meeting.

Kresák reported that Commission 20 had supported a proposal to hold a conference on comets, asteroids, and interplanetary matter at Nice in 1972. The emphasis is expected to be on orbital studies.

The Commission approved the appointment of R. K. Soberman as Vice-Chairman of its new Committee on Interplanetary Dust.

The President announced that A. Hajduk and M. Šimek were preparing a catalogue of radar meteor equipment to be published in *Bull. astr. Inst. Czechoslovakia*. The deadline for inclusion in the list is December 31, 1970.

Ceplecha extended an invitation to all meteor astronomers to submit meteor papers for publication in the *Bull. astr. Inst. Czechoslovakia*. He also reported that the International Union of Amateur Astronomers had asked for the co-operation of Commission 22 in their programs and the Commission welcomed this request.

The President called upon the meeting to rise in a tribute to one of the Commission's consulting members, Prof. Dr J. Zähringer, whose death occurred shortly before the General Assembly.

The following short communications were presented before the close of the second session:

1. R. E. McCrosky: Photographic Meteorite Fall, Lost City.
2. I. Halliday: The Canadian Meteorite Camera Network.
3. P. M. Millman: Meteor Spectra Observed with the Image Orthicon.
4. A. F. Cook: Discrete Levels of Beginning Heights of Meteor Streams.
5. L. Kresák: On The Origin of Short-Period Meteors.

Third Session

Further short papers were presented at this session as follows:

6. Z. Sekanina: Statistical Model of Meteor Streams and its Application to the Association of a Meteor Stream with Asteroid Adonis.
7. B. A. McIntosh: Radar Observations of the 1969 Leonids.
8. D. W. Hughes, D. G. Stephenson: Diurnal Variations in the Mass Distribution of Sporadic Meteoroids.
9. E. Anders: Origin and Formation of Meteorites.
10. E. Anders: Meteorite Influx Rate on the Moon from Trace Element Studies on Apollo 11 and 12 Lunar Samples.
11. B. Hellyer: Mass Distribution in Meteorite Showers.
12. P. W. Hodge, D. E. Brownlee: Interplanetary Dust Detection and Collection.
13. J. A. O'Keefe: Tektites.
14. J. Hunter: The Production of Artificial Meteor Trains by Means of Shaped Charges.

The Vice-President, R. E. McCrosky, expressed the thanks of the entire Commission to the retiring President for his work on behalf of Commission 22 during the past three years, particularly for the preparation of the Report of the Commission and for arranging the sessions at the General Assembly. Dr McCrosky also expressed the appreciation of the Commission for the work of Drs Yavnel and Anders on the Committee on Meteorites.

COMMISSION 23: CARTE DU CIEL

Report on Meeting, 20 August 1970

PRESIDENT: W. Dieckvoss.
SECRETARY: H. Eichhorn.

Dieckvoss opened the meeting with the following statement:

J'ai préparé un texte français pour payer hommage aux contributions magnifiques de nos collègues français. Grâce aux activités des présidents français (Baillaud, Couderc, Sémirot) de la commission et en général de la dévotion exceptionelle des quatre observatoires en France qui participent à la formation du Catalogue Astrographique la tâche principale est finie: Nous possèdons un véritablement grand catalogue des positions précises des époques anciennes, données supérieures pour la détermination des mouvements propres des étoiles faibles comme une possibilité d'étudier la kinématique de la voie galactique et peut-être assister à fixer un système inertial. M. Lacroute vous donnera une résumé historique des travaux sur la Carte du Ciel.

Beaucoup de travaux sont à faire à l'avenir, mais, dans les activités des astronomes individuels avec l'assistance de la commission 24. Des constantes nouvelles des clichés publiés sont en progrès, les résultats du groupe à Bergedorf seront publiés à la fin de cette année. M. A. Günther vous peut montrer la forme des ces résultats. La coordination de recherche galactique (programme de Groningue) concernant la liste de Plaut sur des étoiles variables intéressantes n'a fait presque aucun progrès. M. Günther a quitté l'observatoire de Hambourg et il est maintenant membre du centre de calcul de l'université de Hambourg. Il n'est pas éligible par l'Union, mais il a plusieurs bandes magnétiques avec beaucoup de données et malgré la fin officielle du travail de la détermination des constantes définitives je pense qu'à l'avenir ces données seront plus valables et je propose que nous demandons M. Günther s'il voudrait accepter le rôle commission membre consultant dans le groupe de travaux dans la commission mère No. 24.

La carte propre et les cuivres de grandaire faits pour imprimer les cartes propres ne sont pas utilisés.

Lacroute recalled the history of the *Astrographic Catalogue* (AC). In 1887, sixty-five astronomers met at Paris when the initiation of the project was resolved. Subsequent meetings were held in 1889, 1891, 1896, 1900 and 1909. The presidents of the commission on the AC were H. H. Turner (1919–30), Esclangon (1930–35), Jules Baillaud (1935–55), P. Couderc (1955–61), Sémirot (1961–67) and W. Dieckvoss (1967–70).

The original project called for a map complete to the 15th photographic magnitude, and a catalogue of positions (rectangular plate coordinates) complete to the 12th.

Definitive plate constants were determined at Hamburg-Bergedorf for the zones $+89°$ to $+32°$, and at Strasbourg for the zones $+35° -2°$.

While the value of the AC as a source for reference positions for minor planet observations has diminished, it is constantly becoming more valuable as first epoch material for proper motion determinations.

Luyten offered the following resolution which was approved by acclamation. "Je voudrais encore exprimer mes félicitations, spécialement aux astronomes français d'avoir réussi et ayant achevé ce magnifique catalogue *La Carte du Ciel*".

Van Herk remarked that one could utilize the plates which were originally taken for the Charts, for determining the proper motions of faint stars.

Lacroute and *Dieckvoss* answered the sky was not completely covered by the Chart plates. A list of those plates that are available was given by Couderc.

Wood suggested that these plates should be repeated and would support such an enterprise.

Dieckvoss reported on the work on the determination of definitive plate constants at Hamburg. In these calculations, the systematic field corrections were assumed to be identical for all plates of a zone. (Tables which contain them will be published at the end of 1970.)

Bouigue presents the results of his investigations: Au cours des études entreprises sur les erreurs de champs en vue de l'exploitation du *Catalogue Photographique*, il a été remarqué que les clichés qui concernent la zône $\delta = +10°$, $0^h < \alpha < 12^h$ ne présentent aucun écart systématique entre les positions mesurées et les positions calculées (avec AGK2 ou *Catalogue de Toulouse*). Dans ces conditions, les courbes de fréquence de ces erreurs permettent de connaître la précision que l'on peut attendre sur les positions déduites du *Catalogue Photographique*. On trouve ainsi les erreurs moyennes quadratiques $0^s.02$ pour μ_α et $0''.2$ pour μ_δ.

Lacroute reported on the computation of right ascensions and declinations by the plate overlap method for the zones in the belt covered by the French observatories. Studies of the field were started for the Paris and the Toulouse zones. Also established were tables of the systematic errors of AGK2 and AGK3. At the epoch 1900, the smallest systematic errors result if the AG catalogues and the AC are reduced by the overlap method. The inclusion of meridian positions, i.e. essentially the AGK1, considerably decreases the local systematic errors.

Eichhorn remarks that the use of the AGK1 in the reductions of the AC plates is bound to reduce the residual systematic errors.

Dieckvoss agrees that this is so, in spite of the low accuracy of most of these catalogues, and suggests in addition the use of the selected areas.

Eichhorn points out that certain types of magnitude errors in the positions of the faint stars can be established (in the absence of grating images) only when the positions of equally faint reference stars are available.

Murray agrees that the Selected Areas, which contain stars down to between 14^m and 16^m, could eventually also be useful toward this purpose.

De Vegt reported on the remeasurement of old AC Vatican plates. Most of these are still in good condition. Using the new measurements, the mean error of one position obtained from averaging data on two plates is $0''.17$. This figure was obtained from comparing positions from overlapping plates. According to Günther, this figure (obtained from comparison with reference stars only) is $0''.32$ for the published positions.

Measuring difficulties occured with large and unsymmetrical images toward the edges of the plates.

Dieckvoss remarks that the remeasurement of the long *and* short exposures of the same stars would also help to establish the magnitude equation of the AC positions.

Lacroute is working on finding a fast and accurate measuring machine for remeasuring the plates taken for the AC. The difficulty is finances, but he hopes that he will have one in one or two years, which will also allow the measurement of plates taken on an astrograph with a large field.

Dieckvoss starts a discussion of the repetition of the old zones with the aim of obtaining proper motions of faint stars. Several observatories (Paris, among others) have already started repeating zones. Speaker thinks it would be better to use modern lenses and larger fields.

Wood feels that, if an instrument were to be designed from the beginning for photographic catalogues to follow the astrographic catalogue, it would be better to seek a long focal length rather than a large field. If the largest size of plate that can be tolerated is decided, say 40 cm, and the smallest field, say 4°, the focal length follows and since a large lens is not necessary the lens designer should not have a difficult problem to give good images over the whole field. The scale could be more than doubled.

Dieckvoss agrees.

Eichhorn suggests that reflection (or at least part reflection) optics would probably be cheaper than and superior to all refraction optics for astrometric purposes.

Luyten directs attention especially to big Schmidt telescopes. He can now provide the proper motions of all stars brighter than 21^m and north of $-33°$ with an accuracy corresponding to a probable error of $0''.007$ per year. These proper motions are relative to the system of the 19^m stars.

Dieckvoss announces that all measured coordinates in all zones are being key punched at some French observatories.

Lacroute states that a center for the collection of astrometric data has been set up at Strasbourg – one for that of photometric data at Lausanne. He plans to order the stars by their coordinates and to establish a table of identical stars, and a comparison of various systems. He invites collaboration to avoid duplication and to coordinate the work.

Dieckvoss points out some problems connected with the storage of data. Cards must be duplicated every 4 yr, tapes every year.

Murray proposed a vote of thanks to Dieckvoss for his efforts in making the results of the AGK3 available in advance of publication. This was seconded by

Mikhailov who stated that Dieckvoss' advance information had been very valuable at Pulkovo.

Dieckvoss announced the construction of a compilation catalogue which will be available on cards, and based on the information from AGK2-AGK3-AC. This catalogue contains BD numbers, positions for the equinox 1950.0, the central date, the proper motion components and their mean errors: for right ascension both in seconds of time per year and seconds of arc times the cosine of the declination.

The discontinuation of Commission 23 was agreed upon.

HISTORIQUE DE LA COMMISSION DE LA CARTE DU CIEL

Cette Commission a son origine dans l'*Enterprise internationale de la Carte du Ciel*. On ne saurait comprendre le caractère de ses travaux, leur difficulté, ses déboires, sans rappeler brièvement l'histoire de cette entreprise au cours des 32 années qui ont précédé la naissance, en 1919, de la Commission 23.

Depuis l'invention de la *photographie*, divers essais avaient été faits pour l'appliquer à l'étude du Ciel. Mais, en ce qui concerne les étoiles, ce sont vraiment les admirables clichés des frères Paul et Prosper Henry, obtenus à l'Observatoire de Paris entre 1880 et 1885, qui révélèrent aux astronomes du monde entier la puissance et les possibilités de la nouvelle technique: la rapidité et la précision de l'enregistrement d'innombrables étoiles sur un seul cliché donnèrent aussitôt envie d'établir une Carte et un Catalogue photographique du Ciel. Appuyé par Sir David Gill et par Otto Struve, l'amiral Mouchez, directeur de l'Observatoire de Paris, organisa en 1887, dans cet observatoire, le premier *Congrès International de la Carte du Ciel*, où seize nations furent représentées par 65 astronomes. Cet événement marque l'introduction systématique de la photographie en Astronomie. On ne doit pas oublier que cette entreprise a permis la mise au point et la consécration des méthodes qui ont rendu possible le développement de l'astronomie sidérale.

Rappelons aussi que l'entreprise de la *Carte du Ciel* a préludé à ces grandes Unions Internationales, qui régissent heureusement aujourd'hui les relations scientifiques dans tous les domaines.

Les buts de la Carte du Ciel sont bien définis par les premières résolutions votées:

(a) Dresser une carte générale du Ciel pour l'époque actuelle et obtenir les données qui permettent de fixer, avec la plus grande précision possible, les *positions* et les *grandeurs* de toutes les étoiles jusqu'à un ordre déterminé (les grandeurs étant entendues dans un sens photographique à définir);

(b) Pourvoir aux meilleurs moyens d'utiliser, tant à l'époque actuelle que dans l'avenir, les données fournies par les procédés photographiques.

Le Congrès de 1887 avait prévu que beaucoup de questions difficiles seraient à résoudre pour mener à bien l'entreprise. Il établit un *Comité permanent*, dont le président n'a cessé d'être le Directeur de l'Observatoire de Paris. Ce Comité s'est réuni en 1889, 1891, 1896, 1900 et 1909. En ce temps-là, devant la faiblesse reconnue des mouvements propres de presque toutes les étoiles et la lenteur des changements d'éclat, on pensait qu'il faudrait des siècles pour *"pénétrer les mystères du monde sidéral"*. C'est pourquoi on voulut *a priori* que les documents recueillis fussent *durables à l'échelle séculaire*: on pouvait soit *conserver l'image* inaltérée, soit *mesurer* toutes les positions et les grandeurs et en dresser le *Catalogue*. Le congrès décida l'emploi des deux procédés: ce furent la Carte et le

Catalogue. Les instruments devaient être des réfracteurs du type réalisés par les frères Henry, donnant des champs de 2° de côté. On avait prévu que la voûte céleste devait être couverte *deux fois* pour l'un et l'autre travail: deux séries de clichés pour la Carte, deux séries pour le Catalogue – les centres des clichés de chaque série devant coïncider avec les sommets des clichés de l'autre.

LA CARTE

Sur la Carte, on devait photographier les étoiles jusqu'à la quatorzième grandeur inclusivement *"selon l'échelle en usage en France"* et sous réserve d'une définition photographique ultérieure.

Pour cette Carte, une pose de 30 min parut suffisante pour fournir les étoiles de quatorzième grandeur. En fait, cette durée a donné, en gros, les images jusqu'à $m = 15,0$. Mais pour éviter de confondre ces images avec les imperfections de la couche sensible du support ou du procédé de reproduction, on jugea nécessaire de faire sur chaque cliché trois poses légèrement décalées, de manière que chaque étoile donne trois images, sommets d'un triangle équilatéral. La durée totale de pose était ainsi portée à 1 h 30, ce qui impliquait environ 2 h de temps clair pour chaque cliché. Sous le climat de la plupart des observatoires interessés, au voisinage ou même à l'interieur de certaines grandes villes comme Londres ou Paris, les intervalles de temps propices égaux au moins à 2 h sont relativement rares (l'éclairage urbain ne rendait pas encore impossibles à cette époque ces longues poses).

Pour la conservation indéfinie des documents, on ne voulut pas se fier à la permanence des couches sensibles ni aux possibilités de renouvellement par contretypage. Le procédé retenu fut l'*héliogravure* sur planches de *cuivre*. Malheureusement le travail est fort coûteux, très difficile en raison de la précision exigée et, en définitive, seuls des artisans français l'ont convenablement exécuté. Les dépenses de la Carte gravée durent donc se faire en France – tandis que les mesures, calculs et publications comportaient des dépenses faites dans le pays même où le cliché était pris.

Nous trouvons là déjà les causes matérielles qui ont empêché la *Carte du Ciel* d'atteindre son but: la *longueur des poses*, prohibitive pour les clichés à prendre dans la mauvaise saison, en des stations souvent mal situées, le malheureux choix de l'*héliogravure sur cuivre* trop coûteuse, moins précise que l'original et impropre aux mesures directes, et finalement exécutée à Paris pour tous les pays qui l'ont réalisée.

Le fait est que *nulle part* la double série des clichés *Carte* (centres de déclinaisons paires et impaires) n'a été exécutée. On s'est contentée d'une seule série à 3 poses, certains pays choisissant des centres pairs, d'autres des centres impairs, comme on le verra dans le tableau récapitulatif. Dans certaines zones, on s'est contenté de clichés à une seule pose (de 40 mn). Enfin, il reste quatre bandes où aucun cliché Carte n'a été pris (de $+47°$ à $+54°$, de $+25°$ à $+31°$, de $-23°$ à $-17°$, de $-40°$ à $-32°$).

Quant aux reproductions héliogravées, elles couvrent un peu plus de la moitié du Ciel. En certaines zones (Greenwich, le Vatican en partie) on a publié des reproductions sur papier photographique fort. Ailleurs, la collection des clichés existe mais n'a pas été reproduite.

La situation de la Carte aurait sans doute été meilleure si, dès la decennie 1920–30, il n'était devenu évident que la structure de l'Univers se découvre grâce à des astres beaucoup plus faibles que ceux que la Carte enregistre: sur les clichés de la Carte on trouve bien peu de galaxies. Enfin, l'invention des télescopes de Schmidt, qui donnent dans des champs de 36 degrés carrés des images meilleures que celles des champs de 4° de la Carte, a permis de publier rapidement des atlas comme le *Sky-Atlas*, où figurent les astres jusqu'à la magnitude 20 au moins, avec indication sommaire de leur spectre grâce à la répétition des clichés en deux couleurs. Ces atlas remplissent les désirs des fondateurs de la Carte du Ciel beaucoup mieux que la Carte initiale.

Heureusement, le Catalogue photographique, qui, au début, paraissait une opération accessoire, a connu un destin meilleur, quoique tardif; il donne aujourd'hui la promesse d'applications nombreuses et la collection de ses clichés anciens est une source de documents encore unique.

LE CATALOGUE

La série de clichés à courtes durées de pose devait "assurer une grande précision dans la mesure micrométrique des étoiles et rendre possible la construction d'un *Catalogue*". On décida de se borner à des poses suffisantes pour enregistrer les étoiles jusqu'à la onzième grandeur. A cette fin, chaque cliché de cette seconde sorte comporterait 3 poses de 6 mn, 3 mn et 20 s, convenablement décalées. En fait, le Catalogue donne des images mesurables jusqu'à $m = 12,0$, en gros.

En 1887, on prévoyait seulement de fournir les (α, δ) des étoiles de repère, sans autre obligation. Les réunions suivantes demandèrent la mesure de *toutes* les étoiles jusqu'à la onzième grandeur et la publication des (x, y) et de (m) seulement. La publication des (α, δ) était laissée à la discrétion des observatoires.

Mais il était prévu, dès cette époque, qu'après l'achèvement du Catalogue, un nouveau catalogue résulterait de la discussion du premier, assurerait l'homogénéité de l'ensemble, combinant en une seule les diverses valeurs trouvées pour la position d'une étoile (deux au moins, et souvent davantage, grâce aux recouvrements entre clichés mitoyens).

C'est en 1891 que le Comité Permanent partagea la tâche entre 18 observatoires volontaires. Mais il y eut quelques défections: La Plata et Rio de Janeiro d'abord, puis Santiago. Plus tard, Potsdam publia un travail hâtif puis se désintéressa de l'entreprise: sa zone dut être redistribuée à d'autres. Ces défections furent compensées par de nouveaux volontaires: Hydérabad, Cordoba, Perth et Uccle, auxquels se joignirent Paris, Edimbourg, Oxford et Hambourg. Enfin l'Observatoire de Melbourne fut désaffecté. Sydney compléta ses manuscrits et Paris les imprima.

L'Union Astronomique Internationale (UAI) et la Commission 23

L'UAI fut formée en juillet 1919 à Bruxelles, lors d'une réunion du Comité International de la Recherche (International Research Council). Elle comprenait alors 32 commissions: celle de la Carte du Ciel prit le numéro 23, qui lui est toujours resté.

Le Comité Permanent de la Carte du Ciel cédait sa tâche et ses pouvoirs à cette Commission 23; le Président de ce Comité, Benjamin Baillaud, Directeur de l'Observatoire de Paris (qui devenait le premier président de l'UAI) recommanda, pour présider la Commission 23, H. H. Turner, directeur de l'Observatoire d'Oxford. Ce choix fut approuvé par l'UAI, et la Commission 23 commença sa tâche.

L'enquête faite par Turner sur l'état de l'entreprise est publiée dans le volume I des *Transactions de l'UAI* relatif à la 1ère Assemblée Générale (Rome, 1922).

Turner constate d'abord des retards considérables: non seulement on avait sous-estimé la longueur du travail, mais dès 1909 les observations et les mesures d'Eros avaient requis les équipes de la Carte, puis de 1914 à 1918 la guerre avait pratiquement suspendu l'activité de nombreux participants. Après la guerre, beaucoup d'astronomes avaient disparu et la pauvreté des laboratoires pesa longtemps sur la reprise des travaux.

Turner insista avec raison sur la priorité à donner au Catalogue et l'entreprise paraissait devoir aboutir enfin lorsque la seconde guerre mondiale ruina à nouveau maints participants.

Après 1945, Jules Baillaud (président depuis 1935 de la C.23) joua le même rôle d'animateur que H. H. Turner après la première guerre. Et le travail du Catalogue reprit, avec des difficultés financières et administratives nombreuses. Il fallut battre le rappel des bonnes volontés en tous pays. Il faut souligner que, depuis 1928, devant le coût de l'entreprise, l'UAI a accordé d'importantes contributions financières pour l'impression des Catalogues, lors de chacune de ses Assemblées. Pratiquement, les 20 derniers volumes imprimés ont été payés par elle. Mais il faut remercier aussi les observatoires et les pays participants qui ont financé, pendant si longtemps, des bureaux de mesures et de calculs pour établir les manuscrits.

C'est en 1964 seulement, à l'Assemblée Générale de Hambourg, que l'achèvement du Catalogue a pu être annoncé.

Le tableau joint à ce texte montrera mieux que de longs discours les aléas du Catalogue et l'état où la Carte a été laissée.

Présidents de la Commission 23

Désigné dès 1919, lors de la fondation de l'UAI à Bruxelles, *H. H. Turner* (Oxford) eut la lourde tâche de recueillir l'héritage du Comité Permanent, de faire le bilan de l'entreprise après la première guerre mondiale et d'en relancer les activités jugées rentables, en particulier le Catalogue. Il demeura président jusqu'à sa mort (1930).

E. Esclangon (Paris) lui succède jusqu'en 1935.

J. Baillaud (Paris) devait ensuite assumer la présidence pendant 20 années, de 1935 à 1955. Il eut le mérite, après la seconde guerre, de rallier les bonnes volontés pour achever le Catalogue et, après sa retraite, il consacra ses forces jusqu'à sa mort (en 1960) à l'impression, à Paris, des volumes tardifs.

Paul Couderc (Paris) de 1955 à 1961, puis Pierre Semirot (Bordeaux) de 1961 à 1967, virent se terminer le Catalogue et s'amorcer son perfectionnement et ses applications.

Depuis 1967, W. Dieckvoss (Hambourg) est le président en exercice.

BILAN ET PERSPECTIVES

Le *Catalogue Photographique* couvre désormais le ciel entier et fournit les mesures de toutes les étoiles jusqu'à la magnitude 12,0, la magnitude de ces astres n'ayant qu'une valeur indicative, suffisante pour les identifications; les constantes des clichés sont qualifiées de provisoires.

On s'est demandé longtemps, devant le travail à fournir, s'il y avait lieu de calculer les constantes *définitives* des clichés, en augmentant le nombre des étoiles-repères, en tenant compte de leurs mouvements propres les plus récemment publiés, et en les rapportant toutes à un même système de référence.

L'opération était nécessaire si l'on voulait transformer les millions d'étoiles du Catalogue, à leur tour, en autant d'étoiles-repères fort utiles, par exemple pour les observateurs de comètes, de petites planètes ou pour les réductions des clichés poussant à des magnitudes bien plus élevées, comme les Cartes ou Atlas modernes.

L'entrée en usage des ordinateurs a tranché la question. Le calcul des constantes définitives a été entrepris déjà pour l'hémisphère Nord et le sera, à brève échéance sans doute, pour le Sud.

Préalablement au calcul de ces constantes, il a paru nécessaire de corriger les x, y mesurés des erreurs systématiques (coma, astigmatisme, courbure de champ, si possible équation de magnitude, etc...). Des études préliminaires (faites à Hambourg, à Paris, à Helsinki, à Bordeaux, à Strasbourg, etc...) ont montré qu'il serait sans doute possible d'établir, pour les diverses zones, ou pour d'importantes séries de clichés, des formules relativement simples, aptes à corriger les imperfections principales des x, y. P. Lacroute (Strasbourg) et l'équipe de Hambourg (Kox, Dieckvoss, Günther) Eichhorn et d'autres ont publié d'intéressantes études, générales ou particulières à ce sujet.

Après l'amélioration de ces données, les constantes définitives, calculées au moyen de l'AGK3 et de ses mouvements propres, en utilisant quand c'est possible plus de vingt étoiles-repères, pourront, semble-t-il, fournir la position d'une étoile *bien mesurée* avec une erreur moyenne ne dépassant pas $\pm 0''.16$, en tenant compte des deux clichés où elle figure. (Pour beaucoup d'étoiles, grâce aux chevauchements de clichés, on pourrait trouver plus de deux images.) Bien entendu, rien ne saurait remédier aux mauvaises mesures, ni aux erreurs fortuites de faible amplitude.

Une heureuse constatation: les clichés à 3 poses de la Carte sont parfaitement mesurables (sauf s'il s'agit de leurs étoiles brillantes, dont les images s'imbriquent) et on pourra les réduire au moyen des étoiles faibles du Catalogue.

Le Catalogue a déjà fait découvrir un grand nombre de *binaires* physiques et l'on va déterminer les mouvements propres de toutes celles dont la séparation est inférieure à une minute de degré (1').

Le Catalogue a déjà conduit à la détermination de nombreux *mouvements propres*. La recherche des mouvements propres d'étoiles d'intérêt *spécial* (comme celles du programme de Groningue, 1953) sera sans doute l'un des premiers fruits du Catalogue.

Bientôt, toutes les étoiles de l'hémisphère Nord auront chacune leur fiche perforée et le calcul de leurs (α, δ) ne soulèvera guère de discussions. De même, lorsque les astrographes à grand champ et des appareils de mesure *automatique* des clichés seront en service régulier, la détermination de *tous*

δ	Observatoire volontaire en 1891	Remarque	Observatoire de remplacement	Catalogue	Carte
90° 65°	Greenwich			Travail complet	Clichés à une pose de 40mn reproduits sur carton, agrandis × 2
64° 55°	Vatican			Travail complet	Héliogravure pour ⅓ des champs + 55°, + 56° + 57° – pour le reste sur carton, agrandis × 2 et distribués
54° 47°	Catane			Travail complet (Collection de clichés en mauvais état)	Clichés non pris
46° 40°	Helsingfors (Helsinki)			Travail complet	Clichés pris mais non publiés
39° 32°	Potsdam	Zone publiée hâtivement puis abandonnée	39°, 38°, 37°, 36° 35° et 34° 33° et 32°	Hydérabad Uccle prend les clichés, Paris: mesures et calculs, Paris imprime Hambourg achève les mesures, Paris imprime	Clichés pris, héliogravés et distribués par Uccle
31° 25°	Oxford			Travail complet	Clichés non pris
24° 18°	Paris			Travail complet	Clichés à 3 poses + 24° + 22° + 20° héliogravés Zones
17° 11°	Bordeaux			Travail complet	reproduits et distribués + 18° + 16° + 14° + 12° paires
10° 5°	Toulouse			Travail complet	reproduits et distribués + 11° + 9° + 7° + 5° Zones
+ 5° − 2°	Alger			Travail complet	reproduits et distribués + 3° + 1° − 1°
− 3° − 9°	San Fernando			Travail complet	reproduits et distribués − 3° − 5° − 7°
− 10° − 16°	Tacubaya			Travail complet	reproduits et distribués − 11° − 13° − 15° − 16° impaires

Zone	Station		Station		Observations
— 17° / — 23°	Santiago	Défection	Hydérabad	Travail complet	Clichés non pris
— 24° / — 31°	La Plata	Défection	Cordoba	Travail complet	Clichés pris mais non publiés (sauf héliogravure de la zone — 25°)
— 32° / — 40°	Rio de Janeiro	Défection	Perth	Prend tous les clichés et publie les zones — 32° à — 37° Edimbourg: mesures et calculs — 38°, — 39°, — 40° Paris imprime Vol I — 40°, Vol II — 39°, Vol III — 38°	Clichés non pris
— 41° / — 51°	Le Cap				Clichés pris (avec une seule pose?)
— 52° / — 64°	Sydney				
— 65° / — 90°	Melbourne	Désaffecté en cours de travail	Sydney	A pris tous les clichés – a publié les Vol I, II, III et VIII Les Vol IV — 71° et — 72°, V — 73° et — 74°, VI — 75°, — 76°, — 77° ont été mesurés à Sydney et imprimés à Paris ainsi que le Vol VII — 78°, — 79°, — 80°, — 81°	mais non publiés.

les mouvements propres du Catalogue, si on l'estime utile, sera beaucoup moins coûteuse et laborieuse qu'on ne pensait naguère.

En ce qui concerne la Carte, nous avons vu que, mises à part 4 zones, de 6° chacune en δ, les clichés à longues poses existent (3 poses en général, parfois une seule). Certaines collections sont sans doute mal conservées. Mais, là où les clichés demeurent, ils portent les imgages jusqu'à $m = 15,0$ (en gros) et fournissent de bonnes possibilités d'astrométrie.

Une bonne moitié du ciel a été reproduite par héliogravure. Une partie du reste a été reproduite sur carton. Ailleurs, enfin, les clichés existants n'ont pas été reproduits. Il a été convenu de laisser la Carte en son état actuel, l'intérêt des héliogravures et des reproductions distribuées ayant en partie disparu, si l'on sait conserver les plaques. Des atlas plus modernes (comme le *Sky Atlas*) suffisent aux identifications, fournissent l'indice de couleur et poussent beaucoup plus loin l'investigation de l'Univers.

L'enterprise de la Carte et du Catalogue n'en a pas moins jeté les bases de la coopération internationale dans la science, introduit dans la pratique l'emploi de la photographie, montré ses difficultés, ses méthodes, ses limites. A ceux qui y ont coopéré avec désintéressement, il convient de penser avec gratitudes et les résultats que fournira, dans un proche avenir, le Catalogue, justifieront assurément leur persévérance.

P. COUDERC

COMMISSION 24: STELLAR PARALLAXES AND PROPER MOTIONS (PARALLAXES STELLAIRES ET MOUVEMENTS PROPRES)

Report of Meetings 19, 20, and 22 August 1970

PRESIDENT: W. J. Luyten.
SECRETARY: C. A. Murray.

The President, after welcoming members to the meeting of the Commission, invited discussion on the draft report. The report from the Leander McCormick Observatory had not been received in time for inclusion in the draft report.

Frederick reported that the parallax programme had been continued at a rate of 4000 plates per year since 1967, dealing mainly with Vyssotsky red dwarfs and members of the Hyades. Proper Motions and Strömgren photometry were being determined for high latitude G and K stars and a device for measuring positions and separations of double stars was put into use.

Giclas reported that all proper motions in the Lowell Survey had been stored on an IBM 1130 disk preparatory to printing in a catalogue which would also give UBV photometry, parallaxes and further references.

The draft report was accepted.

Vasilevskis stated that, with the increasing rate of discovery of faint proper motion stars, there was an increasing difficulty in naming them. He proposed that the President appoints a working group to consider nomenclature for such stars, and to prepare a proposal for the next General Assembly. *Eichhorn* remarked that any new system of identification should depend only on the star's position. *Dieckvoss* suggested that *Gliese* be nominated to serve on this working group, and the President announced that he was appointing *Gliese*, and *Hoffleit* as members of this working group, with *Murray* as chairman.

The President then introduced discussion on the proposed merger with Commission 23. With the completion of the Carte du Ciel the work of that Commission has been successfully completed. He himself had taken the initiative in suggesting the amalgamation, after discussion with the President of Commission 23, and the Vice-Presidents and Organizing Committees of both commissions. *Eichhorn* proposed that Commission 23 be absorbed with the sub-committee of Commission 8 on star catalogues, to which *Vasilevskis* remarked that though he was a member of that sub-committee, he had never attended a meeting. *Murray* expressed the view that the work of Commission 23 and 24 overlapped in many respects with that of Commission 8 and no re-organization of astrometric activities within the Union should be carried out without consultation with that Commission. *Vasilevskis* then moved that Commission 24 was in favor of a merger with Commission 23 – this was carried.

The President announced the names of those nominated as officers of the new, merged, Commission, for approval by the General Assembly.

PRESIDENT: S. Vasilevskis.
VICE-PRESIDENT: P. Lacroute.

The President then proposed that the working group on spectroscopic parallaxes be disbanded. After a brief discussion *Blaauw* remarked that the problem of calibration of luminosity criteria still existed. *Vasilevskis*, while agreeing with this, said that it was invariably decided nowadays not to include spectroscopic parallaxes in a parallax catalogue; it was the business of the spectroscopists to determine luminosities, and the need for calibration could be met by a joint working group of astrometrists and spectroscopists, if required. The President then said that if no formal proposal was made for its retention, the working group would cease to exist.

At the joint meeting of Commissions 23 and 24 the following were nominated to serve on the new organizing committee: A. N. Deutsch, W. Dieckvoss, P. Herget, V. V. Lavdovskij, W. J. Luyten, C. A. Murray, K. Aa. Strand, P. van de Kamp, H. W. Wood. *Lacroute* remarked that the committee, as suggested, did not include anyone with interest specifically directed toward the Southern Hemisphere, to which *Strand* replied that the U.S. Naval Observatory had recently undertaken to set up its own astrometric camera to photograph the southern sky, a project which would be carried out within the next five years. *Vasilevskis* added that there should be no problem as the President could always co-opt new members to the Committee. *Lacroute* spoke about the advantages of using space vehicles for astrometric work, in particular for measuring the absolute rotation of the coordinate frame of reference and in determining absolute parallaxes. *Dieckvoss* reported on some recent work in which he used the réseau measures in the Astrographic Catalogue to reduce the star measures to a strictly orthogonal system, thus reducing the number of plate constants required in making overlap solutions.

At the third meeting *Bidelman* presented the report of the working group on spectroscopic parallaxes, which had been set up at the Commission's meeting at Prague, under the chairmanship of *Morgan*. Several members of the working group had contributed. In particular, the chairman drew attention to the large range in precision between various estimates of 'MK' types by different observers, and to the possibility of systematic differences between the northern and southern hemispheres; since, even under optimum conditions, the absolute magnitudes of early-type stars (09-A2) could only be estimated from MK types to within a probable error of about $0^m.45$ which was of the same order of accuracy as estimates from Hβ and Hγ photometry. *Keenan* reported on the calibration of luminosity classes of late-type giants and supergiants; in particular all late-type stars brighter than $m = 5$ should have trigonometric parallaxes measured. In the brief discussion which followed, *Gliese* supported *Morgan*'s point that MK classifications should be used with care, especially those estimated from objective prism plates.

Fracastoro then proposed that a symposium be held to discuss the 'Astrophysical Needs of Astrometry'. There was general approval of this idea. *Strand* remarked that UBV photometry had been carried out for all the parallax stars and some of the comparison stars in the first list of 1000 parallaxes obtained with the USNO 61-in. reflector, but R-I colors of the red stars would be required to distinguish between main-sequence stars and sub-dwarfs.

Upgren reported on the re-activation of the Van Vleck parallax program which consisted mainly of Vyssotsky stars. *Klemola* reported preliminary results from the Lick pilot programme of measuring proper motions relative to galaxies. Secular parallaxes have been measured for the three magnitude groups $m_B = 9.1, 11.7$, and 15.9. Values of the Oort constants ($A = +4.7 \pm 6.7$, $B = -7.1 \pm 5.7$, m.e.) derived from high-latitude stars were poorly determined, but even so, the smallness of B did not confirm Aoki's value. *Panyatov* presented a communication from *Deutsch*, in which he appealed for photographic observations of a list of 153 faint fundamental stars in the -25 to -90 declination zone, in order to relate meridian proper motions to those derived with respect to galaxies. This was part of the Pulkovo KS7 plan.

Thomas briefly described progress with the Herstmonceux parallax programme, in particular, the procedure for continuously monitoring the progress of the observations which was welcomed by *Vasilevskis*. Finally, *Lacroute* spoke about the systematic errors which are likely to occur in the Lick Program.

In addition to the regular meetings of Commission 24 which are reported here, joint meetings were held with Commission 8 and 40 on 'Fundamental Systems and Radio Astronomy', and with Commissions 27, 33, and 37 on 'The Absolute Magnitude of the RR Lyrae Stars'.

COMMISSION 25: STELLAR PHOTOMETRY
(PHOTOMÉTRIE STELLAIRE)

Report of Meeting, 20 August 1970

PRESIDENT: A. W. J. Cousins.
VICE-PRESIDENT: D. L. Crawford.

Business and Scientific Meeting

Business

The proposed officers of the Commission were approved.

STANDARDS FOR THE UBV SYSTEM

A. W. J. Cousins reported that the Working Group appointed at the Prague Meeting had held no actual meetings but conducted business through correspondence. Its terms of reference had been to arrive at 'a new definition of the UBV-system' (*Proceedings* 1967, 142). Hitherto the system had been defined by the stars in the Johnson and Harris list (1954, *Astrophys. J.* **120**, 196 and 1955, *Astrophys. J.* **121**, 779), but these stars were not all observable from the Southern Hemisphere. Cousins made the following proposal: "that the non-variable HR stars brighter than visual magnitude 5.0 and located between $\pm 10°$ declination be adopted as primary standards for the UBV system, with the fainter HR stars in the same zone as secondary standards to assist in defining the colour system". This was accepted by the meeting.

Copies of a memorandum putting forward the case for the equatorial HR stars may be obtained on application to the Cape Royal Observatory. The available UBV photometry of equatorial stars in the *Bright Star Catalogue* has been collected and mean magnitudes and colours are to be published, probably in Royal Observatory Annals. A list of the stars brighter than 5.0 has already appeared as a Cape Royal Observatory Mimeogram and a further compilation will be found in *Commun. Lunar Planet. Lab.*, No. 63. Of the two, the Cape compilation has appreciably smaller random errors, and Johnson has given his approval to the adoption of the Cape list as standards for V and $B - V$.

POSSIBLE ALTERNATIVES TO THE U SYSTEM

A. W. J. Cousins reported that there have been complaints about difficulties in reproducing $U - B$ with apparently similar equipment and filters. Some observers use the Corning 9863 filter as originally specified for U, others the Schott UG 2 (or UG 1) on account of its lower red leak. Contributory factors are mirror reflectivity, atmospheric transmission and the method used for correcting for the latter. Without knowledge of the wavelength variation of spectral intensity at the shortwave end of the U band outside the atmosphere it is impossible to 'reduce' ground-based observations rigorously to outside the atmosphere. It is simpler to reduce all observations to a fixed air mass. If all observers were to adopt similar procedures, then we should end up with mutually consistent results. Johnson, while admitting deficiencies in the present $U - B$, was opposed to introducing any minor changes.

J. Stock agreed that changes would be premature, and advocated more theoretical and observational work. Johnson's $U - B$ measurements are supposed to be 'extra-atmospheric' but are not in fact completely so, and comparison with the Cape 'inside atmosphere' measurements gave residuals that depend on declination.

LISTING OF SEQUENCES SUITABLE FOR CALIBRATION OF PHOTOGRAPHIC PLATES

The President requested Bok and Argue to contact interested parties with a view to drawing up lists of suitable sequences (cf. *Reports* 1970, 234, 245).

Scientific Reports

V. Straižys – 'The Vilnius System for Photometric Three-Dimensional Classification of Stars'. An eight-colour intermediate-band system obtainable with glass and interference filters and photo-electric or photographic photometry. Region 3000–6600 Å. (cf. *Bulletin Vilnius Astr. Obs*, 1970, Nr. 28, 29).

E. E. Mendoza and T. Gómez: 'BVR, Potsdam and Harvard Photometries'. Intercomparison by multiple regression analysis (*Bol. Tonantzintla y Tacubaya* 1969, **5**, 111).

J. Stock: 'Reduction Procedures'. Broad-band colour transformations derived from observations of unreddened stars are not generally applicable to reddened stars (cf. A. Gutiérrez-Moreno and H. Moreno: 1970, *Astron. Astrophys.* 7, 35). Conventional methods for extinction correction may leave considerable residuals, especially in $U - B$. A more rigorous method has been proposed by Gutiérrez-Moreno et al. (1966, *Publ. Dept. Astr. Univ. Chile* 1, 1).

C. Jaschek: 'Information Problems in Photometry and Possible Solutions'. Discussion followed from Jaschek's contribution to the Report (p. 231). Members highly commended the *U.S. Naval Observatory Photoelectric Catalogue* (V. M. Blanco et al., 1968. *Publ. USNO* **21**) but no formal resolution was passed urging continuation or extension. Stoy urged the need for critical judgment in compiling such data. Hauck referred members to the work of the European Centre de Données Stellaires de Strasbourg.

G. E. Kron: 'Stellar Photometry with the Navy Electronic Camera'.

M. F. Walker: 'Application of Electronography to Stellar Photometry'. (*Sky Telesc*, 1970, **40**, 132).

G. Wlérick: 'Photométrie d'objects très faibles identifiés avec des radiosources du Catalogue 3CR'. (Soumis à Astronomy and Astrophysics).

R. V. Willstrop: 'High Speed Photoelectric Photometry'. (cf. *IAU Symp*. **46**: 'The Crab Nebula').

K. Serkowski: 'Development of Ten-Channel Polarimeter'. A Wollaston prism polariser, dichroic filters and ten photomultipliers and integrators. Tested at Siding Spring.

N. G. Roman: 'Comparison of Roman's Photometry of Moderately Faint Stars with Measurements by other Observers'.

W. Liller: 'Photometry of CH Cygni'. Well known rapid variation in UV gradually diminished and disappeared during 1970.

COMMISSION 26: DOUBLE STARS (ÉTOILES DOUBLES)

Report of Meetings, 25 August 1970

PRESIDENT: P. Couteau.
SECRETARY: S. L. Lippincott.

The session was opened by President Couteau. Greetings were conveyed from Finsen and Arend, who were not able to attend. Names of officers were proposed and accepted by the Commission. New Commission members were accepted. The Draft Report was accepted; several further comments on recent progress were made. *Worley* commented on the double star Card Catalogue deposited in Washington. Since the Draft Report, 18 new lists of measures, involving some 6800 cards have been added to the Catalogue. Although unpublished observations are not added to the Catalogue, it remains of interest to submit lists of observations prior to publication since it serves as a signal to help catch the publication when received in the USNO library.

P. Morel described 'The Atlas of Orbits of the Nice Observatory', which includes a computer traced diagram of the orbit indicating the part of orbit observed. A sample copy was made available. Those further interested should consult P. Couteau or P. Morel.

The Circular of P. Muller giving lists of new orbits has also always been considered to serve as a communication outlet for the Commission. Fuller use of the latter aspect would be achieved by giving such information as number of observations not yet published; bibliography of new publications; changes of addresses, etc.

The concern in the long delays in publication of observations was discussed; a less expensive and expedient method must be found in a publication which will be assured maintanance by libraries. The discussion resulted in the resolution given at the end of this report.

The continuing need for communication with astrophysicists on particular double stars which hold answers to problems otherwise unattainable. Of special interest are binaries containing high luminosity, or subdwarf components, classes for which masses are very uncertain. *Luyten* cited the astrophysical interest in wide doubles containing white dwarf components. He further said it is evident that, at least at present, these constitute the only group of objects from which eventually the masses of white dwarfs can be determined – by statistical methods, using orbital motions, but this requires very large telescopes. Photometry is also an essential parameter for the study of double stars. *Franz* reported recent progress at the Lowell Observatory of magnitude determinations of high luminosity components off the main sequence, and also of systems with variable component. The need for contact with other commissions to set up exchange of demands was expressed by *Dommanget*. *Eggen* pointed out that the astrophysicists have the problem of being able to critically select the proper information from publications on double stars.

The need for high dispersion radial velocity observation of binaries with special interest was discussed. *Batten* proposed that astronomers with high dispersion equipment be encouraged to join Commission 26; *Couteau* suggested that Dommanget work with Batten (Commission 30) to establish a list of individual radial velocities for orbital analysis. *Batten* expressed interest in seeing a new edition of the Dommanget-Nys Catalogue of ephemerides of radial velocities which is now nearing the end of its usefulness.

Specialized methods of double star observations are being developed such as from occultations. *R. Hanbury Brown* reported on observations of the double-lined non-eclipsing spectroscopic binary α Virginis, as an example of the information which can be obtained with an intensity interferometer. The observations have been compared with a theoretical model of a binary star to find the inclination of the orbit, angular size of the semi-axis major and of the primary, brightness ratio, position angle of the line of nodes, and sense of orbital motion. By combining the parameters known spectroscopically with those gained above, the parallax to the system was found with an uncertainty of about 4%, along with other information about the components, including their masses. By building a larger and more sensitive interferometer it would be possible to extend the measurements to fainter binaries and to obtain information about their distances, masses, radius and absolute magnitude.

The President expressed the concern of the competition with other programs on large telescopes. It is imperative to maintain observations both in the Northern and Southern hemispheres. Perhaps a list of telescopes, which could be used for double star measures, would be useful. Refractors should be preserved and not allowed to run down; the restoration of a good refractor (Nice) is less expensive than a new telescope. *Djurkovic* announced the possibility of an exchange of staff between the observatory at Belgrade and an observatory in another country-exchange involving only personnel with no monetary considerations.

The President discussed the interest in having another colloquium on double stars, which had been suggested by Arend and Dommanget in the March 1970 Circulaire of Muller. The following title conveys the scope of the subject matter which should be discussed: 'Orbital and Physical Parameters of Double Stars'. It seems appropriate for this colloquium to be held in 1972, possibly at the Sproul Observatory, or Coïmbra.

The proposal made by Dommanget in the Draft Report (p. 256) suggesting that the methods of orbit calculations and parallax derivations be given in the publications was approved. Dommanget will undertake to carry out proposal no. 2 (p. 256) of the Draft Report concerning a statistical study of couples in the Jeffers-Van den Bos Catalogue as a function of separation, magnitude and Δm.

RESOLUTIONS ADOPTED

1. Commission 26 realizes the necessity in the near future to compile homogeneous data for visual double stars for which the following information is known: all orbital elements, photometric, spectroscopic and astrometric; this data will serve as a basis for research on mass-luminosity relation on stellar evolution and related problems. (This was first proposed at *IAU Colloquium* No. 5 on Visual Double Stars at the Nice Observatory September 1969.)

2. Commission 26, facing similar problems for publishing double star observations as Commission 42, supports its resolution concerning the desirability for editors of all journals to accept and even request for publication complete lists of double star observations.

COMMISSION 27: VARIABLE STARS (ÉTOILES VARIABLES)

Report of Meetings, 19 and 21 August 1970

PRESIDENT: L. Detre.
SECRETARY: W. S. Fitch.

During the Brighton General Assembly, the Commission conducted one administrative session, one general scientific meeting, and two specialized sessions devoted to the spectra of variable stars and to the absolute magnitudes of the RR Lyrae stars.

Business Meeting

Following the introduction of the Secretary and of the Interpreter (J. Jung for French), the President remarked briefly on the replies he had received to his request for information to be included in the draft report, and then opened the meeting to general business.

M. W. Feast requested that the Committee on the Spectra of Variable Stars (Chairman: Feast) be dissolved. After a brief discussion, this request was formally approved by the Commission and has been transmitted to the Executive Committee.

The Commission approved a request from *Mrs H. B. Sawyer-Hogg* that the Committee on Variable Stars in Clusters (Chairman: Mrs Sawyer-Hogg; Members: G. H. Herbig, M. W. Feast, L. Rosino) be renamed the Committee on Variable Stars in Globular Clusters. *Herbig* noted that this action should be brought to the attention of Commission 37.

A brief report of the Working Group on Flare Stars (Chairman: P. F. Chugainov; Members: A. D. Andrews, R. E. Gershberg, Sir Bernard Lovell, V. Oskanian) was read by the President. The Commission then approved the request by Chugainov that this Working Group be continued, with their observing activity being especially concentrated on YZ Cmi, UV Cet, EV Lac, and AD Leo.

The Commission also supported the President's motion that the Working Group on Variable Stars in the Magellanic Clouds (Chairman: S. C. B. Gascoigne; Members: S. Gaposchkin, B. Kukarkin, J. Landi Dessy, P. Th. Oosterhoff, A. Thackeray, Sir Richard Woolley), presently engaged in the preparation of a Catalog of Magellanic Cloud variables, be continued.

Zwicky's request that the Working Group for Research on Supernovae (Chairman: F. Zwicky; Members: V. A. Ambartsumian, Ch. Bertaud, A. A. Boyarchuk, L. Detre, E. A. Dibaj, G. Haro, B. V. Kukarkin, H. Lambrecht, L. Rosino, K. Rudnicki, J. L. Sersic, A. D. Thackeray, B. E. Westerlund, P. Wild) be moved from Commission 28 to Commission 27 was supported by *Rosino*, and then approved by Commission 27.

The President raised the question as to whether pulsars should be included in the domain of specific interest of Commission 27. After a discussion in which it was pointed out that there is currently only one known optical pulsar, whose variability may be due entirely to rotation and/or a binary state, it was unanimously agreed that no action should be taken on this question until future developments in pulsar research clearly indicate an appropriate decision.

The President presented to Commission 27 a proposal submitted to the Executive Committee of the Union by the U.S.S.R. National Committee for Astronomy, requesting the creation within the IAU of a Sub-commission on the Physics of Unstable Stars. *Eggen* stated that this appeared to be a new name for the present Commission 27. After vigorous discussion, it was unanimously agreed that the membership of Commission 27 opposed the formation of such a Sub-commission, pending full clarification of the distinction between the physics of variable stars and the physics of unstable stars.

Plans for two new colloquia were announced, pending final approval by the Executive Committee of the IAU. The Vth Variable Star Colloquium, on 'New Directions and Frontiers in Variable Star

Research', is planned to be held at Bamberg, D.B.R. in late August or early September, 1971. A Colloquium on 'Observational Aspects of Novae and Supernovae' is tentatively scheduled for 10–15 September 1971, at Padua and Venice, Italy.

The President read a brief report from Professor Kukarkin, and then he moved the following resolution:

"We, the members of Commission 27, express our great admiration and deep appreciation for the difficult and extremely important work being carried out by our colleagues in the Moscow Bureau, and we request that the IAU grant an annual subvention of $ 1000 toward publication costs of the *General Catalog of Variable Stars*, edited by B. V. Kukarkin, P. N. Kholopov, and Yu. N. Efremov."

The Commission unanimously endorsed this resolution.

The President brought to the attention of the Commission the following requests:

1. *Van Herk* reported experiencing difficulties in a program for improving proper motions of long period variables, due to poor photometric elements. *Mrs Mayall* was requested to send to the *Information Bulletin* for rapid circulation any new photometric data on these stars.

2. *Van Hoof* requested that cooperative observing programs on β CMa stars be carried out at observatories situated in different longitudes, especially in the southern hemisphere.

3. *Cesevich* requested simultaneous photometric and spectroscopic observations of two RW Aurigae-type stars, FG and FH Aql.

4. *Opolski* requested simultaneous UBV and velocity measures of δ Cephei-type variables, and urged those who plan to make velocity measures to circulate their plans as early as possible via the *Information Bulletin*, so that photometric observers can arrange to observe at the same time.

5. *Mumford* requested finding charts for old novae, and the Commission expressed its approval of the work on the preparation of finding charts which is being carried out by Dr. Bertaud.

A letter from the late Dr A. V. Nielsen was read by the President. Dr Nielsen had requested Commission 27 to publish the long series of observations of southern variables by the late A. W. Roberts. *Detre* will publish in the Information Bulletin the list of stars observed by Roberts. *Feast* suggested that photocopies of the observations be made available to interested workers.

The President reopened a proposal initiated by W. J. Miller at the Prague meeting, to the effect that a new scheme be devised for the rapid naming of newly discovered variables. After a short discussion in which *Herbig* stated that he and Kukarkin opposed any present change, the proposal was tabled.

In conclusion, the President proposed and the Commission approved that names of new officers of Commission 27 be forwarded to the Executive Committee of the Union.

JOINT SESSION OF COMMISSIONS 24, 27, 30, 33 AND 37

CHAIRMAN: B. J. Bok.
SECRETARY: W. S. Fitch.

The proceedings of this Joint Meeting will be published in the volume *Highlights of Astronomy* 1970. The following papers were presented:

Sir Richard Woolley: On the Determination of the Absolute Magnitudes of the RR Lyrae Stars.
R. F. Christy: Absolute Magnitudes of RR Lyrae Stars.
G. van Herk: Review of Observational Data on RR Lyrae Stars.
S. V. M. Clube: Absolute Magnitudes of RR Lyrae Stars.
A. R. Klemola: The Lick Observatory Program of Proper Motions of RR Lyrae Stars.

COMMISSION 28: GALAXIES

Report of Meetings, 19 and 20 and 21 August 1970

PRESIDENT: G. C. McVittie
SECRETARY: E. B. Holmberg

The first session dealt with business matters, the second and third sessions with presentation and discussion of a series of papers on extragalactic research. In addition, a joint meeting with Commission 40 was held on 20 August.

Business Meeting

The printed Report presented by the President on research work in the extragalactic field during the past three years was gratefully acknowledged by the Commission. A number of members expressed the wish that in future reports the abbreviations of titles of scientific periodicals should conform to the rules previously established by the IAU.

The President announced that in its first session on 18 August the General Assembly had decided to create two new Commissions: No. 47 on Cosmology, and No. 48 on High Energy Astrophysics. The Commission expressed its approval of this decision, on the condition that, if possible, separate meetings of Commissions 28 and 47 never be arranged simultaneously.

The President proposed, and the Commission approved, the list of names to be sent to the Executive Committee of the Union and those of persons who should form the Organizing Committee of the Commission.

On account of the difficulties in defining the boundary-lines between Commissions 28 and 47, the members of Commission 28 expressed the wish that the new President ask each individual member about his (her) preference as regards membership of the two Commissions.

The Commission accepted the proposal of Dr Zwicky that the Working Group on Supernovae should transfer from Commission 28 to Commission 27.

The business meeting was concluded by a discussion of general rules of proceedings that might possibly be established for future meetings of the Commission.

First Scientific Meeting

At this session the following seven papers were presented:
1. F. Zwicky: Compact Galaxies.
2. V. A. Ambarcumjan: Galaxies with Ultra-Violet Excess.
3. G. O. Abell: Giant Galaxies in Clusters.
4. J. Heidmann: Diameter-Luminosity Relation and Distance of the Virgo Cluster.
5. A. K. Alksnis: Novae in Outer Regions of M 31.
6. J. D. Wray: Velocity Field of Ionized Gas in M 82.
7. H. Oleak: Radio Emission from Supernova Remnants in Distant Galaxies.
Each presentation was followed by a short period of discussion.

Second Scientific Meeting

At the final session the following eight papers were presented:
8. W. A. Baum: Diameter-Redshift Relation.
9. K.-H. Schmidt: On the Mass Discrepancy of Clusters of Galaxies.
10. L. Bottinelli: The Large-Scale Spatial Distribution of Neutral Hydrogen in Galaxies.

11. J. E. Ejnasto: Proposal for 'Complex Investigation of the Galaxy and Nearby External Galaxies'.

12. M. H. J. Demoulin: Remarks on Markarian Galaxies.

13. H. C. Arp and F. Bertola: The Large Optical Extensions of Elliptical Galaxies.

14. G. de Vaucouleurs: Inclination Effects in Apparent Diameters of Galaxies.

15. A. Przybylski: Analysis of HD 32034 in LMC.

As before, each paper was followed by a short period of discussion.

Reports of Working Groups

The three Working Groups of Commission 28 have submitted the following reports on their activities.

The Working Group on Galaxy Photometry, with G. de Vaucouleurs as chairman, held two separate meetings in which reports and papers were presented by J. D. Wray, E. Vandekerkhove, P. Wild, G. de Vaucouleurs, J. E. Solheim, G. O. Abell, C. Fraser, F. Bertola, A. G. Pacholczyk, and P. D. Usher. The reconstituted membership consists of the following: G. O. Abell, H. D. Ables, G. F. Benedict, F. Bertola, J. H. Bigay, T. M. Borchkadze, W. Bronkalla, G. de Vaucouleurs, S. D'Odorico, C. Fraser, P. W. Hodge, E. B. Holmberg, E. K. Kharadze, E. J. Kibblewhite, I. R. King, G. E. Kron, C. D. Mackay, B. E. Markarjan, R. H. Miller, P. Nilson, H. Oleak, G. Paal, N. Richter, H. J. Rood, J. L. Sĕrsic (chairman), C. D. Shane, R. R. Shobbrook, J. E. Solheim, D. W. N. Stibbs, P. D. Usher, E. Vandekerkhove, S. Van den Bergh, B. A. Voroncov-Vel'jaminov, P. Wild, J. D. Wray.

The Working Group on Supernovae, with F. Zwicky as chairman, held one meeting, at which the decision was taken to accept the invitation of Commission 27 to transfer the Working Group from Commission 28 to Commission 27. Short reports were presented by L. Rosino, A. D. Thackeray, L. Detre, J. A. Hynek, and P. Wild; recent results and future plans were briefly summed up by F. Zwicky. The reconstituted membership consists of the following: V. A. Ambarcumjan, Ch. Bertaud, F. Bertola, A. A. Boyarchuk, E. Chavira, L. Detre, G. Haro, J. A. Hynek, E. K. Kharadze, B. V. Kukarkin, H. Lambrecht, J. B. Oke, L. Rosino, W. L. Sargent, J. L. Sĕrsic, A. D. Thackeray, B. E. Westerlund, P. Wild, F. Zwicky (chairman).

The Working Group on the Magellanic Clouds met on 22 August 1970 and short contributions were read from the following authors: J. Graham, P. A. Wayman, W. Tifft, T. Schmidt, Ch. Fehrenbach, A. Ardeberg, M. Walker, M. W. Feast, T. Lloyd Evans and J. Dachs.

The reconstituted membership of the Group is as follows: A. D. Thackeray (chairman), V. M. Blanco, J. Landi Dessy, Ch. Fehrenbach, S. C. B. Gascoigne, P. W. Hodge, G. de Vaucouleurs, P. A. Wayman, B. E. Westerlund.

With regard to future work, the need for regular nova searches in the Clouds, especially with objective prism, was stressed. It is probable that many novae occur without detection.

COMMISSION 29: STELLAR SPECTRA (SPECTRES STELLAIRES)

Report of Meetings, 19 and 25 August 1970

PRESIDENT: M. W. Feast.
SECRETARIES: W. P. Bidelman and L. Houziaux.

Business

The President announced with regret the deaths of two commission members: Armin J. Deutsch and Marianne C. Bretz.

Approval of the Draft Report (which few members had as yet seen) was asked and duly granted. The meeting considered that the reference problem had been satisfactorily handled in the Draft Report.

The newly-proposed Organizing Committee for the Commission was approved by acclamation.

The meeting continued with short reports by G. Cayrel on the working group concerned with equivalent-width measurement by R. Griffin on prospects for increased accuracy in this work, by C. M. Sitterly on matters connected with astronomical data centres, by G. Elste on a new compilation of results of stellar abundance analyses, and by E. Muller on the question of making available spectral Atlases and tracings of representative stellar spectra to the smaller and newer schools of astronomy.

The following resolution was proposed by W. P. Bidelman and carried with acclamation:

"Commission 29 of the International Astronomical Union, mindful of her great service to all astronomical spectroscopists, highly commends Mrs. Charlotte Moore Sitterly of the U.S. National Bureau of standards for her devoted labours in the correlation and dissemination of spectroscopic data, and expresses the earnest hope that her work in the preparation of tables of atomic energy levels and, especially, tables of the multiplet structure of the spectra of the various elements, will be vigorously prosecuted and the results made generally and inexpensively available in the future as in the past."

The following papers were read:

A. B. Severny: Measurement of Weak Magnetic Fields.

R. A. Bell: Spectrum Synthesis with a Computer.

P. S. Conti: Spectroscopic Studies of O-type Stars.

J. L. Greenstein: Helium Lines at the Hot End of the Horizontal Branch.

H. J. Wood: Balmer Line Behaviour of Magnetic and Related Stars.

Cape-Radcliffe Group (read by M. W. Feast): RY Sgr, Evidence for Pulsation and Variable Circumstellar Extinction.

T. D. Fay and A. A. Wyller: Photoelectric Observations of Stellar Line Profiles with a Pressure Scanning Spectrometer.

C. Van 't Veer: Travaux de L'équippe des Étoiles Am à Paris.

J. Hutchings: The Rotation and Extended Envelopes of Be Stars.

A. Ringuelet: Short Period Velocity Variation in πAgr and the Group of V/R Variables.

W. P. Bidelman: Some Interesting Stars Discovered in the Michigan Southern Hemisphere Survey.

V. L. Khokhlova: Silicon Abundances in Si Ap Stars, Especially HD 124227.

J. R. Angel, J. D. Landstreet: Polarization Studies of White Dwarfs.

In addition the following joint meetings with other Commissions took place.

Joint Meeting with Commission 36: 'How to Determine Abundances'. Chairman: Greenstein.

A. B. Underhill: What Does One Need to Know to Determine Abundances?

R. F. Griffin: How Accurate are Measurements of Equivalent Widths and Line Profiles?

B. E. J. Pagel: Uses and Limitations of the Differential Curve of Growth Method.

L. H. Aller: Use of Line Profiles in Abundance Work.

R. Cayrel: Comparison of High Dispersion Studies with Scanner Work.

A. Unsöld: General Survey of Recent Experimental and Theoretical Work on the Abundance Problem.

Joint Meeting with Commission 27: 'The Spectra of Variable Stars'

The meeting was devoted chiefly to Eta Carinae.

B. E. J. Pagel: Spectral Analysis of Eta Carinae.

A. Feinstein: The Stars Near Eta Carinae.

L. H. Aller: A Comparison of the Shell and Core Spectra of Eta Carinae and RRTel.

R. Viotti: The Excitation of the Emission Lines and Some Consideration of the Nature of Eta Carinae.

A. D. Thackeray: The Spectrum of Eta Carinae 8000–11 000 Å and a Comparison of RRTel and S. Dor with Eta Carinae.

In addition the following papers were read:

H. Maehara, Y. Fujita: Spectral Analysis of Long Period Variables.

G. H. Herbig, R. R. Zappala: The Infra-Red Spectra of NML Cygnus and the Leo Infra-Red Object IRC + 10° 216.

WORKING GROUP ON LINE INTENSITY STANDARDS

To increase the efficiency of this working group we suggest that somebody (e.g. Dr R. F. Griffin, if willing) should go to the McMath Solar Telescope and take photoelectric tracings with double pass of bright standard stars – (e.g. αBoo, εVir, αCMi).

The resulting reductions can then be used as a criterion by which to judge the accuracy of given telescope-spectrograph-observer reduction combinations at various observatories.

Suitable limited spectral regions should be investigated (the choice of which could be left to Dr Griffin). Tracings of these regions and tables of equivalent widths could then be published.

The efficiency of the subcommittee on line intensity standards will be substantially enhanced by a reduction of its membership.

We propose that the following should be members: R. F. Griffin, Koelbloed, Pagel, Wright.

GIUSA CAYREL DE STROBEL
Chairman of the Working Group

COMMISSION 30: RADIAL VELOCITIES (VITESSES RADIALES)

Report of Meeting, 19 August 1970

PRESIDENT: D. S. Evans.
SECRETARY: A. H. Batten.

The President in the chair.

Administrative:

Dr H. A. Abt, currently Vice-President, desired by letter to withdraw from office because of other heavy commitments. The Commission endorsed the proposed names of new officers.

No objection was raised to the text of the printed Commission report.

Scientific Communications:

Beardsley described observations of standard velocity stars at 40 Å mm^{-1}. No systematic errors exceeding accidental errors of measurement were found. Personal and flexure errors were under examination. The scarcity of standard velocity stars earlier than F2 and in declinations suited to Allegheny should be remedied.

Heard seeks faint standard velocity stars suitable for calibration of objective-prism observations. A total of 122 spectra at 12 Å mm^{-1} and nine at 15 Å mm^{-1} from Dunlap and Victoria cover 24 stars, 21 of which have at least four spectra each. Seventeen seem suitable for adoption as standards but seven spectra of each star are desirable. *Fehrenbach* has spectrograms of some. *Heard* and *Fehrenbach* will prepare a combined list.

G. de Vaucouleurs with his wife continues interference filter observations (as reported at *IAU Symp.* No. 30) of clusters, H II regions, and galaxies. For recession velocities less than 1100 km sec^{-1} results agree well with 21-cm values, but, thereafter blending of galaxy absorption H and K with night-sky H and K leads to underestimation of optical radial velocities by up to 30%. This has important practical implications for estimation of the Hubble constant.

Davis Philip based his conclusions on the relative accuracy of image-tube and conventional spectrographs on observations of faint A-stars at high galactic latitudes (Table 1).

Table 1

Spectrograph and Telescope	Dispersion at Hγ (Å mm^{-1})	Limiting mag. (Exp. 1^h, width 0.2 mm)	P.E. (km sec^{-1})
Mt. Wilson B (100-in.)	86	11.5	± 3
Westinghouse Fiber Optics (84-in. KPNO)	69	14	± 15
Carnegie Tube (84-in. KPNO)	50	15	± 9

Vera Rubin used an image tube in observing galaxies from a list by *M. S. Roberts* for which optical and 21-cm velocities differed appreciably. The new optical measures agree much better. Published optical velocities can be appreciably in error.

Thackeray described tests of the 2-prism Cassegrain spectrograph with slit rotated in the direction of atmospheric dispersion. Measures of faint standard velocity stars have been made. *Lloyd Evans* continues observation of Cepheid binaries (frequency 15%). Orbital elements of the single-lined binary hot sub-dwarf HD 49798 have been determined.

Batten described *Fletcher's* effort to obtain maximum precision from the Victoria coudé spectrograph at 2.5 Å mm^{-1}. A systematic difference of 0.1 km sec^{-1} has been found between velocities obtained using iron-arc comparison and discharge tube, the former being, apparently, correct. *Batten* will communicate this result to Commission 14.

Pedoussaut studies statistics of spectroscopic binary stars. Of 858 known, 260 have UBV measures. Distribution in the galactic plane is nearly uniform, but there is an asymmetry perpendicular to the plane. Additional photometry is needed to show if the effect is real.

Fehrenbach measures radial velocities in the Large Magellanic Cloud region. The programme covers 494 member stars between magnitudes 9 and 13.5, and 1890 galactic stars. UBV photometry and spectroscopy at 73 Å mm^{-1} are well advanced.

Hube recalled that at Hamburg *Bonneau* reported that stars of types B8 and B9 with GCRV entries of *d* and *e* quality showed an apparent systematic error. On *Thackeray*'s initiative eleven such stars had been observed at Pretoria, Toronto and Victoria. *Hube*'s measures of all the plates showed no significant differences between the observatories. The effect is not explained by measurement difficulties due to diffuse spectrum lines.

D. S. Evans described the arrangement of the pages in the revision of the GCRV, now complete to 17 h R.A. Many colleagues had helped in proof-reading typescript copies. He hoped to finish early in 1971.

A letter from *Abt* suggested that the Commission concern itself with minimizing instrumental and guiding errors, with the encouragement of more observational work, and with improvement of storage and retrieval of numerical values of known velocities. This last project would be closely connected with any astronomical data center that may be created.

COMMISSION 31: TIME (L'HEURE)

Report of Meetings, 19, 21, 24 and 26 August 1970

ACTING PRESIDENT: G. M. R. Winkler.
SECRETARY: C. J. A. Penny.

First Session

The session was opened by the Vice-President, *G. M. R. Winkler*, who had been asked to preside in the absence of the President, *F. Zagar*.

The President appointed Miss Penny, RGO, as Secretary.

The proposed names for officers of the Commission, as well as of new members, were approved by the members.

The President proposed that in the future the Director of the Bureau International de l'Heure (BIH) should be an ex officio member of the Organizing Committee. This was agreed.

The *President* reported that the following names had been proposed as consulting members: G. Becker and A. R. Robbins. They were approved by the members.

The *President* expressed regret that J. P. Blaser had resigned from the Commission.

The Draft Report was approved.

The Director of the BIH, B. Guinot, reported on matters of general interest and of an administrative character. The BIH is one of the permanent services of the Federation of Astronomical and Geophysical Services (FAGS), it is sponsored by the IAU, UGGI and URSI; the parent union is the IAU. The Directing Board presently includes H. M. Smith (IAU) as Chairman, F. Zagar (IAU), W. Markowitz (UGGI), B. Decaux (URSI, CCIR), P. Giacomo (CIPM), and B. Guinot, Director of the BIH.

A minor change of statutes had been made to make it clear that the BIH is only concerned with the scientific problems of time and not adopted legal times.

UT1 and the co-ordinates of the pole are now computed by the simultaneous use of time and latitude observations obtained from 76 instruments and are published in Circular D. Some improvements have been made such as weighting the observations and publication of the raw data as well as the smoothed data for every five days. Work has commenced on reduction of the observations prior to 1964 in order to obtain a homogeneous set of data.

Research centred on two main topics:

The best use to be made of the polar motion determinations deduced from the U. S. Navy Doppler satellite observations by the Dahlgren Polar Monitoring Service and the systematic differences between these results and the IPMS and the BIH results; and secondly, how to keep the reference system, '1968 BIH System', in the long term.

Computation of an international atomic time scale was considerably improved at the end of 1968 when international time comparisons became possible with the use of Loran-C. AT (BIH) is now uniform to about 1×10^{-13} over several years. It is now possible to maintain clocks in agreement with UTC, defined by an exact relationship with AT (BIH) to ± 1 microsecond. The Director of the BIH thanked the U.S. Naval Observatory and the U.S. Coast Guard for loans of equipment and for visits with travelling clocks and also the Paris Observatory and other French organizations for support.

The Director sought views on the need to continue publishing all the observed values of UTO and of the latitudes and their residuals from the BIH solution.

The Director expressed concern about the financial position of the BIH which receives an annual grant of 6000 U.S. $ from FAGS, but the real expenses of the computing service alone are more than 100000 U.S. $, and the difference in the income and expenditure is borne entirely by the Paris

Observatory. The BIH could not continue its work without some other financial support.
The following discussion on the report of the Director of the BIH took place:

In reply to a question from *J. Bonanomi*, the Director said that most of the BIH expenditure is due to salaries and that the costs of the determination of UT and AT are approximately equal.

In reply to a question from *I. Shapiro*, B. Guinot said that he was looking into the differences in the polar coordinates obtained by the three different organizations. He noted that the drift of the mean pole obtained by the IPMS was different from that of the BIH.

After some discussion on the need to publish information on erroneous time signal emissions *W. Markowitz* moved that the BIH should publish a list of agencies responsible for the various services; this was seconded by *H. Enslin* and approved by the members.

H. Enslin proposed that the material in the Annual Report of the BIH should be reduced by publishing figures at $\frac{1}{20}$ yr intervals only. B. Guinot said that it would be necessary to consider also the advice of Commission 19 on this matter. In reply to *R. G. Hall*, B. Guinot said that all observations were available at the BIH on magnetic tape. It was agreed by a small number of members to recommend that the Annual Report of the BIH be condensed.

The *President* urged Commission 31 to consider the scope of its activities. The Commissions of URSI, CCDS, and the IAU are, in many cases, attended by the same people, but the discussions are held in very different climates. Commission 31 should ensure that its views are made known to the Directing Board of the BIH.

The Chairman of CCIR IWP VII/I, H. M. Smith, explained that problems arose from the need for two time systems: UT based on the rotation of the Earth and a uniform atomic time scale based on the second. The control of radio time signals is the responsibility of the CCIR, and at Boulder, 1968, a Working Party had been set up to look into the problem of improving the system for radio time signals which would better suit the needs of users of both UT and AT. The Working Party had proposed that the signals should be emitted without offset, but that there should be steps of 1 second to keep the signals in general agreement with UT and that this system should commence on 1.1.72. The Working Party sought the views of Commission 31 on details of implementation.

During discussion on these points it was noted that if the 1 second steps were made on fixed dates then the signals could diverge from UT by up to 0.7 second. *W. Markowitz* said that he did not believe that a ship navigating by celestial means needed greater accuracy than one second. The *President* postponed discussion on this point until the joint discussion with Commission 4.

A. A. Mikhailov said that a correction is something that must be added so that he proposed:
Time signal + correction = UT

H. M. Smith thought that since $\Delta T = \text{ET} - \text{UT}$ there would also be great advantage in an analogous convention.

The meeting adjourned.

Second Session

At the request of the President, *J. Terrien* reported on the CCDS Meeting of 18–19 June, 1970. He commenced by reporting that the CCDS had expressed great appreciation of the work of B. Guinot as Director of the BIH. He then went on to explain the organization of which the CCDS is a part. The Conférence Générale des Poids et Mesures is a Governmental organization of which the CIPM is the executive body. The Bureau International des Poids et Mesures (BIPM) is a scientific laboratory of which he is the Director, nominated by the CIPM. The CCDS is one of seven consultative committees whose membership is decided by the CIPM. In 1967, when the second was defined in terms of an atomic transition, the CCDS had restricted themselves to defining the interval of time. Since then several international organizations had pressed the CIPM to define a scale of atomic time.

In reporting informally on the draft recommendations, J. Terrien stressed that the CCDS had still to report to the CIPM. He added that although the IAU had no power to change the recommendations he would note any views expressed by Commission 31 and report them at the October meeting of the CIPM.

CCDS recommendation S1 (1970) stressed the need for the adoption of an atomic time scale. Recommendation S2 (1970) reads as follows:

'Le Comité Consultatif pour la Définition de la Seconde propose de définir le Temps Atomique International (TAI) comme suit:

Le Temps Atomique International est la coordonnée de repérage temporal établie par le Bureau International de l'Heure sur base des indications d'horloges atomiques fonctionnant dans divers établissements conformément à la definition de la seconde, unité de temps du Système International d'Unités.'

Recommendation S3 (1970) which is concerned with the practical realization of the atomic time scale states that the duration of the interval unit must be as close as possible to the duration of the second of the SI at a fixed point on the Earth at sea level, that the scale must be as uniform as possible and that the scale unit would only be intentionally altered if it differed significantly from the duration of the second.

J. Terrien reported that the majority of members of the CCDS were in agreement that the BIH should continue to form the atomic time scale, but FAGS had already asked for financial support for the work from adhering governments.

The *President* opened the discussion by remarking that it was essential to know what was meant by the fixed point and suggested that the pole might be suitable. *G. Becker* said that it was not necessary to refer to a fixed point, that any point on an equipotential surface was as good as another. *The President* agreed that he would expect the same rates but he wondered if time would vary from one equipotential point to another. *J. Terrien* said the definition had been phrased for practical reasons and could very easily be changed.

G. Becker noted that the definitions clarified the origin and the scale unit, but did not say how to designate the scale values or the origin; he wondered if dates should be given according to the Gregorian calendar. *W. Markowitz* thought it better to use the Julian date. The *President* said the question of date should be discussed at the joint meeting with Commission 4; the question of a better definition of the origin had been discussed by the Organizing Committee and they had decided that nothing could be gained by trying to improve on the present one. *B. Guinot* said the origin was fixed by reference to the Earth itself and could not be defined more precisely and *N. Stoyko* confirmed that UT was not known to great accuracy.

G. Becker proposed that the next step should be to interest the CIPM in legal time scales. *H. Enslin* endorsed this and said it was important that all countries adopt the new system simultaneously and that it was fully publicized in all countries. The *President* remarked that adjustments to time signals had already been made without difficulty and that publication of information was an internal problem for countries; however he agreed that there would be an increasing need for better communications.

The *President* said that the Commission should review the scope of its activities and interests, for example, representation of the IAU on the CCDS was of increasing importance.

The *President* thanked J. Terrien and wished to join in the complimentary remarks to the BIH, who had done a marvellous job with very little funds. The Commission formally expressed its gratitude to the BIH.

P. L. Bender emphasized the importance of an international atomic time scale in new fields of astronomy, for example, lunar laser ranging, timing pulsars, etc., which already utilized the full accuracy of the atomic time scale.

The *President* reported that unfortunately the IAU had received no official communication from the CCIR of the recommendations and resolutions made at New Delhi. *D. H. Sadler* expressed amazement since a CCIR resolution stated specifically that the IAU should be informed. After further discussion *H. M. Smith* said that although the Commission could not comment on documents which had not been received he would like views on the six points he had raised in an earlier meeting.

The *President* proposed that the discussion on reaction to the CCIR's recommendations be dropped and that the Commission proceed to hold a free discussion giving their own views and this was agreed. The President then read out a draft of the resolution on time signal emissions which had been prepared by the Organizing Committee. After clarification of a few points the meeting adjourned.

Third Session

The *President* reopened discussion on the draft Resolution by remarking on the need to reach an acceptable compromise. The following points were made:

that the difference given should be UTC-UT1 and not UTC-UT2,

that the maximum tolerance had been increased from $0^s.5$ to $0^s.7$ because it would be impossible to keep to a tolerance of $0^s.5$, with 1^s steps on fixed dates and the requirement to give advance notice,

that all time signal emissions *must* include the information necessary to obtain UT1 to at least $0^s.1$.

At the request of *D. H. Sadler*, the President agreed to put to the vote the sentiment expressed in the final paragraph that Commission 31 considered the document as the optimum solution. This was agreed unanimously. The amended Resolution was agreed nem. con.

H. M. Smith proposed the following statement:

"Commission 31 received with interest an informal report by the Director of the BIPM on the proceedings of the meetings of the CCDS (Paris 1970), and noted with satisfaction that the CCDS endorsed the atomic scale of time proposed by the IAU (Prague 1967) and currently operated by the BIH.

Commission 31 would welcome the appointment of a representative nominated by the CIPM as a member of the Directing Board of the BIH."

After translation into French by B. Guinot this was agreed unanimously.

The *President* reported that Commissions 7 and 19 had asked that a representative of Commission 31 be appointed for discussions on the formation of a Working Group with COSPAR on lunar laser ranging. He proposed that Dr Guinot be the representative and this was agreed.

W. Markowitz reported that from 1820 to date there were only two periods, about 1870 and 1910, when the rate of rotation of the Earth would have required more than a single 1 second step a year to be made in time signals if the proposed rules had been in operation, and that in no case would more than two such steps have been needed.

W. Markowitz reported on the use of Loran-C for synchronization of clocks at various laboratories for satellite tracking, and for lunar laser ranging. An experiment with Loran-C conducted at Marquette University indicated that the frequency of electromagnetic radiation is not affected by mass.

B. Guinot presented a slide illustrating the difference TUC (OP)-TUC (USNO) obtained via Loran-C and with portable clocks for a period of fourteen months and noted the remarkable agreement between these two methods.

H. M. Smith showed a similar slide giving comparisons between UTC (RGO) and several international laboratories.

H. M. Smith took the chair whilst *G. M. R. Winkler* reported on portable clocks and the problems of time synchronization. Two series of portable clock trips, of greater than $2\frac{1}{4}$ days duration, had been analyzed, the first with cesium standards, model 5060 A, showed a mean closure error of $+0.1 \pm 1.1$ μsec and the second with standards, model 5061, showed a mean closure error of -0.5 ± 1.5 μsec. *G. M. R. Winkler* presented a slide showing the variations in frequency of typical cesium standards over periods up to 100 days; whilst some did not improve, the frequency of the best clock was determinable to 3×10^{-14} within 40 d.

On time synchronization *G. M. R. Winkler* said that for purposes of economy time signals were superposed on navigational and communication systems. Of the different systems in use HF radio time signals gave global distribution to 1 msec, portable clocks to 0.5 μsec, VLF-OMEGA 1–3 μsec by relative phase tracking and LORAN-C 0.5 μsec in the northern hemisphere (except in the western United States). Communication satellites, such as Telstar and Relay gave time to 0.1 μsec, the navigational satellite, Transit, gave 10 μsec, the experimental satellite GEOS gave 5–10 μsec. Exotic systems such as very long baseline interferometry (VLBI) gave 1 nsec and TV gave 10 nsec locally and 1 μsec regionally. He also reported on the work of *M. Lefebvre* (ONERA) who had achieved an accuracy of 10 μsec using TRANSIT. TV had been used very successfully in Prague and Potsdam and in the U.S.

H. M. Smith thanked G. M. R. Winkler for his review of methods of precision time comparison and the United States Naval Observatory for their comprehensive service with travelling clocks.

J. Bonanomi reported a portable clock experiment lasting 5–7 d with closure errors of 0.2 $\pm$ 1 μsec or better.

The meeting adjourned.

Fourth Session

The President opened the session.

E. Proverbio discussed radio wave propagation. HF and VLF signals are still used for time synchronization and this calls for a knowledge of the velocity of radio propagation to obtain travel times. A great deal of work has been done on the effects of diurnal and seasonal variations. Anomalies due to solar disturbances were also mentioned.

The *President* confirmed that VLF signals are still useful for world-wide synchronization and mentioned two ways of determining path delays: one is to develop a theoretical model ionosphere and produce tables giving path delays for different heights; a second is to measure phase differences between a number of synchronized transmitters at the same site with differing frequencies and produce an empirical model by assuming the path delay from the transmitters to the receiving station to be a function of distance and the phase differences.

R. G. Hall reported on a meeting of IAU Working Group No. 1 on Numerical Data which he had attended and which was enquiring into the feasibility of setting up a data storage centre. Commission 31 was already catered for with the BIH and BIPM, but it had been suggested that the BIH might transfer some of its data to the centre. A matter of direct concern to Commission 31 was the recommendation that all stars be referred to by the Durchmusterung BD number where possible.

B. Guinot reported on the meeting of representatives of Commissions 4, 7, 9, 17, 19 and 31 which had agreed to send representatives from each of these Commissions to the COSPAR Working Group on lunar laser ranging experiments.

The *President* said that there is a need for a precision of from 10–100 μsec for space tracking. In theory it is possible to use a spacecraft as a fixed point for radio tracking and to derive also information on x, y and UT1; but in practice it is better to obtain these data by conventional means and use them to improve the tracking. For missions which will take place during the changeover of the UTC scale the President advised anticipating the new scale.

The *President* made some comments on Loran-C measurements. The phase values reported by the USNO were unsmoothed. Attempts were being made in Washington to measure the sky wave transmissions on two chains at one station. On cycle identification the *President* described difficulties that were encountered and said that it was of paramount importance to use the standard bandwidth of $\pm$ 20 kcs.

The *President* gave his sincere thanks to the very excellent co-operation that the USNO had received from a number of laboratories and mentioned in particular BIH, PTB, DHI, RGO and RRL (Japan).

The *President* asked Commission 31 to consider what interest it had in regard to very long base line interferometry. *W. Markowitz* reported only a resolution of Commission 19 which noted the advantages of VLBI for measurements of the rotation of the Earth.

Finally the *President* acknowledged with pleasure the interest of the members and closed the meeting.

RESOLUTION ADOPTED BY COMMISSION 31

Commission 31 makes the following recommendations:
1. That the frequency offset of UTC be made zero, effective 0^{h}, 1 January 1972.

2. That *step adjustments* shall be exactly 1^s. When a step adjustment is made it shall be at 0^h on the first day of a month with preference for 1 January or 1 July. These step adjustments will be decided upon and announced as early as possible by the BIH.

3. The maximum difference UT1–UTC will be less than $0^s_.7$ unless there are exceptional variations in the rotation of the Earth.

4. *Special adjustment.* The BIH will also announce a unique fraction of a second adjustment to be made at 0^h 1 January 1972, so that UTC and the International Atomic Time Scale (IAT, in French TAI) will differ by an integral number of seconds.

5. The *emission times* of time signals from co-ordinated stations shall be kept as close to UTC (BIH) as feasible with a maximum tolerance of 1 ms.

6. *Nomenclature*

6.1. Clocks in common use will indicate the minutes, seconds and fractions of UTC (French: TUC).

6.2. The terms 'G.M.T.' and 'Z' are accepted as the general equivalents of UTC in navigation and communications.

7. The term ΔUT is defined by: ΔUT = UT1–UTC. Extrapolated and final values of ΔUT will be issued by astronomical observatories and the BIH, and will be given the widest possible distribution.

8. All standard time signal emissions must include information which will enable a user to obtain UT1 with a precision of at least $0^s_.1$.

9. *Designation of the epoch of steps in UTC*

9.1. If UTC is to be advanced, then second 00 will follow 23^h 59^m 58^s of the previous day.

9.2. If UTC is to be retarded, then the second of the previous day 23^h 59^m 58^s will be followed by the next second 0^h 00^m 00^s of the first day of the month.

9.3. The stepped second will be commonly referred to as a "leap" second (in French: intercalaire).

9.4. The time of an event given in the old scale, before the leap second, will be given as a date in the previous month, exceeding 24^h if necessary. The time of an event given in the scale after the step will be given as a date in the new month, with a negative time, if necessary.

Note: Commission 31, taking into account the conflicting requirements of the various users of UTC, including the large number of those requiring immediate knowledge of hour angle, considers that the above represents the optimum solution.

JOINT MEETING OF COMMISSIONS 4 AND 31, ON TIME SCALES, 25 AUGUST 1970

CHAIRMAN: G. A. Wilkins.
SECRETARIES: C. J. A. Penny and A. T. Sinclair.

UNIVERSAL TIME

G. A. Wilkins drew the attention of members of Commission 31 to the resolution of D. H. Sadler (see pp. 60-63) which had been approved at the previous meeting of Commission 4. This resolution requested that adequate means should be provided for making the difference UT1–UTC available to a precision of $0^s_.1$ before UTC is permitted to depart from UT1 by more than about $0^s_.1$. W. Markowitz asked why a precision of $0^s_.1$ was necessary since he doubted whether it was possible to determine positions to the corresponding accuracy of about 100 m. In reply, R. L. Duncombe said that observational errors by navigators were unavoidable, but the time errors should be kept below the level where they would contribute to the result. R. F. Haupt said that the almanacs were designed to allow the determination of positions to 0.1 min of arc, and for this a precision in time of 0.25 sec was required.

The meeting then discussed Resolution No. 1 of Commission 31 (see p. 123). R. L. Duncombe asked why the maximum permissible difference between UTC and UTl had been fixed at $0^s\!.7$, instead of $0^s\!.5$. G. M. R. Winkler replied that it was expected that a maximum tolerance of $0^s\!.7$ could be adhered to by making step changes on the first day of a month only, whereas a maximum tolerance of $0^s\!.5$ could necessitate changes on any day at short notice. After further discussion the resolution was approved by the meeting.

H. M. Smith asked for suggestions as to the best way of transmitting in coded form the corrections to UTC to obtain UTl, and if any greater accuracy than $0^s\!.1$ was required in these corrections. G. E. Taylor thought that the corrections should not be incorporated in the time signal emissions, as those who needed to know the corrections immediately were frequently working in conditions of poor radio reception. Also observers who used the time signals by ear could be upset by a variation of the rythmic beat of the signals due to the coded corrections. He agreed with an idea that had been suggested of broadcasting the corrections with the shipping forecast. H. Enslin had had experience of using a coding system, consisting of an extra signal either shortly before or shortly after the minute signal. He had found it easy to use in practice. A. M. Sinzi commented that in this sort of system signals denotating a negative correction which come just before the minute signal are easy to miss. G. A. Wilkins pointed out that the navigator would only have to note the correction once a month, as the resolution proposes that step changes should only be made on the first day of a month. Everyone was agreed that no greater precision than $0^s\!.1$ was needed in the correction UT1–UTC at this time.

EPHEMERIS TIME

G. A. Wilkins stated that a working group on units and time scales had been set up by Commission 4, with himself as convener. Part of its work would be to look into the possible need for a new definition of the ephemeris time scale.

T. C. Van Flandern gave a summary of his paper entitled 'The need for a new ephemeris time scale', in which he discussed evidence that the present Ephemeris Time scale is not supported by observations of the Sun, Moon, or inner planets. The present Ephemeris Time scale depends on the adopted values for certain parameters, such as the equinox location and the tidal acceleration of the Moon. In the light of modern revisions of the values of these parameters he suggested several possibilities for the revision of the definition of Ephemeris Time.

B. Guinot thought that the working group set up by Commission 4 should consider using Atomic Time in ephemerides, so that Ephemeris Time would no longer be required. T. C. Van Flandern agreed with this for future ephemerides, but said that the best way to extend a time scale backwards was by using the Moon to give Ephemeris Time. I. I. Shapiro said that the variation of the time scale used in the past could be taken as an unknown to be determined by the comparison of theories with observations. A. Stoyko presented figures for the variation of Ephemeris Time from Atomic Time apparently contradictory to those of T. C. Van Flandern, but it was agreed that the two sets of figures were not comparable.

RELATIVISTIC EFFECTS ON TIME SCALES

G. Becker gave a short description of the nature of time in relativity theory. He said that time scales vary with the gravitational field, so to define a time scale it is necessary to specify the gravitational field. On the Earth we should take the gravitational field on the geoid surface as the reference field, and the clocks used to determine the time scales must be at rest. Such a time scale would be coordinate time. The variations in time scales due to changes in the gravitational potential are smaller than the errors in the best clocks available today, and so he suggested that there was no need at present to talk about coordinate time.

I. I. Shapiro attempted to clarify the difference between Ephemeris Time and coordinate time. He said that Ephemeris Time only had a meaning in Newtonian theory, and the coordinate time of relativity theory could not be compared with it. G. M. R. Winkler said that what was required was a simple system for practical time keeping on the Earth's surface.

COMMISSION 33: STRUCTURE AND DYNAMICS OF THE GALACTIC SYSTEM (STRUCTURE ET DYNAMIQUE DU SYSTÈME GALACTIQUE)

Report of Meetings, 20, 21, 22 and 26 August 1970

PRESIDENT: G. Contopoulos.
SECRETARY: F. J. Kerr.

Business Session

Following a request to all Commissions by the Executive Committee, a set of by-laws was prepared by the Organizing Committee and adopted by the Commission.

1. A member of the Organizing Committee serves for two consecutive terms and then retires unless he becomes President or Vice-President. A retiring President stays one more term as member of the Organizing Committee. Any retiring member can be reelected at a later time.

2. The President serves for one term and is then normally succeeded by the Vice-President. Proposals for the new Vice-President and new members of the Organizing Committee to replace those retiring are made by the Organizing Committee at the first meeting of Commission 33 during a General Assembly. If any member of the Commission proposes other names, a secret ballot is taken of all members of the Commission present.

3. New members of Commission 33 are proposed by at least one member of the Commission. The Organizing Committee decides on each proposal.

The proposed names of Commission officers were accepted.

During the last triennium, the Commission report in the *IAU Transactions* was severely reduced in size because of financial difficulties. However, the Commission produced on its own initiative a longer report which appeared as a publication of the University of Thessaloniki. Opinions were sought on the value of this dual form. After several very favourable comments it was agreed that two reports should again be attempted, in spite of the extra work involved.

The following resolution was proposed by the Organizing Committee:

We propose that the superscript II be omitted from the new galactic coordinates (those adopted by the IAU in 1958); henceforth they should be simply l and b. We recommend that authors adopting this practice should indicate in the beginning of each paper that l and b are the new galactic coordinates. The old coordinates should retain the superscript I.

Ogorodnikov said he was not against the proposal, but emphasized that it should be regarded as an exceptional case, as changes are always confusing. Bok pointed out that the change is being carried out slowly, three years having elapsed since an earlier motion to the same effect. Blaauw stated that a major reason for the proposal was to recognize a trend already in effect.

The resolution was carried.

Contopoulos pointed out that the Selected Areas Committee has worked for many years – in fact, its field of work was originally the central subject of the Commission. He then proposed that there should no longer be a separate committee, but instead its work should be taken over by the full Commission. Elvius had been included in the new Organizing Committee to facilitate this change.

The meeting approved the dissolution of the Committee, adding a vote of thanks to the Committee and in particular to Elvius for their very good work over the years.

PAPERS

J. Einasto: A Proposal for Galactic Research.
K. Lodén: The Stockholm Survey of the Southern Milky Way.
J. D. Wray and B. E. Westerlund: An Atlas of the Southern Milky Way.

C. B. Stephenson and N. Sanduleak: The Cleveland Southern O-B Star Survey.

J. A. Agekjan: The Third Integral of Motion in a Stationary Stellar System.

J. Einasto: On the Structure and Evolution of the Galaxy.

Second Meeting

RR LYRAE STARS

Joint Meeting with Commissions 27, 24, 30, 32 and 37.

R. v.d. R. Woolley: On the Determination of the Absolute Magnitudes of the RR Lyrae Stars.

R. F. Christy: Absolute Magnitudes of RR Lyrae Stars.

G. van Herk: Review of Observational Data on RR Lyrae Stars.

S. V. M. Clube: Absolute Magnitudes of RR Lyrae Variables.

A. R. Klemola: The Lick Observatory Program on Proper Motions of RR Lyrae Stars.

Third Meeting

SPIRAL STRUCTURE AND RELATED PROBLEMS

M. Lecar: Report on the *IAU Colloquium* No. 10 'Gravitational *n*-Body Problem'.

F. H. Shu: Observational versus Theoretical Spiral Patterns in External Galaxies and in our Galaxy.

C. Yuan: On the Kinematics of Nearby Stars.

A. Kalnajs: Non-Linear Effects in Axisymmetric Waves.

F. J. Kerr: Local Spiral Structure.

W. Iwanowska: A Statistical Approach to the Problem of Stellar Populations.

L. Martinet and A. Hayli: On the Dynamics of the RR Lyrae Variables and the Existence of a Third Integral of Motion in the Galaxy.

Fourth Meeting

THE DISTANCE SCALE OF OUR GALAXY

B. J. Bok: Report on the RR Lyrae Meeting.

O. J. Eggen: The Distance of the Hyades.

P. A. Wayman: The Error Limits in the Convergence Method for Determining the Distance of the Hyades.

A. Behr: The Ratio of Interstellar Extinction to Reddening.

J. D. Fernie: The General Consequences of Uncertainty in the Ratio of Total to Selective Interstellar Absorption.

J. H. Oort: Summary and Conclusions.

Fifth Meeting

SELECTED AREAS

CHAIRMAN: T. Elvius.
SECRETARY: K. Lodén.

There was no discussion about the printed reports (full version in *Contr. astr. Dep. Univ. Thessaloniki* **53**, 111, 1970; short version in *Transactions IAU* **14A**, 383, 1970), which were accepted.

At the business meeting of Commission 33 it had been decided to dissolve the special Committee for Selected Areas and to include the Chairman in the Organizing Committee of the Commission, which should report directly on Selected Area matters.

Kerr reported on a 21-cm hydrogen-line survey of all the Selected Areas. Profiles have been observed for the central point of each field and at points north, south, east, and west of this point, stepped off 21'. The observations for Selected Areas 1–163 were made at Green Bank with the NRAO 140-ft telescope and the 413-channel receiver. The velocity resolutions are 0.69 and 2.78 km/sec. A catalogue of calibrated profiles will be produced as a first result, and the reduction is under way. Results for individual areas can be sent to those who are interested. Future work will include studies of the velocity field and comparisons of radio and optical data. The southern Selected Areas were observed at Pereyra, Argentina.

Ardeberg reported on the following work performed at La Silla, Chile, in Selected Areas 165, 188, and 205:

1. SA 165: Photoelectric UBV photometry of all stars brighter than $V = 10.5$ (and some fainter) and of all stars of spectral types earlier than G0 and with $V \leqslant 12.5$; photographic UBV photometry for all stars brighter than $V = 15$; photoelectric polarimetry for the 20 brightest stars; slit spectra (dispersion 70 Å/mm) for some bright stars; objective prism spectra (dispersion 110 Å/mm).

2. SA 188: The same photoelectric photometry as in SA 165; objective prism spectra obtained.

3. SA 205: Photoelectric photometry for all stars brighter than $V = 10.5$; photoelectric polarimetry for the 20 brightest stars; objective prism spectra obtained.

Reductions are now under way.

Griffin has measured radial velocities with a photoelectric technique for about 500 stars in Selected Areas 68 to 91 ($+15°$ declination zone). The stars range from G5 to M in spectral type and from 7 to 10 in V. The standard error is about 1 km/sec. The work is to be published in Monthly Notices.

Terzan spoke on his project of constructing maps for each of the 206 Selected Areas and of compiling a reference catalogue from all existing published catalogues and investigations, with keys to the numbering in the various catalogues of the stars considered. Terzan demonstrated maps for SA 68, derived by means of a computer programme due to Ounnas. The technical and financial problems were discussed. It was further mentioned that the long exposures of southern Selected Areas taken at the Uppsala Station at Mount Stromlo could be used for making photographic maps available. The possibility of using inexpensive overlays with printed numbers on them to be placed upon already existing charts was mentioned by Bok. Roman reported that she had maps available for a number of fields. Murray and Bok proposed that the data compiled by Terzan should be available on magnetic tape instead of being printed, copies of which could be obtained relatively cheaply by interested astronomers.

The chairman concluded this last Selected Areas meeting by thanking all who during a long series of years had contributed to the work of the Selected Areas Commission, Sub-Commission, and Committee.

A more extensive report of the Commission meetings, including abstracts of the paper, can be obtained from the secretary.

COMMISSION 34: INTERSTELLAR MATTER AND PLANETARY NEBULAE (MATIÈRE INTERSTELLAIRE ET NÉBULEUSES PLANÉTAIRES)

Report of Meeting, 19, 20 and 25 August 1970

PRESIDENT: D. E. Osterbrock.
SECRETARY: T. K. Menon.

Business Meeting

The draft report was approved subject to minor corrections or additions.

The following resolution was adopted unanimously: "Commission 34 requests all future discoverers of planetaries to send the necessary data about new objects (at least R. A. 1950, Dec. 1950, identication charts) to the Organizing Committee of the Commission. All observers of planetary nebulae are requested to send one copy of their observing data to the Organizing Committee and one copy to Dr L. Kohoutek. This will make it easy to collect all observing material either for the Supplement of the *Catalogue of Galactic Planetary Nebulae* or for preparing a second edition of the 'catalogue'."

Scientific Meetings

PLANETARY NEBULAE

L. H. Aller reviewed the work on planetary nebulae since the 1967 IAU Meeting. Most of the material is included in the Draft Reports. He particularly stressed the outstanding problems such as the distance scale and model atmospheres for central stars.

Y. Terzian mentioned that the NRAO interferometer measurements show a source of size less than 1″ at the center of NGC 7027 with a brightness temperature in excess of 100000 K.

W. Liller and B. Lasker reported that their independent measurements have shown no evidence for variability of any planetary nebula.

L. Higgs reported that out of 400 objects radio fluxes have been measured for about 120 planetaries.

M. Peimbert reported on his scanner measurements of the Balmer continuum to $H\beta$ ratio and the $\lambda 5007/4363$ ratio for 15 planetaries. He found a value of $T_e = 7000 \pm 1000$ K from the first ratio and about 11000 K from the second ratio. He suggested that the two temperatures refer to two different regions. He also found the ratio $He/H \approx 0.115$. He attributed the discrepancy from earlier work to overestimate of weak lines.

INTERNAL MOTIONS IN HII REGIONS

M. G. Smith reported on Fabry-Pérot interferometer measurements of the nebulae M 17, η Carina, M 8 and M 20. The data suggest both expansion of these nebulae and simple doubling of line profiles in some cases. Because of the doubling effects the temperatures determined from line widths may not be relevant. J. Meaburn reported on his search for high velocity features in emission nebulae using single and double Fabry-Pérot interferometers. He suggested that many of the nebulae searched had either systematic or turbulent velocities in the range of 30 to 60 km/sec.

P. Mezger on the other hand reviewed the radio recombination line data and found no evidence for any expansion. He attributed the observed large scale motion to possible rotation of the nebulae.

In the ensuing discussion several speakers stressed the effects of angular resolution on the observed velocities.

R. Louise reviewed the French work on line profiles and radial velocities of HII regions. He suggested the possible existence of two kinds of HII regions, one a low-temperature type and the other a high-temperature type. Some have small variations of electron temperature inside, whereas others have large radial velocity and temperature variations inside.

In a session on new and important results the following papers were presented:

H. M. Johnson and R. H. Rubin: Observation and Classification of the Nebula YM 29.

Y. Terzian: New Low Frequency Observations of the Orion Nebula from the Arecibo Observatory.

L. Goldberg and A. K. Dupree: Radio Recombination Lines and Their Use in Inferring the Properties of Nebulae.

W. Liller: Planetary Nebulae and Their Exciting Stars.

C. R. O'Dell: Dust in Planetary Nebulae.

K. Serkowski and G. V. Coyne: Recent Advances in the Field of Interstellar Polarization.

K. Nandy: Polarization in the Diffuse Interstellar Bands.

J. M. Greenberg: The Distribution of Dust in the Merope Nebulae (By Title Only).

J. M. Greenberg: Extinction by Roughened Particles, Dielectric and Metallic (By Title Only).

B. H. Zellner: Polarization in Reflection Nebulae.

H. Habing: X-Ray Heating and the Ionization of Interstellar He (By Title Only).

J. S. Mathis: Internal Dust in Diffuse Nebulae.

H. Zimmermann: The Dynamical Separation of Dust Particles During Cloud Collisions.

K. Rohlfs: Association of Gas and Dust.

R. Sancisi and H. Van Woerden: Motion of Neutral Hydrogen Connected with the OB Association in Lacerta.

T. De Jong: H^- Densities and Interstellar Diffuse Bands.

R. Henrikson: Pulsars in a Spiral Arm?

ENERGY CONTENT OF INTERSTELLAR GAS AND ITS LARGE-SCALE DYNAMICS

G. Haslam reviewed the radio and optical evidence regarding the structures known as spurs from radio studies. He suggested that the ridges of peak radio emission of the four well established spurs are in the form of small circles in the sky. Some of the spurs appear to have associated faint optical features. He suggested that all of them are shell structures viewed from various angles.

Elly Berkhuijsen reported on the distribution of neutral hydrogen in the vicinity of the continuum spurs. She suggested that the spur structures have associated with them neutral hydrogen clouds following closely the continuum emission ridges.

P. Seymour discussed the harmonic analysis of the optical polarization data, dividing the stars into five distance groups. His analysis suggested the existence of a magnetic field structure of the nature of a spur for the three lower distance groups.

W. Zuzak discussed the model of a spherical source of shocked relativistic electron gas expanding into the interstellar medium with magnetic field. He finds that at the inside surface of the shell the magnetic field varies very sharply and hence the synchrotron radio emission from such a shell can be expected to have sharp gradients at the edges. He suggested that the radio spurs may be such shells.

D. Wentzel discussed a mechanism by which the streaming of cosmic rays along magnetic fields can produce hydromagnetic waves which in turn can transfer their momentum to the surrounding gas. By this process cosmic ray energy can be used directly for the purposes of acceleration and heating of the gas without ionization. This will be particularly true near the edge of the galactic disk.

H. Habing reviewed the processes of heating and cooling of the interstellar medium. He pointed out that the cooling curves are still uncertain. However we probably still must have as yet unknown sources of heating in order to balance the cooling.

J. Bergeron and S. Souffrin described their new computations of heating of interstellar medium by hard UV radiation and compared it with that produced by fast particles. No decisive observational tests are available yet to distinguish between the two processes.

W. Roberts reported on his work on shock waves, spiral structure and formation of stars.

REVISED CATALOG OF 2592 STARS OBSERVED FOR POLARIZATION

This catalog, published in 1958 (*USNO Publ.* **17**, 285–331), provides 1900 positions and the old system of galactic coordinates. Thanks to the efforts of G. A. H. Walker, this polarization catalog has now been revised to contain 1975 positions, l^{II} and b^{II}. When available, detailed spectral types have been added. A limited number of printed copies of the revised catalog are now available free of charge. Send request to John S. Hall, Lowell Observatory, Flagstaff, Arizona, U.S.A.

COMMISSION 35: STELLAR CONSTITUTION
(CONSTITUTION DES ÉTOILES)

Report of Meetings, 19 and 22 August 1970

PRESIDENT: A. Massevich.
SECRETARIES: G. Ruben, R. C. Smith.

Paczynski discussed the final evolution of Population I low- and moderate-mass stars, using a numerical technique designed to deal with the dynamical instabilities found; a semi-convective region appears in the model, similar to the results of Uus. He also described the evolution of helium stars which reached the R CrB part of the HR diagram only if neutrino emission was included. *J. P. Cox* described a survey (with King) of the Cepheid instability strip, using linear and non-linear calculations, aimed at determining Cepheid masses. The left-hand edge of the observed strip is well-described if the masses are those given by evolution without mass-loss. *Von Sengbush* showed that the evolution to the giant branch of a 4 $M_\odot$ star with $X = 0.36$ was very similar to that of a 7 $M_\odot$ star with $X = 0.6$. *Christy* surveyed the non-linear aspects of pulsation over a range of models, including Cepheids and extending to RV Tau stars. Masses of about 0.6 times the evolutionary mass seemed indicated for Cepheids. The disagreement with Cox could be due to uncertainties in colour/temperature conversions or in the opacities. Giannone (with Giannuzzi) reported Case B evolution of close binaries with mass exchange. For the cases studied, the mass of the final white dwarf is in the range 0.2 $M_\odot$ to 0.4 $M_\odot$, depending on the initial separation and mass ratio. Work on post-helium-flash models (in collaboration with Castellani and Renzini) shows the presence of a semi-convective region which increases the time-scale of evolution. *Demarque* showed that the assumptions of relativistic and non-relativistic degeneracy have quite different effects on post-helium flash evolution. It also appears to be necessary for horizontal branch stars to have a range of masses. *Refsdal* showed that moderate mass loss suppressed the loops in the HR diagram in the central-helium-burning stage of evolution. *Schwarzschild* remarked that his models also possessed semi-convection zones. He suggested that details of horizontal branch models might be very sensitive to opacity. *Massevich* reported on late evolution with stellar wind mass loss, which may be important for certain T_{eff} and g, and on binary evolution with mass exchange. Mass loss from the whole system speeds up the evolution.

Second Scientific Session – Joint Meeting with Working Group on Be Stars of Commission 29

CHAIR: R. Herman, E. Schatzman.

Herman described measurements of radial velocity variations in certain Be stars, with timescales of 20 to 30 yr. The timescale for α Dra is about an hour. *Coyne* showed that the polarisation in Be stars is intrinsic and irregularly variable. A list was given of stars whose spectra and polarisation should be simultaneously measured to clarify the nature of the variation. *H. C. Thomas* described computations of the evolution of 5 $M_\odot$ and 9 $M_\odot$ models with rotational mass loss. Unlike some other workers, they found no mass loss in the contraction phase. *Bernacca* discussed the statistics of rotation velocities of Be stars, and showed that not all Be stars need be rotating at break-up.

In discussion, *Schwarzschild* stressed that in rotational mass loss there is no mechanism for removing the mass from near the star, a point that was also stressed by *Cowling*. *Underhill* pointed out that the lines discussed by the observers were formed in regions which are completely outside the star from the point of view of stellar interior calculations.

Third Scientific Session – Solar Neutrinos and Rotation

CHAIR: L. Mestel.

Shaviv discussed the discrepancy between solar models and the observed neutrino flux, from the point of view of mixing in the Sun. He pointed out that the Sun could not be mixed through more than about a third of its mass if evolutionary calculations were to give ages for globular clusters consistent with cosmological ideas. In that case, the theoretical flux could be reduced by no more than 40%.

R. C. Smith compared the effects of rotation in the colour-magnitude diagrams of Praesepe and the Hyades with theoretical predictions and concluded that uniform rotation was excluded for both clusters, but that the data are too uncertain to support any particular law of non-uniform rotation. *H. C. Thomas* described the use of spherically averaged models for evolution with rotation. Rotation lengthens the timescales for H- and He-burning and lengthens the loops in the HR diagram. The carbon and oxygen cores become rotationally unstable before C-burning starts. *Roxburgh* described calculations on convection in rotating shells, assuming a latitude dependence for the efficiency of convection. It was possible to reproduce the solar equatorial acceleration with equatorial heating and an oblateness less than that observed by Dicke.

Fourth Scientific Session – Instabilities

CHAIR: P. J. Ledoux.

Schatzman proposed the mechanism of turbulent diffusion as a means of mixing Li, Be and B in the Sun down to a depth where they could be burned. The depletion of these elements at the surface can be explained with plausible values of the diffusion coefficient, but the generating mechanism of the turbulence was uncertain, possibly being shear instabilities. *James* discussed the role of circulation when the Goldreich-Schubert-Fricke instability is damped by a dynamical instability. The resulting timescale for redistribution of angular momentum is certainly less than the Eddington-Sweet circulation time. *Ruben* described calculations, by the PIC method, of shock wave propagation in stars. Mass loss occurred, the amount depending on the energy in the initial disturbance. Agreement was found with the results of Sparks for lower energies. *Aizenman* reported that a study of the complex roots of the secular stability eigenvalue problem showed the possible existence of slow decaying pulsations, with a period of about 10 million years, in stars just leaving the main sequence. In discussion, *J. P. Cox* and *Schwarzschild* reported the presence of complex roots in later stages of evolution. *Paczynski* remarked that in difference equation schemes the determinant of the equations of stellar structure sometimes changes sign. *Christy* discussed the disagreement in pulsation calculations – different authors make contact with observations at different points, and not all computed quantities agree with observation for any model. He also speculated on shear instabilities in semi-convection zones as a source of the tidal bulge wave proposed to explain βCMa stars.

Business Meetings

CHAIR: A. G. Massevich.

The President's report was presented at the beginning of the first scientific session. On the proposal of Professor Schwarzschild, the report was passed without discussion.

The nominations for the new President and Vice-President were accepted. The list of new members of the Commission was presented. It was agreed to add three names which had been inadvertently omitted. It was decided not to fix by-laws at the moment, but to experiment for the next three years. The President and Vice-President were asked to propose at least three names for the next Vice-President and to communicate them to the Commission members at least 6 weeks before the Assembly.

Additional names shall be added if they are proposed by not less than three Commission members. A secret ballot shall be taken. The President shall report to the General Secretary the three top names from the ballot (in order of preference). It was agreed to form a small organising committee to help the President. Members should serve not more than two terms in the Organising Committee. The list of members for the first Organising Committee was approved.

It was pointed out that work on neutron stars, supernova explosions and similar topics was poorly represented on the Commission. It was decided to discuss these questions with members of Commissions 47 and 48. The question of inviting theoretical physicists working in these fields to be members of the Commission was discussed, and it was suggested that they might be made consultants. Professor Cowling congratulated Professor Massevich on her able organization of the Commission and in particular on her excellent circulars.

COMMISSION 36: THEORY OF STELLAR ATMOSPHERES
(THÉORIE DES ATMOSPHÈRES STELLAIRES)

Report of Meetings, 19, 20, 25 and 26 August 1970

PRESIDENT: A. B. Underhill.
SECRETARY: A. L. T. Powell.

Business Meeting

The members learned with regret of the deaths of L. G. Henyey, J. Q. Stewart and M. H. Wrubel since the last General Assembly. The following additions were made to the Report of the Commission: A-1: – A set of models for F stars has been published by T. Kipper. A-8: – The nature of P Cygni stars has been discussed by L. Luud. C-4: – A more exact equation of radiative transfer has been derived from the basic equations of quantum electrodynamics by A. Sapar who has discussed the implications in detail. Computational techniques for plane-parallel radiative transfer based on mathematical semigroup properties of the operators for diffuse reflection and transmission in scattering atmospheres have been developed by Grant and Hunt. C-7: – Problems of extended stellar atmospheres and of stationary mass loss are being studied by Pustylnik and by Sapar and Viik. In addition analytical expressions for the redistributed intensity as a result of multiple Compton scattering in a spherical atmosphere is under study by Viik. As with the work reported previously, reference must be made to standard abstracting journals to find where these studies are published.

The new Organizing Committee drafted a set of by-laws for the Commission. These are given at the end of this report.

Plans for colloquia to be held in the next three years were discussed. Choice will be made by the new Organizing Committee from among the topics: The transfer of radiation and the formation of spectral lines in the presence of a magnetic field; Theoretical considerations of spectral line profiles; Interdisciplinary aspects of stellar atmospheres. The venues of Boulder, Nice and Istanbul were mentioned and either late spring or early autumn were considered to be the best time for such intensive working meetings.

In response to a resolution submitted by the Executive Committee to Commission 36 for consideration, the following resolution was approved: Commission 36 considers the physics of unstable stars to be a most important subject in which it is deeply interested. However, because the physics of unstable stars is basically the same as that of stable stars, it does not recommend the creation within one of the IAU Commissions of a subcommission on the physics of unstable stars. Commission 36 suggests that colloquia or symposia be held every few years to summarize the problems and results in the field of unstable stars.

The following resolution drafted by R. N. Thomas and J. C. Pecker was also approved for submission to the IAU Executive Committee:

For some years Commission 36 has realized the vital importance of small working colloquia and symposia on topical problems in stellar atmospheres to delineate problems and continually to re-examine the basis of theoretical models and diagnostic spectroscopy. A major part of such colloquia is lost without the prompt preparation and distribution of a set of inexpensive proceedings so that persons unable to attend the colloquia may still share their stimulus. This is particularly true for students, young researchers and mature persons just shifting their attention to another field.

Therefore, Commission 36 proposes to implement the following policy and it recommends and invites other Commissions of the IAU to join it:

(a) to encourage the Organizing Committee and the Editor of the Proceedings of each symposium and colloquium to utilize the staff and secretarial facilities of their institution to prepare quickly camera copy of the major summary papers and discussions for photo-offset reproduction. Paper quality and artistic appearance are negligible considerations;

(b) to encourage collaborative efforts between governmental and institutional publication offices to publish simultaneous English and Russian language editions at a price such that each student can afford to purchase his own working copy.

To ease the problem of literature monitoring, the meeting approved the proposal by R. N. Thomas that a preprint, summary and reprint of all papers relevant to stellar atmospheres should be sent to the Commission President. These would then be collated yearly and a bibliography would be sent to each member and to anybody making a request to receive such information.

COMMISSION 36 BY-LAWS

1. The Organizing Committee is elected by members by ballot, from a list of candidates proposed by the members. The number of candidates is substantially larger than the number to be elected.

2. The possible subjects for next colloquia, symposia or other types of gathering sponsored by Commission 36 are proposed by the members and selected by the Organizing Committee.

3. Questions calling for decisions or statements on behalf of Commission 36 are normally first brought to the attention of the President, secondly evaluated by the Organizing Committee and then submitted to the members of Commission 36 for their approval.

Scientific Meeting

The first topic to be discussed was '*The problem of spectrum formation without the hypothesis of LTE: physical principles and mathematical techniques*'. R. N. Thomas introduced the subject with a review of the physical principles. A summary of his remarks follows:

A step-by-step summary of the physical principles introduced is given. Spectrum formation results from the interaction between energy states of matter and of the radiation field as the radiation diffuses outward through the stellar atmosphere. Such interaction is described by the structure and variation through the atmosphere of the source-function. Section I summarizes the general structure of the source-function, S_ν, in terms of energy states of matter, their interaction among themselves and with the radiation field, and the explicit use of population and conservation equations in making specific the structure and variation of S_ν. It also summarizes the physical and mathematical simplification resulting from imposing the conditions characterizing the classical atmosphere. Section II summarizes the conditions under which astronomical LTE, or LTE-R, results even when the condition of LTE-R is not a priori imposed; viz, collision domination, or homogeneity of radiation field under additional conditions. Section III summarizes the principles upon which a construction of the currently-used form of S_ν rests; its ν-dependence; the relation between S_ν, J_ν, and so-called source- and sink-terms in thick, effectively-thin, and thin atmospheres; and the idea of the New Spectroscopy. Section IV discusses the computation of the values of state-parameters, or equivalently source-sink terms, throughout the atmosphere, trying carefully to distinguish between population and conservation-equation effects. The use of the Temperature-Control-Brackets, TCB, is briefly outlined, in connection with obtaining a physical picture.

A review of the mathematical techniques was then given by L. H. Auer. In summary he said the following:

The mathematical difficulty of the non-LTE problem lies in the fact that we must find simultaneously and self-consistently: (1) the radiation field at each frequency, (2) the electron velocity distribution, and (3) the population of each level. Because the radiation field is not locally controlled, solution must proceed simultaneously at all depths. Modern methods of treating transfer problems enable one to solve any case in which the equation of transfer is linear in the radiation field. There are two classes of such techniques: those based on the integral equation formulation and those based on the differential equation formulation. Both reduce the continuous equations to a set of linear equations which may be directly solved for the source function (or radiation field) at all depths. Since not only the transfer equation but also the equation of statistical equilibrium must be satisfied, iteration is always necessary. In the simplest approach, the transfer equation is linearized and solved one transition at a time. With the new radiation field, new populations are computed and the process iterated to convergence. Alternatively, the entire set of equations, both transfer and statistical equilibrium, may be completely linearized, and the resulting system solved. To the extent that the equations are really linear, this scheme simultaneously and consistently finds all quantities. In complex problems it is easier to apply and more strongly convergent than the former method; however, each iteration requires much more time. It is possible that a mixed scheme involving linearization of several but not all transitions is optimal from the standpoint of computing time.

Four contributed papers were presented. A. G. Hearn considered the effects of gradients of the temperature and electron density on the departures from LTE for stellar absorption lines. The importance of considering the population of many levels in the atom was stressed.

The empirical determination of S_L and b_L as functions of τ for infrared solar lines was considered by C. de Jager and L. Neven for C I lines in the Sun. The source function was found to follow the LTE value, but for some lines, the value of b_L was much greater than unity.

Results of non-LTE atmosphere calculations given by L. H. Auer showed that the line profiles of He II lines computed by non-LTE techniques developed by himself and Mihalas improved the agreement with observation. The discussion was closed by a contribution from A. Skumanich on the application of the Generalized Newton-Raphson method to non-LTE problems.

The second topic for this session was '*The Atmospheres of Cool Stars*'. Y. Fujita discussed the theoretical problems of the atmospheres of cool stars and showed results of opacity calculations done at Tokyo for a wide range of molecules. Recent calculations of the opacity of CN in cool stars were presented by H. R. Johnson. These showed that this source of opacity was very significant when the CN molecule was dominant.

Joint Scientific Meeting with Commission 29

In this session, arranged by Commission 29, the following papers were read on the topic 'How to Determine Abudances':

A. B. Underhill: 'What Does One need to know to Determine Abundances'.

R. F. Griffin: 'How Accurate are Measurements of Equivalent Widths and Line Profiles'.

B. E. J. Pagel: 'Uses and Limitations of the Differential Curve-of-Growth Method'.

L. H. Aller: 'Use of Line Profiles in Abundance Work'.

R. Cayrel: 'Comparison of High Dispersion Studies with Scanner Work'.

A. Unsöld: 'General Survey of Recent Experimental and Theoretical Work on the Abundance Problem'.

The speakers reviewed the field of abundance determination outlining the errors involved and validity of the physical arguments used to interpret the observations as well as the effect of systematic errors in the observations. The accuracy obtained from curve-of-growth methods appears to match the deficiencies in model atmospheres, spectroscopic data and observations. Scanner measurements give results consistent with those from other methods but with some loss in accuracy. Movement towards agreement in the determination of the solar iron abundance from the different available methods has followed from a revision of the Fe I and Fe II f-values. The ramifications of this change on the physical properties of the solar atmosphere were outlined by Unsöld.

Scientific Session

The first half of this session was devoted to a discussion of solar abundances. L. H. Aller, the discussion leader, confined his introductory remarks mainly to the subject of spectrum synthesis. He outlined the methods, assumptions, limitations and results obtained by members of his group at the University of California. By judicious adjustment of certain free physical parameters in a model atmosphere, the computed spectrum can be made to fit the observations in most cases. This approach can be used to determine the abundance of low-abundance elements that only occur in blended lines. Discussion followed concerning the justification of adjusting the various parameters which in some cases gave unrealistic values; in other cases, it was impossible to reproduce all the observed features of the spectrum. J. C. Pecker then gave an account of his recent work on the polarization variation of the solar continuum with wavelength. The results, independent of f-values and damping constant, indicated that the metal composition of the Sun was at least five times the GMA values. The recent upward revision of the magnesium, silicon and iron abundances makes this result more plausible. G. Elste showed how the use of high excitation Fe II lines for the curve-of-growth analysis of A stars has revealed a need for a revision of the f-values. These had given a large scatter on the curve and an anomalously low excitation temperature. A correction, dependent on excitation potential, was found by comparing the original f-values used in the analysis with those given by Garz and Kock. Application of this correction removed the scatter and temperature anomaly.

The second half of this session was devoted to contributions on 'The Problem of Spectrum Formation without the Hypothesis of LTE' that had been received since the announcement of the first session. R. C. Altrock outlined a method of 'Empirical Analysis of Line Profiles'. This was based on the solution for the source function from three lines of one multiplet. Preliminary results on the infrared oxygen triplet showed the source function to be in agreement with those of BCA. The theory of radiative transfer in a two component atmosphere was presented by E. H. Averett. This involved the calculation of the source function in the presence of material motion in hexagonal columns. P. R. Wilson described work undertaken by some of his colleagues on the vertical and horizontal fluctuations of temperature, density and opacity in the solar photosphere. Solution for the source function followed the Feautrier method. From a comparison of old and young Of stars, S. R. Heap described the discrepancy that had been found between the Zanstra temperatures and those derived from model atmospheres. This cannot be resolved by the use of classical models. In a discussion of the formation of the Lyman continuum, W. Kalkofen presented the results of his and R. W. Noyes' work on the construction of a model solar atmosphere that predicted the dissimilar colour and brightness temperatures observed. This model is based on the detailed balancing of three bound levels of hydrogen and the continuum. Recent work on the computation of a non-LTE line-blanketed model was given by R. G. Athay. Line blanketing is taken into account by using a representative atom with three levels and two line transitions. The model atmosphere was found to give good agreement with solar observations.

COMMISSION 37: STAR CLUSTERS AND ASSOCIATIONS
(AMAS STELLAIRES ET ASSOCIATIONS)

Report of Meetings, 19 and 21 August 1970

PRESIDENT: M. Golay.
SECRETARY: G. Lyngå.

A. Administrative Sessions

1. The Draft Report was approved without corrections.
2. S. Van den Bergh presented to the Commission a copy of the David Dunlap atlas of colour-magnitude diagrams for clusters on the UBV system.
3. J. Jung announced a French data centre for stellar data. Reference was made to Working Group 1 where a more detailed description was to be given.
4. A letter from B. V. Kukarkin informed the Commission about the preparation of the 'General Catalogue of Globular Clusters'. Erroneous data in the literature are corrected and properties of globular clusters are redetermined in a uniform system. Indices of richness (IR) and of metallicity (IM) are also determined.
5. G. A. Alter presented the second edition of the *Catalogue of Star Clusters and Associations*, which has been published in Budapest by the Publishing House of the Hungarian Academy of Sciences. Alter described the numbering system of clusters according to galactic coordinates, which was accepted at the XIIth General Assembly and now is used for the ordering of the catalogue. A new system for insertion of supplement data is used and the question was put, whether subscribers would be prepared to pay a nominal sum for this service. No objection was raised.

B. Scientific Sessions

The following papers were given:

Gonzalo Alcaino: 'A UBV Photometric Research on Ten Previously Unstudied Southern Globular Clusters'.

During 1968–70 the 36-in. and 60-in. reflectors at Cerro Tololo Inter-American Observatory have been used to undertake a photometric research on the previously unstudied globular clusters: NGC 1261, NGC 1851, NGC 2808, NGC 4372, NGC 4833, NGC 5286, NGC 6352, NGC 6362, NGC 6541 and NGC 6752.

The following results have been obtained from published photoelectric sequences:

Cluster	$(m-M)_{app}$	$E_{(B-V)}$	$(m-M)_0$
NGC 1851 (48 stars)	15.15	0.24	14.43
HGC 2808 (50 stars)	15.85	0.28	15.01
NGC 6362 (61 stars)	14.70	0.15	14.25
NGC 6752 (71 stars)	13.50	0.00	13.50
From calibrated photographic results:			
NGC 1261 (206 stars)	16.1	0.00	16.1
NGC 6352 (220 stars)	15.40	0.44	14.10

Complete photoelectric plus photographic results for all ten clusters down to $V \approx 17$ are expected to be completed by 1971.

O. Maeder: 'Influence of Axial Stellar Rotation on Age Estimates of Open Clusters'.

Individual rotational corrections on the colour-indices of stars depend only on the product

$v_R \cdot \sin i$. Orientation of the axes do not intervene. The absence of these corrections brings about an age overestimate of 60–70 % for Pleiades and α Persei clusters.

G. Janin and P. Bouvier: 'Numerical Experiments on the Disruption of Star Clusters Through Passing Interstellar Clouds'.

The tidal influence on a small stellar cluster of passing interstellar clouds is studied by means of numerical experiments.

The disruption of the cluster occurred over a time scale of 450×10^6 yr with a cloud concentration of 40 per million cubic parsec.

It is essentially the close encounters which are effective and in particular those of impact parameter smaller than the cloud radius, which had been neglected by Spitzer in his first attempt (1958) to solve this problem.

This paper was illustrated by a film.

P. Pishmish: 'UBV Photometry of the New Open Cluster NGC 2175s, of NGC 2175 and T9'.

UBV magnitudes are determined for stars in the clusters, down to magnitude $V \approx 15.^m5$, using photoelectric determinations as standards. The colour-magnitude and color-color diagrams are discussed. The distances are based on the absolute magnitudes, estimated from the spectral types of the brightest stars, and diameters are derived from star-counts. The full text is published in English in the *Bol. Obs. Tonantzintla y Tacubaya*, 1970 (T9 to be published).

M. Walker: 'Electronographic Photometry of Globular Clusters in the Magellanic Clouds'.

During 1968–69 electrographs of 14 clusters in the Magellanic Clouds were obtained with the spectrocon attached to the $f/7.5$ focus of the 60-in. Tololo reflector. Colour-magnitude diagrams for two clusters, Kron 3 and NGC 2209, have been derived and were shown. Photometry to nearly 23rd magnitude is possible by this method, making it possible to use the Magellanic Clouds as a laboratory in which to study the change in the C-M diagram as a function of age and chemical composition. Details of the instrument and technique will appear in the Sept. 1970 issue of *Sky and Telescope*, and a discussion of the C-M diagram of Kron 3 will appear in the Sept. 1970 issue of the *Astrophys. J.*

R. Wielen: 'On the Life-Times of Galactic Clusters'.

From the observed age distribution of galactic clusters within 1 kpc we deduce that the typical total life-time of a galactic cluster is about 2×10^8 yr. The individual life-times vary between 10^8 and 10^{18} yr. The observed life-times are compared with the evaporation times which are found from numerical experiments with star cluster models containing up to 250 stars. The effect of the galactic tidal field is taken into account and this enhances the rate of escape significantly. The agreement between the resulting theoretical life-times and observed values is good.

COMMISSION 38: EXCHANGE OF ASTRONOMERS (ÉCHANGE DES ASTRONOMES)

Report of Meeting, 19 August 1970

ACTING PRESIDENT: A. Reiz.
SECRETARY: F. B. Wood.

A memorial tribute to Dr H. D. Babcock was read.

The Draft Report was approved without modification. It was suggested that any proposals for new members be presented to the Organizing Committee.

It was reported that a sum of $ 22000 for the next three years to aid in the exchange of astronomers had been approved by the Executive Committee of the IAU. Since this was not felt to be adequate, the President was requested to ask that at least additional $ 4000 be obtained from ICSU or other sources.

Professor M. V. Migeotte had expressed his wish to resign as chairman of the sub-committee charged with travel grants to graduate students. At the suggestion of Dr Minnaert, it was decided to abolish this committee and to have its functions exercized by the Commission President.

A summary was given of the replies to a request from the President to countries which might wish visits of an astronomer from one of the active scientific centres. Only three replies have been received and in each case it has not yet been possible to supply a visiting astronomer. The suggestion that this sort of exchange might properly lie in the domain of Commission 46 was noted. Dr J. Sahade suggested that when experts in particular areas were required, advice of the Presidents of the appropriate IAU Commissions might be sought.

COMMISSION 40: RADIO ASTRONOMY (RADIO ASTRONOMIE)

Report of Meetings, 19 and 25 August 1970

PRESIDENT: J. P. Wild.
SECRETARY: S. F. Smerd.

Business Meetings

The President welcomed members of the Commission and the large gathering of non-members present. He recorded with deep regret the death in July 1968 of Professor S. E. Khaikin of Pulkovo Observatory.

The President then reported on two proposals submitted by the Executive Committee for consideration by the Commissions:

(a) Whether the new Organizing Committee should be elected or appointed; the meeting favoured the former.

(b) That the Commission be encouraged to draw up such 'by-laws and guiding principles' as may be useful to future Committees.

The President drew attention to an innovation introduced by the Organizing Committee, namely the preparation of 'Reports on Work in Progress' by radio astronomy observatories; 34 such reports became available for distribution during the Assembly.

The following matters were then discussed:

1. THE REPORT OF THE COMMISSION

The President had received overwhelming support for his objection to the apparently arbitrary basis on which the Executive Committee had allocated the maximum length of the Report to each Commission. A resolution had already been submitted to the General Assembly expressing this objection and the General Secretary had foreshadowed the need for new arrangements. The question is to be referred to the incoming Executive Committee.

The Secretary had been notified of one minor amendment; on page 468, line 3 from the bottom of the page: read (96) for (95).

2. MEMBERSHIP

The President reviewed the exchange of suggested deletions and additions between members of the Organizing Committee, the General Secretary and himself which led to the proposed list of Commission-40 members as submitted to the General Secretary prior to the Assembly. Since then he had, with the agreement of the Organizing Committee, proposed that membership should be open to all IAU-members who were active radio astronomers. This proposal would ensure that a wide and representative body of radio astronomers would receive advance notice of the Commission's programmes at forthcoming meetings. This principle was approved by the meeting. A Subcommittee consisting of Westerhout (convener), Parijskij, Shakeshaft and Smerd was set up to receive nominations. This resulted in the new list of members of Commission 40 as published on page 299 of these *Transactions*.

3. INTERFERENCE AND FREQUENCY ALLOCATION

F. G. Smith reported on the activities of the Inter-Union Commission for the Allocation of Frequencies for Radio Astronomy and Space Science (IUCAF) particularly in preparation for the World Administrative Radio Conference (WARC) under the International Telecommunication

Union (ITU) in June 1971. Smith has published a summary of the present status of the allocation of frequencies to radio astronomy and of proposed changes in *Nature* **228**, 419, 1970.

It was agreed that two further requirements should be added to this list:

(a) Some protection should be given to observations in the band 235–240 MHz. Radio astronomy was using this band effectively at some locations despite the fact that it is allocated to other services, but proposals now current to allow the use of transmissions from satellites in this band would make it impossible to continue.

(b) Astronomical observations from the far side of the Moon should provisionally be allocated half the electromagnetic spectrum, by concentrating all transmission into selected bands.

4. ORGANIZATION OF LONG-BASELINE INTERFEROMETRY

Kellermann suggested that it may be some years before agreement could be reached on the most suitable system and frequency bands for long-baseline interferometry.

5. COORDINATION OF SOURCE SURVEYS

Shakeshaft suggested that a Subcommittee of Commission 40 could help in achieving better coordination between radio surveys and between radio and optical surveys; after discussion it seemed agreed that at this stage such coordination may be better achieved through direct cooperation between two or more observatories. However, Shakeshaft and Kellermann were appointed as a Subcommittee to investigate and report on existing surveys and cooperation; M. M. Davis agreed to collate the information and will circulate copies to each observatory.

6. COORDINATION OF FLUX-DENSITY CALIBRATIONS OF STANDARD SOURCES

Shakeshaft suggested that Commission 40 should publish flux-density values to an accuracy of, say, 5% in the range 10 MHz–10 GHz for a number of standard sources. The Shakeshaft-Kellermann Subcommittee will prepare such a list for the next General Assembly.

7. FUNDAMENTAL POSITIONAL SYSTEMS FOR RADIO ASTRONOMY

Ryle pointed out that, with radio astronomy positions now being given to better than 1″ arc, there was a need for a number of calibrators of small angular size; some of their requirements, i.e. appreciable flux density, constancy with time and no low-frequency cut-off, suggest that some sources may be suitable both as position and flux-density calibrators. A Working Group consisting of Ryle (convener), Burke, Christiansen, Cohen, Gent, Palmer and Raimond was constituted to pursue this matter.

8. TEMPERATURE STANDARDIZATION OF H I SURVEYS

Van Woerden, the convenor of a Working Group – or Subcommittee – submitted the following report:

"The Subcommittee of Commission 40 on Standardization of H I surveys recommends that 21-cm line observers compare their results with those of others, and determine the ratio of their brightness-temperature scales to those of one or more other surveys. For such comparisons, we recommend the standard field S8 ($l^{II} = 207°$, $b^{II} = -15°$), which has been used in several surveys. Some observers have used as standard fields S7 ($l^{II} = 132°$, $b^{II} = -1°$) or S9 ($l^{II} = 356°$, $b^{II} = -4°$); we recommend those as secondary standards.

The authors of some of the major surveys are at present intercomparing their scales and expect to publish the results of this in the literature. In view of this, we recommend that the Subcommittee be now dissolved."

The report was adopted and the Subcommittee dissolved.

9. The unit of flux density

The President suggested that the quantity 10^{-26} W m^{-2} Hz^{-1}, given a suitable name and symbol, would – in conjunction with the usual decimal prefixes (μ, m, k, M etc.) – be a desirable unit to express the flux densities throughout the range required in radio astronomy. The proposal lapsed for want of support.

10. Relation between IAU commission 40 and URSI commission V

Christiansen pointed out that (a) radio astronomy was in an anomalous position both in IAU and URSI; (b) there was a need to avoid duplication; (c) the IAU had no wish to abolish Commission 40 as long as radio astronomers wanted the latter; (d) URSI was likely to remain a Union for radio science only in spite of UGGI proposals for a merger; and (e) radio astronomers should continue to emphasize techniques at URSI and astronomy at the IAU although there was a clear need for an exchange flow.

11. Nomenclature for sources

Andrew drew attention to the confusion caused by the use of different nomenclatures in different source surveys; he suggested the general adoption of the Parkes system and pleaded that a source name should not change with improved position determinations.

12. Contamination of space

Lovell suggested that the continued and increased use of space for scientific and other purposes called for a re-affirmation of the appeal to all Governments concerned to consult with the IAU, as expressed in the IAU Resolution No. 1 of 1961. Lovell and Smerd were asked to draft the new resolution: the form in which this was approved by the General Assembly is published as Resolution No. 10 elsewhere in this volume.

Ryle stressed the specific need (see also item 3) for new ITU regulations to ensure that out-of-band transmissions by space vehicles be reduced well below the level currently required for ground-based transmitters. A resolution along these lines as drafted by Hagen and F. G. Smith was approved by the meeting; the form in which the resolution was affirmed by the General Assembly is published as Resolution No. 11 elsewhere in this volume.

13. Organization of commission-40 meetings

Baldwin pointed out that the tight schedule of Commission-40 meetings made the Chairman's job unpleasant and necessitated overflow meetings. He proposed the following guide-lines to improve future meetings: the three-yearly meetings should not be the place to present (a) 'last week's' work; (b) papers that appeared in publications more than, say, 3 months before the Assembly, and (c) papers also presented at other meetings during the Assembly.

Christiansen drew attention to the Executive's recommendation to limit future Commission meetings to 6 sessions.

Westerhout would be sorry if a re-organization of Commission-40 meetings prevented young astronomers from presenting their work and becoming known. *Lovell* drew attention to the need for greater inter-Commission liaison. Following *J. A. Robert's* complaint that these positive suggestions may be lost, it was agreed to send copies of the minutes to the General Secretary and to members of the new Organizing Committee

14. Election of the organizing committee

The President presented the proposed list of nominations for the Organizing Committee. The Commission approved this Committee without further nominations.

Scientific Meetings

In addition to the Joint Discussions 'Helium in the Universe', 'Interstellar Molecules', 'Photoelectric Observations of Stellar Occultations' and 'Pulsars and Cosmic Rays' of which accounts are given elsewhere in this volume, Commission 40 had the following scientific meetings:

1. A Joint Meeting organized by Commission 10 on 'Mapping the Sun in X, UV and Radio Wavelengths'; for details see the report of Commission 10.

2. *Extra-Galactic Work*
   (with Commission 28) August 20 and 21; organized by J. Lequeux.
   P. A. G. Scheuer: 'Extragalactic Radio Sources'.
   R. J. Allen: 'Extragalactic Hydrogen Line'.
   J. E. Baldwin: 'HI in M 33'.
   J. R. Shakeshaft: '5C Survey of the Coma Cluster'.
   C. D. Mackay: 'Orientation of the Axes of Radio Galaxies'.
   P. P. Kronberg: 'Polarization Measurements'.
   H. P. Palmer: 'Structure of Radio Sources'.
   P. K. Wraith: 'Structure of Unidentified Radio Sources'.
   K. I. Kellermann: 'Very Long Baseline Interferometry'.
   J. L. Locke and B. H. Andrew: 'Observation of Variable Sources'.
   E. G. Daintree: '11-cm Survey'.

3. *Supernovae Remnants and Pulsars*
   (with Commission 34) August 20 and 21; organized by G. Westerhout.
   J. A. Roberts: 'Summary of the Crab Nebula Symposium (No. 46) Held at Jodrell Bank'
   J. R. Dickel and B. R. Hermann: 'High Resolution Observations of Supernova Remnants'.
   M. R. Kundu: 'Polarization Observations of Supernova Remnants at 3 and 6 cm Wavelengths'.
   I. Rosenberg: 'Distribution of Brightness and Polarization in Cassiopeia A at 5.0 GHz'.
   D. E. Hogg and R. G. Conway: 'The Polarized Brightness Distribution for the Crab Nebula'.
   L. Matveyenko: 'The Brightness Distribution of the Crab Nebula at 3.5 mm'.
   G. S. Downs: 'Interstellar Scintillations in Pulsar Signal at 2388 MHz'.
   K. R. Lang: 'Interstellar Scintillation of Pulsar Radiation'.
   W. C. Erickson *et al.* (read by G. Westerhout): 'VLBI Observations of NP 0531 and Other Small Angular Diameter Radio Sources at 121.6 MHz'.
   M. Cohen: 'VLBI Observations'.
   I. S. Shklovsky: 'High-Frequency Flux of the Crab Nebula'.
   R. M. Price *et al.*: 'High Time-Resolution Structure in Pulsars'.
   R. Schonhart: 'Polarization and Other Characteristics of NP 0532'.
   M. I. Large (read by F. G. Smith): 'Pulsar Luminosity Function'.
   J. G. Davies: 'Pulsar Searches'.
   T. W. Cole: 'Pulsar Sub Pulses and the Emission Process'.
   A. T. Moffet: 'Pulsar Intensities at 3.8 cm'.
   D. C. Backer and F. D. Drake: 'Non-Random Phenomena in Pulsar Pulse Shapes'.
   J. M. Rankin: 'Dispersion Measure Change in the Crab Nebula Pulsar'.
   C. Heiles: 'Strong Pulses in the Crab Nebula Pulsar'.
   M. Grewing and M. Walmsberg: 'On the Interpretation of the Dispersion Measure'.
   Y. Terzian: 'Dispersion Measures'.
   I. Milogradov-Turin: 'A Survey of the Background Radiation at 38 MHz'.
   E. Berkhuysen: 'Background Survey at 820 MHz'.
   R. Wielebinski: 'Background Surveys at 150 and 85 MHz at Parkes, Australia'.

4. *Interstellar Radio Spectroscopy* (with Commission 34) August 21 and 22; organized by G. Westerhout and Yu. N. Parijskij'.

H. J. Habing, C. E. Heiles *et al.*: 'The Hat Creek 21-cm Line Survey at Intermediate and High Latitudes'.

A. Sandqvist and F. J. Kerr: 'Lunar Occultations of the Galactic Center Region in the Hɪ-, OH- and CH_2O-lines'.

G. R. Knapp (read by F. J. Kerr): 'High Frequency Resolution Observations of 21-cm Profiles in the Directions of Dense Interstellar Dust Clouds'.

R. Sancisi: 'Neutral Hydrogen in the Region of the Dust Cloud L 134'.

S. J. Goldstein: 'The 21-cm Line Spectrum of the North Galactic Pole'.

M. P. Hughes *et al.* (read by A. R. Thompson): 'Interferometer Observations of Galactic 21-cm Line Absorption'.

V. Radhakrishnan *et al.*: '21-cm Absorption Measurements at Parkes'.

F. Sato (read by title only): 'Interstellar Absorption in the Direction of Emission Nebula IC 1795

V. A. Hughes: 'Absorption Measurements in the Direction of W 49'.

K. W. Riegel: '21-cm Absorption Measurements in the Direction of W 31'.

L. E. Snyder and D. Buhl: 'Detection of Emission from an Unidentified Interstellar Molecule'.

B. E. Turner and C. E. Heiles: 'Nonthermal OH Emission from Interstellar Dust Clouds'.

R. D. Davies: 'Mapping of OH Sources Using Long Baseline Interferometry'.

R. S. Booth: 'OH Variability of VYCMa'.

B. F. Burke *et al.*: 'Long Baseline Interferometry of H_2O Line Sources'.

W. T. Sullivan: 'Intensity and Velocity Variations in Galactic Sources of 1.35-cm H_2O Line Radiation'.

5. Hɪɪ *Regions*

(with Commission 34) August 22; organized by G. Westerhout and R. D. Davies.

C. G. Wynn-Williams: 'Maps of the Condensations in W 49A and W3'.

L. A. Higgs: 'A Microwave Survey of Planetary Nebulae'.

Y. Terzian: 'Radio Emission from Small Nebulae'.

R. M. Hjellming and C. M. Wade: 'Radio Emission from Novae'.

R. S. Booth: 'Hɪɪ Region Interferometry'.

A. Pedlar: 'Stark Broadening in Hɪɪ Regions'.

6. *The Sun and Interplanetary Space*

(with Commission 10) 25 August; organized by M. R. Kundu.

J. P. Hagen: (a) 'Observation of a "Smoke Ring" Associated with a Flare'. (b) 'Radio and Optical Observation of Two Nearly Simultaneous Limb Flares'. (c) 'Radio and Optical Observation of Flare Showing Mass Motion'.

S. Enome and H. Tanaka (read by T. Takakura): 'Expanding Limb Burst at 3.75 GHz'.

P. Kaufmann: 'Permanent Changes of the State of Polarization after Microwave Bursts'.

H. Urbarz: 'Magnetic Tape Record of Spectral Development of a Narrow Band Type IV Burst'.

J. Roosen: 'Dynamical Spectral Observations of Bursts in the 160–320 MHz Range'.

A. Maxwell: 'New Spectral Observations of Decimetre Bursts'.

G. A. Dulk: 'Radioheliograph Observations of Harmonic Burst Radiation and Implication on Coronal Structure'.

Y. Leblanc: 'Recent Results at Decametre Wavelengths at Nançay and Arecibo'.

R. G. Stone and J. Fainberg: 'RAE Results on Low-Frequency Solar Bursts'.

P. A. Dennison: 'Interplanetary Scintillation and Solar Wind'.

V. V. Vitkevich (read by N. Lotova): 'On Solar Wind Characteristics'.

N. Lotova: 'On the Effective Scale of Irregularities from Scattering and Scintillation Observations'.

R. Ekers: 'Observation of the Acceleration of Solar Wind near the Sun'.

L. T. Little: 'Is the Scale of Interplanetary Medium as Deduced from Interplanetary Scintillations Compatible with that Derived from Space Probe Observations'.

C. Caroubalos: 'Stereo Project'.

7. *Radiation Theory and Astrophysical Plasmas*
   (with Commission 43) August 21; organized by S. F. Smerd.
V. L. Ginzburg: 'Radiation Mechanisms'.
D. F. Smith: 'Stream Stabilization'.
R. Ramaty: 'Gyro-Synchrotron Radiation and Solar Radio Bursts'.
H. C. Ko: 'Radiation in Magnetoplasma'.
J. P. Wild: 'Simplified Formulae for Synchrotron Radiation'
V. V. Zheleznyakov: 'On the Polarization of Radio Emission from Pulsars'.
V. L. Ginzburg and L. M. Eruchimov: 'The Role of Electron Density Fluctuations in the Determination of the Dispersion and Rotation Measures'.

8. *Fundamental Positional Systems for Radio Astronomy*
   (with Commissions 8 and 24) August 21; organized by C. A. Murray and J. R. Shakeshaft.
M. Ryle: 'The Measurement of the Accurate Positions of Radio Sources'.
W. Fricke (read by C. A. Murray): 'A Fundamental System for Planets, Stars, Galaxies and Radio Sources'.
I. I. Shapiro: 'Pulsar Observations and Spherical Astronomy'.
Short contributions were also presented by M. H. Cohen and H. Gent.

9. *Planetary and Lunar Radio and Radar Astronomy*
   (with Commissions 16 and 17) August 24; organized by G. H. Pettengill

Radio Studies

D. O. Muhleman: 'Interferometer Measurements of Venus and Mercury'.
C. H. Mayer: 'Observations of Venus, Uranus and Neptune'.
A. C. E. Sinclair: 'Interferometer Observations of Venus and Jupiter'.
E. E. Epstein: 'Microwave Spectrum of Mars and 3-mm Data on Mars, Jupiter, Saturn and Uranus'.
A. D. Kuzmin: 'Measurements of Mars Radio Emission at 8.22 mm and Evaluation of Thermal and Electrical Surface Properties'.
A. D. Kuzmin: 'Measurements of the Radio Emission of Uranus at 8.22 mm'.
S. Gulkis: 'Jupiter Observations'.
R. G. Conway: 'Polarization of Jupiter's Decimetric Emission'.
G. L. Berge: 'The Position of Jupiter's Radio Emission Centroid at 21 cm'.

Radar Studies

G. L. Tyler: 'Bistatic Lunar Observations'.
J. E. B. Ponsonby: 'Mapping the Moon by CW Radar at 162 MHz'.
I. I. Shapiro: 'Topography of the Moon, Mercury and Venus'.
C. C. Counselman: 'Topography of Mars'.
G. H. Pettengill: 'Venus Radar Maps'.

10. *Miscellaneous Papers*
    August 25; organized by M. Ryle.
C. G. T. Haslam: '408 MHz Survey'.
J. R. Baker: '408 MHz Polarization'.
R. S. Le Poole: 'Search for Radio Quiet BSO's'.
D. A. Graham and F. G. Smith: 'Pulsar Faraday Rotations'.
J. W. Brooks and J. D. Murray (read by V. Radhakrishnan): 'Zeeman Measurements of Orion A'.
F. D. Drake: 'Performance and Potential of Spherical Reflectors'.
H. D. Greyber: 'Acceleration and Confinement of Plasma Clouds'.

COMMISSION 41: HISTORY OF ASTRONOMY
(HISTOIRE DE L'ASTRONOMIE)

Report of Meetings, 21 and 24 August 1970

PRESIDENT: E. Rybka.
SECRETARY: M. A. Hoskin.

After welcoming the 70 members and guests, the President asked those present to stand in memory of the late Aleksander Birkenmajer and A. V. Nielsen.

The following matters were then discussed:

1. CONSIDERATION OF MEMBERSHIP OF ORGANIZING COMMITTEE

The Commission approved the proposed names for the new officers of the Commission.

2. ADOPTION OF REPORT OF COMMISSION 41, REPORT OF THE WORKING GROUP ON ASTRONOMICAL MICROFILMS AND MANUSCRIPTS, AND ANNOUNCEMENT CONCERNING 'JOURNAL FOR THE HISTORY OF ASTRONOMY'

The Commission Report was approved. *Professor Gingerich* then reported on the Working Group on Astronomical Microfilms and Manuscripts, citing several important microfilm projects completed or underway, including the Lowell Observatory and the Paris Observatory papers.

Dr. Hoskin, Editor of *Journal for the History of Astronomy*, announced that in order to cope with the number of suitable papers offered for publication, the journal would appear three times annually from February 1971.

3. DISCUSSION OF COOPERATION WITH THE INTERNATIONAL UNION FOR THE HISTORY AND PHILOSOPHY OF SCIENCE AND THE INTERNATIONAL UNION OF AMATEUR ASTRONOMERS

(a) *Cooperation with the International Union for the History and Philosophy of Science (IUHPS)*

Ways of improving cooperation between Commission 41 and IUHPS were discussed. After it was explained that IUHPS had no Commission structure comparable with that of the IAU, the Commission agreed it would be desirable to invite IUHPS to appoint an additional member of the Organizing Committee of Commission 41, who would become a Consultant of the Commission if not already a member of the IAU. The Commission agreed to recommend this to the General Secretary and the Executive Committee of IAU.

(b) *Cooperation with the International Union of Amateur Astronomers (IUAA)*

Commander Derek Howse, Chairman of the Historial Commission of the newly-founded IUAA, invited Commission 41 to forward suggestions to him of useful work which his Historical Commission might undertake. He already intended to offer the services of the National Maritime Museum, Greenwich, as a clearing house for manuscript and other information. Welcoming the creation of the Historical Commission, *Professor Rybka* suggested liaison might best be effected by Commander Howse's becoming a Consultant of Commission 41.

4. Consideration of Arrangements for the General History of Astronomy

Professor Rybka reminded the Commission of Resolution No. 4 of the XIII General Assembly endorsing the preparation of such a *History*, and *Professor Gingerich* outlined some tentative suggestions which had been made. Cooperation with IUHPS was especially desirable, and there was reason to think the proposal would have warm support from IUHPS. The cooperation might take the form of Commission 41 and IUHPS each appointing five members to an editorial board; such a board would need to meet for a short but intensive working session to thrash out the structure of the *History* and possible contributions, and arranging for any necessary refereeing might then become the responsibility of the editor, on the advice of the Editorial Board, and a Managing Director might then undertake responsibility for seeing the material through the press. *Dr M. A. Hoskin* and *Dr E. G. Forbes* had tentatively indicated their willingness to act as Editor and Managing Editor respectively.

Dr Whitrow urged that contributions should be issued initially in separate fascicles. *Dr Hoskin* agreed that this was attractive to contributors, but thought that such a procedure might prove inconvenient to both publishers and booksellers.

Dr Fleckenstein spoke of the need of a single master-mind of the calibre of Cantor if a unified history was to result.

Dr Van't Veer expressed the fear that the emphasis would be on facts rather than interpretation, but the President considered the fear unjustified.

The Commission approved without dissent the actions taken and proposed by the Organizing Committee.

5. Preparation of By-laws for Commission 41

The President announced that all Commissions had been asked to prepare By-laws to regulate their own conduct.

6. Discussion of Arrangements for Copernican Celebrations in 1973

The President announced a proposal that Commissions 41 and 47 should sponsor Symposia in Poland in 1973, the 500th anniversary of the birth of Copernicus. This was approved without dissent.

The Commission also noted with approval the possibility of a meeting being arranged by Italian scholars in 1973 in celebration of Copernicus.

7. Discussion of Kepler Celebrations in 1971

Professor Fleckenstein reported on the work of the Kepler Committee founded by IUHPS, in respect of the symposium planned for Leningrad in 1971, the 400th anniversary of the birth of Kepler. He gave a detailed list of speakers and titles arranged by Professor Mikhailov. Professor Fleckenstein explained that of all the celebrations planned, this alone was of truly international character. In particular, contributions would be in various major languages.

Report of Second Meeting

8. Announcements of Forthcoming Meetings and Publications

Professor G. Righini announced that the centenary in 1972 of Arcetri Observatory near Florence was to be marked by a meeting, probably in July, to discuss minor figures in the history of astronomy in the 17th, 18th and 19th centuries. The Commission fully supported these proposals.

A catalogue of rare 15th and 16th century astronomical books is being prepared by *Signora Giovanna Grassi Conti*. The catalogue is at present limited to the libraries of European observatories, but it was hoped that it would be possible in the future to extend the range of the catalogue both geographically and to a more recent date.

It was announced that *Professor D. J. Price* of Yale University is planning a series of facsimile editions of celestial charts, atlases and similar material, to be published by the Theatrum Orbis Terrarum Publishing Company in Amsterdam. Suggestions for possible contributions would be welcomed.

Professor R. B. Berendzen of Boston University is organizing a conference on the education of astronomers to be held in New York beginning 30 August 1971, wherein special reference will be made to the use of material from the history of astronomy.

D. J. Schove announced that a conference on eclipses and chronology since 1375 B.C. would be held in London on 6 November 1970.

Scientific Meetings

The following short papers were presented:

A. Beer: 'The Warburg Museum in Hamburg'.

J. O. Fleckenstein: 'Astronomy in Basle'.

D. J. Schove: 'Sunspot Cycles and the Spectrum of Time'.

Miss A. V. Douglas spoke of her research and the consequences of the closing of St. Helena Observatory and *Commander Derek Howse* described instruments erected at the Royal Navy School Greenwich.

COMMISSION 42: PHOTOMETRIC DOUBLE STARS
(ÉTOILES DOUBLES PHOTOMÉTRIQUES)

Report of Meetings, 19, 20 and 24 August 1970

PRESIDENT: F. B. Wood.
SECRETARY: M. W. Ovenden.

Business Sessions

The business meetings of Commission 42 at the XIV General Assembly of the IAU were held on 1970 August 19 and 24, Dr F. B. Wood presiding. This summary has been compiled from the minutes of the separate meetings.

The present arrangements for the preparation and distribution of the 'Bibliography and Program Notes for Eclipsing Binaries' (B.P.N.) will continue under the direction of Dr G. Larsson-Leander. The Commission agreed that present and past numbers could be made available to libraries and other institutions, at Dr Larsson-Leander's discretion, an adequate charge being made for the services involved.

The work of the 'Times of Minima' Committee, under Dr T. Herczeg, will continue to be based on the Plavec list, which is, however, being updated; a revised list will be circulated to Commission members in due course. It was agreed that a selected list of those stars suitable for visual observation be also made and maintained.

Dr Katherine Kron reported on the coordinated programmes for GL Car and AR Cas. A composite light-curve of AR Cas produced at Catania was exhibited. The Committee on Coordinated Programmes recommended several stars for coordinated observation (see Appendix). The Commission gave unanimous approval to the proposal that it authorize the request for reservation of special nights far in advance at observatories with guest-investigator facilities.

It was agreed that the Joint Working Group of Commissions 30 and 42 be continued for a further three years, with a view to completing the list of binaries in need of observation.

It was agreed to re-establish the Committee for 'Observations from Outside the Earth's Atmosphere'. While recognizing that the IAU already has such a Commission, it was felt that a subcommittee of Commission 42 was required to bring their special needs to attention.

The Commission passed the following resolution:

"While appreciating the financial difficulties of publishing extensive tabular matter, Commission 42 believes that the individual original observations of eclipsing variables are frequently the most important contributions to new knowledge in papers in this field, and should always be published in full. The deposit of observations in some central store does not appear, to this Commission, to be a fully satisfactory alternative, and it therefore urges the editors of all journals to accept, and even request, for publication the complete lists of individual photometric and/or radial-velocity observations."

The Commission voted to approve a request for IAU subvention to the amount of $ 200 towards partial support for the B.P.N.

It was voted to accept the Draft Report, as published.

The Commission approved the nominations of Dr M. Plavec as President and Dr T. Herczeg as Vice-President for 1970–73.

The Commission elected the following Organizing Committee for 1970–73: A. H. Batten, M. G. Fracastoro, K. Gyldenkerne, M. Kitamura, G. Larsson-Leander, D. M. Popper, V. P. Tsesevich, F. B. Wood.

The Commission elected the following Sub-committees for 1970–73, the Chairman having power to co-opt:

1. *Cooperative Programmes*: K. Gyldenkerne (Chairman), M. Kitamura (Vice-chairman), M. G. Fracastoro, G. F. G. Knipe, N. L. Magalashvili, J. Sahade, K. Serkowski, F. B. Wood.

2. *Times of Minima*: T. Herczeg (Chairman), B. Cester, G. F. G. Knipe, A. Landolt, Mrs. M. W. Mayall, R. Szafraniec.

3. *Bibliography*: G. Larsson-Leander (Chairman), B. Cester, J. Grygar, O. Gunther, D. Hall, M. Kitamura, K. K. Kwee, J. Sahade, A. R. Shulberg, S. D. Sinvhal.

4. *Observations from Outside the Earth's Atmosphere*: Y. Kondo (Chairman), J. Grygar, D. Ya. Martynov, J. Sahade, S. Sobieski.

5. *Working Group, Commissions* 30 + 42: From Commission 42, A. H. Batten, M. Plavec, D. M. Popper, J. Sahade.

The business sessions concluded with a brief description, by Dr Fracastoro, of his Catalogue of the Light-curves of Eclipsing Variable Stars.

Scientific Sessions

The following papers were read:

R. H. Koch: 'New Graded Photometric Catalogue of Eclipsing Binaries'.

R. E. Wilson (and E. S. Devinney): 'Synthetic Light Curves'.

J. B. Hutchings: 'Reflection and Other Effects on Light Curves'.

J. E. Merrill: 'Comments on Solution Procedures'.

K. Serkowski: 'Polarization.'

D. Lauterborn: 'Evolution with Mass Loss from System'.

J. Hazlehurst: 'Contact Binaries'.

K. O. Wright: 'Ultraviolet Spectra of VV Cephei'.

Additional short contributions were also made by E. G. Reuning, K.-Y. Chen, A. D. Thackeray, H. Mauder, K.-C. Leung, J.-P. Zahn and Y. Sobouti.

APPENDIX. REPORT OF THE COMMITTEE ON COORDINATED PROGRAMMES

The following stars are recommended for observation in a coordinated campaign:

1. HD 35921, 6th. mag, 09.5III dbl-lined sp. binary, dec. $= +34°$, and above all P$=4.0025$ d. At present there is half a light curve, obtained by Dr Mayer, and some sp. observations by Dr Batten.

2. Beta Lyrae. Drs Sahade and Batten recommended good coverage of two, not necessarily successive, cycles by coordinated spectroscopic and photometric observations with interference filters or scanner observations. All orders for filters should be placed at the same time as they will be expensive.

3. γ_2 Velorum at dec. $= -60°$ and a period of 78 days was recommended for sponsorship in a coordinated programme of photometric observations only. These would be in 6 narrow-band regions, 3 emission-rich and 3 emission-poor. Dr Sahade has already corresponded with Infrared Industries and hopes to place a wholesale order soon. The European Southern Obs. has pledged cooperation and it is expected that telescope time at Cerro Tololo will be obtainable.

4. The long-period eclipsing binaries 31 Cyg, 32 Cyg and Zeta Aur have been put into one campaign because the common denominator is telescope time on critical nights. (See resolution under Business Sessions.)

The Sub-committee felt that there was still a deplorable lack of coordination between observers of eclipsing binaries in general, and recommended increased use of and attention to the reports of programmes planned and in progress as published in the semi-annual bibliography.

The Sub-committee discussed former coordinated programmes and the problem of how to make the best use of the data obtained. It recommends that the coordinator of each campaign follow up on the publication of results and their evaluation and make a status report to Commission 42 at appropriate intervals.

KATHERINE G. KRON, Chairman

COMMISSION 43: PLASMAS AND MAGNETOHYDRODYNAMICS IN ASTROPHYSICS (PLASMAS ET MAGNÉTO-HYDRODYNAMIQUE EN ASTROPHYSIQUE)

Report of Meeting, 24 August 1970

PRESIDENT: Leverett Davis, Jr.
SECRETARY: J. R. Jokipii.

Business Meeting

It was proposed by the President and the Organizing Committee that R. Lüst and R. J. Tayler be nominated as future President and Vice-President, respectively, of Commission 43, and that the Organizing Committee consist of Leverett Davis, Jr., S. A. Kaplan and B. P. Lehnert. This was approved.

The draft report of the commission was approved.

The question of bylaws for Commission 43 was briefly discussed. It was suggested that the method of selection of officers and organizing committee was a matter to be covered by by-laws. Perhaps the list of prospective officers and Organizing Committee could be circulated to the membership before selection.

A number of joint discussions, specifically with Commissions 10, 34, 35, and 43, were proposed for the next meeting.

It was urged that the stringent length restrictions on Commission Reports be relaxed.

Scientific Meeting

The following short papers were presented:
L. Mestel: 'Review of Recent Work on Stellar and Solar Interiors'.
Friedrich Meyer: 'Solar Atmosphere'.
A. B. Severny: 'Stellar Magnetic Fields; New Measurements'.
K. Runcorn: 'Planetary Dynamo Mechanisms'.
J. Wilcox: 'Solar Rotation'.
D. Smith: 'Radiation from Collisionless Shocks'.

COMMISSION 44: ASTRONOMICAL OBSERVATIONS FROM OUTSIDE THE TERRESTRIAL ATMOSPHERE (OBSERVATIONS ASTRONOMIQUES AU-DEHORS DE L'ATMOSPHÈRE TERRESTRE)

Report of Meetings, 19, 20, 22, 25 and 26 August 1970

PRESIDENT: R. Wilson.
SECRETARY: C. Jordan.

Business Meeting

The President announced that the Joint IAU/COSPAR Symposium on 'New Techniques in Space Astronomy,' proposed by Commission 44, had been held in Munich at the Max-Planck Institute from 10–14 August, 1970. The Commission 44 proposal that an IAU Symposium on 'UV and X-Ray Spectroscopy of Astrophysical and Laboratory Plasmas' should be held at Utrecht in 1971 has been accepted.

The President described how the Commission 44 report for 1967–70 had been split into sections according to several topics and thanked the authors of the different sections for their collaboration. The sections were on Solar Physics, The Interplanetary Medium, Cosmic X and γ-ray Sources, the Cosmic X and γ-ray Background, UV Stellar Astronomy and Radio Astronomy. They were written by Dr R. Tousey, Dr M. Neugebauer, Professor L. Gratton, Dr W. Kraushaar, Dr H. E. Butler and Professor F. G. Smith respectively.

The President explained that the scientific sessions of Commission 44 would follow the plan of the Report, with each session being introduced by a review paper, given where possible by the author of the relevant section.

Each Commission is free to arrange its own system of operation and make its own By-laws, but the President pointed out that the General Secretary and Executive Committee wished to know how each Commission intended to organize itself, preferably by the end of the General Assembly. Following a description of how the Commission had been run in the past three years, with the Organizing Committee consulted on executive matters and the full commission membership consulted on scientific matters, Professor Gratton proposed that the present system seemed satisfactory and that it was unnecessary to define any specific rules of operation.

The President listed the proposals received for the new President, Vice-President and Organizing Committee and invited further proposals.

The membership of the Commission has been reduced by removing the names of those members regularly failing to reply to circulars.

The President invited those who wished to join the Commission to add their names to the list posted on the main notice-board.

Representation of the IAU on other international organizations was discussed. Professor C. W. Allen was proposed as the IAU representative on the Solar Particles and Radiations Monitoring Organization. Professor Dollfus agreed on representing Commission 44 as well as Commission 17 on the Intercommission Committee on the Exploration of the Moon.

Commission 44 wishes to note the letter of Professor F. Zwicky, which draws attention to the importance of astronomers taking an active interest in the work leading to an International Treaty on the Use of Outer Space.

The relationship between Commissions 44, 47 and 48 and between Commissions 44, 10 and 12 was discussed. It was agreed that overlap between commissions should be minimized, but no clear way of dividing interests between the above commissions was apparent. There was a general feeling, expressed particularly by Professor L. Gratton and the President, that Commission 44 should limit itself to techniques and observations rather than include theoretical analyses of data. The President thought that the Executive Committee of the IAU should examine the role of all the commissions to avoid proliferation.

Regarding future symposia, Dr Giaconni pointed out that by 1972 many data are expected from new satellites devoted to X and γ-ray Astronomy. He suggested that an IAU symposium on Cosmic X and γ-ray Astronomy should be held in the eastern United States in 1972 or 1973 and that this should be similar to the 1969 Rome Symposium. It was agreed that such a meeting should be held jointly with Commission 48 and that the President should initiate a joint proposal.

Scientific Meetings

The first solar meeting opened with R. Tousey reporting some of the highlights of the past year. The eclipse of 7 March 1970, during which two rockets were launched from Wallops Island, West Virginia, to reach apogee at totality and so record the flash spectrum in the ultraviolet, was considered as the main event.

C. Jordan then reported results obtained from a recent Skylark rocket flight instrumented by the Culham group. These have been used to derive the temperature structure of the chromosphere-corona transition region.

K. Evans presented results from recent rocket-borne crystal spectrometers. Models of active regions were constructed and abundances of elements in the corona were determined.

L. Goldberg presented results from a multigroup experiment designed to observe the XUV flash spectrum of 7 March 1970. H-$L\alpha$- was observed between 1 and 1.3 solar radii and it has been suggested that this is due to scattering of H-$L\alpha$ by neutral hydrogen.

The meeting concluded with G. S. Vaiana presenting X-ray images of the Sun obtained by the A.S. and E. group over the last two years. The images have a resolution of 5 arc sec for active regions and 1 arc sec for flares.

Further results were reported during the second solar session.

K. Frost presented results from the GSFC hard X-ray experiment on OSO 5. Various flare profiles were presented and their correlation with Type III radio bursts was discussed.

J. H. Parkinson described results from a joint experiment between the UCL and Leicester groups, also on OSO 5. Soft X-ray maps of the Sun were presented and models of active regions were constructed as a function of the photospheric magnetic field.

Results from a third experiment on OSO 5 were presented by W. M. Neupert. The spectra from two crystal spectrometers were discussed, with special emphasis on the feature at 1.86 Å and its resolution into several iron lines.

L. W. Acton reported the analysis of X-ray spectra above 3 keV obtained on three rocket flights. The spectra were assumed to be thermal so electron temperatures and emission measures were derived.

Finally, L. van Speybroeck considered X-ray photographs obtained on the same day as the 7 March 1970 eclipse. The relations between the white light coronograph observations and the X-ray emitting regions on the solar disk were discussed.

Scientific Meetings, Second Session

The first meeting consisted of a series of papers concerning the physics of the interplanetary medium.

J. C. Brandt read the Draft Report prepared by Dr M. Neugerbauer who was unable to attend the meeting. Interplanetary space has been continuously monitored over the last five years from space vehicles, and important advances have been made in our understanding of the solar wind. It has been shown to be electrically neutral, for example, and to have only a small variability with solar cycle.

J. C. Brandt reported that the two-fluid model of the solar wind cannot explain the observed properties even at 1 AU from the Sun. The inclusion of viscosity as a method of heating the proton component is essential to reduce the ratio T_e/T_p to a value near to the measured ones of between 1.5 and 5.0.

T. Gehrels reported work on polarization measurements on the atmosphere of Venus, using the polariscope balloon-borne instrument. Measurements at 2150 Å and 2800 Å show 22.8% linear polarization which can be attributed to the molecular atmosphere above the Venusian clouds.

T. Gehrels then gave a paper on the Pioneer F + G missions to Jupiter. These will carry an imaging polarimeter with which the polarization of Jupiter will be mapped, along with measurements on any asteroids to which the spacecraft may pass sufficiently close.

F. Zwicky made a plea for more widespread understanding of existing international treaties governing the peaceful uses of space, and suggested that for the purposes of future legislation, a more flexible definition of where a country's 'airspace' ends and where 'space' begins, should be made.

The second meeting of the day was concerned with recent results in the field of stellar UV spectroscopy.

Dr H. E. Butler (Edinburgh, U.K.) began the meeting with a review of progress in UV astronomy between 1967 and 1970, pointing out that many of the new observations have resulted from advances in space instrumentation technology.

Recent observations were described and the importance of reliable photometric calibration was stressed.

Dr A. D. Code (Wisconsin, U.S.A.) described the scanning spectrometer experiment carried on OAO-A2 and summarized the observed stellar spectra. Spectra have been recorded between 1100 Å and 3500 Å with a spectral resolution of about 10 Å. The observed continuum fluxes give reasonable agreement with model atmosphere calculations for spectra types earlier than AO, but departures are found for later types. Adjustments of the metal abundances help to match the observed and theoretical curves. Stars having the same spectral type and luminosity class have similar UV spectra, but the line spectra of supergiants and main sequence stars show distinct differences. The instrument intensity calibration has proved to be extremely stable throughout the first 15 months in orbit.

Dr B. D. Savage (Wisconsin, U.S.A.) described measurements of UV interstellar extinction obtained from the OAO-A2 stellar spectra. The observed extinction curves have a peak near 2200 Å and a minimum in the region 1800–1300 Å, followed by an increase in extinction at shorter wavelengths. Theoretical curves for graphite particles could not be matched with the observed extinction, some other constituent being needed to explain the peak near 2200 Å.

The results obtained by the OAO-A2 'Celescope' experiment were described by Dr W. A. Deutschman (Smithsonian, U.S.A.). Data analysis is now in progress and a catalogue listing the UV magnitudes (± 0.2) of 10000 stars is expected to be ready before 1971.

Mr J. W. Campbell (Edinburgh, U.K.) described results obtained using broad-band UV photometers on unstabilized 'Skylark' rockets. Measurements of 94 early-type stars had been made in wavebands centred near 1450 Å, 2150 Å and 2550 Å.

Scientific Meetings, Third Session

The meetings consisted mainly of contributed papers in the field of cosmic X-ray and γ-ray astronomy.

L. Gratton opened the first meeting with a review of cosmic X-ray sources, and listed eleven sources which have been identified with optical counterparts.

K. A. Pounds reviewed the X-ray and γ-ray background; the break in the X-ray spectrum was discussed, together with the isotropy of the general background radiation and the enhancement along the Galactic plane.

B. A. Cooke reported a new survey of the Centaurus-Norma region in the 2–16 keV range, discussing the changes in the intensities of sources observed in this region over the past few years, particularly Lupus X-1 and Cen X-3. An unusual spectrum was presented for the latter source.

J. H. M. Bleeker described the results of a 20–80 keV search for X-rays from the Galaxy using 4 identical independent scintillation detectors. No discrete sources were observed in the Galactic centre region. The spectrum of the Galactic enhancement, incorporating all published observations of the excess radiation in the X- and γ-ray region, was discussed in the light of theoretical considerations.

V. L. Ginzburg discussed the chemical composition of quasars in relation to the effect of γ-radiation. No chemical changes have been observed in the envelopes of quasars – the possible reasons for this were presented.

P. Gorenstein described a recent X-ray observation of the Cygnus Loop using a focusing collector. The structure of the source in X-rays was discussed, the resolution of the detector being $\frac{1}{2}°$. The spectrum of the source was best fitted by thermal bremsstrahlung with line emission from O VIII contributing $\frac{1}{3}$ of the total emission in the observational energy range.

The second meeting continued with a description by R. E. Griffiths of the results of a recent rocket experiment to search for iron line emission from Sco X-1. A narrow line was not observed with the LiF crystal spectrometer, but the overall spectrum in the 4–15 keV region from proportional detectors showed an excess around the expected positions of the emission lines from Fe XXV and XXVI. It was concluded that the emission lines are broadened by the process of electron-scattering in a dense source.

L. Acton presented the results of a proportional counter experiment on Sco X-1 with the detection of iron line emission at 6.6 keV.

W. A. Hiltner reported the results of optical studies of the intensity variations of Sco X-1 and the correlation of X-ray and optical fluctuations.

H. M. Johnson discussed a programme for the optical identification of X-ray sources, using the Mount Palomar Schmidt telescope.

J. E. Grindlay considered flare stars as candidates for X-ray sources. Intensities were derived, assuming a synchrotron model, for X-rays of energies greater than 10 keV.

In the discussion following Grindlay's paper, Lovell commented on the very large energy release during the 3–4 h flaring of YZ Can. Maj.

T. A. Clark concluded the meeting with some results from the Radio Astronomy Explorer satellite on observations of the galactic background continuum radiation in the Local System, and discussed the possible H II distributions.

Scientific Meetings, Fourth Session

The second meeting on stellar UV spectroscopy began with a report by Dr D. C. Morton (Princeton, U.S.A.) on recent ultraviolet stellar spectra of high resolution (0.3 Å) obtained by an objective grating Schmidt camera flown in a stabilized rocket. Interstellar absorption lines were recorded in spectra of ζ-Oph and δ-Sco between 1100 Å and 1500 Å. The interstellar lines include H I 1216 Å, C II 1334 Å and O I 1302 Å from which an abundance ratio [O I]/[H I] of about 7×10^{-3} can be estimated. As the solar ratio is 6×10^{-4} an overabundance of O I in the interstellar cloud is implied.

The interpretation of the interstellar hydrogen absorption in these spectra was discussed by E. B. Jenkins (Princeton, U.S.A.) who also reviewed interstellar H I absorption data obtained on earlier rocket flights. Using a statistical method to determine the correct continuum level, equivalent widths were estimated for the 1216 Å line and values for the interstellar atomic hydrogen density were derived. The estimated density (~ 0.2 cm^{-3}) is significantly lower than the value obtained from 21 cm radio data or the density derived from OAO-A2 observations.

Ultra-violet spectrophotometric measurements of bright stars obtained by balloon-borne instrumentation were described by C. Navach (Geneva, Switzerland). An objective prism Schmidt camera was used to record spectra between 2000 Å and 5000 Å and the observed intensities agree well with those obtained by Stecher.

G. C. Sudbury (Edinbrugh, U.K.) described spectrophotometric measurements made from unstabilized rockets using photoelectric scanning techniques. Photometric data were obtained for 100 stars in the wavelength range 1850–3000 Å. Measurements of the background sky brightness at 2425 Å have provided new data on the distribution of radiation from the galactic plane, which can be interpreted to give information about the location of dust clouds in the Galaxy.

C. M. Humphries (Edinburgh, U.K.) presented photographic spectra recently recorded using an objective prism Schmidt camera carried on a stabilized Skylark rocket. Spectra between 1700 Å and 5000 Å were recorded for stars brighter than $M_v \simeq 5$ in Centaurus and Lupus.

During the third meeting on UV stellar spectroscopy Dr A. Boggess (GSFC Greenbelt, U.S.A.) described the design and operations plan for OAO-B, the second astronomical observatory satellite which will be launched late in 1970. This satellite will have a scanning spectrometer fed by a Cassegrain telescope system. The importance of detailed planning to optimize the observing programme was stressed. A Guest-observer programme will provide ten percent of the available time for observations by invited astronomers.

In a second paper, Dr A. Boggess gave details of a geosynchronous astronomical satellite which is currently being studied for a possible launch in 1974. This satellite would use a Cassegrain telescope and a spectrometer with crossed echelle and grating dispersion, followed by image storage tube detection with T.V. data read-out. The high efficiency data-collection system and the geosynchronous orbit would provide a space observatory with virtually direct ground control.

The complete list of papers presented during the Commission 44 sessions is as follows:

Thursday, 20 August 1970

SOLAR MEETING I

R. Tousey, 'Solar Astronomy, Some Highlights from the Past Year'.

W. M. Burton, C. Jordan, A. Ridgeley and R. Wilson. 'The Structure of the Chromosphere – Corona Transition Region from Limb and Disk Intensities'.

R. M. Batstone, K. Evans, J. H. Parkinson and K. A. Pounds, 'New X-Ray Spectral Data on Solar Active Regions from a Recent Rocket Experiment'.

L. Goldberg, 'XUV Flash Spectrum of the 1970 Eclipse'.

G. S. Vaiana, 'Quiescent and Active X-Ray Corona; Observations and Results'.

SOLAR MEETING II

K. J. Frost: 'Observations of Hard Solar X-Rays by OSO 5'.

J. H. Parkinson, K. A. Pounds, and J. R. Herring: 'Time Variations in the Solar X-Ray Flux from the OSO 5 Data'.

W. M. Neupert: 'X-Ray Spectra Associated with Solar Activity'.

L. W. Acton, R. C. Catura, J. L. Culhane, P. C. Fisher, and A. J. Meyeropp: 'High Resolution Measurements of Solar X-Ray Spectra above 3 keV'.

L. P. van Speybroeck, A. S. Krieger, and G. S. Vaiana: 'X-Ray Observations of the 1970 Eclipse'.

Saturday, 22 August 1970

INTERPLANETARY MEETING

M. Neugebauer: IAU Draft Report "The Interplanetary Medium", read by J. C. Brandt.

J. C. Brandt: 'The Mean Free Path in the Solar Wind Plasma'.

T. Gehrels: 'Results from Polariscope Balloon Flights'.

T. Gehrels: 'The Pioneer F + G Missions to Jupiter'.

STELLAR MEETING I

H. E. Butler: 'Review of Ultra-Violet Stellar Astronomy'.

A. D. Code: 'Ultra-Violet Stellar Spectra from OAO II Observations'.
B. D. Savage: 'Interstellar Extinction Measurements Based on OAO II Observations'.
W. A. Deutschman: 'Status Report on Project Celescope'.
J. W. Campbell: 'Photoelectric UV Observations'.

Tuesday, 25 August 1970

COSMIC X AND γ-RAY MEETING I

L. Gratton: 'X-Ray Sources'.
K. A. Pounds: 'The X and Gamma Ray Background Radiation'.
B. A. Cooke: 'A New X-Ray Survey of the Centaurus-Norma Region'.
J. H. M. Bleeker: 'A Search for Hard X-Ray Emission from the Galaxy'.
V. L. Ginzburg: 'The Effect of Gamma Radiation on the Chemical Composition of Quasars'.
P. Gorenstein: 'X-Ray Observations of the Cygnus Loop'.

COSMIC X AND γ-RAY MEETING II

R. E. Griffiths: 'X-Ray Line Emission from Sco X-1'.
L. Acton: 'X-Ray Line Emission from Sco X-1'.
W. A. Hiltner: 'Optical Variations of Sco X-1'.
H. M. Johnson: 'A Programme for the Optical Identification of X-Ray Sources'.
J. E. Grindlay: 'Flare Stars as X-Ray Sources'.
T. A. Clark: 'Observations of the Galactic Background Continuum Radiation in the Local System'.

Wednesday, 26 August 1975

STELLAR MEETING II

D. Morton: 'Results of Recent Rocket Experiments'.
E. B. Jenkins: 'Results of Recent Rocket Experiments'.
C. Navach: 'Ultra-Violet Spectrophotometry of Stars Based on Balloon Borne Observations'.
G. C. Sudbury: 'Low Resolution Photoelectric UV Stellar Observations'.
C. M. Humphries: 'Objective Prism Spectrophotometry from a Stabilized Rocket'.

STELLAR MEETING III

A. Boggess: 'The Observing Programme for OAO-III'.
A. Boggess: 'Proposed Astronomical Satellite in Geosynchronous Orbit'.

Additional Solar Papers

V. A. Krat: 'Recent Balloon Observations of the Sun'.
R. M. Bonnet: 'Balloon Observations of the Solar MgII Lines'.
J. F. Vesecky: 'The Cooling of Solar Flare Plasmas, as Observed by OSO 4'.

COMMISSION 45: SPECTRAL CLASSIFICATIONS AND MULTI-BAND COLOUR INDICES (CLASSIFICATIONS SPECTRALES ET INDICES DE COULEUR A PLUSIEURS BANDES)

Report of Meetings, 19, 22 and 26 August 1970

PRÉSIDENT: Ch. Fehrenbach.
SECRÉTAIRE: B. Hauck.

Séance administrative

Le rapport du Président est accepté à l'unanimité. Deux adjonctions sont demandées (Voir Appendice A). L'organisation future de la Commission est acceptée également à l'unanimité.

W. P. Bidelman et B. Hauck représenteront la Commission 45 dans le groupe de travail inter-commissions.

Le Président souligne l'importance du problème posé par la collection des données numériques et invite les membres de la commission à renoncer à la prochaine séance afin de pouvoir participer à celle du groupe de travail, présidé par Mrs Dr. Sitterly et consacré à ces questions.

W. Buscombe annonce la préparation d'un catalogue indiquant les types MK et les mesures UBV pour près de 10000 étoiles par Mrs P. Kennedy (Mt Stromlo). B. Hauck rapporte ensuite sur l'état des travaux du fichier photométrique entrepris à Genève et à Lausanne. Deux fichiers, l'un contenant les mesures publiées dans la littérature et le second les moyennes pondérées, sont prévus pour chaque système. En fin de séance, D. J. Mac Connell présente le rapport concernant le 'Michigan Southern Spectral Survey'. Ce travail est effectué avec le télescope de Schmidt de l'Université de Michigan à Cerro Tolo, Chili. Des plaques courant 65 % du ciel sud ont déjà été obtenues (dispersion de 108 A/mm à Hγ). De nombreuses étoiles particulières ont été découvertes et les résultats sont publiés (H. E. Bond et al., A. J. Juin 1970).

Première séance scientifique

Le Président propose la création d'un groupe de travail consacré aux données numériques qui sont du ressort de notre commission. A l'unanimité, la commission accepte un groupe composé de B. E. Westerlund, W. P. Bidelman, Mme M. Barbier et B. Hauck, ce dernier étant président.

Communication:

Mlle L. Divan présente l'important travail fait dans le système Barbier, Chalonge, Divan (BCD) et les possibilités offertes par ce système pour la classification stellaire. (Présentation d'un catologue de données spectrophotométriques pour environ 400 étoiles et de l'étalonnage en magnitude absolue de la classification BCD). Mme G. Cayrel (Spectral classification of some halo population stars) montre que les étoiles froides déficientes en métaux ont un type spectral trop précoce et que l'analyse de spectres à grande dispersion permet d'obtenir une classification correcte. Elle propose un symbole pour identifier dans un catalogue cette catégorie d'étoile.

R. M. West présente ensuite une méthode de classification automatique des étoiles G5-K5 à partir de spectres obtenus au prisme-objectif (170 Å/mm). Le programme permet d'obtenir les résultats avec une précision de ± 1 sous-classe pour le type spectral, ± 1.2 en Mv et ± 0.15 en [Fe/H] dans le cas des géantes.

A. K. Alknis lit ensuite une communication de V. Straižys montrant la possibilité d'obtenir une représentation stellaire tri-dimensionnelle à partir du système photométrique de Vilnius.

A. G. D. Philip (classification of giant A stars far of the galactic plan by means of uvby β photometry) montre qu'il est possible de distinguer les étoiles de la branche horizontale des étoiles nor-males en étudiant les diagrammes $c_1/b - y$ et $m_1/b - y$.

Seconde séance scientifique

D. L. Crawford (standard stars for uvby photometry) présente une liste de 304 étoiles devant servir de standard lors de mesures faites dans le système uvby. Cette liste sera publiée prochaine-ment et il est souhaitable que ceux qui font des observations dans ce système l'utilisent.

P. W. Hill communique ensuite les résultats provisoires obtenus au Cape Observatory pour l'étoile $CD - 42° 14462$, suspectée d'être une naine blanche, soit $V = 10.30$, $B - V = + 0.01$ et $U - B = - 0.79$.

Ch. Fehrenbach présente alors les résultats spectroscopiques et photométriques obtenus par l'équipe de Marseille pour le Grand Nuage de Magellan. Des spectres à 73 Å/mm ont permis la classification spectrale de 142 étoiles. Une liste de 33 étoiles est proposée comme étoiles standards pour la classification des étoiles supergéantes. Des mesures photométriques UBV de 317 étoiles ont également été obtenues. L'absorption moyenne (galactique et dans le Nuage) a été trouvée égale à $V = 0,30$. Les diagrammes couleur – couleur établis par séquence de magnitude – permettent d'établir les couleurs intrinsèques des étoiles supergéantes et montrent une variation très nette avec la lumi-nosité.

A. Ardeberg (A study of NGC 1910, LMC) a également obtenu de nombreuses mesures spectro-scopiques et photométriques des étoiles de cette association à l'aide des télescopes de l'ESO. Ses résultats suggèrent que l'association est extrêmement jeune et que l'absorption interstellaire interne est de 0.7 magnitude pour V.

W. P. Bidelman donne ensuite lecture des rapports de W. W. Morgan et de Ph. C. Keenan à la commission des parallaxes trigonométriques. L'importance de la classification spectrale pour l'ob-tention de bonnes parallaxes spectroscopiques est évidente et les erreurs systématiques (notamment entre les hémisphères nord et sud) devraient être éliminées.

J. A. William, avec M. Aller et G. Elste (University of Michigan), a entrepris l'établissement d'un catalogue d' étoiles pour lesquelles les abondances sont connues.

P. M. Williams (Cambridge Observatories) a étudié une représentation tri-dimensionnelle (T_{eff}, g et [Fe/H]) pour les étoiles froides à partir de mesures d'indices spectrophotométriques et de couleur infrarouge.

E. K. Kharadze et R. A. Bartaya (on the identification of peculiar stars by low dispersion spectra) ont obtenu la classification d'environ 10000 étoiles dans les Selected Areas 2 à 43. 134 étoiles particulières ont été trouvées, 97 sont nouvelles.

W. Seitter annonce en fin de sécance que l'*Atlas spectral de Bonn* est publié et elle le présente.

APPENDICE A. ADDENDA AU RAPPORT DE LA COMMISSION

(a) *R. F. Garrison*

Hiltner, Garrison and Schild (*Astrophys. J.* 1969) have published spectral classifications on the MK system of all O and B type stars earlier than B8, brighter than 6.5. and south of $-20°$. They are now extending this survey to all OB stars brighter than 10th magnitude with stars chosen from the Heidelberg objective prism survey. The observations should be completed early in 1971.

(b) *W. P. Bidelman*

The Statement in paragraph 220 for the Draft Report concerning the work of H. E. Bond on the relation between chemical composition and eccentricity of stellar galactic orbits is a bit misleading. His precise conclusion (1969) is that "Among the stars with more than one-tenth the metal content of the Sun ($\Delta m < 0.^m08$), there now appears to be little or no correlation between kinematics and chemical composition". For larger metal deficiencies, such a correlation does indeed exist, though it shows considerable scatter.

COMMISSION 46: TEACHING OF ASTRONOMY
(ENSEIGNEMENT DE L'ASTRONOMIE)

Report of Meetings, 19 and 21 August 1970

PRESIDENT: E. A. Müller.
SECRETARY: T. L. Swihart.

First Session

Membership

The *President* explained the policy of the Commission concerning membership. Each country adhering to the IAU should have a representative in Commission 46 in charge of making the liaison between the activities of the Commission and the astronomy teachers in the various institutions and all what concerns the teaching of astronomy in his (or her) own country. Every member representative should be designated or approved by his (or her) National Committee of Astronomy. Two countries previously without representation were added to the Commission: *Austria*, represented by *H. F. Haupt*, and *Brazil* by *S. Ferraz-Mello*. By personal contacts during the Brighton meetings the President added the following countries to the Commission: *Indonesia*, represented by B. Hidayat; *Iran* by *A. Therian*, *Finland* by K. A. Hämeen-Anttila, and *Yugoslavia* by F. Dominko.

The *Presidnet* then proposed that the Organizing Committe of the Commission should consist of the President, the Past-President, the Vice-President, and of those members in charge of continuing or developing one of the projects of the Commission, and such others as considered necessary by the President.

Commission Report

No special remark was made concerning the report prepared for the *IAU Transactions* **XIVA**.

National Reports

The *President* remarked that a mimeographed copy of the reports on national activities in astronomy education could be consulted in the book exhibition at the University of Sussex on the occasion of the IAU meeting.

G. O. Abell gave a summary of activities of the Committee for Education in Astronomy (CEA) in the U.S. The CEA is an official organ of the American Astronomical Society and its purposes are similar to those of Commission 46 except on a national rather than international level. *R. E. Berendzen* announced that an International Conference on Astronomy Education and on the History of Modern Astronomy will be held from August 30 to September 1, 1971, in New York City. Participation from as many countries as possible is desired. Interested persons unable to attend are encouraged to send relevant material or information to Berendzen (Dept. of Astronomy, Boston University, Boston, Mass. 02215, U.S.A.)

The IAU/UNESCO International Schools for Young Astronomers

J. Kleczek reported on the previous schools held in England (1967), Italy (1968), and India (1969). The 1970 ISYA is to be held in Argentina during October and November. The purpose of the schools is to give a concentrated expert instruction and training in special topics of modern astronomy for a small number of selected young astronomers or physicists, with or without Ph.D., who otherwise would not have such opportunities available to them. If funds are available future ISYA will be planned, one every year. *J. Kleczek* had agreed to continue serving as Secretary of the ISYA for another three-years term. It was felt that an Assistant Secretary of the ISYA should be designated to help Kleczek in his important task. As a result of numerous private consultations before and during the Brighton meetings the President invited *I. Atanasijević* to serve as Assistant Secretary of the ISYA. He was introduced to the Commission in the beginning of the fourth session.

The Visiting Professors Project

Speaking for *M. Minnaert*, who most regrettably was absent due to illness, *E. A. Müller* reported on this project. The aim is to provide contacts between places remote from main centres of astronomical research which could benefit by the visit of an astronomer teaching a course of particular interest to them and possibly helping them develop a research and/or teaching programme on the one hand, and, on the other, astronomers who would be willing and able to make such visits. India, Indonesia and Uruguay had previously indicated a desire for such visits, and Brazil also responded at the present meeting. One of the problems is to find the capable astronomers who can devote some of their time to go and teach, stimulate, and encourage astronomical research at one of the interested institutions. Another problem is the funding of such visits. In some cases living expenses for the duration of the visit can be provided by the host institution, but travel expenses are quite another matter. It was suggested that such visits could perhaps be combined with other sanctioned travel of the interested astronomers. For example, a visit to Brazil and Uruguay could be accomplished by an astronomer whose observing schedule takes him to one of the observatories in Chile. It was also pointed out that the solar eclipse in Africa in 1973 could provide for such visits to that continent. The President proposed that *M. Minnaert* should continue to be in charge of this project.

After the session *R. Steinitz* informed the Commission Secretary that he would be willing to devote some time as Visiting Professor in one or the other of the South American countries.

The ICSU Committee on Science Teaching

As representative of the ICSU Committee on Science Teaching (CST) *E. A. Müller* mentioned the formation, the purpose, the activities and the plans of the CST which she had described briefly in the Assembly Times No. 2 of 19th August 1970. The main objectives of the CST are (a) to further on an international scale progress in the teaching of science at all levels, (b) to co-operate with other organizations concerned with any aspect of the teaching of science, (c) to facilitate co-operation among the teaching committees of the individual scientific unions. The Committee consists of the Chairman and six members appointed by ICSU and of one representative of each Scientific Union, member of ICSU, interested in the teaching of science. Some observers of the UNESCO Division of Science Teaching are invited to attend the Committee's meetings and to collaborate with the Committee in the programmes which are of mutual interest. A major conference on science education is envisaged for 1972 or 1973 which will be concerned with the various aspects of the training and re-training of science teachers for the teaching of integrated science. Before this major conference one or two symposia on specific topics are planned for 1971 and 1972. *E. A. Müller* agreed to continue to represent the IAU on the ICSU Committee of Science Teaching.

Integrated Sciences

E. A. Müller pointed out that in many countries astronomy education in primary and secondary schools is quite inadequate. Since only few schools can offer a separate course in astronomy, astronomy should be integrated into courses, in particular physics. She suggested that perhaps Commission 46 should start a project of preparing leaflets or teacher's guides on selected topics of modern astronomy which would explain the topics and show how they relate to topics in physics, for example. Some material already in existence, which may at least partially achieve these aims, were mentioned. *E. Schatzman* stated that he thought the best approach is to work directly with the physicists. If we can educate them so that they understand astronomy adequately, physics courses will automatically give a more balanced approach in which astronomy is given proper coverage. The decision of the Commission on the suggested project was postponed to a later session.

The International Exchange and/or Unilateral Gifts of Teaching Material in Astronomy

D. McNally brought up the problem of the large expenses of many essential educational materials for developing countries and the situation of countries with currency exchange problems. One possible answer is that those institutions, organizations or individuals which can afford to do so, make materials available as gifts. *B. Bok* mentioned how he goes about sending scientific literature directly to developing institutions. He also suggested that the Commission contact the UNESCO representatives to see what they would suggest. It was agreed that *D. McNally* would be in charge of this project. The *President* mentioned that UNESCO issues coupons for developing countries which can be exchanged for books published in any country. The conditions are different for each country. All countries eligible for UNESCO assistance have a UNESCO representative who can be contacted directly or through the local Ministry of Education or its equivalent.

The Contratype Project (CP)

E. K. Kharadze described a project proposed by himself, *E. V. Kononovich* and *E. K. Straut* which consists of providing a clearing house where slides, pictures, etc., of reasonable current interest could be made available for reproduction to anyone interested. This is for educational rather than research purposes. All interested observatories would provide materials (suitably standarized if necessary) to the central headquarters of the CP. Anyone interested could have the desired materials sent to them for reproduction, and the originals would then be returned to CP. As *D. Wentzel* had been involved in a similar type project it was suggested that he meet Kharadze to discuss details. The President proposed that *E. K. Kharadze, E. V. Kononovich* and *D. Wentzel* should be in charge of the CP. Miss *M. Gerbaldi* agreed to act as secretary of this group.

Astronomy Educational Material

It was announced that in order to be useful the list of AEM should be updated every three years for each General Assembly. *T. L. Swihart* would collect the English materials and *V. Kourganoff* would be in charge of compiling the materials of the non-English countries by asking the Commission members of all countries for collaboration.

Second Session

This session was devoted to the discussion of the list of *Astronomy Educational Material* which had been prepared by Commission 46 for this General Assembly. *E. A. Müller* explained that the AEM list was prepared for the benefit of teachers and students of all levels and, in particular, it was aimed to serve as guideline to those institutions and persons all over the world who were trying to build up an astronomical library and/or get some astronomy going. She pointed out that a strict and careful selection had to be made as to the usefulness of each item for the teachers and the students. Of the language that she could judge she felt that the items in French, German and Italian had been very well selected, but that the books in English had not been carefully selected. It was decided that as far as the items in English go, some weeding out was to be done to make the list more useful and the US members of the Commission were invited to do this. *P. V. Sudbury* proposed that the AEM should include a list of planetarium manufacturers, the planetaria being a very useful astronomy teaching aid for schools and colleges. The Commission decided that such a list should be included in the AEM, and Sudbury agreed to compile the list.

The main problem is the publication, the price and the distribution of the AEM compilation. In order to be really useful, it should have a large, world-wide distribution. The question is how to go about this. Some members of the audience felt that the present form of the AEM was satisfactory, others indicated that the publication of the AEM in its present form would be too expensive. *B. Bok* suggested that the Commission should contact UNESCO who might be the agency for a worldwide distribution of the AEM. The *President* agreed to do so.

It was decided that for the time being the AEM would be kept in its present form. Each Commission member has a copy of the AEM and has the task to publicize the existence of this compilation to astronomy or science institutions, to teachers associations and others in his (or her) own country. Orders for copies could be sent to the President who could provide mimeographed copies for a modest price covering the paper and mailing costs.

Third Session

At the beginning of this session *D. Wentzel* reported on the meeting he had with *E. K. Kharadze* and *M. Gerbaldi* concerning the Contratype Project. They decided on the following: The major observatories would be contacted and asked to supply contact prints of current interest to the CP file. These will then be sent to any institution on demand for reproduction; the originals will then immediately be returned to the CP. Costs will thus be limited essentailly to postage. The file will constantly be updated and information on what is available will be periodically circulated. It was again stressed that this is for educational not research purposes. The present secretary for the CP is Miss *M. Gerbaldi*, Institut d'Astrophysique, Paris, France. Here the file will be established, at least for the time being. Notice will be given when the CP is ready to receive requests. It was emphasized that southern hemisphere material should be included. *D. McNally* also stressed that care should be taken that no copyright problems might arise, i.e. to ascertain that the user is free to reproduce the material in any way he wishes.

The session was then devoted to discussions concerning *astronomy education at the primary and secondary school level.* Everyone realised the importance of giving some basic knowledge of modern astronomy to the general public, the school children and, in particular, to the science teachers. The discussions mainly dealt with the basic questions of how astronomy should be introduced in schools and how the science teachers should be guided and instructed in modern astronomy. Since most schools do not offer a separate course in astronomy, the importance of integrating astronomy in science courses, mainly physics, was stressed. *D. McNally* pointed out that the students at this age group, growing up in the space age, are much more aware of the recent advances than their science teachers and, therefore, the teachers must currently be brought up-to-date. *F. Egger* said that astronomy should not be introduced simply as an application of mathematics and physics but students actually need some practical experience in astronomy (simple observational experiments, school telescopes, planetaria). He thinks that the curriculae introducing astronomy should be prepared by the teachers themselves with the aid of professional astronomers, and not vice versa. *W. Buscombe* gave an example of a text written by physicists who knew little astronomy thus giving a misleading view of the subject. However, the project was so expensive that the schools concerned are stuck with the text for years to come. *R. A. Coutrez* suggested that astronomy should be integrated into the methodology of science at high schools. The *President* mentioned a paper prepared by *E. V. Kononovich* and *E. K. Straut* on the Astronomy education in Soviet schools. In the U.S.S.R. the methodology of school astronomy has been worked out as described in the paper. Due to absence from Brighton of both authors, their paper was not further discussed. A few recent teachers' guides and projects of integrated sciences were then mentioned. *P. V. Sudbury* called attention to the book *Learning about space* (published by H. M. Stationary Office, London) as a teacher's guide for children aged 7–13. *E. A. Beet* mentioned the *Nuffield Physics Project* in England for ages 11–16, and he said that the teachers hesitate to introduce the project because they are not trained in this system. *R. Berendzen* described a programme called *Search for life in the universe* designed for ages 13–15. It unifies many sciences and is inquiry oriented. Text and teachers' guides are available as are consultants (in the Boston area only, at present). He also mentioned that pre-college level books, slides, filmloops etc. are available from NASA Bibliography, Washington, D.C. *O. Gingerich* described the *Harvard Project Physics* (published by Holt, Rinehart and Winston), a secondary school physics course which in its second unit introduces in a historical sequence (from Greek astronomy, through Copernicus, Kepler and Newton) selected astronomical materials. Good radio and television programmes in Astronomy can be very useful. *R. Steinitz*, for example, reported that the Astronomy programmes on the Israel radio were very successful. The teachers were forced to keep up on the subject because the students had followed the radio course and then came up with many further questions to their teachers. *C. Titulaer* described a course in Astronomy prepared for the Netherlands television starting in early 1971. It consists of a series of 12 lessons each of 25 minutes duration. Information is available from TELEAC, Postbus 2414, Utrecht, The Netherlands.

Coming back to the question from the first meeting, the *President* asked whether the Commission should actively prepare materials to aid astronomy education in schools. *G. O. Abell* felt that this is properly the job of individuals rather than of the Commission itself. The concensus of the audience was that the Commission should not prepare aids for astronomy teaching but that it should encourage persons who have abilities in preparing such materials to do so. The task of the Commission is to disseminate the available information on astronomy teaching aids.

Fourth Session

The *President* opened the session by presenting to the Commission a set of 10 *Rules of Operation* governing the Commission. After a brief discussion and some change in wording the proposed rules were accepted by a vote of the Commission members present. *J.-C. Pecker* asked that the rules should be made available to the various National Committees. The Commission rules then were handed over to the IAU General Secretary for further handling.

R. A. Coutrez then read a recommendation by the Belgium National Committee on astronomy. After some discussion on the wording the following recommendation was put to a vote and accepted by the Commission members present: "Commission 46 recommends that a sufficiently extended course on basic modern astronomy, which might be optional, should be maintained or incorporated within the programmes of the first years of university teaching for the undergraduate students of colleges of arts and science as well as colleges of engineering."

The French version of the recommendation reads as follows:

"La Commission 46 recommande qu'un cours de base suffisamment étendu sur les éléments d'astronomie moderne, cours eventuellement à option, soit maintenu ou incorporé dans les programmes des premières années de l'enseignement universitaire donné aux étudiants des facultés des sciences exactes et appliquées et des écoles d'ingénieurs."

The Commission's recommendation was passed on to the IAU General Secretary for further handling.

The rest of the session was devoted to discussions on the *education in astronomy* at the general university level and on the *education of astronomers*. Concerning astronomy for *non-sciences student* *R. Berendzen* and *O. Gingerich* described projects which emphasize the historical development of astronomy. It was pointed out, however, that the historical approach requires very experienced teachers. *D. Wentzel* remarked that often astronomy is the only science course a student takes at the University. Therefore, in such a course things should be stressed that are important for science in general and not only for astronomy. How this is done at the University of Maryland is described by Wentzel in an article to appear in the *American Journal of Physics* in early 1971. *R. Leclaire* mentioned the pioneering course SPACE AND MAN introduced by the Department of Education in Ontario giving guidelines to teachers and students for studies centred on space. It is an interdisciplinary course which extends through the humanities and the sciences, thus giving the student a general education rather than a training in any one discipline.

A discussion followed dealing with the question of when to introduce astronomy into the curriculum of *science students*. Various opinions were presented and no final conclusion was reached. Whereas *R. Steinitz* thinks that astronomy might be more effectively taught in later rather than earlier years when students have better backgrounds, *D. McNally* said that astronomy should be taught soon to physics and other science students so as not to postpone their exposure. As much as possible astronomy should be integrated with other science subjects – the barriers between them should be cut down. This is important in view of the fact that some of these science students will be future high school teachers. *E. Schatzman* stated that often physics students resist taking separate courses in astronomy and, therefore, they should get their astronomy education directly in the physics courses. To this end more astronomy oriented persons must get into physics chairs. *J.-C. Pecker* stressed that in training scientists one should not neglect to teach the students how to use a library, look up references, etc. For the practical training *D. Wentzel* announced the availability at low cost of selected prints from the *Palomar Sky Survey* and certain Moon photographs released by NASA. The six *Palomar prints for use in laboratory exercises* may be ordered at the Caltech Bookstore, CALTECH, Pasadena, California 91109, U.S.A. The *Lunar Photographs for Lecture and Laboratory* may be ordered at Bara Photographic Inc., 4805 Frolich Lane, Hyattsville, Maryland 20781, U.S.A.

Concerning the *education of astronomers U. Steinlin* stressed the importance of laboratory exercises and practical training of the students. He described inexpensive equipment constructed at the Observatory in Basel with students' needs specifically in mind. Details could be made available on request. Many members of the audience expressed their hope that the detailed information and description would soon be published. On the *graduate level E. Schatzman* strongly recommended that students take part in research projects as early as possible. Taking formal courses and examinations is not the only or the best way of finding the best future research workers. *E. A. Müller* commented on the need and the usefulness of concentrated schools of one to two weeks duration on some selected subject for advanced astronomy students of smaller institutions.

Due to lack of time the training of space astronomers was not discussed. *Th. Page* had sent a paper on this subject, but he was not present at Brighton to present and discuss it.

COMMISSION 47: COSMOLOGY (COSMOLOGIE)

Report of Meetings, 19, 20, 21, 24 and 25 August 1970

ACTING PRESIDENT: W. Bonnor.
SECRETARY: M. S. Longair.

In the absence of Ya. B. Zeldovich, W. B. Bonnor was nominated Acting President of Commission for the duration of the XIVth General Assembly.

The new commission met on the 19, 20, 21 and 25 August. The first meeting began with the nomination of the Vice-President and Organizing Committee.

The meeting of 19th August then continued with the following scientific programme under the chairmanship of W. B. Bonnor.

THE MICROWAVE BACKGROUND RADIATION

R. B. Partridge: 'New Experimental Limits on the Anisotropy of the Cosmic Microwave Background'.

E. K. Conklin: 'Observations of Large-Scale Anisotropy in the 3K Background Radiation'.

T. L. May and G. C. McVittie: 'A General Theory of Universes Containing Matter and Black-Body Radiation'.

R. A. Sunyaev and Ya. B. Zeldovich (read by M. S. Longair): 'Distortions of the Relict Radiation Spectrum'.

THE ORIGIN OF GALAXIES

CHAIRMAN: V. A. Ambartsumian.

R. A. Sunyaev and Ya. B. Zeldovich (read by M. S. Longair): 'Fluctuations in the Relict Radiation – Fluctuations of Density and Antimatter'.

E. R. Harrison: 'Comments on the Origin of Galactic Spin'.

P. J. E. Peebles: 'Recent Work on Galaxy Formation at Princeton'.

E. Schatzman: 'Cosmology and Antimatter'.

L. M. Ozernoi and G. V. Chibisov (read by M. J. Rees): 'The Parameters of Galaxies as a Result of Cosmological Turbulence'.

OPEN OR CLOSED UNIVERSE, MISSING MASSES, GRAVITATIONAL RADIATION AND BLACK HOLES

CHAIRMAN: P. J. E. Peebles.

W. A. Fowler: 'The Microwave Background Radiation and Helium Production in the Canonical Big-Bang'.

M. Rowan-Robinson: 'Do Dead Galaxies Outnumber the Living?'

A. M. Wolfe: 'The X-Ray Background and the Structure of the Universe'.

J. Pachner: 'Comment on the Paper "The Origin of Galactic Spin" by E. R. Harrison'.

R. Ruffini: 'Gravitational Radiation and Black Holes'.

R. B. Partridge: 'A Search for Pulses of Intense Electromagnetic Radiation'.

S. van den Bergh: 'Limits to the Number of Black Holes in the Virgo Cluster'.

P. D. Noerdlinger and V. Petrosian: 'The Effect of Cosmological Expansion on the Sizes of Galaxy Clusters'.

The sessions scheduled to take place on 24th August were cancelled because they clashed with the Joint Discussion on Helium.

MIXMASTER UNIVERSES – QUANTUM MECHANICS AND COSMOLOGY

CHAIRMAN: G. R. Burbidge.

I. D. Novikov: 'Mixmaster Universe Singularities – Physical Processes near a Singularity'.

M. A. H. McCallum: 'Evolution of Anisotropy and Observational Characteristics of some Simple Bianchi Cosmologies'.

R. Matzer: 'Mixmaster Universes'.

B. K. Harrison: 'Notes on Quantum Cosmology'.

F. Hoyle: 'Quantum Mechanics and Cosmology'.

Y. Ne'eman: 'Comments on Time Asymmetry'.

J. V. Narlikar: 'Quantum Mechanics and Cosmology'.

V. L. Ginzburg: 'On the Role of Quantum Fluctuations of the Gravitational Field in Cosmology'.

It was noted that at the extraordinary assembly of the IAU to be held in Poland in 1973, it is intended to hold symposia on Cosmology and a topic from relativistic astrophysics.

COMMISSION 48: HIGH ENERGY ASTROPHYSICS
(ASTROPHYSIQUE DE GRANDE ÉNERGIE)

PRESIDENT: H. Friedman.
SECRETARY: M. J. Rees.

No formal meeting report was received.

NUMERICAL DATA FOR ASTRONOMERS AND ASTROPHYSICISTS (DONNEÉS NUMÉRIQUES POUR LES ASTRONOMES ET LES ASTROPHYSICIENS)

PRESIDENT: Ch. M. Sitterly.
SECRETARY: C. W. Allen.

GENERAL COMMENTS

During the two sessions held by this Group, a lively and widespread interest in data handling was indicated. More than 80 persons attended each session and 16 countries were represented. The comments of 14 speakers, most of whom are operating data centers in various fields of research, stimulated active discussion. The serious need for coordinated effort and constructive planning with regard to Data Centers was striking. Ignorance of existing centers was apparent. Inter-Commission cooperation in publicizing all centers within the Union must be stimulated.

DATA CENTERS

An appraisal of the purpose, method of operation, personnel and future outlook of each Data Center is a basic step toward constructive future planning. Within a given Center three types of work are involved:

(a) *Input.* This usually requires a continuing literature search – past, present and future. Existing abstracting services supplemented by individual surveys of current periodicals relevant to each field provide this information.

(b) *Critical Analysis of Data Obtained from the Literature Search.* Here, there is not unanimous agreement, but, in general, a Center that produces reliable critical data is more promising and more secure than one whose output is completely uncritical. Unreliable data will tend to discredit the Data Center, and the expense of operating the Center is justifiable only for reliable data.

D. S. Evans provided a model example of Numerical Data Procedure in describing the preparation of The Radial Velocity Catalogue. He appraised the data with extreme care and judgment before accepting them and drew the moral 'Do not go to a computer until the data are correct'. Errors propagate.

(c) *This leads to the third process – Output.* Publication of critically analyzed data in standard form at regular intervals is desirable. Response to inquiries, as far as is feasible, increases the justification for maintaining the Center. Distinction must be made between the Data Center (for analyzed data only) and the Information Center (for uncritical information).

Much emphasis was placed on the necessity of assembling the data in machine-readable form. This is well-illustrated in connection with Star Catalogues, which were widely discussed and which include a number of specialized topics for which data are needed by astronomers. The catalogues must be key-punched and kept up to date. Codification for machines is needed and a single system for the identification of a star is desirable. All agreed that the combining of catalogues presents a serious problem. This requires critical appraisal of data. Cross-indexing demands a skill beyond that of a computer. Equally important consideration must be given to the question of what data to include for a star. It was again stressed that the propagation of errors in the various catalogues constitutes a serious problem.

The specific problems connected with Star Catalogues are embodied in a more general report prepared by H. Abt. He is Chairman of a special Committee appointed by the Astronomical Society of the Pacific, to consider the feasibility and desirability of establishing an Astronomical Data Center. His report contains a number of pertinent suggestions on recovering published data about individual stars, star clusters and the like, so that statistical studies of objects with specific charac-teristics can be made. This study extends the scope to the study of galactic structure. This report will published in the *Publ. Astron. Soc. Pacific.*

Attention must, also, be given to the formation of new centers as the need develops. It was sug-gested that one might be planned on numerical experiments and theoretical work. For example, data on stellar models could be put in machine-readable form and stored.

The need for a catalogue of pulsars was mentioned, along with the desirability of forming a center to deal with radio and X-ray sources.

In general, the handling of data in any center must be systematic and follow sound general guidelines that have been well tested.

METHOD OF PROCEDURE

The suggestions for handling numerical data were varied, but the willingness to discuss the prob-lems openly, and to cooperate in an effort to solve them, was reassuring. Although a completely mechanized center with a hook-up among observatories is visualized, the trend appeared to favor starting on a small scale. Coordination is urgently needed, and a center for centers may be desirable, but there is the possibility of having an overall central agency functioning as a broker for individual centers. To some it appeared to be too soon for repositories.

With regard to mechanization, it was proposed that astronomers attempt to use University com-puters that operate on a non-profit basis. It is desirable to use the same type of machines, so that programs are interchangeable. A circular letter requesting information on the type of machine for-mat used in various centers might be useful.

The important task assigned to the Group can possibly be handled by a Commission, or by sub-committees in each Commission concerned. At least four such groups under the Commissions are now actively engaged in this type of work.

As a result of the present meeting, 16 centers, representing eight countries, have reported to the Working Group, which is an encouraging start.

In view of the ideas suggested during the course of these meetings, the Working Group made the following recommendations:

1. That a Permanent Working Group be established;
2. That the first task of the Permanent Working Group shall be to centralize information on existing Data Centers and distribute this information through the IAU Information Bulletin. Information on existing lists of errors in Catalogues should, also, be distributed.

The first recommendation was accepted by the General Assembly. Details are being worked out regarding the membership of this Permanent Working Group.

CHARLOTTE MOORE-SITTERLY, Chairman

Washington, D.C., U.S.A. 24 *September,* 1970

SUPPLEMENT TO REPORT

Working Group 1 (Brighton): C. W. Allen, T. Lederle, M. Migeotte, E. L. Schatzman; C. Moore-Sitterly, Chairman.

Speakers at Two Sessions held in Brighton:

The Members of the Group, C. W. Allen, T. Lederle, M. Migeotte, E. L. Schatzman, C. Moore-Sitterly and W. P. Bidelman, D. S. Evans, Ch. Fehrenbach, A. Hynek, J. Jung, P. Lacroute, W. J. Luyten, J. B. Sykes, G. A. Wilkins

Recommendation 3 (Brighton). That the main Data Center Leaders shall constitute the Perma-nent Working Group. The following names are proposed:

G. A. Wilkins, Chairman   C. Jaschek
H. Abt   J. Jung
W. P. Bidelman   T. Lederle]
D. S. Evans   V. B. Nikonov
R. H. Garstang   N. G. Roman
R. Giacconi   F. G. Smith
B. Hauck   J. B. Sykes
J. A. Hynek   H. M. Van Horn
                                 R. Wilson
Additional members may be co-opted.

IAU SUB-COMMITTEES AND INTER-COMMISSION COMMITTEES

Commissions 4, 7, 8, 26
  Working Group within Commission 4 – Ephemerides
  Working Group with IAU, COSPAR and Paris Office
  Combined Data and Information Center
Commissions 5, 9, 38, 46
  At Joint Session agreed to form an information center with A. G. Velghe Director
Commission 14
  Committee 2   Transition Probabilities
  Committee 4   Structure of Atomic Spectra
  Committee 5   Diatomic Molecules
Commission 34
  Data on Planetary Nebulae – will centralize.
Commission 45
  Spectral Classifications and Multi-band Color Indices – Ch. Fehrenbach.
  Working Group

DATA CENTERS REPORTED TO WORKING GROUP 1 SEPTEMBER 1970

Atomic Energy Levels, W. C. Martin, Natl. Bur. Std., Washington, D.C., 20234, U.S.A.
Cross Sections for Collisions of Electrons and Photons with Atoms, Ions and Small Molecules,
L. J. Kieffer, Joint Institute for Laboratory Astrophysics, Boulder, Colorado, 80302, U.S.A.
Eclipsing Binaries, Photoelec. Obs. publ., continuing file of data, F. B. Wood, Univ. of Florida,
Gainesville, Florida, 32601, U.S.A.
Ephemerides, Planetary Data, Star Catalogues, G. A. Wilkins, Royal Greenwich Obs., Herst-
monceux Castle, Hailsham, Sussex, U.K. – R. L. Duncombe, U.S. Naval Obs., Washington, D.C.,
20390, U.S.A. – Paris center
Extra-Galactic Objects, J. D. Wray, Dearborn Observatory, Evanston, Illinois, 60201, U.S.A.
Globular Clusters, Helen Sawyer-Hogg, David Dunlap Obs., Richmond Hill, Ontario, Canada
Observatories, Instruments (Computer listings) etc., A. G. Velghe, Obs. Royale de Belgique,
Uccle-Brussels, Belgium
Planetary Research Center, W. A. Baum, Lowell Observatory, Flagstaff, Arizona, 86001, U.S.A.
Pulsars-Center needed, Consult R. Giacconi, 11 Carleton St., Cambridge, Massachusetts, 02142,
U.S.A.
Star Catalogues, J. Jung, Stellar Data Center – Strasbourg
                 Radial Velocity Center – Marseille
            T. Lederle, Astronomisches Rechen-Institut, Heidelberg, G.F.R.
            C. O. R. Jaschek, La Plata, Argentina
            P. W. Hill, University Obs., St. Andrews, Fife, U.K.
            See, also, Ephemerides (above).

Radio Sources, W. N. Brouw, Sterrewacht Leiden, Leiden, The Netherlands
         Master List – Ohio State Univ.
         Radio Astronomy – California Institute of Technology.
         Data on stars and galaxies.
Spectroscopic Binaries, A. H. Batten, Dominion Astroph. Obs., Victoria, B.C.
Transition Probabilities, W. L. Wiese, Natl. Bur. Std., Washington, D.C., 20234, U.S.A.

PART 4

ASTRONOMER'S HANDBOOK

ASTRONOMER'S HANDBOOK

The Astronomer's Handbook (*IAU Transactions* **XIIC**, edited by J.-C. Pecker and *IAU Transactions* **XIIIB**, edited by L. Perek) contains documents of permanent value. Only the parts changed after the XIVth General Assembly are supplemented here. The numbering of paragraphs is the same as in **XIIC**.

PART 1 OF ASTRONOMER'S HANDBOOK

I. SHORT HISTORY OF THE INTERNATIONAL ASTRONOMICAL UNION

D. Table 2 should be supplemented as follows:

XIVth General Assembly, 1970, Brighton, Sussex, England

Members of the Union, after the General Assembly	2590
Participants (850 Members)	2255
Number of Commissions	39
Volumes of *Transactions* A	566 pages
B	378 pages
Highlights 1970	805 pages
President: B. Strömgren	1970–73
General Secretary: C. de Jager	1970–73
Assistant General Secretary: G. Contopoulos	1970–73
Vice-Presidents: B. J. Bok	1970–76
B. Lovell	1970–76
E. R. Mustel	1970–76

II. ADMINISTRATION AND FINANCES OF THE UNION

A. ADMINISTRATION

The address of the Secretariat is:

International Astronomical Union
Space Research Laboratory
The Astronomical Institute
21, Beneluxlaan
UTRECHT
The Netherlands

The Secretariat is headed by the General Secretary, C. de Jager. The staff of the office consists of the Executive Secretary, A. Jappel, the Secretary of the office, Mrs J. Daňková, and a typist.

All problems related to Symposia and Colloquia are dealt with by the Assistant General Secretary

> Prof. G. Contopoulos
> Astronomical Department
> University of Thessaloniki
> THESSALONIKI
> Greece

He also maintains the contact with the publisher.

B. FINANCES

(a) The XIVth General Assembly resolved to maintain the unit of contribution at 900 gold francs, that is 293,94 U.S. dollars, for the years 1971, 1972, 1973.

III. THE INTERNATIONAL ASTRONOMICAL UNION AND THE OTHER INTERNATIONAL SCIENTIFIC ORGANIZATIONS

A. THE INTERNATIONAL COUNCIL OF SCIENTIFIC UNIONS

The terms of office of other international bodies differ from those of the IAU. Information in this respect should be looked up in the Yearbook of the ICSU for the current year. Only IAU representatives are mentioned in the following paragraphs.

The IAU is represented by the General Secretary, C. de Jager, on the Executive Committee of the ICSU.

B. IAU REPRESENTATIVES ON SCIENTIFIC AND SPECIAL COMMITTEES AND IN COMMISSIONS OF ICSU

1. *Committee on Space Research (COSPAR)*: C. de Jager
2. *International Abstracting Board (IAB)*: J. B. Sykes
3. *Federation of Astronomical and Geophysical Sciences (FAGS)*: B. Guinot and G. H. Wilkins
   IAU representatives in FAGS services:
   Bureau International de l'Heure (BIH): H. M. Smith and G. M. R. Winkler.
   B. Guinot is Director of the Bureau
   International Polar Motion Service (IPMS): E. P. Fedorov and B. Guinot
   Quarterly Bulletin on Solar Activity: Organizing Committee of IAU Commission No. 10
   International Ursigrams and World Day Service (IUWDS): R. Michard
   Solar Particles and Radiation Monitoring Organization (SPARMO): C. W. Allen, C. de Jager and P. Simon

C. IAU REPRESENTATIVES ON INTER-UNION COMMITTEES AND IN COMMISSIONS OF ICSU

1. *Comité International de Géophysique (CIG)*: discontinued
2. *Inter-Union Committee on Frequency Allocations for Radio Astronomy and Space Science (IUCAF)*: O. Hachenberg, D. S. Heeschen, V. A. Sanamian and F. G. Smith
3. *Inter-Union Commission on Solar Terrestrial Physics (IUCSTP)*: Z. Švestka
4. *Inter-Union Commission for Science Teaching (IUCST)*: Edith A. Müller
5. *Committee for Data on Science and Technology (CODATA)*: G. A. Wilkins
6. *Inter-Union Commission on Spectroscopy (IUGS)*: B. Edlén, G. Phillips and M. J. Seaton

D. IAU REPRESENTATIVES IN OTHER ORGANIZATIONS

1. *La Fondation Internationale du Pic-du-Midi*: A. Lallemand
2. *Le Comité Consultatif pour la Définition de la Seconde (CCDS) du Bureau International des Poids et Mesures:* H. Enslin, B. Guinot, Wm. Markowitz and H. M. Smith
3. *Le Comité Consultatif pour la Définition du Mètre (CCDM) du Bureau International des Poids et Mesures*: A. H. Cook
4. *International Radio Consultative Committee (CCIR)*: F. G. Smith and H. M. Smith
5. *European Physical Society (EPS)*: G. Contopoulos

IV. SERVICES AND FUNCTIONS OF THE IAU

No substantial changes as against *Transactions* **XIIC**.

As for the abstracting journal *Astronomischer Jahresbericht* – now *Astronomy and Astrophysics Abstracts* – see *Information Bulletin* No. 23 p. 62.

V. PUBLICATIONS OF THE INTERNATIONAL ASTRONOMICAL UNION

Reference is made to the Report of the Executive Committee, pp. 26–28 and to *Information Bulletin* No. 25.

VI. SYMPOSIA OF THE INTERNATIONAL ASTRONOMICAL UNION

Reference is made to the Report of the Executive Committee, p. 6, 7 and to *Information Bulletins* Nos. 24 and 25.

PART 2 OF ASTRONOMER'S HANDBOOK

STYLE BOOK

A. HOW TO PREPARE MANUSCRIPTS FOR PRINTING

I. GENERAL CONSIDERATIONS

The *presentation* of a scientific article is, for the reader, almost as important as its scientific content. Therefore some rules of presentation, based both on experience and logic, should be followed.

A manuscript passes through the following channels:

Author — Prepares manuscript according to rules recommended by the IAU.

Editor — Modifies, if necessary, the manuscript as regards its scientific and formal contents, in such a way as to satisfy the requirements of editorial clarity.

Publisher — Takes responsibility for all sub-editing, printing and distributing of the publication.

II. RULES FOR AUTHORS

(a) *Presentation*

The first imperative requirement is that the manuscript be submitted before the deadline set by the editor, in duplicate; one copy has to go to the referee.

The manuscript should include: title (brief), full name of the author, the author's 'home' institution, country of the author, abstracts (in the case of contributions to a Symposium, but not in the case of reports of Presidents of Commissions), text of the article or report, references, tables, figures, captions.

Texts should be typed double- or triple-spaced, so as to permit the editors and referees to make the necessary corrections in the manuscript. Ample margins of 3 cm should be provided at both sides of the page.

Typing on both sides of the paper is strongly discouraged.

(b) *Abstracts*

Articles submitted for publication in a Symposium volume should be accompanied by an abstract in the language used by the author. An English abstract is also recommended if the article is not in English.

An abstract should be concise, clear and comprehensive, in continuous text. It is not a table of contents; it should summarize the substance of the conclusions.

(c) *Figures*

(i) Figures should be drawn in Indian ink on white mat paper, or tracing paper; hatching or cross hatching is permissible but any shading can only be done by indicating the area to be dotted on a separate trace.

(ii) Figures should include accurate and clear lettering.

(iii) *The width of the drawing lines should be calculated in such a way that, after photographic reduction, the figure is, at most, about* 12 *cm wide and* 8 *cm deep.* Only very detailed figures can be enlarged to a whole page (i.e., 19 cm × 12 cm).

(iv) The author should write his name and the number of each figure lightly in soft pencil in the margin or in the back.

(v) *Captions should be typed on a separate sheet of paper.*

(vi) Plates should be provided on glossy paper, glazed, with normal range of contrast. The top must be indicated.

(vii) The author will mark on his manuscript the location he suggests for figures.

(viii) In the text the reference 'Figure' is to be written in full. In captions the abbreviation 'Fig.' is to be used.

(d) *Tables*

(i) Tables should have titles; they should be numbered by roman numerals.

(ii) Tables should not be too large in width (take into account decimal points, intervals, units, etc). Authors should prepare tables in such a way that the editor is not obliged to suggest a break in the table; such a break is not desirable and it involves an expensive correction. In order to save space, column headings should be abbreviated; lower case letters might be used to refer to useful explanations given as notes at the bottom of the table.

(e) *References*

(See also Section C. III for 'Abbreviations of titles of Scientific Periodicals')

(i) Authors should verify all references by referring back to the original publication and should avoid quoting second-hand references without checking them.

(ii) *References should be in alphabetical order at the end of the article*, as follows:

A single author. If several references are given, they should appear in chronological order. In the text the name of the author appears in parenthesis, followed by the year. If more than one reference correspond to the same year the letters a, b,... follow the year, both in the text and in the list of references.

Two authors. References should appear alphabetically, according to the spelling of the name of the first of the two authors, as they are given in the original article; in each alphabetical group chronological order should be observed. In the text one should mention the names of the two authors, and the date, and should avoid replacing the name of the second author by *et al.*

Three or more authors. Only the name of the first author should be quoted in the text (followed by *et al.*). All the names of authors should appear in the bibliography.

"... The abundances derived from the chromosphere and lower corona (Pottasch 1963a, 1963b) are of the order of 20 times that given for the photosphere by Goldberg *et al.* (1960). However, Goldberg and Müller (1959), Goldberg and Aller (1960), found that..."

Goldberg, L. and Aller, L. H.: 1960, *Astrophys. J.* **131**, 213.
Goldberg, L. and Müller, E. A.: 1959, *Monthly Notices Roy. Astron. Soc.* **121**, 733.
Goldberg, L., Müller, E. A., and Aller, L. H.: 1960, *Astrophys. J. Suppl.* **5**, 1.
Pottasch, S. R.: 1963a, *Astrophys. J.* **137**, 945.
Pottasch, S. R.: 1963b, *Monthly Notices Roy. Astron. Soc.* **125**, 543.

(iii) A reference contains:

(α) for a *periodical*: surname and initials of the author(s), date, abbreviation of the periodical, volume, page.

(β) for a *book*: surname and initials of the author(s), date, title (in the original language and possibly translated into the language of the author; in the latter case, the translated title should be put in parentheses), publisher, place of publication, and possibly chapter and page.

This reference should be typed as follows, noting carefully the punctuation to be used:

Chandrasekhar, S.: 1961, *Hydrodynamic and Hydromagnetic Stability*, Clarendon Press, Oxford.

In the case where the quoted work is part of a collection, this can be indicated after the title.

Thomas, R. N., Athay, R. G.: 1961, *The Solar Chromosphere*, Vol. VI in the series: Interscience Monographs and Texts in Physics and Astronomy, Interscience Publ., New York.

In case where a work is published in translation, and where the reference is made to the translation, this should be noted.

Spitzer, L.: 1959, *Physique des gas complètement ionisés*, Dunod, Paris, transl. from the English by J.-E. Blamont (*Physics of Fully Ionized Gases*, Interscience Publ., New York, 1956).

(γ) For an *article in a book*: surname of the author(s), initials, date, initials and surname of the scientific editor of the book, title of the book (in the original language), publisher, place of publication, and, possibly, chapter and page.

Kuiper, G. P.: 1951, in J. A. Hynek (ed.), *Astrophysics*, McGraw-Hill Co., New York, Toronto, London, Ch. 8, p. 128.

Some series of books are so well-known that they can be referred to only by the Volume and page, as in the case of a periodical.

Becker, W.: 1963, *Stars and Stellar Systems* 3, 241.
Urey, H. C.: 1959, *Handbuch der Physik* 52, 363.

(δ) For an *unpublished article* or an *article in preparation* or *in course of publication*: give the year, the title of the paper, possibly the journal, or the Institution, and add any further information in parenthesis, like: (in preparation), (in press), (preprint), (Ph.D. Thesis), etc.

Kalnajs, A.: 1965, 'The Stability of Highly Flattened Galaxies', Harvard University (Ph.D. Thesis).

(iv) References in 'Reports on Astronomy'
Because of the need to shorten the 'Reports on Astronomy', various shorthand methods of references have been proposed. It is recommended to include in the text, after the name, the Astronomy and Astrophysics Abstracts year and number of reference:

Wilson, P. R. (AAA 1, 071.026)

(f) *Footnotes*

Footnotes should be avoided as far as possible. They appear at the bottom of the page and are to be indicated by one or two asterisks, daggers, etc.

(g) *Notations and formulae*

(i) The international rules should be generally applied for the *abbreviation of units, for numerical and mathematical formulae, and for notations which are strictly astronomical.*
It is preferable to use the non-abbreviated name of a quantity or unit rather than an obscure or uncertain abbreviation.
Especially dangerous are notations for units such as 'cc' (instead of cm^3), the use of non-metrical units (inches, pounds, etc., which must be written out), the units 'km/sec/sec' which should be written $km\ s^{-2}$ (because of the ambiguity of fraction lines).

(ii) *Formulae and miscellaneous symbols*

These should be written clearly. If necessary one can add explanations in the margin in black pencil (for example: Greek kappa).
Some symbols that may cause confusion are:
Capital Z, lower-case z, and number 2; K, k and Greek κ; a and α; r and γ; r and v; v and ν; w and ω; X, x, $\times$ and χ; z (script) and 3; C and c; l (l.c.) and e; u and μ; n and η; Q (script)

and 2; S, s and 5; V and U; q and g; $_1$ (subscript) and , (comma); ^{1} (superscript) and $'$ (prime).

Most typewriters do not distinguish between l and 1, or between 0 (zero) and O. Also distinction should be made between hyphens (-) and dashes, or minus signs ($-$).

It is recommended to write all mathematical formulae carefully by hand. Sometimes it is necessary to be precise in noting what is in exponent and what is in subscript.

In cases of ambiguity formulae should be presented by the author as follows (all indications not to be printed should be encircled and marked in black pencil):

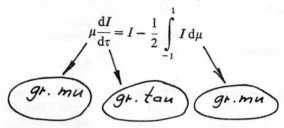

$$\mu \frac{dI}{d\tau} = I - \frac{1}{2} \int_{-1}^{1} I \, d\mu$$

Accumulations of exponents should be avoided: e^{-t} is correct, but $e^{-(t_1)^2}$ is difficult to read without error of interpretation. Instead, it should be written: $\exp(-t_1{}^2)$.

Vectors should be written clearly and indicated by wavy underlining (to point out that they should be printed in bold face), or this indication may be given in a marginal note.

The arguments of operators: exp, sin, etc. when they contain several terms, should be placed in brackets.

In complicated formulae, use

$$(x/y) \quad \text{which is preferable to} \quad \frac{x}{y} \, .$$

It is also recommended to use

$$\left(\frac{\sin(a+b)}{a^2+b^2+c^2}\right)^{1/2} \quad \text{instead of} \quad \sqrt{\frac{\sin(a+b)}{a^2+b^2+c^2}}$$

$$x^{1/n} \quad \text{instead of} \quad {}^n\sqrt{x}$$

(iii) *Numbers*

Groups of three figures, before or after the decimal point, are separated by a space and never by a period or a comma.

Correct: 38 932.071 172.

To be avoided: 38,932.071,172, or 38.932.071.172, or 38932.071172.

An exception is made for groups of four figures; one has to write 3759 Å but 12 133 Å.

Only small numbers (smaller than 12), when they do not appear in formulae or are not followed by the name of a unit, should be written in full: three observatories, but 3 cm, or 3 kT.

(iv) *Algebraic symbols*

All symbols are in italics; the operators are in roman:
Example: x, y, t, z, etc. but $\cos t$, $\exp(a+bt)$, $\mathrm{tg}\, z$, dx/dt, etc.

Vectors are in bold-face type: **B**, **H** $\cos wt$, etc.

Thus, it is unnecessary to underline x, a and y in $\cos(ax+y)$; however in ambiguous cases it is necessary to indicate italics in black pencil in the margin.

(v) *Brackets*

The normal order of brackets is as follows: parentheses (), the usual brackets [], braces {}, angular brackets ⟨⟩, followed by large parentheses, etc.

$$\left\langle \left\{ \left[\left(\langle \{ [( )] \} \rangle \right) \right] \right\} \right\rangle$$

(vi) *Units and prefixes*

For physical symbols and units the rules given by the International Union of Physics should be adhered to. For a complete list see the *Document UIP* **11** (SUN 65-3) 1965. A few examples and *specific* astronomical units are given below. Note that symbols are never followed by a period, neither are they given in the plural: 7 cm and NOT 7 cm. or 7 cms. or 7 cms

ångström unit	Å	hertz (not c/s)	Hz
Astronomical Unit	AU		
		gram	g
bar	bar	kilogram	kg
atmosphere (pressure)	atm		
		dyne	dyn
second	s		
ephemeris time, universal time	ET, UT	degree centrigrade	°C
steradian	sterad	degree Fahrenheit	°F
		degree Kelvin	K
		gauss	G

The use of the following prefixes is acceptable:

T	tera	$=10^{12}$		d	deci	$=10^{-1}$
G	giga	$=10^9$ (not B = Billion)		c	centi	$=10^{-2}$
M	mega	$=10^6$		m	milli	$=10^{-3}$
k	kilo	$=10^3$		μ	micro	$=10^{-6}$
h	hecto	$=10^2$		n	nano	$=10^{-9}$
D	deca	$=10^1$		p	pico	$=10^{-12}$
				f	femto	$=10^{-15}$
				a	atto	$=10^{-18}$

The symbol μm for 10^{-6} m is preferred to μ (micron).

(vii) *Chemical and spectroscopic symbols*

Chemical elements: roman capitals Cu; H; O; N, etc.
Ions: the degree of ionization in small capitals: CaII; FeXIV, etc.
Spectral lines: in general roman capitals: H, Hα, Hβ, K, Lα, etc.
The symbols α, β are not in index.
Energy levels: in italics: *s*, *p*, *S*, etc.
Atomic weight: The atomic number is placed as a left subscript: ^{198}Hg.
Wavelengths: to be indicated in a homogeneous fashion within an article, either by λ 5303 or by 5303 Å.

(viii) *Astronomical symbols*

Spectral classifications: in roman: B 5, cG2, Me 5, etc.
Abbreviations such as: pe (photo-electric), pg (photographic): in roman.
Magnitudes: the abbreviation 'mag' can be used, or the word 'magnitude' written out in full; another alternative is to use the superscript *m*
To note: $B = 12.3$ mag or $B = 12^m.3$.
An hour can be written as 14^h36^m or 14 36 or 14:36.

(ix) *Constellations*

The three-letter abbreviation adopted by the IAU* should be used, preceded by the name of the star (Greek letters, numbers, etc.).

* *Trans. IAU* **4** (1932), 221.

(h) *General organization of articles*

It is important to respect the hierarchy in the numbering of paragraphs. Numbering is as follows: 1, A, I, a, i, α. It is important to use this sequence in order to avoid errors of subordinations from one paragraph to another. The use of the symbols 1 and A should be reserved, in principle, for parts, sections, or chapters of a volume. The symbols I, a, i and α should generally be used within individual articles.

B. SPELLING AND TRANSLITERATION

I. GENERAL CONSIDERATIONS

In *Transactions of the IAU*, the scientific and administrative reports are in English or in French. However, certain speeches (official opening ceremonies, in particular) can be in other languages (Russian, German, etc.). In the Symposium volumes the same rule is generally followed.

II. ACCENTS AND DIACRITICAL MARKS

In French, a careful use of accents is recommended, even in the use of capitals.
As far as possible, diacritical marks in Czech, Turk, Spanish, etc., proper names will be included.

III. USE OF CAPITALS

Initial capitals are common in English and are the rule for all German substantives; they are, on the other hand, very rare in French.

The rules below will be applied in both English and French texts. The following words should begin with a capital and used as a name or title: *President, Vice-President, Commission, Committee, Members* (of the Union) (but *members* of Commissions), *Resolution* (when specifically designated: 'Resolution No. 17', but 'the *resolutions* voted by the General Assembly'), *Appendix*.

The names of individual objects (Sun, Moon, Galaxy) are to begin with a capital. On the other hand, one should speak of *minor planets*, of *spiral galaxies*, without capitals. *The* Galaxy should not be confused with *a* galaxy.

The physical effects named after scientists are written with capitals: Rayleigh, Doppler, etc. The units named after scientists are not, however, capitalized: ampere, joule, watt, ångström, etc.

IV. USE OF NUMBERS

Texts published by the IAU are almost exclusively of a scientific nature; therefore numbers should be written in figures. An exception is made for numbers expressed in a single word, provided that the unit is also expressed without abbreviation: *three dimensions, three cubic centimetres*, (but 3 cm^3); *millions of stars* (but some 10^6 cm); *six per cent* (but 6%), etc.

The Commissions of the IAU are designated by arabic numerals (Commission 36), the General Assemblies of the Union by roman numerals, or written in full: *Xth General Assembly or Tenth General Assembly*.

V. SCHEME OF TRANSLITERATION OF THE CYRILLIC ALPHABET

IAU Commission No. 5 recommended the following scheme for transliteration of the Cyrillic alphabet which has been endorsed by the XIVth General Assembly (Resolution No. 12).

а	a	к	k	х	kh
б	b	л	l	ц	ts
в	v	м	m	ч	ch
г	g	н	n	ш	sh
д	d	о	o	щ	shch
е	e	п	p	ъ	″
ё	e	р	r	ы	y
ж	zh	с	s	ь	′
з	z	т	t	э	eh
и	i	у	u	ю	yu
й	j	ф	f	я	ya

Authors are asked to use this scheme in all IAU publications.

C. ABBREVIATIONS

I. MISCELLANEOUS ABBREVIATIONS

(a) There is no separation between letters in such cases as UNESCO, IAU, ESO, ESRO, FK_3, PZT, IGY, BD, etc.

The names of countries are written with separating periods: U.S.A., U.K., U.S.S.R. For East and West Germany D.D.R. (G.D.R.) and B.R.D. (F.R.G.) should be used.

(b) Proper names of persons and countries: titles are omitted in scientific reports, but may be included in formal and administrative reports. Usual abbreviations are:

Prof, Dr, Mr, Messrs, Mrs, Miss (without a period in English); M., MM., Mme, Mlle (in French). Initials are given in the bibliographies. In the text (and except in cases of possible confusion, such as authors bearing a very common name) they are omitted if an author is quoted, but retained when the reference is related to a definite action. Example: "In his presidential address, R. W. Jones referred to..."; "the partition function has been computed by Matsushima and by J. J. Smith..."

In the text, names of cities and countries should be expressed in the language of the author (London or Londres, Lyons or Lyon, etc.); in mailing addresses the local spelling of the city should be used (London, Lyon, etc.).

II. LIST OF COMMON ABBREVIATIONS

The list below gives a table of the principal abbreviations found in publications of the IAU and of the ICSU. Obviously such a list cannot be comprehensive; it does not include abbreviations used in restricted fields by specialists only.

AAA	Astronomy and Astrophysics Abstracts
AAS	American Astronomical Society
AGK	Astronomischer Gesellschaft Katalog
APFS	Apparent Places of Fundamental Stars
BD	Bonner Durchmusterung
BIH	Bureau International de l'Heure
CCIR	Comité Consultatif International des Radiocommunications
CETEX	Committee on Contamination by Extraterrestrial Exploration
CIG	Comité International de Géophysique
CNRS	Centre National de la Recherche Scientifique
CODATA	Committee for Data for Science and Technology
COSPAR	Committee on Space Research
CPD	Cape Photometric Durchmusterung
CST	Committee on Science Teaching
EPS	European Physical Society
ESO	European Southern Observatory
ESRO	European Space Research Organization

FAGS	Federation of Astronomical and Geophysical Services
FK	Fundamental Katalog
GC	General Catalogue
GCVS	General Catalogue of Variable Stars
HD	Henry Draper Catalogue
HR	Hertzsprung-Russell
IAB	ICSU Abstracting Board
IAG	International Association of Geodesy
IAGA	International Association of Geomagnetism and Aeronomy
IAU	International Astronomical Union
ICSU	International Council of Scientific Unions
IPMS	International Polar Motion Service
IQSY	International Quiet Sun Years
ISO	International Standardization Organization
ITA	Institute of Theoretical Astronomy (Leningrad)
ITU	International Telecommunication Union
IUCAF	Inter-Union Committee on Frequency Allocations for Radioastronomy and Space Science
IUCI	Inter-Union Committee on the Ionosphere
IUCS	Inter-Union Commission on Spectroscopy
IUCSTP	Inter-Union Commission on Solar-Terrestrial Physics
IUGG	International Union of Geodesy and Geophysics
IUHPS	International Union of the History and Philosophy of Science
IUPAP	International Union of Pure and Applied Physics
IUTAM	International Union of Theoretical and Applied Mechanics
IUWDS	International Ursigram and World Days Service
JOSO	Joint Organisation for Solar Observations
NASA	National Aeronautics and Space Administration
NRL	National Research Laboratory
PZT	Photographic Zenith Tube
RAS	Royal Astronomical Society
SPARMO	Solar Particles and Radiation Monitoring Organization
SUN	IUPAP Commission for Symbols, Units and Nomenclature
UIP	Union Internationale de Physique
UNESCO	United Nations Educational, Scientific and Cultural Organization
URSI	Union Radio-Scientifique Internationale
WDC	World Data Center
WMO	World Meteorological Organization

III. ABBREVIATIONS OF TITLES OF SCIENTIFIC PERIODICALS

For abbreviations of titles of scientific periodicals *Access* should be followed. *Access* is being published by Chemical Abstracts Service and has replaced the former *Chemical Abstracts' List of Periodicals*.

(a) *Principal Abbreviations*

- *Abstr.*: Abstract
- *Accad.*: Accademia
- *Acad.*: Academy, Academie, Academia, Academica
- *Ann.*: Annalen, Annaler, Annales, Annali, Annals, Annuae, Annual, Annuel
- *Anuar.*: Anuario
- *Annu.*: Annuaire
- *Astron.*: Astronomia, Astronomie, Astronomy, Astronomical, etc.
- *Astrofis.*, *Astrofiz.*, *Astrophys.*: Astrofisica, Astrofizika, Astrophysics, etc.
- *Ber.*: Bericht
- *Bol.*: Boletim, Boletin
- *Boll.*: Bollettino
- *Bull.*: Bulletin
- *Compt. Rend.*: Compte(s) Rendu(s)
- *Commun.*: Communications

- *Contr.*: Contributions
- *Geofis., Geofiz., Geofys., Geophys.*: Geofisica, Geofizika, Geofysik, Geophysics, etc.
- *Inst.*: Institut, Institute, Institution
- *Int.*: International, etc.
- *Ist.*: Istituto
- *J.*: Journal
- *Jahrb.*: Jahrbuch
- *Jahresber.*: Jahresbericht
- *Mem.*: Memoirs, Mémoires, Memorias, Memorie, etc.
- *Monthly Notices:* Monthly Notices
- *Nachr.*: Nachrichten
- *Natl.*: National
- *Naut.*: Nautical
- *Obs.*: Observatoire, Observatory
- *Oss.*: Osservatorio
- *Proc.*: Proceedings
- *Pubbl.*: Pubblicazione
- *Publ.*: Publicaciones, Publicationes, Publications, Publikationer
- *Quant.*: Quantitative
- *Quart.*: Quarterly
- *Rend.*: Rendiconto, Rendu
- *Rept.*: Report
- *Repr.*: Reprint
- *Res.*: Research
- *Result.*: Resultados, Resultats
- *Rev.*: Review, Revue, Revista
- *Roy.*: Royal
- *Schr.*: Schriften
- *Sci.*: Science, Sciencia, Scienze, etc.
- *Ser.*: Serie, Series, Serija, etc.
- *Soc.*: Societa, Société, Society, etc.
- *Supl.*: Suplemento
- *Suppl.*: Supplément, Supplement, Supplemento, etc.
- *Trans.*: Transactions
- *Transl.*: Translation
- *Un.*: Union, etc.
- *Univ.*: Universidad, Université, University, etc.
- *Z.*: Zeitschrift
- *Ztg.*: Zeitung
- *Zh.*: Zhurnal

(b) *Some Important Journal Abbreviations*

Abastumansk. Obs. Bull.
Acad. Roy. Belg. Bull. Cl. Sci.
Acta Astron.
Acta Astron. Sinica
Adv. Astron. Astrophys.
Am. J. Math.
Am. J. Sci.
Am. Scientist
Ann. Astrophys.
Ann. Geophys.
Ann. Obs. Roy. Belg.
Ann. Harv. Coll. Obs.
Ann. N.Y. Acad. Sci.
Ann. Phys. N.Y.
Ann. Physik
Ann. Phys. Paris

Ann. Tokyo Astron. Obs.
Ann. Rev. Astron. Astrophys.
Arkiv Astron.
Arkiv Fysik
Astrofizika
Astronaut. Acta
Astron. Astrophys.
Astron. J.
Astron. Mitt. Zürich
Astron. Nachr.
Astron. Zh.
Astrophysics
Astrophys. J.
Astrophys. J. Letters
Astrophys. J. Suppl.
Astrophys. Letters

Astrophys. Norv.
Astrophys. Space Sci.
Atti Accad. Nazl. Lincei. Rend.
Australian J. Phys.
Australian J. Phys. Astrophys. Suppl.

Bol. Obs. Tonantzintla Tacubaya
Bull. Am. Astron. Soc.
Bull. Astron.
Bull. Astron. Inst. Czech.
Bull. Astron. Inst. Neth.
Bull. Astron. Obs. Roy. Belg.
Bull. Classe Sci. Acad. Roy. Belg.
Bull. Geodes.
Bull. Soc. Roy. Sci. Liège
Bull. Signal.

Can. J. Phys.
Celes. Mech.
Coll. Astrophys. Liège
Comments Astrophys. Space Phys.
Comm. Pure Applied Math.
Compt. Rend. Acad. Sci. Paris
Contr. Oss. Milano-Merate
Cosmic Electrodyn.
COSPAR Symp.

Dokl. Akad. Nauk

Earth Extraterrest. Sci.
Earth Planet. Sci. Letters

Geochim. Cosmochim. Acta
Geomagnetizm i Aeronomiya
Geophys. J.

IAU Circ.
IAU Inform. Bull.
IAU Symp.
Icarus
Izv. Krymsk. Astrofiz. Obs.

J. Astronaut. Sci.
J. Atmospheric Sci.
J. Atmospheric Terrest. Phys.
J. Geophys. Res.
J. Obs.
J. Opt. Soc. Am.
J. Phys. A General Phys.
J. Physique
J. Phys. Soc. Japan
J. Plasma Phys.
J. Res. Natl. Bur. Std.
J. Roy. Astron. Soc. Can.

Komety i Meteory
Koninkl. Ned. Akad. Wetenschap.

Math. Rev.
Mem. Br. Astron. Assoc.
Mem. Roy. Astron. Soc.
Mem. Soc. Astron. Ital.
Meteoritics
Meteoritika
Mitt. Astron. Ges.
Monthly Notices Roy. Astron. Soc.
Moon

Natl. Bur. Std. U.S. Monograph
Nature
Naturwissenschaften
Nucl. Fusion
Nuovo Cimento
Nuovo Cimento Letters

Observatory
Oss. e Mem. Oss. Arcetri

Phil. Mag.
Phys. Abstr.
Phys. Ber.
Phys. Blätter
Phys. Earth Planet. Interiors
Phys. Rev.
Phys. Rev. Letters
Phys. Fluids
Phys. Today
Planetary Space Sci.
Proc. Inst. Elec. Engrs. London
Proc. Inst. Elec. Electron. Engrs.
Proc. Nat. Acad. Sci.
Proc. Phys. Soc. Japan
Proc. Phys. Soc. London
Proc. Roy. Soc. Edinburgh
Proc. Roy. Soc. London
Prog. Theor. Phys. Kyoto
Publ. Astron. Inst. Prague
Publ. Astron. Soc. Japan
Publ. Astron. Soc. Pacific
Publ. Dominion Astrophys. Obs.

Quart. Bull. Solar Activ.
Quart. J. Roy. Astron. Soc.

Ref. Zh.
Rev. Geophys.
Rev. Mod. Phys.
Roy. Observ. Bull.

Science
Sci. Am.
Sitzber. Deut. Akad. Wiss. Berlin
Sky Telesc.
Smithsonian Contrib. Astrophys.
Solar Phys.

Soobshch. Byurakansk. Obs. Akad. Nauk Arm.
  S.S.R.
Soviet Astron.
Soviet Phys. Dokl.
Soviet Phys. JETP
Soviet Phys. Usp.
Space Sci. Rev.
Stockholm Obs. Ann.

Tokyo Astron. Bull.
Trans. IAU
Trans. Roy. Soc. Edinburgh
Trudy Inst. Teor. Astron.

U.S. Naval Obs. Repr.
Usp. Fiz. Nauk

Veroeffentl. Astron. Rechen-Inst. Heidelberg
Vestn. Leningr. Gos. Univ.
Vistas Astron.

Z. Angew. Phys.
Z. Astrophys.
Z. Geophys.
Z. Naturforsch.
Z. Phys.
Zh. Eksperim. Teor. Fiz.

D. SIGNS USED IN CORRECTING PROOFS

ℬ	Delete; take out.
⊙ (inverted)	Turn inverted letter right side up.
stet · · · · ·	Let it remain; change made was wrong.
◻	Indent one *em*.
⊙	Insert a period.
‖	The type line is uneven at the side of the page; straighten.
✗	A broken letter.
-	A hyphen.
ital.	Use italics.
⌒	Join together.
ℬ̃	Take out letter and close up.
center	Put in middle of page, or line.
≣	Straighten lines.
ⱽ	Insert an apostrophe.
⋀	Insert a comma.
⌐¬	Raise the word or letter.
⌊⌋	Lower the word or letter.
⊏	Bring matter to the left.
⊐	Bring matter to the right.
#	Make a space.
ⱽᵃ	Insert a superior letter or numeral.
⋀̂	Insert an inferior letter or numeral.
lead	A thin metal strip used to widen the space between the lines.
space out	Spread words farther apart.
¶	Make a paragraph.
no ¶	Run on without a paragraph.
cap.	Use a capital.
l.c.	Use the lower case (small type), *i.e.*, not capitals.
s.c.	Small capitals.
w.f.	Wrong font—size or style.
font	Kind of type.
tr. ∩∪	Transpose.
rom.	Use roman letter.
⋀	Indicates where an insertion is to be made.
Qy. or (?)	Doubt **as** to spelling, etc.
≡	Indicates CAPITAL letters.
═	Indicates SMALL CAPITAL letters.
—	Indicates *italic* letters.
∿	Indicates **boldface** letters.
≋	Indicates **BOLDFACE CAPITALS**
∿	Indicates ***Boldface italic***.

PART 3 OF ASTRONOMER'S HANDBOOK

EXECUTIVE COMMITTEE

STATUTES AND BY-LAWS, WORKING RULES

RULES FOR SCIENTIFIC MEETINGS

MEMBERSHIP OF COMMISSIONS

ALPHABETICAL LIST OF MEMBERS

LIST OF COUNTRIES ADHERING TO THE UNION

EXECUTIVE COMMITTEE

1970–73

PRESIDENT

Professor B. Strömgren, Observatoriet, Østervoldgade 3, Copenhagen K, Denmark

VICE-PRESIDENTS

Professor M. K. V. Bappu, Director of the Astrophysical Observatory, Kodaikanal, India
Professor B. J. Bok, Steward Observatory, University of Arizona, Tucson, Arizona 85721, U.S.A.
Professor L. Gratton, Laboratorio di Astrofisica, Casella Postale 67, 00044 Frascati (Rome), Italy
Sir Bernard Lovell, Nuffield Radio Astronomy Laboratories, Jodrell Bank, Macclesfield, Cheshire, U.K.
Professor E. R. Mustel, Astronomical Council, USSR. Academy of Sciences, Vavilov Str. 34, Moscow V-312, U.S.S.R.
Professor J. Sahade, 47–848, 9°A, La Plata, Argentina

GENERAL SECRETARY

Professor C. de Jager, c/o Space Research Laboratory of the Astronomical Institute, 21 Beneluxlaan, Utrecht, The Netherlands

ASSISTANT GENERAL SECRETARY

Professor G. Contopoulos, Astronomical Department, University of Thessaloniki, Thessaloniki, Greece

UNION ASTRONOMIQUE INTERNATIONALE

STATUTS

I. DÉNOMINATION, BUTS ET DOMICILE

1. L'Union Astronomique Internationale (ci-après dénommée l'Union) est une organisation non-gouvernementale, qui a pour buts de:
   (a) développer l'astronomie par la coopération internationale,
   (b) encourager l'étude et le développement de l'astronomie sous tous ses aspects,
   (c) servir et sauvegarder les intérêts de l'astronomie.

2. L'Union a son siège légal à Bruxelles.

II. AFFILIATION DE L'UNION

3. L'Union adhère au Conseil International des Unions Scientifiques.

III. MEMBRES DE L'UNION

4. L'Union a pour membres:
   (a) des personnes morales (Pays adhérents)
   (b) des membres individuels (Membres)

IV. ORGANISATIONS AFFILIÉES

5. L'Union peut accepter l'affiliation d'organisations internationales non-gouvernementales qui contribuent au développement de l'astronomie.

V. PAYS ADHÉRENTS

6. Les pays adhèrent à l'Union
soit:
   (a) par l'intermédiaire de l'organisation par laquelle ils adhèrent au Conseil International des Unions Scientifiques, ou par l'intermédiaire d'un Comité National d'Astronomie approuvé par cette organisation,
soit:
   (b) s'ils n'adhèrent pas au Conseil International des Unions Scientifiques, par l'intermédiaire d'un Comité National d'Astronomie reconnu par le Comité Exécutif de l'Union.
   (c) Les Organisations ou Comités mentionnés à l'article 6a et les Comités Nationaux d'Astronomie mentionnés à l'article 6b sont dénommés ci-après organismes adhérents.

7. L'adhésion d'un pays à l'Union est proposée par le Comité Exécutif et approuvée par l'Assemblée Générale; elle prend fin si le pays se retire de l'Union.

8. Les Pays Adhérents sont répartis en catégories. Le nombre des catégories est fixé par le Règlement. Un pays qui sollicite son adhésion indique la catégorie dans laquelle il désire être classé. La proposition peut être refusée par le Comité Exécutif si la catégorie est manifestement inadéquate.

INTERNATIONAL ASTRONOMICAL UNION

STATUTES

I. DENOMINATION, OBJECTS AND DOMICILE

1. The International Astronomical Union (referred to as the Union) is a non-governmental organization, whose objects are
   (*a*) to develop astronomy through international co-operation,
   (*b*) to promote the study and development of astronomy in all its aspects,
   (*c*) to further and safeguard the interests of astronomy.

2. The legal domicile of the Union is Brussels.

II. ADHERENCE OF THE UNION

3. The Union adheres to the International Council of Scientific Unions.

III. COMPOSITION OF THE UNION

4. The Union is composed of
(*a*) corporate members (Adhering Countries)
(*b*) individual members (Members).

IV. AFFILIATED ORGANIZATIONS

5. The Union may admit the affiliation of international non-governmental organizations which contribute to the development of astronomy.

V. ADHERING COUNTRIES

6. Countries adhere to the Union either
   (*a*) through the organization by which they adhere to the International Council of Scientific Unions, or through a National Committee of Astronomy approved by that organization,
or
   (*b*) if they do not adhere to the International Council of Scientific Unions, through a National Committee of Astronomy recognized by the Executive Committee of the Union.
   (*c*) The Adhering Organizations and National Committees of Astronomy are referred to as adhering bodies.

7. Adherence of a country to the Union is approved, on the proposal of the Executive Committee, by the General Assembly; it terminates if the country withdraws from the Union.

8. Adhering Countries are classified in categories. The number of categories shall be specified in the By-laws.
   A country requesting adherence shall specify the category in which it desires to be classed. The specification may be declined by the Executive Committee if the category proposed is manifestly inadequate.

VI. MEMBRES

9. Les Membres sont admis dans l'Union par le Comité Exécutif, sur proposition de l'un des organismes adhérents mentionnés à l'article 6, en considération de leur activité dans une branche de l'astronomie.

VII. ASSEMBLÉE GÉNÉRALE

10. (*a*) L'activité de l'Union est dirigée par l'Assemblée Générale des représentants des Pays Adhérents et des Membres. Chaque Pays Adhérent nomme un représentant autorisé à voter en son nom.

(*b*) L'Assemblée Générale rédige un Règlement qui précise les modalités d'application des Statuts.

(*c*) Elle nomme un Comité Exécutif chargé d'exécuter les décisions de l'Assemblée Générale, et d'administrer l'Union pendant la période séparant les réunions de deux Assemblées Générales ordinaires successives. Le Comité Exécutif rend compte de sa gestion à l'Assemblée Générale. L'Assemblée Générale, en acceptant le rapport du Comité Exécutif, le décharge de sa responsabilité.

11. (*a*) Sur les questions concernant l'administration de l'Union, sans implication budgétaire, le vote à l'Assemblée Générale a lieu par Pays Adhérent, chaque pays disposant d'une voix. Les Pays Adhérents qui ne sont pas à jour de leurs cotisations annuelles au 31 décembre de l'année précédant l'Assemblée Générale ne peuvent pas participer aux votes.

(*b*) Sur les questions engageant le budget de l'Union, le vote a lieu de même par Pays Adhérent, dans les conditions et avec les réserves prévues à l'article 11 (*a*), le nombre de voix de chaque Pays Adhérent étant égal à l'indice de sa catégorie, définie conformément à l'article 8, augmenté d'une unité.

(*c*) Les Pays Adhérents peuvent voter par correspondance sur les questions figurant à l'ordre du jour de l'Assemblée Générale.

(*d*) Un scrutin n'est valable que si au moins deux tiers des Pays Adhérents disposant du droit de vote en vertu de l'article 11 (*a*) y prennent part.

12. Sur les questions scientifiques n'engageant pas le budget de l'Union, les Membres de l'Union disposent chacun d'une voix.

13. Sur toutes les questions prévues aux articles 11 et 12, les décisions sont prises à la majorité absolue des suffrages. Cependant, une décision de modification des Statuts n'est valable que si elle a été prise à la majorité des deux tiers des voix des Pays Adhérents qui disposent du droit de vote en vertu de l'article 11 (*a*).

14. Une proposition de modification des Statuts ne peut être discutée que si elle figure, en tant que telle, à l'ordre du jour de l'Assemblée Générale.

VIII. COMITÉ EXÉCUTIF

15. Le Comité Exécutif se compose du Président de l'Union, de six Vice-Présidents, du Secrétaire Général et du Secrétaire Général Adjoint, élus par l'Assemblée Générale sur la proposition du Comité Spécial des Nominations.

IX. COMMISSIONS DE L'UNION

16. L'Assemblée Générale crée des Commissions en vue d'assurer la réalisation des buts qu'elle se propose.

X. REPRÉSENTATION LÉGALE DE L'UNION

17. Le Secrétaire Général est le représentant légal de l'Union.

VI. MEMBERS

9. Members are admitted to the Union by the Executive Committee, on the proposal of an adhering body referred to in article 6, with regard to their achievements in some branch of astronomy.

VII. GENERAL ASSEMBLY

10. (*a*) The work of the Union is directed by the General Assembly of representatives of Adhering Countries and of Members. Each Adhering Country appoints a representative authorized to vote in its name.

(*b*) The General Assembly draws up By-laws governing the application of the Statutes.

(*c*) It appoints an Executive Committee to implement the decisions of the General Assembly, and to direct the affairs of the Union in the interval between meetings of two successive ordinary General Assemblies. The Executive Committee reports to the General Assembly. The General Assembly, in accepting the report of the Executive Committee, discharges it of liability.

11. (*a*) On questions concerning the administration of the Union, not involving its budget, voting at the General Assembly is by Adhering Country, each country having one vote. Adhering Countries which have not paid their annual contributions up to 31 December of the year preceding the General Assembly may not participate in the voting.

(*b*) On questions involving the budget of the Union, voting is similarly by Adhering Country, under the same conditions and with the same reservations as in article 11 (*a*), the number of votes for each Adhering Country being one greater than the number of its category, as defined in article 8.

(*c*) Adhering Countries may vote by correspondence on questions on the agenda for the General Assembly.

(*d*) A vote is valid only if at least two thirds of the Adhering Countries having the right to vote by virtue of article 11 (*a*) participate in it.

12. On scientific questions not involving the budget of the Union the Members of the Union each have one vote.

13. On all questions in articles 11 and 12, decisions are taken by an absolute majority of the votes cast. However, a decision to change the Statutes is only valid if taken with the approval of at least two thirds of the votes of the Adhering Countries having the right to vote by virtue of article 11 (*a*).

14. A motion to change the Statutes can only be discussed if it appears, in specific terms, on the agenda for the General Assembly.

VIII. EXECUTIVE COMMITTEE

15. The Executive Committee consists of the President of the Union, six Vice-Presidents, the General Secretary and the Assistant General Secretary elected by the General Assembly on the proposal of a Special Nominating Committee.

IX. COMMISSIONS OF THE UNION

16. The General Assembly forms Commissions for such purposes as it may decide.

X. LEGAL REPRESENTATION OF THE UNION

17. The General Secretary is the legal representative of the Union.

18. (*a*) Pour chaque Assemblée Générale ordinaire, le Comité Exécutif prépare un projet de budget pour la période à courir jusqu'à la prochaine Assemblée Générale ordinaire. Il soumet le projet, avec les comptes de la période précédente, à un Comité des Finances nommé par l'Assemblée Générale. Le Comité des Finances examine les comptes et le projet de budget, sur lesquels il présente deux rapports distincts, qu'il soumet à l'approbation de l'Assemblée Générale.

(*b*) Chaque Pays Adhérent verse annuellement à l'Union un nombre d'unités de cotisation qui est fonction de sa catégorie. Le nombre d'unités de cotisation pour chaque catégorie est fixé par le Règlement.

(*c*) Le montant de l'unité de cotisation est fixé par l'Assemblée Générale, sur la proposition du Comité Exécutif et avec l'avis du Comité des Finances.

(*d*) Le paiement des cotisations est à la charge des organismes adhérents. La responsabilité de chaque Pays Adhérent envers l'Union est limitée au montant des cotisations dues par ce pays à l'Union.

(*e*) Un Pays Adhérent qui cesse d'adhérer à l'Union renonce de ce fait à ses droits sur l'actif de l'Union.

19. La décision de dissoudre l'Union n'est valable que si elle est prise à la majorité des trois quarts des voix des Pays Adhérents qui disposent du droit de vote en vertu de l'article 11 (*a*).

20. Si, par suite d'événements indépendants de la volonté de l'Union, des circonstances apparaissent qui rendent impossible le respect des clauses de ces Statuts et du Règlement établi par l'Assemblée Générale, les organes et membres du Comité Exécutif de l'Union, dans l'ordre fixé ci-dessous, prendront toutes dispositions qu'ils jugeront nécessaires pour la continuation du fonctionnement de l'Union. Ces dispositions devront être soumises à une autorité supérieure dès que cela deviendra possible, jusqu'à ce qu'une Assemblée Générale extraordinaire puisse être réunie. L'autorité est dévolue dans l'ordre ci-dessous: L'Assemblée Générale; une Assemblée Générale extraordinaire; le Comité Exécutif, réuni ou par correspondance; le Président de l'Union; le Secrétaire Général; ou, à défaut de la possibilité de recourir à l'une de ces autorités ou de leur disponibilité, un des Vice-Présidents.

21. Ces Statuts entrent en vigueur le 1er Septembre 1970.

22. Les présents Statuts sont publiés en versions française et anglaise. En cas d'incertitude, la version française fait seule autorité.

REGLEMENT

1. Les demandes d'adhésion des pays à l'Union Astronomique Internationale (ci-après dénommée l'Union) sont examinées par le Comité Exécutif et soumises à l'approbation de l'Assemblée Générale.

2. Les propositions de modifications de la liste des Membres sont, après examen attentif des suggestions des Présidents de Commissions, soumises pour avis au Comité des Nominations,

XI. BUDGET AND DUES

18. (*a*) For each ordinary General Assembly the Executive Committee prepares an estimate of the budget for the period to the next ordinary General Assembly. It submits this estimate, together with the accounts for the preceding period, to a Finance Committee appointed by the General Assembly. The Finance Committee examines the accounts and the estimate of the budget, on which it presents two separate reports, and submits them to the General Assembly for approval.

(*b*) Each Adhering Country pays annually to the Union a number of units of contribution according to its category. The number of units of contribution for each category shall be specified in the By-laws.

(*c*) The amount of the unit of contribution is determined by the General Assembly, on the proposal of the Executive Committee and with the advice of the Finance Committee.

(*d*) The payment of contributions is the responsibility of the adhering bodies. The liability of each Adhering Country in respect of the Union is limited to the amount of that country's dues to the Union.

(*e*) An Adhering Country that ceases to adhere to the Union resigns at the same time its rights to a share in the assets of the Union.

XII. DISSOLUTION OF THE UNION

19. The decision to dissolve the Union is only valid if taken with the approval of three quarters of the votes of the Adhering Countries having the right to vote by virtue of article 11 (*a*).

XIII. EMERGENCY POWERS

20. If, through events outside the control of the Union, circumstances arise in which it is impracticable to comply with the provisions of these Statutes and of the By-laws drawn up by the General Assembly, the organs and officers of the Union, in the order specified below, shall take such actions as they deem necessary for the continued operation of the Union. Such action shall be reported to a higher authority immediately this becomes practicable until such time as an extraordinary General Assembly can be convened. The following is the order of authority:

The General Assembly; an extraordinary General Assembly; the Executive Committee in meeting or by correspondence; the President of the Union; the General Secretary; or failing the practicability or availability of any of the above, one of the Vice-Presidents.

XIV. FINAL CLAUSES

21. These Statutes enter into force on 1 September 1970.

22. The present Statutes are being published in French and English version. In case of doubt, the French version is the only authority.

BY-LAWS

I. MEMBERSHIP

1. Applications of countries for adherence to the International Astronomical Union (referred to as the Union) are examined by the Executive Committee and submitted to the General Assembly for approval.

2. Proposed changes in the list of Members are, with due regard to the suggestions of the President

composé d'un représentant de chaque Pays Adhérent désigné par l'organisme adhérent habilité, avant la décision du Comité Exécutif.

3. Les Commissions peuvent, avec l'approbation du Comité Exécutif, coopter des consultants qu'elles jugent en mesure d'apporter une contribution utile à leur travail. L'adhésion des consultants a pour terme le dernier jour de la première Assemblée Générale ordinaire qui suit leur admission, à moins qu'elle ne soit renouvelée.

4. Une organisation affiliée peut participer au travail de l'Union dans les conditions fixées par accord entre l'organisation et le Comité Exécutif.

II. L'ASSEMBLÉE GÉNÉRALE

5. L'Union se réunit en Assemblée Générale ordinaire régulièrement une fois tous les trois ans. Si le lieu et la date de l'Assemblée Générale ordinaire n'ont pas été décidés lors de la précédente Assemblée Générale, ils sont fixés par le Comité Exécutif et communiqués aux organismes adhérents au moins six mois à l'avance.

6. Le Président peut convoquer, avec l'accord du Comité Exécutif, une Assemblée Générale extraordinaire. Il est tenu de le faire à la demande du tiers des Pays Adhérents.

7. L'Ordre du Jour de chaque Assemblée Générale ordinaire est arrêté par le Comité Exécutif et communiqué aux Organismes Adhérents au moins quatre mois avant le premier jour de la réunion. Il devra inclure la proposition du Comité Exécutif concernant le montant de l'unité de cotisation qui permet l'application de l'article 24.

8. (a) L'Ordre du Jour doit inclure toute motion ou proposition reçue par le Secrétaire Général au moins cinq mois avant le premier jour d'une Assemblée Générale ordinaire, qu'elle émane d'un organisme adhérent, d'une Commission de l'Union, ou d'une Commission mixte dans laquelle l'Union est représentée.

(b) Une motion ou proposition concernant l'administration ou le budget de l'Union qui ne figure pas à l'Ordre du Jour préparé par le Comité Exécutif, ou tout amendement à une motion qui figure à l'Ordre du Jour, ne peut être discuté qu'avec l'accord préalable des deux tiers au moins des voix des Pays Adhérents représentés à l'Assemblée Générale et disposant du droit de vote en vertu de l'article 11 (a) des Statuts.

9. S'il y a doute sur le caractère administratif ou scientifique d'une question donnant lieu à un vote, l'avis du Président est prépondérant.

10. En cas de partage égal des voix, le Président a voix prépondérante.

11. Le Président peut inviter des représentants d'autres organisations, des scientifiques et de jeunes astronomes à participer à l'Assemblée Générale. Avec l'accord du Comité Exécutif, il peut déléguer ce privilège au Secrétaire Général en ce qui concerne les représentants d'autres organisations, aux organismes adhérents en ce qui concerne les scientifiques et les jeunes astronomes.

III. LE COMITÉ SPÉCIAL DES NOMINATIONS

12. (a) Les propositions pour les élections du Président de l'Union, des six Vice-Présidents, du Secrétaire Général et du Secrétaire Général Adjoint sont soumises à l'Assemblée Générale par le Comité Spécial des Nominations. Ce Comité se compose du Président en fonction et du Président sortant, d'un membre proposé par le Comité Exécutif sortant et n'appartenant ni au Comité Exécutif actuel ni au Comité Exécutif précédent, et de quatre membres choisis par le Comité des Nominations

of Commissions, submitted for advice to the Nominating Committee, consisting of one representative of each Adhering Country designated by the appropriate adhering body, before decision by the Executive Committee.

3. Commissions may, with the approval of the Executive Committee, co-opt consultants whom they consider may contribute to their work. The adherence of consultants expires on the last day of the ordinary General Assembly next following their admission, unless renewed.

4. An affiliated organization may participate in the work of the Union as mutually agreed between the organization and the Executive Committee.

II. GENERAL ASSEMBLY

5. The Union meets in ordinary General Assembly, as a rule, once every three years. The place and date of the ordinary General Assembly, unless determined by the General Assembly at its previous meeting, shall be fixed by the Executive Committee and communicated to the adhering bodies at least six months beforehand.

6. The President, with the consent of the Executive Committee, may summon an extraordinary General Assembly. He must do so at the request of one third of the Adhering Countries.

7. The agenda of business for each ordinary General Assembly is determined by the Executive Committee and is communicated to the adhering bodies at least four months before the first day of the meeting. It shall include the proposal of the Executive Committee in regard to the unit of contribution as called for in article 24.

8. (a) Any motion or proposal received by the General Secretary at least five months before the first day of an ordinary General Assembly, whether from an adhering body, from a Commission of the Union, or from an Inter-Union Commission on which the Union is represented, must be placed on the agenda.

(b) A motion or proposal concerning the administration or budget of the Union which does not appear on the agenda prepared by the Executive Committee, or any amendment to a motion that appears on the agenda, shall only be discussed with the prior approval of at least two thirds of the votes of Adhering Countries represented at the General Assembly and having the right to vote by virtue of Statute 11 (a).

9. If there is doubt as to the administrative or scientific character of a question giving rise to a vote, the President determines the issue.

10. Where there is an equal division of votes, the President determines the issue.

11. The President may invite representatives of other organizations, scientists and young astronomers to participate in the General Assembly. Subject to the agreement of the Executive Committee he may delegate this privilege concerning representatives of other organizations to the General Secretary, and concerning scientists and young astronomers to the adhering bodies.

III. SPECIAL NOMINATING COMMITTEE

12. (a) Proposals for elections to the President of the Union, six Vice-Presidents, the General Secretary and the Assistant General Secretary are submitted to the General Assembly by the Special Nominating Committee. This consists of the President and past President of the Union, a member proposed by the retiring Executive Committee from Members not belonging to the Executive Committee, both immediately past and present, and four members selected by the Nominating

parmı douze membres proposés à la réunion des Présidents de Commissions. Les Membres du Comité Spécial des Nominations doivent tous appartenir à des pays différents. Le Secrétaire Général et le Secrétaire Général Adjoint participent au travail du Comité Spécial des Nominations à titre consultatif.

(*b*) Le Comité Spécial des Nominations est nommé par l'Assemblée Générale et est responsable directement devant elle. Il reste en fonction jusqu'à la fin de l'Assemblée Générale ordinaire qui suit immédiatement sa nomination, et il peut combler toute vacance survenant parmi ses membres.

IV. LE COMITÉ EXÉCUTIF ET SES MEMBRES

13. (*a*) Le Président de l'Union reste en fonction jusqu'à la fin de l'Assemblée Générale ordinaire qui suit immédiatement celle de son élection; les Vice-Présidents restent en fonction jusqu'à la fin de la deuxième Assemblée Générale ordinaire qui suit celle de leur élection. Ils ne sont pas rééligibles immédiatement pour les mêmes fonctions.

(*b*) Le Secrétaire Général et le Secrétaire Général Adjoint restent en fonction jusqu'à la fin de l'Assemblée Générale ordinaire qui suit immédiatement celle de leur élection.

Normalement, le Secrétaire Général Adjoint succède au Secrétaire Général, mais l'un et l'autre peuvent être réélus aux mêmes fonctions pour une seconde période consécutive.

(*c*) Les élections ont lieu au cours de la dernière réunion de l'Assemblée Générale, les noms des candidats proposés ayant été annoncés au cours d'une réunion antérieure.

14. Le Président sortant et le Secrétaire Général sortant deviennent conseillers du Comité Exécutif jusqu'à la fin de l'Assemblée Générale ordinaire qui suit immédiatement celle de la fin de leur mandat. Ils participent au travail du Comité Exécutif et assistent à ses réunions sans droit de vote.

15. Le Comité Exécutif peut combler toute vacance survenant en son sein. Toute personne ainsi nommée reste en fonction jusqu'à l'Assemblée Générale ordinaire suivante.

16. Le Comité Exécutif peut rédiger et publier des Directives pour expliciter les Statuts et le Règlement.

17. Le Comité Exécutif nomme le représentant de l'Union qui doit siéger au sein du Comité Exécutif du Conseil International des Unions Scientifiques; si ce représentant n'est pas déjà un membre élu du Comité Exécutif, il en devient conseiller.

18. (*a*) Le Secrétaire Général est responsable auprès du Comité Exécutif des dépenses qu'il engage, qui ne doivent pas dépasser le montant des fonds mis à sa disposition.

(*b*) Un bureau administratif, sous la direction du Secrétaire Général, est chargé de la correspondance, de la gestion des fonds de l'Union, et de la conservation des archives.

V. COMMISSIONS

19. (*a*) Les Commissions de l'Union poursuivent les buts scientifiques de l'Union par des moyens tels que l'étude de domaines particuliers de l'Astronomie, l'encouragement de recherches collectives et la discussion de questions relatives aux accords internationaux et à la standardisation.

(*b*) Les Commissions de l'Union établissent des rapports sur les sujets qui leur ont été confiés.

20. Chaque Commission se compose de:

(*a*) un Président et au moins un Vice-Président élus par l'Assemblée Générale sur la proposition du Comité Exécutif. Ils demeurent en fonction jusqu'à la fin de l'Assemblée Générale ordinaire qui suit immédiatement celle de leur élection. Ils ne sont pas normalement rééligibles.

(*b*) un Comité d'Organisation, dont les membres sont désignés par la Commission sous réserve de l'approbation du Comité Exécutif. Le Comité d'Organisation assiste le Président et le(s) Vice-

Committee from among twelve Members proposed at the meeting of Presidents of Commissions. No two members of the Special Nominating Committee shall belong to the same country. The General Secretary and the Assistant General Secretary participate in the work of the Special Nominating Committee in an advisory capacity.

(b) The Special Nominating Committee is appointed by the General Assembly to which it reports direct. It remains in office until the end of the ordinary General Assembly next following that of its appointment, and it may fill any vacancy occurring among its members.

IV. OFFICERS AND EXECUTIVE COMMITTEE

13. (a) The President of the Union remains in office until the end of the ordinary General Assembly next following that of his election; the Vice-Presidents remain in office until the end of the second ordinary General Assembly following that of their election. They may not be re-elected immediately to the same offices.

(b) The General Secretary and the Assistant General Secretary remain in office until the end of the ordinary General Assembly next following that of their election.

Normally the Assistant General Secretary succeeds the General Secretary though both officers may be re-elected for another term.

(c) The election takes place at the last session of the General Assembly, the names of the candidates proposed having been announced at a previous session.

14. The retiring President and the retiring General Secretary become advisers to the Executive Committee until the end of the ordinary General Assembly next following that of their retirement. They participate in the work of the Executive Committee and attend its meetings without voting right.

15. The Executive Committee may fill any vacancy occurring among its members. Any person so appointed remains in office until the next ordinary General Assembly.

16. The Executive Committee may draw up and publish Working Rules to implement the Statutes and By-laws.

17. The Executive Committee appoints the Union's representative on the Executive Committee of the International Council of Scientific Unions; if not already an elected member of the Executive Committee, this representative will become its adviser.

18. (a) The General Secretary is responsible to the Executive Committee for not incurring expenditure in excess of the funds at his disposal.

(b) An administrative office, under the direction of the General Secretary, conducts the correspondence, administers the funds, and preserves the archives of the Union.

V. COMMISSIONS

19. (a) The Commissions of the Union shall pursue the scientific objects of the Union by activities such as the study of special branches of astronomy, the encouragement of collective investigations, and the discussion of questions relating to international agreements or to standardization.

(b) The Commissions of the Union shall prepare reports on the work with which they are concerned.

20. Each Commission consists of:
(a) A President and at least one Vice-President elected by the General Assembly on the proposal of the Executive Committee. They remain in office until the end of the ordinary General Assembly next following that of their election. They are not normally re-eligible.

Président(s) dans leur tâche. Une Commission peut décider qu'elle n'a pas besoin de Comité d'Organisation.

(c) des membres de l'Union, nommés par les Président, Vice-Président(s) et Comité d'Organisation, en considération de leurs spécialités; leur désignation est soumise à confirmation par le Comité Exécutif.

21. Entre deux Assemblées Générales ordinaires, les Présidents de Commissions peuvent coopter, parmi les Membres de l'Union, de nouveaux membres des Comités d'Organisation et des Commissions elles-mêmes.

22. Les Commissions rédigent leur propre règlement. Les décisions sont prises, à l'intérieur des Commissions, par un vote de leurs membres.

VI. ORGANISMES ADHÉRENTS

23. Le rôle des organismes adhérents est d'encourager et de coordonner, sur leurs territoires respectifs, l'étude des diverses branches de l'astronomie, particulièrement en ce qui concerne leurs besoins sur le plan international. Ils ont le droit de soumettre au Comité Exécutif des propositions pour discussion par l'Assemblée Générale.

VII. FINANCES

24. Chaque Pays Adhérent verse à l'Union une cotisation annuelle, qui est un multiple de l'unité de cotisation en fonction de sa catégorie, comme suit:

Catégories définies conformément à l'article 8 des Statuts:	1	2	3	4	5	6	7	8
Nombre respectif d'unités de cotisations:		1	2	4	6	10	14	20	30

25. Les ressources de l'Union sont consacrées à la poursuite de ses buts, y compris:
(a) les frais de publication et les dépenses administratives;
(b) l'encouragement des activités astronomiques qui nécessitent la coopération internationale;
(c) la cotisation due par l'Union au Conseil International des Unions Scientifiques.

26. Les ressources provenant de dons sont utilisées par l'Union en tenant compte des vœux exprimés par les donateurs.

VIII. PUBLICATIONS

27. L'Union a la propriété littéraire de tous les textes imprimés dans ses publications, sauf accord différent.

28. Les Membres de l'Union ont le droit de recevoir les publications de l'Union gratuitement ou à prix réduits, à la discrétion du Comité Exécutif qui décide en fonction de la situation financière de l'Union.

IX. CLAUSES FINALES

29. Ce règlement entre en vigueur le 1er septembre 1970.
Il peut être modifié avec l'approbation de la majorité absolue des voix des Pays Adhérents qui disposent du droit de vote en vertu de l'article 11 (a) des Statuts.

30. Le présent règlement est publié en versions française et anglaise. En cas d'incertitude, la version française fait seule autorité.

(b) An Organizing Committee, whose members are appointed by the Commission subject to the approval by the Executive Committee. The Organizing Committee assists the President and Vice-President(s) in their duties. A Commission may decide that it needs no Organizing Committee.

(c) Members of the Union, appointed by the President, Vice-President(s) and the Organizing Committee, in consideration of their special interests; their appointment is subject to the confirmation by the Executive Committee.

21. Between two ordinary General Assemblies, Presidents of Commissions may co-opt, from among Members of the Union, new members to the Organizing Committees and to the Commissions themselves.

22. Commissions draw up their own rules. Decisions within Commissions are taken according to the vote of their members.

VI. ADHERING BODIES

23. The functions of the adhering bodies are to promote and co-ordinate, in their respective territories, the study of the various branches of astronomy, more especially in relation to their international requirements. They are entitled to submit to the Executive Committee motions for discussion by the General Assembly.

VII. FINANCE

24. Each Adhering Country pays annually to the Union a number of contributions according to its category as follows:

Category as defined in Statute 8: 1 2 3 4 5 6 7 8
Number of units of contribution: 1 2 4 6 10 14 20 30

25. The income of the Union is to be devoted to its objects, included

(a) costs of publication and expenses of administration
(b) the promotion of astronomical enterprises requiring international co-operation
(c) the contribution due from the Union to the International Council of Scientific Unions.

26. Funds derived from donations are used by the Union in accordance with the wishes expressed by the donors.

VIII. PUBLICATIONS

27. The Union has the copyright to all materials printed in its publications, unless otherwise arranged.

28. Members of the Union are entitled to receive the publications of the Union free of charge or at reduced prices at the discretion of the Executive Committee taking due regard of the financial situation of the Union.

IX. FINAL CLAUSES

29. These By-laws enter into force on 1 September 1970.
They can be changed with the approval of an absolute majority of the votes of the Adhering Countries having the right to vote by virtue of Statute 11 (a).

30. The present By-laws are being published in French and English version. In case of doubt, the French version is the only authority.

UNION ASTRONOMIQUE INTERNATIONALE

DIRECTIVES

Adoptées par le Comité Exécutif le 1er Septembre 1970

I. PUBLICATIONS

1. Les publications de l'Union Astronomique Internationale approuvées, dans le budget, par l'Assemblée Générale, sont préparées par le Bureau Administratif de l'Union.

2. Les Commissions de l'Union peuvent, avec l'approbation du Comité Exécutif, avoir leurs propres publications.

3. Le Comité Exécutif décide sur la proposition du Secrétaire Général, des modalités de distribution des publications de l'Union.

4. Les Membres de l'Union peuvent acquérir les publications de l'Union à un prix réduit.

II. APPARTENANCE À L'UNION

A. *Pays Adhérents*

6. Les demandes d'adhésion à l'Union formulées par les pays sont examinées par le Comité Exécutif compte tenu des points suivants:

(*a*) justesse du choix de la catégorie dans laquelle le pays souhaite être classé

(*b*) situation actuelle de l'astronomie dans le pays formulant la demande, et ses possibilités de développement

(*c*) mesure dans laquelle le futur organisme adhérent est représentatif de l'activité astronomique de son pays.

7. Les demandes proposant une contribution annuelle appropriée seront soumises pour décision à l'Assemblée Générale, avec la recommandation du Comité Exécutif.

B. *Membres*

8. Les personnes proposées pour devenir Membres de l'Union doivent en principe être choisies parmi des astronomes et des chercheurs dont les activités sont étroitement liées à l'astronomie, compte tenu de

(*a*) la qualité de leur œuvre scientifique

(*b*) la mesure dans laquelle leur activité scientifique implique des recherches astronomiques

(*c*) leur désir de contribuer à la poursuite des buts de l'Union.

9. Les jeunes astronomes doivent être considérés comme pouvant devenir Membres de l'Union dès qu'ils ont fait la preuve de leur capacité (en principe par une thèse de doctorat ou son équivalent) et de leur aptitude (quelques années d'activité fructueuse) à mener une recherche personnelle.

10. Pour les astronomes professionnels, leur contribution à l'astronomie peut consister soit en des recherches personnelles, soit en une collaboration assidue à des programmes importants d'observations.

11. Les autres personnes ne peuvent devenir Membres de l'Union que si certains de leurs travaux originaux concernent étroitement la recherche astronomique.

12. Huit mois avant une Assemblée Générale ordinaire, il sera demandé aux organismes adhérents

INTERNATIONAL ASTRONOMICAL UNION

WORKING RULES

adopted by the Executive Committee on 1 September 1970

I. PUBLICATIONS

1. The publications of the International Astronomical Union approved, in the budget, by the General Assembly are prepared by the Administrative Office of the Union.

2. Commissions of the Union may, with the approval of the Executive Committee, issue their publications independently.

3. The distribution of publications of the Union is decided, on the proposal of the General Secretary, by the Executive Committee.

4. Members may purchase the publications of the Union at reduced prices.

II. MEMBERSHIP

A. *Adhering Countries*

6. Applications of countries for adherence to the Union are examined by the Executive Committee for

   (*a*) the adequacy of the category in which the country wishes to be classified

   (*b*) the present state and expected development of astronomy in the applying country

   (*c*) the degree to which the prospective adhering body is representative of its country's astronomical activity.

7. Applications proposing an adequate annual contribution to the Union shall, with the recommendation of the Executive Committee, be submitted to the General Assembly for decision.

B. *Members*

8. Individuals proposed for Union membership should, as a rule, be chosen from among astronomers and scientists, whose activity is closely linked with astronomy taking into account

   (*a*) the standard of their scientific achievement

   (*b*) the extent to which their scientific activity involves research in astronomy

   (*c*) their desire to assist in the fulfilment of the aims of the Union.

9. Young astronomers should be considered eligible for membership after they have shown their capability (as a rule Ph.D. or equivalent) of and experience (some years of successful activity) in conducting original research.

10. For full-time professional astronomers the achievement in astronomy may consist either of original research or of substantial contributions to major observational programmes.

11. Others are eligible for membership only if they are making original contributions closely linked with astronomical research.

12. Eight months before an ordinary General Assembly, adhering bodies will be asked to propose

de proposer de nouveaux Membres. Les propositions devront parvenir au Secrétaire Général au moins cinq mois avant la première session de l'Assemblée Générale. Les propositions reçues après cette date limite ne seront prises en considération que si des circonstances exceptionnelles justifient le retard.

13. Chaque proposition du nouveau Membre doit être présentée séparément et indiquer le nom, les prénoms et l'adresse postale du candidat (de préférence celle de son Institut ou Observatoire), ses date et lieu de naissance, l'Université devant laquelle il a soutenu sa thèse ou le diplôme équivalent, la date de soutenance, la situation actuelle du candidat, les titres et renseignements bibliographiques de deux ou trois de ses articles ou publications les plus significatifs et, s'il y a lieu, tous les renseignements susceptibles d'être pris en considération par le Comité des Nominations.

14. (a) Les Présidents de Commissions qui désirent suggérer de nouveaux Membres doivent adresser leurs suggestions au Secrétaire Général au moins cinq mois avant la première session d'une Assemblée Générale ordinaire. Les propositions devront fournir les mêmes renseignements que ceux mentionnés à l'article 13.

(b) Le Secrétaire Général fait part de ces suggestions aux organismes adhérents intéressés.

15. Le Secrétaire Général préparera deux listes pour le Comité des Nominations

(a) l'une contenant les noms des candidats proposés par les organismes adhérents

(b) l'autre contenant les noms des candidats proposés par les Présidents de Commissions, mais qui ne sont pas déjà inclus dans les propositions des organismes adhérents.

16. A partir des deux listes mentionnées à l'article 15, le Comité des Nominations prépare les propositions définitives de nouveaux Membres de l'Union.

17. Les organismes adhérents peuvent proposer la radiation de Membres ayant abandonné le domaine de l'astronomie pour d'autres activités, à moins qu'ils ne continuent à apporter une contribution à l'astronomie. Ces propositions doivent être portées à la connaissance du Secrétaire Général et du Membre concerné.

18. Le Secrétaire Général publiera la liste alphabétique des Membres de l'Union dans les Transactions de chaque Assemblée Générale ordinaire.

III. MEMBRES DES COMMISSIONS

19. Les membres des Commissions de l'Union sont cooptés par les Commissions. Cette procédure est régie par des règles établies par les Commissions elles-mêmes.

20. Les Commissions devraient choisir, ou approuver, la liste des membres de leurs commissions compte tenu de la spécialité de ces personnes, en particulier de leur activité scientifique dans le domaine de recherches de la Commission, et de leur contribution au travail de la Commission. Elles peuvent

(a) inviter des Membres de l'Union à devenir membres de la Commission

(b) radier les membres de la Commission qui n'ont pas contribué à son activité

(c) accepter ou refuser les demandes présentées par des Membres de l'Union, ou par des personnes proposées comme tels, en vue d'appartenir à la Commission

(d) suggérer l'élection comme Membres de l'Union de personnes n'y appartenant pas, ce qui leur permettrait alors de devenir membres de la Commission.

21. Les Membres de l'Union ne peuvent pas, en règle générale, appartenir à plus de trois Commissions.

22. Les Membres de l'Union peuvent demander à être admis dans une Commission en écrivant au Président de cette Commission. Ils ne devraient faire cette demande que si leur propre activité rentre dans le cadre des recherches de la Commission et s'ils sont décidés à contribuer au travail de la Commission.

23. Les membres des Commissions peuvent se retirer d'une Commission en écrivant à son Président.

24. En envoyant leurs propositions de nouveaux Membres, les organismes adhérents peuvent également suggérer le choix d'une Commission pour chaque candidat.

new Members. The proposals should reach the General Secretary not later than five months before the first session of the General Assembly. Proposals received after the closing date will only be taken into consideration if the delay is justified by exceptional circumstances.

13. Each proposal shall be written separately. It should include the name, first names and postal address of the candidate, preferably that of his Institute or Observatory, his place and date of birth, the University and the year of his Ph.D. or equivalent title, his present occupation, titles and bibliographic data of two to three of his more important papers or publications, and details, if any, worthy to be considered by the Nominating Committee.

14. (a) Presidents of Union Commissions wishing to suggest new Members for admittance should address their suggestions to the General Secretary five months before the first session of an ordinary General Assembly. The proposals should contain particulars as in article 13.
(b) The General Secretary notifies the adhering bodies in questions of such suggestions.

15. The General Secretary shall prepare two lists for the Nominating Committee
(a) one containing the candidates proposed by the adhering bodies,
(b) the other containing those suggested by Presidents of Commissions, but not included among the proposals of adhering bodies.

16. The Nominating Committee prepares the final proposals for Union membership from the two lists as mentioned in article 15.

17. Adhering bodies should propose cancellation of Members who have left the field of astronomy for other interests, unless they continue to contribute to astronomy. Such proposals should be announced to the Member concerned and to the General Secretary.

18. The alphabetical list of Union Members will be published by the General Secretary in the Transactions of each ordinary General Assembly.

III. COMMISSION MEMBERSHIP

19. Members of Union Commissions are co-opted by Commissions. The rules governing the procedure of such co-option are drawn up by the Commissions themselves.

20. Commissions should choose, or approve of, Commission members taking into account their special interests, in particular their scientific activity in the appropriate fields of research and their contribution to the work of the Commission. They may

(a) invite Members to become members of their Commission

(b) remove members who have not contributed to the work of Commission

(c) accept or reject applications for membership from existing or proposed Members

(d) suggest non-Members for election as Members, thus enabling them to become members of the Commission.

21. Members may not, as a rule, be members of more than three Commissions.

22. Members may apply for Commission membership by writing to the President of the Commission concerned. Such applications should only be made if the Member is actively engaged in the appropriate field of research and is prepared to contribute to the work of the Commission.

23. Members of Commissions may resign from a Commission by writing to its President.

24. Adhering bodies, in sending in their proposals for new Members, may also suggest one Commission for each candidate.

25. Le Secrétaire Général enregistrera et analysera la liste des membres des Commissions; si cela est nécessaire, il tentera de trouver une solution aux anomalies évidentes.

26. Le Secrétaire Général publiera la liste des membres des Commissions dans les Transactions de chaque Assemblée Générale ordinaire.

IV. CONSULTANTS

27. Peuvent être élus Consultants des personnes qui ne sont pas astronomes, mais qui sont susceptibles de servir les intérêts de l'astronomie.

28. Les Commissions doivent en principe envoyer, pour approbation, leurs propositions de consultants au Secrétaire Général au moins cinq mois avant la première session d'une Assemblée Générale ordinaire.

29. Le Secrétaire Général préparera une liste des personnes proposées comme consultants et la soumettra pour approbation au Comité Exécutif.

30. Le Bureau Administratif établira une liste alphabétique des consultants.

31. Les consultants peuvent participer aux réunions de l'Union. Ils peuvent avoir droit de vote dans leurs Commissions respectives. Ils reçoivent gratuitement le Bulletin d'Information de l'Union.

25. The General Secretary will record and analyse the list of members of Commissions; if necessary he will try to resolve any outstanding anomalies.

26. The list of Commission members will be published by the General Secretary in the Transactions of each ordinary General Assembly.

IV. CONSULTANTS

27. Eligible as Consultants are non-astronomers in a position to further the interest of astronomy.

28. Proposals of Commissions for the approval of consultants should, as a rule, reach the General Secretary not later than five months before the first session of an ordinary General Assembly.

29. The General Secretary shall prepare a list of those proposed for admission as consultants and submit it to the Executive Committee for approval.

30. The Administrative Office will maintain an alphabetical list of consultants.

31. Consultants may participate in the meetings of the Union. They may have voting right in the respective Commission. They receive, free of charge, the Information Bulletin of the Union.

RULES FOR SCIENTIFIC MEETINGS

A. GENERAL

One of the essential tasks of the Union is to encourage circulation of ideas and to disseminate information on scientific results by organizing scientific meetings.

Such meetings are (a) Symposia (b) Colloquia (c) Commission Meetings and (d) Joint Discussions during General Assemblies.

Each meeting has to be proposed or sponsored by one or more Commissions. It is organized either by the Organizing Committee of a Commission (case c), or by an Organizing Committee appointed by the Executive Committee (cases a, b, d).

Symposia and Joint Discussions cover broader fields and have larger attendance. Their Proceedings are published by the IAU.

Colloquia and Commission Meetings deal with somewhat narrower topics and last a shorter time. Their Proceedings are not published by the IAU, except for short summaries of the Commission Meetings which appear in the Proceedings of the General Assemblies.

The publication of Proceedings of Colloquia is decided and arranged by their Organizing Committees.

Commission Meetings held in the course of General Assemblies are of two kinds: (a) Business Meetings, including organization of specific projects, and (b) Scientific Meetings on specific subjects. Such meetings can be organized jointly with other Commissions. Meetings of Working Groups deal, as a rule, with business.

Attendance of Symposia and Colloquia is by invitation from the President of the Scientific Organizing Committee only.

Symposia and Colloquia held in conjunction with a General Assembly should preferably take place in the same country as the General Assembly.

Proposals for Symposia and Colloquia are addressed to the Executive Committee through the Assistant General Secretary (copies should be sent to the General Secretary). Such proposals should be accompanied by letters of Presidents of Commissions sponsoring them. The proposal should contain (1) the title of the Symposium or Colloquium, (2) the date and place, (3) the contact address, (4) the suggested composition of the (Scientific) Organizing Committee, (5) the composition of the Local Organizing Committee, (6) the approximate number of participants, (7) any financial support other than from the IAU, (8) the IAU support asked for, (9) the suggested Editor(s) of the Proceedings (if there will be Proceedings) and, (10) a preliminary scientific programme, or simply the topics to be covered. The proposal should also state why the meeting is useful or necessary at the time proposed.

Proposals for Meetings of Commissions or Joint Discussions during a General Assembly are addressed to the General Secretary, according to the time-table set by him.

The agenda of scientific meetings may consist of (a) Invited Papers, (b) Contributed Papers, and (C) Discussions.

The Organizing Committee decides on the invited speakers, accepts or rejects contributed papers, decides on the subjects, programme, way of discussion, translation, recording of discussions, organization of the secretariat, etc. Invited speakers should, as a rule, review a certain topic and not restrict themselves to their own work.

It is advisable to have mimeographed summaries of papers for distribution to all participants before the Meeting.

Any proposal for a Symposium or Colloquium should be made at least one year before the meeting.

Presidents of related Commissions should be informed of the proposals and invited to send in their comments.

Fully documented proposals are submitted by the Assistant General Secretary to the Executive Committee. The Executive Committee either accepts or rejects the proposals. In the former case the Executive Committee decides, after consultation with the authors of the proposal and interested Presidents of Commissions, on the title, date and place, president and members of the Organizing Committee, financial support, Editor(s) of the Proceedings and topics to be covered.

Changes in the original proposals are sometimes necessary in order to ensure the international character and broad scientific coverage in the Organizing Committee or to avoid overlapping with similar meetings. The Executive Committee may ask other Unions to co-sponsor a particular Symposium.

The President of the Organizing Committee

(1) invites participants to the Symposium or Colloquium once the meeting is accepted by the Executive Committee;

(2) accepts or rejects requests for participation;

(3) invites contributions and sets a deadline for submission of abstracts of papers.

(4) informs the Assistant General Secretary as follows:

About 4 months before the meeting:

(a) detailed programme

(b) choice of participants

(c) provisions made for local organization.

About 2 months before the meeting: distribution of the IAU financial allocation.

The Assistant General Secretary should also be informed of any change in the programme of the meeting.

C. PARTICIPATION OF THE IAU IN JOINT SYMPOSIA

The IAU is willing to participate in Symposia, jointly with other Unions or Inter-Union Committees, either as a principal or a subsidiary participant.

As the principal participant, the IAU:

(a) proposes the Symposium and, either alone or in consultation with other Unions, plans the main features such as scope of subject, place and date;

(b) appoints an Organizing Committee and its President;

(c) invites the co-operation and participation of other Unions; such co-operation could include, depending on circumstances:

(i)   discussions of the main features;

(ii) appointment of representatives on the Organizing Committee;

(iii) the drawing up of a list of participants, a scientific programme, methods of presentation, etc., through the representatives on the Organizing Committee;

(iv) discussion on the publication of the Proceedings, together with all associated questions of free distribution, price, payments and subsidies;

(d) would normally expect the other Unions to make some contribution to the expenses of the Symposium;

(e) would normally arrange for publication of the Proceedings as one of the IAU Symposium volumes, but would consult the other Unions as to their wish to have special copies or to contribute financially in return for the volume being included in their series.

As a subsidiary participant, the IAU would:

(a) accept the leading role of the principal participant in all relevant matters;

(b) wish to appoint representatives in the Organizing Committee, to be consulted on all important questions, and to be kept informed of decisions on other matters;

(c) in principle, contribute towards the expenses;

(d) wish to be consulted in regard to the publication of the Proceedings, especially in regard to the distribution of free copies to participants.

The IAU would like to suggest that Symposia organized jointly by one or more Unions should always be announced in the following terms:

Organized by the IAU in co-operation with IUGG and COSPAR (e.g.)

or

Organized by IUHPS (e.g.) in co-operation with the IAU.

D. FINANCIAL PROBLEMS

The Executive Committee of the IAU allocates to each Symposium or Colloquium a limited sum which is meant as a 'catalyzer' to encourage other international or national organizations to subsidize it. Colloquia receive smaller amounts than Symposia. The IAU grant is used exclusively to cover travel expenses of the participants and to meet some of the expenses of the secretariat of the Symposium or Colloquium, if necessary. Other expenses are generally covered as follows:

Subsistence expenses of the participants: by the participants themselves or by grants from their governments or home institutions.

Travel expenses of the participants: partly by grants from the IAU funds, partly by grants from other organizations.

Expenses of the secretariat of the Symposium or Colloquium: partly from the IAU funds, partly by organizations in the host country or by home institutions of the members of the Organizing Committee.

Local organization (receptions, social events, excursions): by the host municipality, country or institute.

The president of the Organizing Committee encourages each participant to apply for financial aid from his home institute or country.

He invites applicants to specify their requests for IAU support (the form 'Application for Travel Grant' can be obtained from the Executive Secretary).

After the President of the Organizing Committee establishes the financial support available to participants from other sources, he suggests to the Assistant General Secretary how to distribute the IAU funds.

E. LOCAL ORGANIZATION

As soon as one of the members of the Organizing Committee is put in charge of the local organization, he appoints a Local Organizing Committee and informs the President of the Organizing Committee of its composition.

The Local Organizing Committee takes care of the smooth running of the Symposium or Colloquium. It does not receive any financial help from the IAU, the necessary expenses being covered by local funds or by contribution from the participants.

The requirements of local organization are generally as follows:

1. Meeting rooms suitable for the expected number of participants should be located and reserved.

2. Arrangements should be made to mimeograph and distribute papers presented by the participants, if desirable.

3. Sufficient secretarial and technical assistance should be made available for typing and duplicating, and for operating tape recorders, microphones, projection equipment, etc.

4. If the Organizing Committee decides to have the discussions recorded, arrangements should be made either (a) to have all discussions recorded on tape, or (b) to provide each speaker with a sheet of paper on which he is to write out his remarks immediately after he has delivered his contribution. In the first case it is highly desirable to have the tapes transcribed before the end of the meeting and ask the speakers to edit their remarks. This will avoid lengthy correspondence between Editor and contributor.

5. Information on accommodation (hostels, hotels, etc.) should be sent to the President of the Organizing Committee and to prospective participants as early as possible.

6. All participants should be asked to send their wishes as regards accommodation, excursions, and social events to the Local Organizing Committee.

7. Receptions and excursions can be organized during the free periods of the meeting, or just before and after the meeting. A ladies programme is a welcome courtesy.

8. Participants should be informed of the reservations made for them, and how to proceed to their hotel and to the meeting place on arrival. Arrangements to meet each participant at the airport or the railway station are welcome.

It is customary for the Local Organizing Committee to print or mimeograph the final programme, including any other useful information, to be distributed to participants on arrival.

F. PUBLICATION

The IAU believes that the Proceedings of Symposia are of such general interest that their early publication is desirable. Their publication and distribution have therefore been entrusted to a commercial publisher.

The main requirements for Proceedings volumes are:
their early publication;
their scientific value and originality.

It occurred in the past that a paper, first published in a Symposium volume was later offered to a regular journal, and *vice versa*. Thus it may happen that the paper in the Symposium volume lags behind the very same paper published in the journal.

Many Editors find it difficult to refuse publication of papers once accepted for oral presentation. This policy, however, reduces the quality of the published Proceedings, and increases their costs. Moreover, it leads to unnecessary duplication.

This can be avoided if Editors adhere to the following rules:

(1) Normally only invited papers should be included in Symposium volumes.

(2) Contributed papers should not be accepted for publication unless they are of high quality and contain new material (relevant to the subject of the Symposium) not to be published elsewhere. In general, authors should be advised to send their contributed papers to regular journals, and only publish summaries in the Proceedings, with reference to the original papers.

(3) It may sometimes be advisable to have a number of contributions critically summarized by an eminent scientist into a single short paper.

Editors should not hesitate to discard, or reduce the length of, contributed papers in order to add to the scientific value of the Proceedings. Detailed theoretical derivations, lists of observations and tabular data should be avoided.

The Editor of Symposium Proceedings is appointed by the Executive Committee, as a rule from among the members of the Organizing Committee, on the proposal of its President. It is desirable that the Editor, or one of the Editors, should be from an English-speaking country.

The Editor is responsible both for the scientific value and the appearance of the Proceedings. His main tasks are:

(a) To inform the participants in ample time before the meeting in what general form their contributions should be submitted and what is the deadline for submitting the final version of the manuscripts. This deadline should never be longer than a few days after the meeting.

(b) To inform the participants about the IAU rules (see the IAU Style Book, pp. 254–264.) for publication of the IAU Proceedings, and to emphasize that all contributed papers should be refereed before publication.

(c) To discuss with the Organizing Committee and Local Organizing Committee the method of recording the discussions. In some Proceedings, discussions have a most prominent place, in others only some of the most important discussions are included, and emphasis lies on speedy publication.

(d) To arrange the refereeing of contributed papers. This may be done by the Editor himself or by another member of the Organizing Committee. For the sake of early publication prospective referees should be asked how many papers they can referee within a given time. The Editor should either accept a paper, or reject it. No time should be allowed for revising a paper.

(e) To reduce the length of papers and discussions, to avoid duplication, and to improve the language whenever necessary.

(f) To check whether the IAU rules have been applied in each paper. For instance a paper should have an abstract, good figures with captions, references in correct form, correct abbreviations and transliterations, clear mathematical symbols, numbered sections, etc. The same applies to the Discussions. The Editor should try to make all necessary corrections himself. In some cases he may have to ask the authors to provide missing abstracts, better figures, etc. This should be done at the earliest possible date and a deadline should be fixed, usually after previous consultation with the publisher. If the material requested does not arrive within the deadline, the whole paper should be omitted. The Proceedings should not be delayed because of a few incomplete papers.

(g) To write an Introduction, Table of Contents, etc. In order to secure uniformity the Editor may consult recent IAU Symposium Proceedings. However, the new IAU Style Book (p. 254-264) supersedes any previous practice by authors, or Editors.

(h) To send copies of the edited manuscripts to the authors in cases of substantial changes by the Editor and to inform them that excessive author's corrections will not be acceptable.

(i) *To send the edited manuscript to the publisher not later than two months after the Symposium.* Copies of the Introduction, Table of Contents, and abstracts of the papers should be sent to the Assistant General Secretary.

(j) To be in close contact with the Assistant General Secretary of the IAU and the publisher, especially as regards deadlines and doubtful cases.

In order to obtain a presentation that will indicate that the volumes of IAU Symposia form a series, the Union requests that the Scientific Editors adhere to the following recommendations:

Title page. The title page should include explicitly, above the title of the Symposium and the date on which it was held, the words 'IAU Symposium No...', followed, where appropriate, by the words 'Organized by the IAU in co-operation with...' in which the list of organizations is limited to the Scientific Unions, the Scientific Committees (such as COSPAR), and Inter-Union Commissions of the ICSU (such as IUCSTP). The participation of UNESCO will be acknowledged by the following wording: 'Published for the International Council of Scientific Unions with the financial assistance of UNESCO'.

Table of Contents

Introduction. The introduction by the Editor should mention the various aspects of the organization of the Symposium, the members of the Scientific Organizing Committee and of the Local Organizing Committee, and express appreciation of their work.

The support of the IAU and of other Unions, of the UNESCO, of international, national, or local organizations should also be recognized.

Proceedings of the Symposium

Alphabetical Index of names and subjects, if practicable.

About two months after receipt of the manuscript the first proofs will be sent to the authors, and copies to the Editors.

The Editors should set a deadline for receiving the corrected first proofs back. Proofs not returned in time should be corrected by the Editor; in this case belated author's corrections are ignored. About one month after the corrected first proofs are in, the second proofs are sent to the Editors. Second proofs are not sent to the authors.

Printing and binding will take another two months.

Thus, Symposium volumes may be on the market 8 months after the Symposium.

Symposium participants shall obtain, free of charge, 25 copies of their contributions. Further copies can be ordered when the first proofs are returned. All IAU members can purchase Symposium volumes at reduced prices.

COMPOSITION DES COMMISSIONS

MEMBERSHIP OF COMMISSIONS

4. Commission des Ephémerides. (Ephemerides)
PRÉSIDENT: J. Kovalevsky.
VICE-PRÉSIDENT: R. L. Duncombe.
COMITÉ D'ORGANISATION: V. K. Abalakin, W. Fricke, A. M. Sinzi, G. A. Wilkins.
MEMBRES: Bec (Miss) A., Clemence G. M., Chebotarev G. A., de Greiff J. A., Fursenko (Mrs) M. A., Gondolatsch F., Haupt R. F., Herrick S., Klepczynski W. J., Lahiri N. C., Lederle T., Morrison L. V., Morando B., Mulholland J. D., O'Handley D. A., Rodriques M., Sadler D. H., Sadler (Mrs McBain) F. M., Sato Y., Seidelmann P. K., Shapiro I. I., Taylor G. E., van Flandern T. C.

5. Commission de Documentation. (Documentation)
(Comité du Comité Exécutif)
PRÉSIDENT: J. B. Sykes.
VICE-PRÉSIDENT: K. F. Ogorodnikov.
Sans Comité d'Organisation.
MEMBRES: Alter G., Aoki S., Beer A., Belorizky D., Bourgeois P., Bouška J., Chandrasekhar S., Dewhirst D. W., Fricke W., Heinemann K., Heintz W. D., Henn (Miss) F., Hirose H., Kleczek J., Luplau-Janssen C. E., McNally D., Martynov D. Ya., Meadows A. J., Mein P., Pecker J.-C., Sticker B., Velghe A. G., Weidemann V., Wempe J., Wilkins G. A., Witkowski J., Zinner E.

6. Commission des Télégrammes Astronomiques. (Astronomical Telegrams)
(Comité du Comité Exécutif)
PRÉSIDENT: P. Simon.
VICE-PRÉSIDENT: J. Hers.
Sans Comité d'Organisation.
MEMBRES: Candy M. P., Cunningham L. E., Kozai Y., Marsden B. G. (Secrétaire de la Commission) Martynov D. Ya., Mrkos A., Roemer (Miss) E., Whipple F. L.

7. Commission de la Mécanique Céleste. (Celestial Mechanics)
PRÉSIDENT: G. N. Duboshin.
VICE-PRÉSIDENT: P. J. Message.
COMITÉ D'ORGANISATION: G. A. Chebotarev, W. J. Eckert, Y. Hagihara, J. Kovalevsky, Y. Kozai, K. Stumpff, V. C. Szebehely, F. Zagar.
MEMBRES: Antonacopoulos G., Acki S., Aksenov E. P., Batrakov Yu. V., Belorizky D., Böhme S., Bozis G., Broucke R., Brumberg V. A., Cesco R. P., Chapront J., Clemence G. M., Contopoulos G., Cook A. H., Cox J. F., Cunningham L. E., Danby J. M. A., Deprit A., Duncombe R. L., Fabre H., Ferraz-Mello S., Galibina I. V., Garfinkel B., Gaska S., Giacaglia G., Goldreich P., Goudas C. I., Hadjidemetriou J. D., Hamid S. E., Henon M., Herget P., Herrick S., Hertz H., Hori G., Jefferys W. H., Jeffreys H., Katsis D. N., Kaula W. M., Kustaanheimo P. E., Lundquist C. A., Magnaradze (Miss) N. C., Marsden B., Meffroy J., Merman G. A., Morando B., Musen P., Mulholland J. D.,

Nacozy P. F., Orlov A. A., Peale S. J., Petrovskaya M. S., Popović B., Rabe E., Ryabov J. A., Roy A. E., Sadler D. H., Schubart J., Sconzo P., Sehnal L., Sharaf S. G., Shapiro I. I., Sibahara R., Siegel C. L., Slavenas P. V., Stiefel E., Thiry Y., Vinti J. P., Wilkens A., Williams C. A., Yarov-Yarovoy M. S.

8. **Commission de l'Astronomie de Position. (Positional Astronomy)**
PRÉSIDENT: W. Fricke.
VICE-PRÉSIDENT: G. van Herk.
COMITÉ D'ORGANISATION: B. L. Klock, P. Lacroute, A. A. Nemiro, J. L. Schombert, R. H. Tucker, J. E. B. von der Heide, H. W. Wood, H. Yasuda.
MEMBRES: Adams A. N., Anguita C. A., Argyrakos J. G., Atkinson R. d'E, Bacchus P., Barros M. G. P. de, Bohrmann A., Dambara T., Débarbat (Miss) S., De Moraes A., De Sousa Pereira Osório J. J., De Vegt Ch., Dieckvoss W., Duncombe R. L., Eichhorn H. K. v. W., Gliese W., Gordon Ya. E., Guinot B., Haas J., Hansson N., Harris B. J., Heintz W. D., Hoffleit (Miss) E. D., Høg E., Hughes V. A., Jackson P., Korol' A. K., Kosin G. S., Larink J., Laustsen S., Lévy J. R., Maître V., Marcus (Miss) E., Melchior P. J., Mitić L. A., Murray C. A., Nefed'eva A. I., Novopashennyj B. V., Okuda T., Petrov G. M., Pilowski K., Podobed V. V., Polozhentsev D. D., Postoiev A., Proverbio E., Quijano L., Reiz A., Requième Y., Rybka P., Sanding H.-U., Scheepmaker A. C., Schmeidler F., Schuler W., Scott F. P., Sémirot P., Slaucitajs S., Stoy R. H., Symms L. S. T., Ševarlić B. M., Tavastsherna K. N., Teleki G., Thomas D. V., Tsubokawa I., Tuzi K., Verbaandert J., Woolsey E. G., Zimmerman G. K., Zverev M. S.

9. **Commission des Instruments Astronomiques. (Astronomical Instruments)**
PRÉSIDENT: V. B. Nikonov.
VICE-PRÉSIDENT: A. B. Meinel.
COMITÉ D'ORGANISATION: W. A. Baum, W. C. Livingston, J. D. McGee, N. N. Mihelson, J. Ring, J. Rösch, B. Valníček.
MEMBRES: Aly M. K., Andrews D. H., Atkinson R. d'E., Babcock H. W., Bakos G. A., Baranne A., Barros M. G. P. de, Beck H. G., Behr A., Bhattacharyya J. C., Blitzstein W., Bowen I. S., Boyce P., Brahde R., Brealey G. A., Brosterhus E. B. F., Brückner G., Chincarini G., Connes P., Couder A., Cousins A. W. J., Crawford D. L., de Jonge J. K., Delbouille L., Delsemme A. H., Dobronravin P. P., Dollfus A., Duchesne M., Dunham T. Jr., Dunkelman L., Einasto J. E., Evans J. V., Fellgett P. B., Fischer R. R., Fitch W. S., Ford W. K. Jr., Giovanelli R. G., Gottlieb K., Gratton L., Gunn J. E., Hall J. S., Hallam K. L., Hilliard R., Høg E., Hooghoudt B. G., Ingrao H. C., Ioannisiani B. K., Jayarajan A. P., Jensch A., Karandikar R. V., Kissel K. E., Köhler H., Kopylov I. M., Krasovskij V. I., Kron G. E., Kurt V. G., Lallemand A., Lasker B. M., Linfoot E. H., Linnik V. P., McMullan D., Melnikov O. A., Merman N. V., Minett H. C., Mohler O. C., Monin G. A., Moroz V. I., Odgers G. J., Petford A. D., Pierce A. K., Purgathofer A., Rakos K. D., Richardson E. H., Roddier F., Schmahl G., Schönaich W., Schroeder D. J., Seddon H., Shcheglov P. V., Sisson G. M., Smyth M. J., Strand K. Aa., Suharev L. A., Taylor D. J., Texereau J., Tull R. C., Ulrich B. T., Väisälä Y., Walker M. F., Wallenquist A., Walraven Th., Wayman P. A., Wellgate G. B., Wilcock W. L., Wlerick G., Wynne C. J., Zacharov I.

10. **Commission de l'Activité Solaire. (Solar Activity)**
PRÉSIDENT: J. T. Jefferies.
VICE-PRÉSIDENT: K. O. Kiepenheuer.
COMITÉ D'ORGANISATION: L. D. de Feiter, R. G. Giovanelli, V. A. Krat, R. Michard, S. Nagasawa, G. Newkirk, P. Simon, Z. Švestka.
MEMBRES: Acton L. V., Allen C. W., Altschuler M., Aly M. K. M., Anderson K. A., Athay R. G., Babcock H. W., Ballario M. C., Balli E., Banin V. G., Beckers J. M., Bell B.,

Bhatnagar A., Billings D. E., Bonov A. D., Bray R. J., Brückner G., Bruzek A., Bumba V., Carlquist P., Cimino M., Coutrez R. A. J., Covington A. E., de Groot T., de Jager C., Deubner F. L., Dezsö L., Dizer M., Dodson-Prince H. W., Dollfus A., Dubov E. E., Dunn R. B., Elste G., Fokker A. D., Fortini T., Fracastoro M. G., Friedman H., Fritzová-Švestková L., Gaizauskas V., Gleissberg W., Gnevyshev M. N., Gnevysheva R. S., Godoli G., Goldberg L., Gopasyuk S. I., Gurtovenko E. A., Hagen J. P., Hanasz J., Hansen R., Hedeman E. R., Hiei E., Hirayama T., Howard R. F., Hyder Ch. L. (Secrétaire de la Commission), Ioshpa B. A., Jäger F. W., Jakimiec J., Jakovkin N. A., Jensen E., Kai K., Kjeldseth Moe O., Kleczek J., Kopecký M., Křivský L., Krüger A., Kuklin G. V., Kundu M. R., Künzel H., Lehnert B. P., Leighton R. B., Leroy J. L., Letfus V., Loughhead R. E., McKenna-Lawlor S. M. P., Macris C. J., Makita M., Malitson H. H., Maltby P., Malville J. M., Mandel'shtam S. L., Martres M. J., Mattig W., Maxwell A., Menzel D. H., Mergentaler J., Mogilevskij E. I., Mohler O. C., Moiseev I. G., Moreton G. E., Moriyama F., Mustel E. R., Nakagawa Y., Neupert W. M., Nishi K., Notuki M., Obridko V. N., Öhman Y., Orrall F. Q., Parkinson W. H., Pick-Gutmann M., Pneuman G. V., Popovici C., Rayrole J., Razmadze T. S., Reeves E. M., Rigutti M., Ringnes T. S., Roberts W. O., Rompolt B., Rösch J., Růžičková-Topolová B., Saito K., Sakurai K., Sarabhai V. A., Schlüter A., Schmidt H. U., Schröter E. H., Servajean R., Severny A. B., Shapley A. H., Sheeley N. R., Shklovski J. S., Sitnik G. F., Slonim E. M., Smerd S. F., Smith E. v. P., Smith Henry J., Stenflo J. O., Stepanov W. E., Steshenko N. V., Sturrock P. A., Suemoto Z., Takakura T., Tanaka H., Tandberg-Hanssen E., Trellis M., Tuominen J. V., Valníček B., Van Allen J. A., Van 't Veer F., Vassiljeva G. Y., Vitinskij Yu. I., von Klüber H., Waldmeier M., Warwick C. S., Wilcox J. M., Wild J. P., Winckler J. R., Xanthakis J., Zirin H., Zwaan C.

12. Commission de la Radiation et de la Structure de l'Atmosphère Solaire. (Radiation and Structure of the Solar Atmosphere)

PRÉSIDENT: R. G. Athay.

VICE-PRÉSIDENT: R. G. Giovanelli.

COMITÉ D'ORGANISATION: L. Delbouille, M. N. Gnevyshev, R. Michard, Miss E. A. Müller, A. K. Pierce, Z. Suemoto.

MEMBRES: Adam (Miss) M. G., Allen C. W., Bappu M. K. V., Beckers J. M., Billings D. E., Blackwell D. E., Blaha M., Blamont J. E., Böhm K.-H., Böhm-Vitense E., Bray R. J. De, Brück H. A., Bruzek A., Chamberlain J. W., Christiansen W. N., De Jager C., Dubov E. E., Edlén B., Edmonds F. N., Elste G., Evans J. W., Friedman H., Gingerich O. J., Gökdogan (Mrs) N., Goldberg L., Gopasyuk S. I., Grevesse N., Haupt R. F., Hiei E., Hotinli M., Houtgast J., Hubenet H., Jordan C., Jäger F. W., Jefferies J. T., Kanno M., Kato S., Kawaguchi I., Kiepenheuer K. O., Kononovich E. V., Kopecký M., Krat V. A., Kuklin G. V., Laborde G., Labs D., Leighton R. B., Leroy J.-L., Locke J. L., Lopez Arroyo M., Loughead R. E., Lüst R., Mathias O. F. W., Matsushima S., Mattig W., Mein P., Mergentaler J., Migeotte M. V., Moore-Sitterly Ch., Mouradian Z., Mugglestone D., Namda O., Neven L., Newkirk G. A., Nicolet M., Nishi K., Nikolsky G. M., Orrall F. Q., Oster L., Pande C., Pajdušáková L., Parkinson W. H., Pecker J.-C., Pecker-Wimel (Mrs) C., Peyturaux R., Redman R. O., Reeves E. M., Righini G., Rigutti M., Roland G., Schröter E. H., Seaton M. J., Severny A. B., Sitnik G. F., Sobolev V. M., Švestka Z., Swensson J. W., Tandberg-Hanssen E., Teplitskaya R. B., Thomas R. N., Tousey R., Uchida Y., Unsöld A., Voigt H. H., Von Klüber H., Waldmeier M., Warwick J. W., Wilson P. R., Wlérick G., Zirin H., Zirker J. B.

14. Commission des Données Spectroscopiques Fondamentales. (Fundamental Spectroscopic Data)

PRÉSIDENT: A. H. Cook (1) (2).

VICE-PRÉSIDENT: R. H. Garstang (2)

COMITÉ D'ORGANISATION: G. Herzberg (5), W. Lochte-Holtgreven (2), M. V. Migeotte, Mrs Ch. Moore-Sitterly (4), J. G. Phillips (5), R. Tousey (4).

MEMBRES: Allen C. W. (2), Andrew K. L. (4), Baird K. M. (1), Bates D. R. (3), Bely O., Blaha M. (3), Branscomb L. M. (3), Burgess A. (2) (3), Carver J. H., Corliss C. H. (2), Dalgarno A. (3), Dobronravin P. P., Dressler K. (5), Dufay M., Edlén B. (1) (4), Engelhard E. (1), Essen L., Felenbok P. (5), Gabriel A. H. (3), Garton W. R. S. (4), Grant I. P., Green L. C. (2), Herman R. (3) (5), Hesser J. E. (2) (5), Hindmarsh W. R. (3), House L. L., Humphreys C. J. (1) (4), Jacquinot P., Jordan H. L., Junkes J., Kessler K. G. (1) (5), King R. B. (2), Lawrence G. (2), Layzer D. (2), Littlefield T. A. (1), Mandel'shtam S. L., Martin W. C. Jr. (4), Mel'nikov O. A., Milford S. N., Mohler O. C., Monfils A. (5), Naqvi A. M. (2), Nevin T. E. (5), Nicholls R. W. (5), Nussbaumer H. (2) (3), Obi S., Oetken L., Peach G. (3), Prokof'ev V. K. (2), Rosen B. (5), Salpeter E. W., Schadee A. (5), Seaton M. J. (3), Shore B. W. (2), Smit J. A., Smith G. (2), Somerville W. B. (3), Swings J.-P. (2), Swings P. (5), Takayanagi K. (3), Tech J. L. (2) (4), Terrien J. (1), Traving G. (3), Trefftz E. (2) (3), Van Bueren H. G., Van Regemorter H. (3), Varsavsky C. M., Wares G. W. (2), Wiese W. L. (2), Wilson R. (4), Zirin H.

(1) Member of Committee 1 (PRESIDENT: B. Edlén)
    Standards of Wavelength
(2) Member of Committee 2 (PRESIDENT: R. H. Garstang)
    Transition Probabilities
(3) Member of Committee 3 (PRESIDENT: H. van Regemorter)
    Collision cross-sections and line broadening
(4) Member of Committee 4 (PRESIDENT: Ch. Moore-Sitterly)
    Structure of Atomic Spectra
(5) Member of Committee 5 (PRESIDENT: J. G. Phillips)
    Molecular Spectra

15. Commission pour l'Étude Physique des Comètes. (Physical Study of Comets)

PRÉSIDENT: V. Vanýsek.

VICE-PRÉSIDENT: A. H. Delsemme.

COMITÉ D'ORGANISATION: L. Biermann, G. Herzberg, B. Levin, N. Richter, E. Roemer, F. Whipple, K. Wurm.

MEMBRES: Arpigny C., Bertaud F., Beyer M., Bobrovnikoff N. T., Bouška J., Brandt J. C., Brown P. Lancaster, Dobrovolskij O. V., Donn B. D., Dossin F., Guiguay G., Harwit M. O., Haser L., Hunaerts J. J., Liller W., Lüst Rhea, Lyttleton A., Malaise D., Martel-Chossat (Mme) M. T., Miller F. D., Mrkos A., Öpig E., Pflug K., Poloskov S. M., Probstein R. F., Rahe J. (Secrétaire de la Commission), Rémy-Battiau L., Riives W. G., Sekanina Z., Sivaraman K. R., Swings P., Vsekhsvyatskij S. K., Waterfield R. L.

16. Commission pour l'Étude Physique des Planètes et des Satellites. (Physical Study of Planets and Satellites)

PRÉSIDENT: G. H. Pettengill.

VICE-PRÉSIDENT: S. K. Runcorn.

COMITÉ D'ORGANISATION: J. Connes, S. Miyamoto, V. I. Moroz, T. C. Owen, B. A. Smith.

MEMBRES: Ashbrook J., Barabashov N. P., Barreto L. M., Baum W. A., Belton M. J. S., Berge G. L., Bobrov M. S., Boyce P. B., Broadfoot L., Brunk W. E., Bullen K. E., Camichel H., Chamberlain J. W., Collinson E. H., Colombo G., De Marcus W. C., de Mottoni G., de Vaucouleurs G., Dickel J. R., Dollfus A., Drake F. D., Evans J. V., Fink U., Fox W. E., Gehrels T., Giclas H. L., Goldstein R. M., Goody R. M., Greyber H. D., Guerin P., Günther O., Hagfors T., Hall J. S., Halliday I., Haupt H. F., Herzberg G., Hunten D. M., Ingrao H. C., Irvine W. M., Jeffreys H., Jelley J., Kaplan L., Karandikar R. V., Kellog W. W., Kondo Y., Kopal Z., Koval I. K., Kuiper G. P., Kuzmin A. D., Leighton R. B., Levin

B. J., Liddel U., Link F., Lipskij Y. N., Luplau-Janssen C. E., Martynov D. Ya., McElroy B., Mayer C. H., Menzel D. H., Middlehurst B., Millman P. M., Narrayana J. V., Öpik E. J., Ottelet I., Rösch J., Safronov V., Sagan C. E., Salisbury J. W., Shapiro I. I., Shimizu T., Shoemaker E., Sinton W., Smith H. J., Smoluchowski R., Spinrad H., Strong J. D., Tombaugh C., Trafton L. M., Van Allen J. A., Wallace L. V., Whitaker E. A., Wildey R. L., Wildt R., Williams I. P., Wilson A. G., Young A. T.

17. Commission de la Lune. (The Moon)

PRÉSIDENT (par intérim): A. Dollfus.
VICE-PRÉSIDENTS: G. P. Kuiper, B. J. Levin.
COMITÉ D'ORGANISATION: E. Anders, G. Fielder, D. H. Menzel, P. M. Millman, T. Weimer.
MEMBRES: Arthur D. W. G., Ashbrook J., Baldwin R., Bell (Miss) B., Boneff N., Brunk W. E., Cameron (Mrs) W., Drofa V. K., Dzapiashvili V. P., Eckert W. J., Elston W. E., Ezerskij V. I., Gavrilov I. V., Gold T., Green J., Guest J. E., Habibullin S. T., Hall R. G., Hirose H., Hopmann J., Jeffreys H., Kopal Z., Koziel K., Link F., Lipskij Y. N., Markowitz Wm., Martynov D. Ya., Middlehurst (Miss) B. M., Mikhailov A. A., Miyamoto S., Morrison L. V., Moutsoulas M., Nefed'ev A. A., O'Keefe J. A., Pettengill G. H., Potter H. I., Rösch J., Runcorn S., Sadler (Mrs McBain) F. M., Sagan C., Sato Y., Shapiro I. I., Shoemaker E., Sinton W. M., Sonett Ch. P., Strom R. G., Troitskij V. S., Urey H. C., van Flandern T. C., Watts C. B., Whipple F. L., Whitaker E. A., Wildey R. L.

19. Commission de la Rotation de la Terre. (Rotation of the Earth)

PRÉSIDENT: H. M. Smith.
VICE-PRÉSIDENT: C. Sugawa.
COMITÉ D'ORGANISATION: E. P. Fedorov, B. Guinot, Wm. Markowitz, P. J. Melchior, R. W. Tanner, R. O. Vicente, S. Yumi (ex officio)
MEMBRES: Abraham H. J., Atkinson R., Bender P. L., Billaud G., Bonanomi J., Brkic Z. M., Buchar E., Cecchini G., Chudovicheva N. A., Dramba C., Enslin H., Essen L., Fichera E., Fleckenstein J. O., Gama L. I., Hall R. G., Hers J., Iijima S., Jeffreys H., Kakuta Ch., Kalmykov A. M., Kulikov K. A., Lagrula J., Lederle T., Levallois J. J., Meinig M., Mikhailov A. A., Milovanović V., Miyadi M., Narayana J. V., Nicolini T., Okuda T., Opalski W., Orlov B. A., Orte A., Oterma L., Paquet P., Parijskij N. N., Pavlov N. N., Popov N. A., Postoiev A., Proverbio E., Randić L., Rice D. A., Sakharov V. I., Scheepmaker A. C., Sekiguchi N., Sevarlić B. M., Stoyko A., Stoyko N., Takagi S., Tardi P., Teleki G., Thomas D. V., Torao M., Tsao Mo, Winkler G., Witkowski J.

20. Commission des Positions et des Mouvements des Petites Planètes, des Comètes et des Satellites. (Positions and Motions of Minor Planets, Comets and Satellites)

PRÉSIDENT: F. K. Edmondson.
VICE-PRÉSIDENT: L. Kresák.
COMITÉ D'ORGANISATION: S. Arend, G. A. Chebotarev, P. Herget, Y. Kozai, E. Roemer, H. W. Wood.
MEMBRES: Ahmed I. I., Bielicki M., Böhme S., Boyer L., Bruwer J. A., Candy M. P., Clemence G. M., Cristescu C., Cunningham L. E., Debehogne H., Evdokimov J. V., Everhart E., Galibina (Mrs) I. V., Gehrels T., Giclas H. L., Gutierrez-Moreno A., Herrick S., Hertz H. G., Hirose H., Itzigson M., Kamienski M., Kazimirchak-Polonskaya E. I., Klepczynski W. J., Kovalevsky J., Kuiper G. P., Marsden B., Michkovitch V. V., Milet B., Missana N., Orlov A. A., Oterma L., Popović B., Porter J. G., Protitch M. B., Rabe E., Rasmusen H. Q., Robertson W. E., Sagnier J. L., Schmitt A., Schrutka-Rechtenstamm G., Schubart J., Sekanina Z., Shapiro I. I., Sitarski G., Steins K. A., Strobel W., Sultanov G. F., Torroja J. M., Väisälä Y., Van Biesbroeck G., Van Houten C. J., Van Houten-Groeneveld I., Whipple F. L., Wild P., Wilkins G. A., Yakhontova N. S., Zadunaisky P. E.

21. Commission de la Luminescence du Ciel. (The Light of the Night Sky)
PRÉSIDENT: H. Elsässer.
VICE-PRÉSIDENT: J. L. Weinberg.
COMITÉ D'ORGANISATION: M. Huruhata, M. F. Ingham, P. V. Kulkarni, F. E. Roach, Yu. L. Trutse, G. Weill.
MEMBRES: Anderson K., Asaad A. S., Bates D. R., Blackwell D. E., Chamberlain J. W., Cook A. F., Dauvillier A., Divari N. B., Dufay M., Dumont R., Fesenkov V. G., Fesenkova E. V., Galperin G. I., Gauzit J., Giese R. H., Grandmontagne R., Harang L., Haug U., Henry R. C., Hunten D. M., Jarrett A. H., Kaplan J., Karandikar R. V., Karyagina Z. V., Kastler A., Krasovskij I., Megrelishvilli T. G., Neužil L., Nicolet M., Pearse R. W. B., Pettit (Mrs) H. B. Knaflich, Robley R., Sanchez-Martinez F., Saxena P. P., Schmidt Th., Steiger W. R., Tanabe H., Van Allen J. A., Vassy E., Wallace L. V., Wolstencroft R. D.

22. Commission des Météores et Météorites. (Meteors and Meteorites)
PRÉSIDENT: R. E. McCrosky.
VICE-PRÉSIDENT: B. A. Lindblad.
COMITÉ D'ORGANISATION: E. Anders, Z. Ceplecha, W. G. Elford, H. Hirose, B. J. Levin, P. M. Millman.
MEMBRES: Abbott W. N., Astapovich I. S., Babadzhanov P. B., Cook A. F., Davies J. G., Davis J., Ellyett C. D., Fedinskij V. B., Fesenkov V. G., Fireman E. L., Gauzit J., Guiagay G., Guth V., Hajduk A., Halliday I., Hawkins G. S., Hemenway C. L., Hey J. S., Hodge P. W., Hoppe J. A., Jacchia L., Javnel' A. A., Kaiser T. R., Katasev L. A., Keay C. S. L., Kizilirmak A., Kostylev K. V., Kramer K. N., Kresák L., Kresáková M., Krinov E. L., Lovell B., McIntosh B. A., Nazarova T. N., O'Keefe J. A., Olivier C. P., Öpik E. J., Plavcová Z., Rajchl J., Russell J. A., Shao C. Y., Šimek M., Soberman R. K., Southworth R. B., Štohl J., Verniani F. F., Whipple F. L., Wood J. A.

24. Astrométrie Photographique. (Photographic Astrometry)
PRÉSIDENT: S. Vasilevskis.
VICE-PRÉSIDENT: P. Lacroute.
COMITÉ D'ORGANISATION: A. N. Deutsch, W. Dieckvoss, P. Herget, V. V. Lavdovskij, W. J. Luyten, C. A. Murray, K. Aa. Strand, P. van de Kamp, H. W. Wood.
MEMBRES: Abhyankar K. D., Blaauw A., Bouigue R., Brosche P., Cecchini G., Churms J., Couderc P., Delhaye J., Dessy J. Landi, Eichhorn H. K., Fatchikhin N. W., Fracastoro M. G., Franz O. G., Fredrick L. W., Gallouët L., Giclas H. L., Gliese W., Goyal A. N., Günther O., Haas J., Harris B. J., Hoffleit (Miss) E. D., Klemola A. R., König A., Kox H., Lippincott (Miss) S. L., Lourens J. v. B., Meurers J., Missana N., Moreno H., Morgan W. W., O'Connel D. J. K., Oja T., Paloque E., Schilt J., Semirot P., Smart W. M., Stearns C. L., Thomas D. V., van Altena W. F., Vyssotsky A. N., Wagman N. E., Wilkens A.

25. Commission de Photométrie Stellaire. (Stellar Photometry)
PRÉSIDENT: D. L. Crawford.
VICE-PRÉSIDENT: M. Golay.
COMITÉ D'ORGANISATION: A. W. J. Cousins, O. J. Eggen, G. E. Kron, K. Osawa, J. Stock, V. Straižys.
MEMBRES: Argue A. N., Bahng J. D. R., Becker W., Behr A., Bigay J. H., Bok B. J., Bok P. F., Borgman J., Brück H. A., Chis G., Dachs J., Eelsalu H. T., Feinstein A., Fernie J. D., Fitch W. S., Gallonët, Graham J. A., Gutiérez-Moreno A., Haffner H., Hardie R. H., Hill P. W., Hiltner W. A., Hoag A. A., Holmberg E. B., Irwin J. B., Joshi S. C., Johnson H. L., Landolt A., Lasker B. M., McCarthy S. J., Masani A., Mayer P., Mendoza V., Mianes P., Mitchell R. I., Moreno H., Morris S. C., Muller A. B., Mumford G. S., Notni P., Philip A. G., Roslund C., Rufener F. G., Rybka E., Sarma M. K. B., Schmidt H.,

Smyth M. J., Steinlin U. W., Strohmeier W., Ulrich B., Velghe A. G., Wallenquist A. E. E., Walraven T., Wesselink A. J., Willstrop R. V., Young A. T.

26. Commission des Étoiles Doubles. (Double Stars)
PRÉSIDENT: J. Dommanget.
VICE-PRÉSIDENT: Miss S. L. Lippincott.
COMITÉ D'ORGANISATION: P. Couteau, A. N. Deutsch, P. Muller, K. Aa. Strand, P. van de Kamp.
MEMBRES: Arend S., Baize P., Batten A. H., Cester B., Da Silva A. V., Djurković P. M., Eggen O. J., Finsen W. S., Fracastoro M. G., Franz O. G., Freitas-Mourao R. R. de, Güntzel-Lingner U., Hadjidemetriou J. D., Heintz W. D., Hidajat B., Holden F., Hopmann J., Jeffers H. M., Johnson M. C., Jonckheere R., Kuiper G. P., Kulikovsky P. G., Kumar S. S., Luplau-Janssen C. E., Luyten W. J., Scarfe C. D., van Albada G. B., van Biesbroeck G., van den Bos W. H., Wierzbiński St., Wieth-Knudsen N. P., Worley C. E., Zagar F.

27. Commission des Étoiles Variables. (Variable Stars)
PRÉSIDENT: O. J. Eggen.
VICE-PRÉSIDENT: M. W. Feast.
COMITÉ D'ORGANISATION: A. A. Boyarchuk, R. F. Christy, L. Detre, G. H. Herbig, K. K. Kwee, W. Wenzel.
MEMBRES: Alania I. F., Andrews A. D., Arp H. C., Ashbrook J., Bakos G. A., Balázs-Detre J., Banerji A. C., Bateson F. M., Bertaud Ch., Bhatnagar P. L., Chavira E., Chis G., Daube-Kurzemniece I. A., de Kock R. P., Dickens R. J., Efremov Yu. I., Fernie J. D., Fitch W. S. (Secrétaire de la Commission), Eskioglu A. N., Gaposhkin S., Gascoigne S. C. B., Geyer E. H., Godoli G., Gursky H., Hansen C. J., Haro G., Harwood M., Heiser A., Herr R. B., Heyden F. J., Hoffleit E. D., Kholopov P. N., Huffer C. M., Huruhata M., Hutchings J. B., Ishchenko I. M., Jarzebowski T., Jones A. F., Kippenhahn R., Kopylov I. M., Kordylewski K., Kraft R. P., Krzeminski W., Kuhi L. V., Kukarkin B. V., Kumsishvili J. I., Kunkel W. E., Landolt A. U., Ledoux P., Leung K.-C., Lortet-Zuckermann M. C., McNamara D. H., Maffei P., Mannino G., Masani A., Mavridis L. N., Mayall M. W., Miller W. J., Mirzoyan L. V., Mumford G. S., O'Connell D. J. K., Odgers G. J., Oosterhoff P. T., Opolski A., Osawa K., Oskanyan V., Payne-Gaposhkin C. H., Peltier L. C., Piotrowski S., Plaut L., Popova M., Rakos K. D., Richter G., Rodgers A. W., Romano G., Rosino L., Sarma M. B. K., Sawyer-Hogg H. B., Shapley H., Sinvhal S. D., Slavenas P. V., Smak J., Smith Harlan J., Stepien K., Strohmeier W., Swope H. H., Szeidl B., Tammann G. A., Tsesevich V. P., Tempesti P., Tremko J., Usher P. D., Vandekerkhove E., Van Genderen A. M., Van Hoof A., Wachmann A. A., Walker M. F., Wallerstein G. N., Walraven T., Wehlau A., Wesselink A. J., Wright F. W., Zwicky F.

28. Commission des Galaxies. (Galaxies)
PRÉSIDENT: Mrs E. M. Burbidge.
VICE-PRÉSIDENT: E. B. Holmberg.
COMITÉ D'ORGANISATION: F. Bertola, G. C. McVittie, M. S. Roberts, M. Schmidt, J. L. Sěrsic, S. Van den Bergh, B. A. Vorontsov-Vel'yaminov.
MEMBRES: Abell G. O., Alfvén H., Ambartsumian V. A., Arp H. C., Bahcall J. N., Baum W. A., Bigay J. H., Bok B. J., Bolton J. G., Bondi H., Bonnor W. B., Braccesi A., Burbidge G. R., Contopoulos G., Courtès G., Davies R. D., Demoulin M.-H. J., Dessy J. Landi, de Vaucouleurs G., Dibaj E. A., Einasto J. E., Elvius A. M., Evans D. S., Field G. B., Ford W. K., Freeman K. C., Gascoigne S. C. B., Haro G., Heckmann O., Heeschen D. S., Heidmann J., Hodge P. W., Hoyle F., Humason M. L., Johnson H. M., Kalinkov M., Kellerman K. I., Kiang T., King I. R., Kundu M. R., Kustaanheimo P. E., Layzer D., Lindblad P. O., Low F. J., Lynden-Bell D., Lynds C. R., McClure R. D., McCrea W. H., Markarian B. E., Mayal N. U., Mills B. Y., Minkowski R. L., Morgan W. W., Narlikar

J. V., Neugebauer G., Neyman J., Oke J. B., Omer G. C., Oort J. H., Osterbrock D. E., Pachner J., Page T. L., Pishmish de Recillas P., Poveda A., Prendergast K. H., Pronik V. I., Reaves G., Reddish V. C., Richter N., Rindler W., Robinson I., Rood H. J., Rosino L., Rubin V. C., Rudnicki K., Sandage A., Saslaw W. C., Schücking E., Schürer M., Sciama D. W., Scott E. L., Shakeshaft J. R., Shane C. D., Shapley H., Spinrad H., Stein W., Stibbs D. W. N., Tammann G. A., Thackeray A. D., Tifft W. G., Treder H.-J. Vandekerkhove E., Visvanathan N., Westerlund B. E., Whitford A. E., Whitrow C. J., Wild P., Wills D., Wilson A. G., Zel'dovich Ya. B., Zwicky F.

29. Commission des Spectres Stellaires. (Stellar Spectra)
PRÉSIDENT: Y. Fujita.
VICE-PRÉSIDENT: O. C. Wilson.
COMITÉ D'ORGANISATION: M. K. V. Bappu, A. A. Boyarchuk, R. Cayrel, M. W. Feast, Mrs M. Hack, G. H. Herbig, L. Searle.
MEMBRES: Abhyankar K. D., Abt H. A., Aller L. H., Andrillat H., Andrillat Y., Asaad A. S., Babcock H. W., Berger J., Bertaud Ch., Bidelman W. P., Bloch (Miss) M., Boesgaard (Mrs) A. M., Boggess A. III, Bonsack W. K., Bouigue R., Brück H. A., Buscombe W., Cayrel-de Strobel G., Chalonge D., Climenhaga J. L., Code A. D., Conti P. S., de Groot T., Divan L., Dobronravin P. P., Dolidze M. V., Dunham Jr. T., Edmonds Jr F. N., Evans D. S., Fringant A. M., Gollnow H., Gorbackij V. G., Gratton L., Greenstein J. L., Griffin R. F., Griffin R. M., Groth H. G., Gathrie B. N. G., Haro G., Heard J. F., Heintze J. R. W., Henize K. G., Herman R., Houziaux L., Huang Su-Shu, Iwanowska W., Jaschek C. Ô. R., Jaschek M., Jugaku J., Keenan P. C., Kienle H., King R. B., Kodaira K., Kogure T., Kopylov I. M., Kraft R. P., Larsson-Leander G., McCuskey S. W., McNamara D. H., Mannino G., Mel'nikov O. A., Milligan J. E., Morton D. C., Mustel E. R., Myerscough (Mrs) V. P., Nicholls R. W., Nikitin A. A., Nishimura S., Oetken L., Oke J. B., Osawa K., Pagel B. E. J., Pasinetti L. E., Payne-Gaposhkin C. H., Pedoussaut A., Preston G. W. III, Przybylski A., Ringuelet-Kaswalder A., Rodgers A. W., Rosen B., Sahade J., Sanwal N. B., Sargent W. L., Schild R. E., Sinnerstad U., Slettebak A., Smak J., Stawikowski A., Stecker T. P., Svolopoulos S. N., Swensson J. W., Swings J. P., Swings P., Taffara S., Thackeray A. D., Thompson G. I., Tsuji T., Underhill A. B., Vandekerkhove E., Van 't Veer-Menneret C., Voigt H. H., Wallerstein G. N., Wehlau W. H., Wellmann P., Wempe J., Willstrop R. V., Wilson R., Wolff S. G., Wood H. J., Wright K. O., Wyller A. A., Yamashita Y.

30. Commission de Vitesses Radiales. (Radial Velocities)
PRÉSIDENT: R. Bouigue.
VICE-PRÉSIDENT: R. F. Griffin.
COMITÉ D'ORGANISATION: D. S. Evans, H. Gollnow, L. Oetken, A. D. Thackeray.
MEMBRES: Abt H. A., Barbier-Brossat (Mrs) M., Batten A. H. (Secrétaire de la Commission), Beardsley W. R., Bertiau F. C., Blaauw A., Boulon J., Buscombe W., Crampton D., de Jonge J. K., de Vaucouleurs G., Edmondson F. K., Eelsalu H. T., Fehrenbach Ch., Fletcher J. M., Garrison R. F., Gratton L., Harding G. A., Heard J. F., Heintze J. R. W., Hube D. P., Kharadze E. K., Kraft R. P., Martin (Miss) N., Perry C. K., Philip A. G. Davis, Preston G. W. III, Rubin (Mrs) V. C., Sahade J.

31. Commission de l'Heure. (Time)
PRÉSIDENT: G. M. R. Winkler.
VICE-PRÉSIDENT: H. Enslin.
COMITÉ D'ORGANISATION: D. J. Belocerkovskij, J. Bonanomi, T. Gökmen, B. Guinot, G. Hemmleb, S. Iijima, Wm. Markowitz, D. A. Orte, H. M. Smith, M. Torao, F. Zagar.
MEMBRES: Abraham H. J., Arbey L. M. J., Bakulin P. I., Bender P. L., Billaud G., Brkić Z. M., Caprioli G., Decaux B., Dingle H., Drâmbă C., Dubois-Chevallier (Mrs) R., Fuchs J.,

Gama L. I., Gougenheim A., Hall R. G., Hers J., Koebcke F., Lacombe C. G., Lederle T., Madwar M. R., Matsunami N., Melchior, P. J., Mikhailov A. A., Miyadi M., O'Handley D. A., Opalski W., Pavlov N. N., Pensado J., Postoiev A., Proverbio E., Randić L., Sadler D. H., Sandig H.-U., Schuler W., Stoyko (Mrs) M., Stoyko N., Shcheglov V. P., Shiryaev A. V., Šternberk B., Takagi S., Tardi P., Thomson N. M., Tsao Mo, Verbaandert J., Von der Heide J. E. B., Yumi S.

33. Commission de la Structure et de la Dynamique du Système Galactique. (The Structure and Dynamics of the Galactic System)

PRÉSIDENT: S. W. McCuskey.

VICE-PRÉSIDENT: L. Perek.

COMITÉ D'ORGANISATION: J. A. Agekian, G. Contopoulos, T. Elvius, F. J. Kerr, G. G. Kuzmin, C. C. Lin, P. Mezger, B. E. Westerlund.

MEMBRES: Aarseth S., Ambartsumian V. A., Baldwin J. E., Barbanis B., Becker W., Beer A., Bidelman W. P., Bigay J. H., Blaauw A., Blanco V. M., Bok B. J., Boulon J. Burke B. F., Courtès G., Crawford D., Cuperman S., Delhaye J., Dickel (Mrs) H. R., Dickel J. R., Dieter (Mrs) N. H., Dzigashvili R. M., Edmondson F. K., Einasto J. E., Elsässer H., Elvius A. M., Fehrenbach Ch., Fenkart R., Freeman K., Fricke W., Fujimoto M., Georgelin Y., Goldreich P., Gyldenkerne K., Haffner H., Haug U., Hayli A., Heckmann O., Hénon M., Hohl F., Hunter C., Idlis G. M., Innanen K. A., Irwin J. B., Iwaniszewska-Lubienska C., Iwanowska W., Jaschek C. O. R., Johnson H. M., Kaburaki M., Kalnajs A., Kharadze E. K., Kholopov P. N., King I. R., Kinman T. D., Klare G., Kukarkin B. V., Kurth R., Lecar M., Lindblad P. O., Lodén K., Lodén L. O., Lunel M., Luyten W. J., Lynden-Bell D., Lyngå G., McCarthy M. F., MacConnell D. J., MacRae D. A., Malmquist K. G., Martinet L., Mavridis L. N., Miller R. H., Mirzoyan L. V., Mohr J. M., Monnet G., Morgan W. W., Münch G., Murray C. A., Nahon M. F., Neckel Th., Ogorodnikov K. F., Ollongren A., Oort J. H., Ostriker J. P., Ovenden M. W., Palmer P., Perry C. L., Pesch P., Philip A. G. D., Pilowski K., Plaut L., Priester W., Ramberg J. M., Riegel K. W., Roman N. G., Rubin (Mrs) V. C., Rudnicki K., Rybicki G. B., Sanduleak N., Schilt J., Schmidt H., Schmidt K. H., Schmidt M., Schmidt-Kaler Th., Seggewiss W., Shane W. W., Sharov A. S., Sher D., Shimizu T., Shu F. H., Slettebak A., Spiegel E. A., Steinlin U. W., Stephenson C. B., Stibbs D. W. N., Sturch C., Svolopoulos S. N., Szebehely V., Tammann G. A., The Pik-Sin, Toomre A., Upgren A., Vanderlinden H. L., Vandervoort P. O., van Hoof A., Varsavsky C. M., Velghe A. G., Verschuur G. D., Vetešník M., Wayman P. A., Weaver H., Westerhout G., White R. E., Wielen R., Woltjer L., Woolley R.

34. Commission de la Matière Interstellaire et des Nébuleuses Planétaires. (Interstellar Matter and Planetary Nebulae)

PRÉSIDENT: F. D. Kahn.

VICE-PRÉSIDENT: H. van Woerden.

COMITÉ D'ORGANISATION: J. E. Baldwin, A. Behr, G. Khromov, T. K. Menon, D. E. Osterbrock, B. J. Robinson.

MEMBRES: Aller L. H., Andrillat H., Andrillat Y., Arny T. T., Boggess A., Bok B. J., Brück (Mrs) M. T., Capriotti E. R., Chopinet (Miss) M., Chvojková E., Courtès G., Coyne G. V., Czyzak S. J., Davies R. D., Demoulin (Miss) M. H. J., Dibaj E. A., Dickel (Mrs) H. R., Dieter N. H., Divan L., Dorschner J., Dunham T., Dyson J. E., Elvius (Mrs) A. M., Felten J. E., Field G. B., Friedemann C., Gardner F. F., Gaustad J. E., Gehrels T., Goldberg L., Goldsworthy F. A., Goldwire H. C., Greenberg J. M., Grewing M., Grubisich C., Grzedielski S., Gurzadyan A., Habing H. J., Hall J. S., Hardebeck (Mrs) E. C., Heiles C. E., Henize K. G., Herbig G. H., Hidajat B., Higgs L. A., Hiltner W. A., Hjellming R. M., Hobbs L. M., Hummer D. G., Jenkins E. B., Johnson H. M., Kaler J. B., Kamijo F., Kaplan S. A., Kerr F. J., Kharadze E. K., Khavtasi J. S., Kipper A. J., Ko Hsien-Ching,

Kogure T., Kohoutek L., Lambrecht H., Lasker B. M., Liller W., Lin C. C., Loden L. O., Louise R., Lynds B. T., McCrea W. H., McGee R. X., McNally D., Martel-Chossat M. T., Mathis J. S., Mathewson D. S., Meaburn J., Mezger P. G., Miller J. S., Minin I. N., Minkowski R. L., Münch G., Nandy K., Okuda H., Osaki T., Ozernoj L. M., Palmer P., Pecker J.-C., Peimbert-Sierra M., Perinoto M., Petrosian V., Pikel'ner S. B., Pottasch S. R., Razmadze N. A., Riegel K. W., Rohlfs K., Rose W. K., Rosino L., Rozhkovskij D. A., Rubin (Mrs) V. C., Safronov V. S., Savage B. D., Savedoff M. P., Schalen C., Schatzman E. L., Schmidt K. H., Schmidt Th., Schmidt-Kaler Th., Seaton M. J., Seddon H., Serkowski K., Shak G. A., Shakeshaft J. R., Shao C. Y., Shapiro M. M., Sharpless S. L., Shklovskij I. S., Smith M. G., Sobolev V. V., Sofia S., Somerville W. B., Souffrin S., Spitzer L., Strömgren B., Syrovatskij S. I., Takakubo K., Terzian Y., Thackeray A. D., Thompson A. R., Treanor P. J., Van de Hulst H. C., Van Horn H. M., Vanýsek V., Verschuur G. L., Wiswanathan N., von Hoerner S., Vorontsov-Vel'yaminov B. A., Walker G. A. H., Wentzel D. G., Westerhout G., Wickramasinghe N. C., Williams D. A., Williams I. P., Williams R., Witt A. N., Wolstencroft R. D., Yabushita S., Zimmermann H., Zuckerman B. M.

35. Commission de la Constitution des Étoiles. (Stellar Constitution)
PRÉSIDENT: E. Schatzman.
VICE-PRÉSIDENT: L. Mestel.
COMITÉ D'ORGANISATION: A.N. Cox, C. Hayashi, R. Kippenhahn, Mrs A.G. Massevich, G. Ruben.
MEMBRES: Anand S. P. S., Arnett W. D., Appenzeller I., Baglin D., Bhatnagar P. L., Bierman L., Bodenheimer P., Bondi H., Boury A., Brownlee R. R., Burbidge G. R., Cameron A. G. W., Carson T. R., Chandrasekhar S., Chiu H. Y., Christy R. F., Cohen J. M., Cowling T. G., Cox J. P., Demarque P. R., Dingens P. S. A., Dluzhnevskaya O. B., Eggleton P. P., Eminzade T. A., Epstein I., Ezer (Mrs) D., Faulkner J., Ferraro V. C. A., Fowler W. A., Forbes J. E., Frantzman Yu. L., Fricke K. J., Gabriel M., Gaughlan (Mrs) G., Giannone P., Gjellestad (Miss) C., Hamada I., Hitotuyanagi Z., Hoshi R., Hoyle F., Iben Jr I., Imshennik V. S., James R. A., King D. S., Kothari D. S., Krook M., Kumar S. S., Kushwaha R. S., Lebovitz N., Ledoux P., McCrea W. H., Marx G., Masani A., Meyer-Hofmeister (Mrs) E., Monaghan J. J., Morris S. C., Nishida M., Ohyama N., Osaki Y., Paczynski B., Popova (Mrs) M. D., Porfir'jev V. V., Poveda A., Reeves H., Reiz A., Rosseland S., Rouse C. A., Roxburg I., Sakashita S., Sakurai T., Salpeter E. E., Savedoff M. P., Schwarzschild M., Sears R. S., Sengbusch K. V., Smeyers P., Souffrin P., Spiegel E., Stibbs D. W. N., Strittmatter P. A., Strömgren B., Suda K., Sugimoto D., Sweet P. A., Tayler R. J., Temesváry S., Thomas H. C., Truran J. W., Tuominen J. V., Tutukov A. V., Uchida J., Van der Borght R., Van Horn H. M., Vardya M. S., Vila S. C., Weigert A., Zahn J. P., Zhevakin S. A., Ziolkowski J.

36. Commission de la Théorie des Atmosphères Stellaires. (The Theory of Stellar Atmospheres)
PRÉSIDENT: R. N. Thomas.
VICE-PRÉSIDENT: R. Cayrel.
COMITÉ D'ORGANISATION: J. T. Jefferies, B. E. J. Pagel, J.-C. Pecker, V. V. Sobolev, Miss A. B. Underhill.
MEMBRES: Abhyankar K. D., Aller L. H., Arpigny C., Athay R. G., Auer L. H., Auman J. R., Avrett E. H., Baschek B., Biermann L., Bless R. C., Böhm K. H., Böhm-Vitense E., Busbridge I. W., Carson T. R., Cayrel-de-Strobel C., Conti P. S., Cuny (Miss) Y., de Feiter L. D., de Jager C., Delache Ph., Dumont (Miss) S., Edmonds F. N., Elste G. H. E., Faraggiana (Mrs) R., Finn G. D., Fischel D., Frisch U., Gebbie K. B., Gingerich O., Gökdogan N., Gordon Ch., Grant I. P., Greenstein J. L., Grevesse N., Groth H. G., Gussmann E. A., Hack (Mrs) M. H., Hardorp J., Hearn A. G., Hitotuyanagi Z., Hotinli M., House L. L., Houtgast J., Houziaux L., Huang S. S., Hummer D. G., Hunger K.,

Ivanov V. V., Johnson H. R., Kalkofen W., Kandel R. S., Kopylov I. M., Krishna Swamy K. S., Kumar S. S., Kushwaha R. S., Matsushima S., Menzel D. H., Mihalas D., Mirzoyan L. V., Miyamoto S., Mugglestone D., Müller (Miss) E. A., Münch G., Mustel E. R., Mutschlecner J. P., Myerscough V. P., Nariai K., Neff J. S., Neven L., Orrall F. Q., Phillips J. G., Pottasch S. R., Praderie F., Rudkjøbing M., Rybicki G. B., Saito S., Sapar A., Schatzman E. L., Schmalberger D. C., Seaton M. J., Skumanich A., Sitnik G. F., Sobouti Y., Spiegel E. A., Spite M., Stibbs D. W. N., Strom S. E., Strömgren B., Swihart T. L., Traving G., Ueno S., Uesugi A., Unno W., Unsöld A., van Regemorter H., van 't Veer (Mrs) C., van 't Veer F., Vardya M. S., Weidemann V., Wellmann P., Wilson P. R., Wright K. O., Wyller A. A., Zirker J. B., Zwaan C.

37. Commission des Amas Stellaires et des Associations. (Star Clusters and Associations)
PRÉSIDENT: G. Larsson-Leander.
VICE-PRÉSIDENT: I. King.
COMITÉ D'ORGANISATION: K. A. Barkhatova, S. B. C. Cascoigne, M. Golay, L. Rosino, J. Ruprecht, S. van den Bergh, H. van Schewick, M. F. Walker.
MEMBRES: Aarseth S., Agekyan T. A., Alter G. A., Artyukhina (Mrs) N. M., Balázs B., Becker W., Blaauw A., Bouvier P., Cuffey J., Eggen O. J., Feast M. W., Feinstein A., Haffner H., Hénon M., Hoag A. A., Johnson H. L., Kadla (Mrs) Z., Kholopov P. N., Lavdovskij V. V., Lodén L. O., Lohmann W., Lynden-Bell D., Lyngå G. (Secrétaire de la Commission), Mavridis L. N., Menon T. K., Meurers J., Morgan W. W., Murray C. A., Pishmish de Recillas (Mrs) P., Poveda A., Reddish V. C., Sandage A., Sawyer Hogg (Mrs) H. B., Serkowski K., Shapley H., Sharov A. A., Smyth M. J., Swope (Miss) H. H., Takase B., Terzan A., The Pik-Sin, Vanderlinden H. L., van Hoerner S., Walker G. A. H., Wallenquist A. A. E., Weaver H. F., Wooley R.

38. Commission pour l'Echange des Astronomes. (Exchange of Astronomers)
(Comité du Comité Exécutif)
PRÉSIDENT: A. Reiz.
VICE-PRÉSIDENT: P. M. Routly.
COMITÉ D'ORGANISATION: D. Ya. Martynov, J. Sahade, J. P. Wild, F. B. Wood.
MEMBRES: Abetti G., Bappu M. K. V., Bok B. J., Bowen I. S., Delhaye J., Keller G., Kienle H., Kourganoff V., MacRea D., Mikhajlov A. A., Miyamoto S., Mohr J. M., Page T. L., Righini G., Rosseland S., Stoy R. H., Swings P., Witkowski J., Wyatt S. P. Jr.

40. Commission de la Radioastronomie. (Radio Astronomy)
PRÉSIDENT: D. S. Heeschen.
VICE-PRÉSIDENT: Yu. N. Parijskij.
COMITÉ D'ORGANISATION: R. D. Davies, O. Hachenberg, Mrs M. Pick-Gutmann, J. R. Shakeshaft, G. Swarup, H. Tanaka, J. P. Wild.
MEMBRES: Akabane K., Alexander J. K., Allen R. J., Altenhoff W. J., Andrew B. H., Argyle P. E., Baart E. E., Baldwin J. E., Barrett A. H., Berge G. L., Blum E. J., Boischot A., Bolton J. G., Bracewell R. N., Broten N. W., Brouw W. N., Brown R. Hanbury, Burbidge G. R., Burke B. F., Carr T. D., Ceccarelli M., Christiansen W. N., Clark B. G., Cocke J., Cohen M. H., Cooper B. F. C., Costain C. H., Coutrez R. A. J., Covington A. E., Daene H., Daintree E. G., Davies J. G., Davis M. M., de Groot T., Demoulin M. H., Denisse J. F., Dent W. A., Dickel H. R., Dickel J. R., Dieter N. H., Douglas J. N., Downs G. S., Drake F. D., Dulk G. A., Dyson F. J., Elgaröy O., Ellis G. R. A., Elsmore B., Elwert G., Epstein E., Erickson W. C., Eriksen G., Eshleman V. R., Feix G., Felton J. E., Field G. B., Findlay J. W., Fleischer R., Fokker A. D., Fomalont E. B., Foster P. R., Friedman H., Galt J. A., Gardner F. F., Gent H., Ginzburg V. L., Gold T., Goldstein S. J., Goldwire H. C. Jr., Gonze R., Gonze-Delys C., Gorgolewski S., Gosachinsky I. V., Goss W. Miller, Gower J. F. R., Grewing M., Gulkes S., Haddock F., Hagen J. P., Hartz T. R., Hazard C.,

Heidmann J., Heiles C., Hewish A., Hey J. S., Higgs L. A., Hill E. R., Hjelming R. M., Hobbs R. W., Högbom J. A., Höglund B., Hogg D. E., Hooghoudt B. G., Howard W. E., Hughes V. A., Ihsanova V. N., Jauncey D. L., Jennison R. C., Joshi M. W., Kaftan-Kassim M. A., Kahn F. D., Kai K., Kakinuma T., Kardasev N. S., Kaufmann P., Kawabata K., Kazes I., Kellermann K. I., Kenderdine S., Kerr F. J., Ko H., Komesaroff M., Kotelnicov V. A., Kraus J. D., Krishnan T., Kundu M. R., Kuzmin A. D., Laffineur M., Lang K. R., Large M. I., Lequeux J., Le Squeren A. M., Lilley A. E., Lindblad P. O., Little A. G., Locke J. L., Lovell B., McAdam W. B., McGee R. X., McKenna-Lawlor S., McLean D., McVittie G. C., MacRae D. A., Maran S. P., Matvejenko L. I., Maxwell A., Mayer C. H., Meeks M. L., Menon T. K., Mezger P., Mills B. Y., Minkowski R. L., Moffet A. T., Moriyama F., Morimoto M., Muller C. A., Nicolson G. D., Oort J. H., Osterbrock D. E., Palmer H. P., Pauliny-Toth I. I. K., Penzias A. A., Pettengill G. H., Ponsonby J. E. B., Price R. M., Priester W., Rabben H. H., Radhakrishnan V., Raimond E., Ramaty R., Rao U. V. Gopala, Razin V. A., Reber G., Riegel K. W., Righini G., Roberts J. A., Roberts M. S., Robinson B. J., Roeder R. C., Roger R. S., Rogers A. E. E., Rogstad D. H., Rohlfs K., Rubin R. H., Rydbeck O. E. H., Ryle M., Salomonovich A. E., Salpeter E. E., Sanamjan V. A., Scheuer P. A. G., Schmidt M., Schwarz U. J., Scott P. F., Seaquist E. R., Seielstad G. E., Sheridan K. V., Shklovskij I. S., Shuter W. L. A., Simon P., Slee O. B., Smerd S. F., Smith A. G., Smith F. G., Smith Harlan J., Soboleva N. S., Sorochenko R. L., Stanley G. J., Steinberg J. L., Stewart P., Stone R. G., Swanson P. N., Swenson G. W. Jr., Takakura T., Terzian Y., Thompson A. R., Tlamicha A., Tolbert C. R., Townes C. H., Troickij V. S., Tsuchiya A., Tsuruta S., Turner B. E., Turner K. C., Turlo Z., Turtle A. J., Ulrich, B. T., Van de Hulst H. C., Van der Laan H., Van Woerden H., Varsavsky C. M., Veron P., Vitkevic V. V., Wade C. M., Walsh D., Warwick J. W., Weaver H. F., Weliachew L., Wendker H. J., Westerhout G., Westfold K. C., Whiteoak J., Wickramasinghe N. C., Wielebinski R., Williams D. R., Wills D., Wilson R. W., Woltjer L., Zhelezniakov V. V., Zuckerman B. M.

41. Commission de l'Histoire de l'Astronomie. (History of Astronomy)
Président: O. Gingerich.
Vice-Président: W. Hartner.
Comité d'Organisation: J. O. Fleckenstein, Miss D. Hellman, Z. Horský, M. A. Hoskin, P. G. Kulikovskij, E. Rybka.
Membres: Abetti G., Argyrakos J. G., Ashbrook J., Baehr U., Beer A., Chenakal V. L., Cimino M., Collinder P., Dadić Z., Deeming T. J., Dewhirst D. W., Dingle H., Dobrzycki J., Douglas (Miss) A. V., Erpylev N. P., Evans D. S., Ferrari d'Occhieppo K., Filliozat J., Freiesleben H. C., Freitas-Mourao R. R. De, Hawkins G. S., Hayli A. F., Hirose H., Kamieński M., Kennedy J. E., Kiang T., King H. C., Kotsakis D., Lévy J. R., Link F., McKenna-Lawlor (Mrs) S. M. P., Meadows A. J., Michel H., Michkovitch V. V., Norlind W., Nørlund N. E., Omer G. C., Pelseneer J., Petri W., Pogo A., Ronan C. A., Rosen E., Samaha A. H. M., Shcheglov V. P., Shukla K. S., Slavenas P. V., Sticker B., Wattenberg D., Whitaker E. A., Whitrow G. J., Wright (Miss) H., Yabuuti K., Zagar F., Zinner E.

42. Commission des Étoiles Doubles Photométriques. (Photometric Double Stars)
Président: M. Plavec.
Vice-Président: T. Herczeg.
Comité d'Organisation: A. H. Batten, M. G. Fracastoro, K. Gyldenkerne, M. Kitamura, G. Larson-Leander, D. M. Popper, V. P. Tsesevich, F. B. Wood.
Membres: Abyankar K. D., Bappu M. K. V., Beer A., Binnendijk L., Blitzstein W., Broglia P., Cester B., Chis G., Cillié G. G., Cousins A. W. J., Cristaldi S., Dadaev A. N., de Kort J., Ebbighausen E. G., Ferrari d'Occhieppo K., Fredrick L. W., Fresa A., Gaposhkin S., Grygar J., Günther O., Gyldenkerne K., Heard J. F., Huang Su-Shu, Huffer C. M., Irwin J. B., Jurkevich I., Kizilirmak A., Koch R. H., Kopal Z., Knipe G. F. G., Krat V. A.,

Kristenson H., Kruszewski A., Kwee K. K., Landolt A. U., Lavrov M. I., Linnell A. P., Magalashvili N. L., Mammano A., Martynov D. Ya., Mayer P., Merrill J. E., Mumford G. S., Nelson B., O'Connell D. J. K., Ovenden M. W., Paczynski B., Piotrowski S., Popovici C., Purgathofer A. Th., Rakos K. D., Reuning E., Sahade J., Schmidt H., Serkowski K., Shapley M. B., Shulberg A. R., Sinvhal S. D., Smak J., Sobieski S., Strohmeier W., Szafraniec R., Todoran I., Tremko J., Van Hoof A., Vetešník M., Walter K., Wehlau W. H., Weigert A., Wellmann P., Wesselink A. J., Whitney B. S., Widorn T., Wright K. O.

43. **Commission des Plasmas et Magnéto-Hydrodynamique en Astrophysique. (Astrophysical Plasmas and Magnetohydrodynamics)**
PRÉSIDENT: R. Lüst.
VICE-PRÉSIDENT: R. J. Tayler.
COMITÉ D'ORGANISATION: L. Davis Jr, S. A. Kaplan, B. P. Lehnert.
MEMBRES: Alfvén H., Axford W. I., Bel N., Bhatnagar P. L., Biermann L., Bumba V., Carlquist P., Chandrasekhar S., Chitre S. N., Cowling T. G., Csada I. K., Ferraro V. C. A., Ginzburg V. L., Godoli G., Jensen E., Jokipii J. R., Kopecký M., Krause F., Kuperus M., Kushwaha R. S., Ledoux P., Mauder H., Mestel L., Michel F. C., Parker E. N., Piddington J. H., Pikel'ner S. B., Pohl E., Sarma M. B. K., Scargle J. D., Schlüter A., Severny A. B., Simon R., Skalafuris A., Smerd S. F., Spitzer L., Sreenivasan S. R., Sturrock P. A., Sweet P. A., Shklovskij I. S., Talwar S. P., Trehan S. K., Tuominen J. V., Wentzel D. G., Westfold K. C., Wilcox J. M.

44. **Commission des Observations Astronomiques au-dehors de l'Atmosphère Terrestre. (Astronomical Observations from outside the Terrestrial Atmosphere)**
PRÉSIDENT: V. K. Prokof'ev.
VICE-PRÉSIDENT: A. D. Code.
COMITÉ D'ORGANISATION: C. de Jager, L. Goldberg, L. Gratton, K. Pounds, R. Tousey, R. Wilson.
MEMBRES: Acton L. W., Alexander J. K., Blamont J. E., Bless R. C., Boggess A. III, Boneff N., Boyd R. L. F., Brandt J. C., Bruckner G. E., Butler H. E., Carver J. H., Chamberlain J. W., Chubb T. A., Clark T. A., Courtes G., Coutrez R., Culhane J. L., Danielson R. E., Davies L., Davis R. J., Dimov N. A., Dollfus A., Fazio G. G., Fichtel C. E., Fisher P. C., Fredga K., Friedman H., Gabriel A. H., Garmire G. P., Gehrels T., Giacconi R., Ginzburg V. L., Glaser H., Gold T., Greisen K. J., Gursky H., Haddock F., Hallam K. L., Hartz T. R., Hayakawa S., Hearn A. C., Helmken H., Henize K. G., Herzberg G., Hinteregger H. E., Houziaux L., Ivanov-Kholodnyj G. S., Jechlel C., Jenkins E. B., Jordan C., Kondo Y., Kraushaar W. L., Kupperian J. E., Lindblad B. A., Lovell B., Lüst R., McCracken K. G., McWhirter R. W. P., Manara A., Mandel'shtam S. L., Massevich A. G., Matsuoka M., Mehltretter J. P., Migeotte M. V., Mikhajlov A. A., Milford S. N., Monfils A., Morton D. C., Müller (Miss) E. A., Ness N. F., Neupert W. M., Newkirk G. A., Nicolet M. Nikolsky G. M., Oda M., Papagiannis M., Parkinson W. H., Rao U. R., Reeves E. M., Rense W. A., Righini G., Rogerson J. B., Roman N. G., Sato H., Savage B., Smith Henry J., Sofia S., Speer R., Stecher T. P., Steinberg J. L., Stone R. G., Takakura T., Van de Hulst H. C., Van Duinen R. J., Walsh D., Weinberg J. L., Zhongolovich I. D., Zwicky F.

45. **Commission des Classifications Spectrales et Indices de Couleur à plusieurs bandes. (Spectral Classifications and Multi-band Colour Indices)**
PRÉSIDENT: B. Westerlund.
VICE-PRÉSIDENT: C. O. R. Jaschek.
COMITÉ D'ORGANISATION: W. P. Bidelman, Ch. Fehrenbach, W. Iwanowska, H. L. Johnson, E. K. Kharadze, W. W. Morgan, U. Sinnerstad.
MEMBRES: Ardeberg A. L., Bahng J. D. R., Bappu M. K. V., Barbier-Brossat M., Bartaya R. A.,

Beer A., Blanco V. M., Boyce P. B., Buscombe W., Cameron R. C., Canavaggia R., Cester B., Cowley A. P., Crawford D. L., Dessy J. L., Divan L., Duflot M., Elvius T., Evans D. S., Feast M. W., Garrison R. F., Geyer E. H., Golay M., Gyldenkerne K., Hack M. H., Hallam K. L., Hauck B., Henize K. G., Herman R., Hiltner W. A., Hoag A. A., Jaschek M., Keenan P. C., Kopylov I. M., Kron G. E., Loden K., Low F. J., Lynga G., McCarthy M. F., McCuskey S. W., McNamara D. H., Mendoza-V. E., Nikonov V. B., Notni P., Oja T., Osawa K., Perry C. L., Philip A. G. D., Preston G. W. III, Reddish V. C., Roman N. G., Rudkjøbing M., Sanduleak N., Sanwal N. B., Schmidt-Kaler Th., Seitter W. C., Sharpless S. L., Sinvhal S. D., Slettebak A., Steinlin U. W., Stephenson C. B., Straižys V. L., Strömgren B., Treanor P. J., Upgren A. R., Van den Bergh S., Walker G. A. H., Warner B., Williams J. A., Wilson O. C., Yoss K. M.

46. Commission pour l'Enseignement de l'Astronomie (The Teaching of Astronomy)
PRÉSIDENT: Miss E. A. Müller.
VICE-PRÉSIDENT: D. McNally.
COMITÉ D'ORGANISATION: E. K. Kharadze, J. Kleczek, E. V. Kononovich, V. Kourganoff, E. Schatzman, T. L. Swihart, D. G. Wentzel.
MEMBRES: Abell G. O., Abhyankar K. D., Andrillat H. L., Asaad A. S., Atanasijević I., de Barros M. G. P., Climenhaga J. L., Coutrez R. A. J., Dinulescu N., Dominko F., Doughty N. A., Elgarøy O., Elvius T., Ferraz-Mello S., Grushinskij N. P., Hack (Mrs) M., Hämeen-Antilla K. A., Haupt H. F., Hazer S., Hidajat B., Iwaniszewska (Mrs) C., Jarret A. H., Jørgensen H. E., Lambrecht H., Marik M., Mavridis L., Moreno H., Nicolov N. S., Pishmish (Mrs) P., Ringuelet-Kaswalder (Mrs) A., Rodgers A. W., Scheffler H., Steinitz R., Takase B., Therian (Mrs) A., Torroja J. M.

47. Commission de la Cosmologie (Cosmology)
PRÉSIDENT: Ya. B. Zeldovich.
VICE-PRÉSIDENT: P. J. E. Peebles.
COMITÉ D'ORGANISATION: M. S. Longair (Secrétaire de la Commission), W. H. McCrea, G. C. McVittie, C. W. Misner, H. Nariai, Y. Ne'eman, A. R. Sandage, D. W. Sciama.
MEMBRES: Aizu K., Alfven H., Ambartsumian V. A., Andrillat H., Bardeen J., Bondi H., Bonnor W. B., Burbidge G. R., Cohen J. M., Dautcourt G., Davidson W., Dicke R. H., Ellis G. F. R., Felten J. E., Field G. B., Fujimoto M., Godart O., Gold T., Harrison E. R., Hawking S. W., Hayakawa S., Hayashi C., Heckmann O., Heidmann J., Hoyle F., Kato S., Kawabata K., Kellermann K. I., Kovetz A., Kozlovsky B. Z., Kustaanheimo P. E., Layser D., Narlikar J. V., Neyman J., Nishida M., Noerdlinger P., Novikov I. D., O'Connel R. F., Oort J. H., Ozernoj L. M., Pachner J., Petrosian V., Rees M. J., Rindler W., Robinson I., Roeder R. C., Pyle M., Sapar A. A., Sato H., Scheuer P. A. G., Schmidt M., Schücking E., Scott E. L., Sersic J. L., Setti G., Shaviv G., Simon R., Tauber G., Tayler R. J., Thorne K. S., Treder H. J., Wheeler J. A., Whitrow C. J., Wilson A. G., Zelmanov A. I.

48. Commission de l'Astrophysique de Grande Energie. (High Energy Astrophysics)
PRÉSIDENT: H. Friedman.
VICE-PRÉSIDENT: M. J. Rees.
COMITÉ D'ORGANISATION: A. G. W. Cameron, V. L. Ginzburg, T. Gold, K. I. Greisen, P. Morrison, F. Pacini, D. W. Sciama, M. M. Shapiro, I. S. Shklovskij, F. G. Smith.
MEMBRES: Alvarez L., Burbidge G. R., Chubb T. A., Clark G., Dautcourt G., Fazio G. G., Felten J. E., Fichtel C. E., Field G. B., Fisher P. C., Fowler W. A., Garmire G. P., Giacconi R., Gratton L., Greyber H. D., Gursky H., Hayakawa S., Hoyle F., Jelley J. V., Kozlovsky B. Z., Kurt V. G., Longair M. S., Matsuoka M., Ne'eman Y., Oda M., Parker E. N., Petrosian V., Pounds K. A., Reeves H., Rossi B., Salpeter E. E., Saslaw W. C., Schatzman E., Scheuer P. A. G., Setti G., Shaviv G., Sofia S., Sturrock P. A., Sunyaev R. A., Syrovatskij S. J., Thorne K., Truran J. W., Wheeler J. A., Woltjer L., Zeldovich Ya. B.

LISTE ALPHABÉTIQUE DES MEMBRES
ALPHABETICAL LIST OF MEMBERS

Names are given under the first letter of the first word: for example, the name Van Agt will be found under V.

The names of new Members, elected at the XIVth General Assembly, are prefixed by an asterisk.

The names of the countries are given in the English form.

Names of astronomers using the Cyrillic alphabet are transliterated according to the rules published in *Transactions* XIVA, 13.

Please inform the Executive Secretary of the IAU, Beneluxlaan 21, Utrecht, The Netherlands, of any errors you may find in this list.

Les noms figurent à la première lettre du premier mot; par exemple, le nom Van Agt se trouve à la lettre V.

Les noms des nouveaux membres élus par la XIVe Assemblée générale sont annotés d'un astérisque.

Les noms des pays sont donnés sous leur forme anglaise.

Les noms des astronomes utilisant l'alphabet cyrillique sont donnés selon les règles de translitération publiées dans les *Transactions* XIVA, 13.

Les Membres de l'Union sont priés de bien vouloir informer le Secrétaire Exécutif de l'UAI, Beneluxlaan 21, Utrecht, Pays Bas, de toutes les erreurs qu'ils pourraient relever dans la présente liste.

AARSETH (Dr S.), Institute of Theoretical Astronomy, Madingley Road, Cambridge, U.K.

ABALAKIN (Dr V. K.), Institute of Theoretical Astronomy, USSR Academy of Sciences, Leningrad, U.S.S.R.

ABBOT (Dr C. G.), Astrophysical Observatory of the Smithsonian Institution, Washington, D.C., U.S.A.

ABBOTT (Prof. Dr W. N.), 42 Leophoros A. Michalacopoulou, Ilisia, Athens 612, Greece

ABDALA (Dr J.), Edificio 'Pio XII' primer piso No. 1, San Francisquito a Puente Ayacucho, Caracas, Venezuela

ABELL (Dr G. O.), Department of Astronomy, University of California, Los Angeles 24, California 90024, U.S.A.

ABETTI (Prof. G.), Instituto Naz. Ottica-Arcetri, Largo Enrico Fermi 7 50125, Firenze, Italy

ABHYANKAR (Dr K. D.), Director, Observatory of Hyderabad, Hyderabad, Deccan, India

ABRAHAM (H. J.), Mount Stromlo Observatory, Canberra, A.C.T., Australia

ABRAMI (Prof. A.), Osservatorio Astronomico, via G. B. Tiepolo No. 11, Trieste (409), Italy

ABREU (Dr J. C. de Brito e), Observatorio Astronomico de Lisboa, Tapada, Lisbon, Portugal

ABT (Dr H. A.), Kitt Peak National Observatory, Box 4130, Tucson, Arizona 85717, U.S.A.

* ACTON (Dr L. W.), Dept. 52–14, Bldg 202, Lockheed Research Lab., 3251 Hanover Street, Palo Alto, Calif. 94304, U.S.A.

ADAM (Miss Dr M. G.), University Observatory, South Parks Road, Oxford, U.K.

ADAMOPOULOS (Dr G.), Rue Lamias No. 8, Athens (6), Greece

ADAMS (A. N.), U.S. Naval Observatory, Washington 25, D.C., U.S.A.

ADEL (Prof. A.), Atmospheric Research Observatory, Arizona State College, Flagstaff, Arizona, U.S.A.

ADOLFSSON (Dr T.), Lund Observatory, S-222 24 Lund, Sweden

* AFANAS'EVA (Dr P. M.), Main Astronomical Observatory, USSR Academy of Sciences, Pulkovo, U.S.S.R.

AGEKJAN (Prof. Dr T. A.), University of Leningrad, Leningrad, U.S.S.R.

AHMED (I.I.), Helwan Observatory, Helwan near Cairo, U.A.R.

AHNERT (Dr P.), Zentralinstitut für Astrophysik, Sternwarte Sonnenberg, DDR 64 Sonnenberg, G.D.R.

* AIZU (Prof. Dr K.), Physics Department, Rikkyo University, Nishi-Ikebukuro 3-chome, Toshima-ku, Tokyo 171, Japan

AKABANE (Dr K.), Tokyo Astronomical Observatory, Mitaka, Tokyo, Japan

AKSENOV (Dr E. P.), Sternberg Astronomical Institute, Moscow, U.S.S.R.

* AKYOL (Dr M. Ü.), Department of Astronomy, P.K. 21, Bornova, Izmir, Turkey

ALANIA (Dr I. F.), Abastumani Astrophysical Observatory, Abastumani, Georgian SSR, U.S.S.R.

* ALEXANDER (J. B.), Royal Greenwich Observatory, Herstmonceux Castle, Hailsham, Sussex, U.K.

ALEXANDER (J. K.), Code 615, Radio Astronomy Section, Ionospheric and Radio Physics Branch, Laboratory for Space Science, NASA-Goddard Space Flight Center, Greenbelt, Maryland 20771, U.S.A.

ALFVÉN (Prof. H.), Royal Institute of Technology, S-100 44 Stockholm 70, Sweden

ALKSNIS (Dr A. K.), Astrophysical Laboratory, Latvian Academy of Sciences, Riga, U.S.S.R.

ALLADIN (S. M.), Department of Astronomy, Osmania University, Hyderabad, India

ALLEN (Prof. C. W.), University of London Observatory, Mill Hill Park, London, N.W. 7, U.K.

* ALLEN (Dr R. J.), Kapteyn Astronomical Laboratory, Groningen, The Netherlands

ALLER (Prof. L. H.), Astronomy Department, University of California, Los Angeles, California 90024, U.S.A.

ALMÁR (Dr I.), Konkoly Observatory, Szabadsághegy, Budapest XII, Hungary

* ALTENHOFF (Dr W.), Max-Planck-Institut für Radioastronomie, Argelanderstrasse 3, 5300 Bonn, F.R.G.

ALTER (Dr G.A.), Bet Yizhaq, Israel

* ALTSCHULER (Dr M.), High Altitude Observatory, Boulder, Colorado 80302, U.S.A.

* ALVAREZ (Dr L.), Lawrence Radiation Laboratory, University of California, Berkeley, California 94720, U.S.A.

ALY (Dr M. K. M.), Director, Helwan Institute of Astronomy and Geophysics, Helwan Observatory, Egypt, U.A.R.

AMBARTSUMIAN (Prof. Dr V. A.), President of the Academy of Sciences of Armenian SSR, Erevan, Armenia, U.S.S.R.

* ANAND (Prof. S. P. S.), Department of Astronomy, University of Toronto, Toronto 5, Ontario, Canada

ANDERS (Prof. E.), Enrico Fermi Institute, University of Chicago, 5640 S. Ellis Avenue, Chicago, Illinois 60637, U.S.A.

* ANDERSON (Dr B.), Nuffield Radio Astronomy Laboratories, Jodrell Bank, Macclesfield, Cheshire, U.K.

* ANDERSON (Dr K.), Space Science Laboratory, University of California, Berkeley, California 94728, U.S.A.

* ANDREW (Dr B.H.), Astrophysics Branch, National Research Council, Ottawa 7, Ontario, Canada

ANDREW (K.L.) address unknown

* ANDREWS (A.D.), Armagh Observatory, Armagh, N. Ireland, U.K.

ANDREWS (D. H.), Dominion Astrophysical Observatory, R.R. 7 Victoria, B.C., Canada

* ANDREWS (Dr P.J.), Radcliffe Observatory, P.O. Box 373, Pretoria, South Africa

ANDRILLAT (Prof. Dr H.), Laboratoire d'Astronomie de la Faculté des Sciences de Montpellier, Hérault, France

ANDRILLAT (Mrs Dr Y.), Laboratoire d'Astronomie de la Faculté des Sciences de Montpellier, Hérault, France

* ANDRLE (Dr P.), Astronomical Institute, Czechoslovak Academy of Sciences, Budečská 6, Prague 2, Czechoslovakia

ANGUITA (Dr C.A.), Universidad de Chile, Observatório Astronómico Nacional, Casilla 36-D, Santiago, Chile

* ANTONACOPOULOS (Dr G.), Laboratory of Astronomy, University of Athens, 57, Solonos Street, Athens, Greece
AOKI (Prof. Dr S.), Tokyo Astronomical Observatory, University of Tokyo, Mitaka, Tokyo, Japan
* AOUDOUZE (Dr J.), Institut d'Astrophysique, 98bis Boulevard Arago, Paris 14e, France
* APPENZELLER (Dr I.), Universitätssternwarte, Geismarlandstrasse 11, 3400 Göttingen, F.R.G.
ARAKELYAN (Dr M. A.), Byurakan Astrophysical Observatory, Byurakan, Armenia, U.S.S.R.
ARBEY (Prof. L. M. J.), Directeur de l'Observatoire, 25 Besançon, Doubs, France
* ARDEBERG (A. L.), Institute of Astronomy, S-222 24 Lund, Sweden
AREND (Prof. Dr S. J. V.), Astronome Honoraire, Observatoire Royal de Belgique, avenue Circulaire, 3, B-1180 Bruxelles, Belgium
ARGUE (A. N.), The Observatories, Madingley Road, Cambridge, U.K.
ARGYLE (Dr P. E.), Dominion Radio Astrophysical Observatory, Box 248, Penticton, B.C., Canada
ARGYRACOS (Prof. Dr J. G.), 193 Patission Street, Athens (816), Greece
ARIAS-DE GREIFF (Prof. J.), Observatorio Astronómico Nacional, Apartado Nacional 2584, Bogotá 1, D.E., Colombia
* ARKHIPOVA (Dr V. P.), Sternberg Astronomical Institute, Moscow, U.S.S.R.
* ARNETT (Dr W. D.), Department of Space Science, Rice University, Houston, Texas 77001, U.S.A.
* ARNOULD (M. L. L.), chaussée de Tubize, 146, B-1430 Wauthier-Braine, Belgium
* ARNQUIST (Dr W. N.), Douglas Advanced Research Laboratories, 5152 Bolsa Avenue, Huntington Beach, California 92647, U.S.A.
* ARNY (Dr T. T.), Department of Physics and Astronomy, University of Massachusetts, Amherst, Massachusetts 01002, U.S.A.
ARP (Dr H. C.), Mount Wilson and Palomar Observatories, 813 Santa Barbara Street, Pasadena, California, U.S.A.
ARPIGNY (Prof. Dr C.), Institut d'Astrophysique, Université de Liège, avenue de Cointe, 5, B-4200 Cointe-Ougrée, Belgium
ARROYO (Dr M. L.), Observatorio Astronómico, Alfonso XII, 3, Madrid 7, Spain
ARTHUR (D. W. G.), Lunar and Planetary Laboratory, University of Arizona, Tucson, Arizona, U.S.A.
ARTIUKHINA (Mrs Dr N. M.), Sternberg Astronomical Institute, Moscow, U.S.S.R.
ASAAD (Dr A. S.), Helwan Observatory, Helwan near Cairo, Egypt, U.A.R.
ASHBROOK (Dr J.), 16 Summer Street, Weston 93, Massachusetts, U.S.A.
* ASLANOV (Dr I. A.), Shemaha Astrophysical Observatory, Azerbajdzhan Academy of Sciences, Shemaha, U.S.S.R.
ASTAPOVICH (Dr I. S.), 15 Nikolsko-Botanicheskaya 17a log. 45, Kiev-33, U.S.S.R.
ATANASIJEVIC (Dr I.), Department of Astronomy, Faculty of Sciences, Driehuizerweg 200, Nijmegen, The Netherlands
ATHAY (Dr R. G.), High Altitude Observatory of the University of Colorado, Boulder, Colorado 80302, U.S.A.
ATKINSON (R. d'E.), Astronomy Department, Indiana University, Swain Hall West, Bloomington, Indiana 47401, U.S.A.
* AUER (Dr L. H.), Department of Astronomy, Watson Astronomy Center, Box 2023 Yale Station, Yale University, New Haven, Connecticut 06520, U.S.A.
AUMAN (Dr J. R.), Department of Geophysics, University of British Columbia, Vancouver 8, B.C., Canada
AVIGNON (Miss Dr Y.), Observatoire de Paris, Section d'Astrophysique, 92-Meudon, France
AVRETT (Dr E. H.), Smithsonian Astrophysical Observatory, 60 Garden Street, Cambridge, Massachusetts 02138, U.S.A.
AXFORD (W. I.), Department of Astronomy, Cornell University, Ithaca, New York, 14850 U.S.A.
* AYDIN (Dr C.), Faculty of Science, Department of Astronomy, Ankara, Turkey

* BAARS (J. W. M.), Dwingeloo Radio Observatory, Dwingeloo, The Netherlands

* BAART (Dr E. E.), Rhodes University, P.O. Box 94, Grahamstown, South Africa

BABADZHANOV (Dr P. B.), Director, Astrophysical Institute, Dushanbe, U.S.S.R.

BABCOCK (Dr H. W.), Mount Wilson and Palomar Observatories, 813 Santa Barbara Street, Pasadena, California 91106, U.S.A.

BACCHUS (Prof. P.), 1, impasse de l'Observatoire, 59-Lille (Nord), France

BAEHR (Dr U.), Astronomisches Rechen-Institut, Quickerstrasse 44a, 69 Heidelberg, F.R.G.

* BAERENTZEN (Dr J.), Ole Roemer Observatory, DK-8000 Aarhus C, Denmark

* BAGILDINSKIJ (Dr B. K.), Main Astronomical Observatory, USSR Academy of Sciences, Pulkovo, U.S.S.R.

* BAGLIN (Miss Dr A.), Institut d'Astrophysique, 98bis Boulevard Arago, Paris 14e, France

BAHCALL (Dr J. N.), Kellog Radiation Laboratory, California Institute of Technology, Pasadena, California 91109, U.S.A.

BAHNER (Dr K.), Landessternwarte, Königstuhl-Heidelberg, F.R.G.

BAHNG (Prof. J. D. R.), Dearborn Observatory, Northwestern University, Evanston, Illinois, U.S.A.

BAIRD (Dr K. M.), Division of Applied Physics, National Research Council, Sussex Drive, Ottawa, Ontario, Canada

BAIZE (Dr P.), 6, Rue Daubigny, 75-Paris 17e, France

BAKER (Dr J. G.), 60 Garden Street, Cambridge, Massachusetts 02138, U.S.A.

BAKER (Dr N.), Astronomy Department, Columbia University, New York, New York 10027, U.S.A.

BAKOS (Prof. G. A.), Department of Physics, University of Waterloo, Waterloo, Ontario, Canada

BAKULIN (Dr P. I.), Sternberg Astronomical Institute, Moscow, U.S.S.R.

BALÁZS (Dr B.), Konkoly Observatory, P.O. 114, Box 67, Budapest XII, Hungary

BALÁZS-DETRE (Mrs Dr J.), Konkoly Observatory, Szabadsághegy, P.O. 114, Box 67, Budapest XII, Hungary

BALDWIN (Dr J. E.), Cavendish Laboratory, Free School Lane, Cambridge, U.K.

* BALDWIN (Dr R.), Oliver Machinery Co., 445 Sixth Street, NW Grand Rapids, Michigan 49502, U.S.A.

BALKLAVS (Dr A. E.), Astrophysical Laboratory, Latvian Academy of Sciences, Riga, U.S.S.R.

BALLARIO (Prof. M. C.), Osservatorio Astrofisico di Arcetri, Via S. Leonardo 75, Firenze, Italy

BALLI (Prof. Dr E.), University Observatory, Beyazit, Istanbul, Turkey

BALTÁ-ELIAS (Prof. J.), Universidad de Madrid, Madrid, Spain

BANERJI (Dr A. C.), Muir Central College, Allahabad University, Allahabad, India

BANIN (Dr V. G.), Post Office Meget, Sibizmiran, Irkutsk, U.S.S.R.

* BANOS (Dr G.), Astronomical Institute, National Observatory of Athens, Athens (306), Greece

BAPPU (Dr M. K. V.), Director, Astrophysical Observatory, Kodaikanal, India. Vice-President of the Union.

BARABASHOV (Prof. Dr N. P.), Director of the Astronomical Observatory, Khar'kov, U.S.S.R.

BARANNE (A.), Observatoire de Marseille, 2 place Le Verrier, 13-Marseille 4e, France

BARBANIS (Dr B.), University of Patras, Department of Astronomy, Patras, Greece

BARBER (D. R.), 'Craigwell', Coulsdon Road, Sidmouth, Devon, U.K.

BARBIER-BROSSAT (Mrs M.), Chargée de Recherche, Observatoire de Marseille, 2, place Le Verrier, 13-Marseille 4e, France

* BARBIERI (Dr C.), Osservatorio Astronomico di Padova, Padova, Italy

* BARBON (Dr R.), Osservatorio Astrofisico Asiago (Vicenza), Italy

* BARDEEN (Dr J. M.), Department of Astronomy, University of Washington, Seattle, Washington 98105, U.S.A.

BARKHATOVA (Mrs Dr K. A.), State University, Sverdlovsk, U.S.S.R.

BARLIER (F.), Observatoire de Paris, Section d'Astrophysique, 92-Meudon, France

BARNEY (Miss Dr I.), Yale University Observatory, Box 2023 Yale Station, New Haven, Connecticut, U.S.A.

* BARNOTHY (Dr J. M.), Department of Mathematics, Barat College, Lake Forest, Illinois 60045, U.S.A.

BARRETO (Prof. L. M.), Observatorio Nacional, Rua General Bruce, 586, São Cristóvão, Rio de Janeiro, Brazil

BARRETT (Prof. A. H.), Research Laboratory of Electronics, Massachusetts Institute of Technology, Cambridge, Massachusetts 02139, U.S.A.

BARTAYA (Mrs Dr R. A.), Astrophysical Observatory, Abastumani, Georgia, U.S.S.R.

* BARTOLINI (Dr C.), Osservatorio Astronomico Universitario, via Zamboni 33, Bologna, Italy

BASCHEK (Dr B.), Lehrstuhl für Theoretische Astrophysik, 69 Heidelberg, Berliner Strasse 19, F.R.G.

* BASTIN (Dr J. A.), Physics Department, Queen Mary College (London University), Mile End Road 1, U.K.

BATES (Prof. D. R.), Department of Applied Mathematics, Queen's University, Belfast, North Ireland, U.K.

BATESON (F. M.), Astronomer-in-Charge, 18 Pooles Road, Greerton, Tauranga, New Zealand

BATRAKOV (Dr J. W.), Institute of Theoretical Astronomy, Leningrad, U.S.S.R.

BATTEN (Dr A. H.), Dominion Astrophysical Observatory, R.R. 7 Victoria, B.C., Canada

BAUER (C. A.), Pennsylvania State University, Department of Physics, State College, Pennsylvania, U.S.A.

BAUM (Dr W. A.), Planetary Research Center, Lowell Observatory, Flagstaff, Arizona 86002, U.S.A.

BAUSTIAN (W. W.), Kitt Peak National Observatory, 950 North Cherry Avenue, Tucson, Arizona, U.S.A.

BEALS (Dr C. S.), Manotic, Ontario, Canada

BEARDSLEY (W. R.), Allegheny Observatory, Riverview Park, Pittsburg 14, Pennsylvania, U.S.A.

BEAVERS (Dr W.), Iowa State University, Ames, Iowa, U.S.A.

* BEC (Miss A.), Bureau des Longitudes, 3, Rue Mazarine, Paris 8e, France

BECK (H.-G.), Carl Zeiss Observatory, Jena, G.D.R.

BECKER (Prof. Dr F.), Klingsorstrasse 3/96, 8 München 61, F.R.G.

BECKER (Dr R. A.), Aerospace Corporation, P.O. Box 95085, Los Angeles, California, U.S.A.

BECKER (Prof. W.), Astronomisches Institut der Universität Basel, Venusstrasse 7, Ch-4102 Binningen, Switzerland

BECKERS (Dr J. M.), Sacramento Peak Observatory, Sunspot, New Mexico 83349, U.S.A.

BEER (Dr A.), The Observatories, Madingley Road, Cambridge, U.K.

BEGGS (D. W.), The Observatories, Madingley Road, Cambridge, U.K.

BEHR (Prof. Dr A.), Director, Hamburg Observatory, 205 Hamburg 80, Gojenbergsweg 112, F.R.G.

BEL (Miss Dr N.), Institut d'Astrophysique, 98bis Boulevard Arago, Paris 14e, France

BELL (Miss Dr B.), Harvard College Observatory, Cambridge, Massachusetts 02138, U.S.A.

BELL (Dr R. A.), Department of Physics and Astronomy, University of Maryland, College Park, Maryland, U.S.A.

BELOTSERKOVSKIJ (Dr D. J.), Institute of Physico-Technical Measurements, Moscow, U.S.S.R.

BELTON (Dr M. S.), Kitt Peak National Observatory, 950 North Cherry Avenue, Tucson, Arizona 85717, U.S.A.

BELY (O.), Observatoire de Nice, Le Mont-Gros, 06-Nice, France

* BENDER (Dr P.), Joint Institute for Laboratory Physics, University of Colorado, Boulder, Colorado 80302, U.S.A.

BENEVIDES-SOARES (Dr P.), Instituto Astronómico e Geofísico, Caixa Postal 30 627, São Paulo OO-SP, Brazil

* BERGE (Dr G. L.), Owens Valley Radio Observatory, California Institute of Technology, Pasadena, California 91109, U.S.A.

BERGER (Dr J.), Institut d'Astrophysique, 98bis Boulevard Arago, Paris 14e, France

BERKNER (Dr L. V.), 3632 N.E. 24th Avenue, Fort Lauderdale, Florida, U.S.A.

* BERNAS (Dr R.), Centre de Spectrométrie Nucléaire et de Spectrométrie de Masse du C.N.R.S., Bat. 108, CNRS 1, 91-Campus, Orsay, France

BERTAUD (Dr Ch.), Observatoire de Paris, Section d'Astrophysique, 92-Meudon, France

BERTIAU (Dr F.C.), (Rév. P.), Specola Vaticana, Castel Gandolfo, Vatican City State

BERTOLA (Dr F.), Osservatorio astrofisico di Asiago (Vicenza), Italy

BEYER (Dr M.), Justus-Brinckmann-Strasse 101, Hamburg-Bergedorf, F.R.G.

BHATNAGAR (A.), Uttar Pradesh State Observatory, Manora Peak, Naini Tal, India

BHATNAGAR (Dr P.L.), Department of Applied Mathematics, Indian Institute of Science, Bangalore 3, South India

BHATTACHARYYA (J.C.), Astrophysical Observatory, Kodaikanal 3, India

BHONSLE (R. V.), Physical Research Laboratory, Ahmedabad, India

BIDELMAN (Dr W.P.), Department of Astronomy, University of Texas, Austin, Texas 78712, U.S.A.

BIELICKI (Dr M.), Astronomical Observatory of Warsaw University, Al. Ujazdowskie 4, Warsaw, Poland

BIERMANN (Prof. Dr L.), Max-Planck Institut für Physik und Astrophysik, Institut für Astrophysik, Föhringer Ring 6, München 23, F.R.G.

BIGAY (Dr J.H.), Directeur de l'Observatoire de Lyon, 69 Saint-Genis-Laval, France

BILLAUD (G.), Observatoire de Paris, 61, avenue de l'Observatoire, Paris 14e, France

BILLINGS (Dr D. E.), High Altitude Observatory of the University of Colorado, Boulder, Colorado, U.S.A.

* BINGHAM (Dr R.G.), Royal Greenwich Observatory, Herstmonceux Castle, Hailsham, Sussex, U.K.

BINNENDIJK (Prof. Dr L.), Flower and Cook Observatory, University of Pennsylvania, Philadelphia 4, Pennsylvania, U.S.A.

* BISNOVATYI-KOGAN (Dr G.S.), Institute of Applied Mathematics, USSR Academy of Sciences, Moscow, U.S.S.R.

BLAAUW (Prof. Dr A.), Director of the Kapteyn Astronomical Laboratory, Broerstraat 7, Groningen, The Netherlands

BLACKETT (Prof. P.M.S.), Imperial College of Science and Technology, South Kensington, London, S.W. 7, U.K.

BLACKWELL (Prof. D.E.), University Observatory, Oxford, U.K.

BLAHA (Dr M.) Goddard Space Flight Center, Code 680, Greenbelt, Maryland 20771, U.S.A.

BLAMONT (Prof. J.E.), Service d'Aéronomie, Réduit de Verrières, 91-Verrières-le-Buisson, France

BLANCO (Dr V.M.), Observatorio Inter-Americano de Corro Tololo, Casilla 63-D, La Serena, Chile

BLESS (Dr R.C.), Washburn Observatory, University of Wisconsin, Madison, Wisconsin 53705, U.S.A.

* BLINOV (Dr N.S.), Sternberg Astronomical Institute, Moscow, U.S.S.R.

BLITZSTEIN (Dr W.), Flower and Cook Observatory, University of Pennsylvania, Philadelphia 4, Pennsylvania, U.S.A.

BLOCH (Miss Dr M.), Observatoire de Lyon, Saint-Genis-Laval (Rhône), France

BLUM (Dr E.J.), Observatoire de Paris, Section d'Astrophysique, 92-Meudon, France

BOBROV (Dr M.S.), Astronomical Council, USSR Academy of Sciences, Vavilov Str. 34, Moscow V-312, U.S.S.R.

BOBROVNIKOFF (Dr N.T.), The Ohio State University, Columbus, Ohio, U.S.A.

* BODENHEIMER (Dr P.), Lick Observatory, University of California, Santa Cruz, California 95060, U.S.A.

* BOESGAARD (Mrs Dr A.M.), Institute of Astronomy, University of Hawaii, 2525 Correa Road, Honolulu, Hawaii 96822, U.S.A.

BOGGESS (Dr A. III), Goddard Space Flight Center, Washington 25, D.C., U.S.A.

BOGORODSKIJ (Dr A. F.), Astronomical Observatory of the Kiev University, Kiev, U.S.S.R.

BÖHM (Prof. Dr K.-H.), Astronomy Department, University of Washington, Seattle, Washington 98105, U.S.A.

BÖHM-VITENSE (Mrs Prof. Dr E.), Astronomy Department, University of Washington, Seattle, Washington 98105, U.S.A.

BÖHME (Dr S.), Astronomisches Rechen-Institut, Mönchhofstrasse 12–14, Heidelberg, F.R.G.

BOHRMANN (Prof. Dr A.), Angelhofweg 31, 6901 Wilhelmsfeld, F.R.G.

BOISCHOT (Dr A.), Observatoire de Paris, Section d'Astrophysique, 92-Meudon, France

BOK (Prof. B. J.), Steward Observatory, University of Arizona, Tucson, Arizona 85721, U.S.A. (Vice-President of the Union)

BOK (Lie Rak), Director of Phyongyang Astronomical Observatory, Phyongyang, Korea, P.D.R.

BOK (Mrs Dr P. F.), Steward Observatory, University of Arizona, Tucson, Arizona 85721, U.S.A.

BOLDT (Dr E. A.), NASA, Goddard Space Flight Center, Greenbelt, Maryland 20771, U.S.A.

BOLTON (J. G.), CSIRO, Division of Radiophysics, Post Office Box 76, Epping, N.S.W. 2121, Australia

BONANOMI (Dr J.), Directeur de l'Observatoire de Neuchâtel, Neuchâtel, Switzerland

BONDI (Dr H.), King's College, Strand, London, W.C. 2, U.K.

BONEFF (Prof. Dr N.), Bulgarian Academy of Sciences, Section of Astronomy, '7 November' 1, Sofia, Bulgaria

BONEV (B.), Observatoire Populaire de Stara-Zagora, Bulgaria

BONNEAU (Prof. M.), Faculté des Sciences, Place Victor Hugo, 13-Marseille, France

* BONNET (Dr R.), L.P.S.P. 91, Verrières le Buisson, France

BONNOR (Prof. W. B.), Queen Elisabeth College, Campden Hill Road, London W. 8., U.K.

BONOV (A. D.), Université de Sofia, Sofia, Bulgaria

BONSACK (Prof. W. K.), Institute for Astronomy, 2840 Koluwalu Street, University of Hawaii, Honolulu, Hawaii 96822, U.S.A.

BORGMAN (Dr J.), 'Kapteyn' Sterrewacht, Mensingenweg 20, Roden (Drente), The Netherlands

BOSMAN-CRESPIN (Mrs D.), place d'Italie, 4, B-4000 Liège, Belgium

BOTELHEIRO (Dr A.), Observatorio Astronomico, Lisboa, Portugal

* BOTEZ (Mrs Dr E.), Pedagogical Institute, Suceava, Roumania

BOUIGUE (Prof. R.), Directeur de l'Observatoire de Toulouse, 1, avenue Camille-Flammarion, 31-Toulouse, France

BOULON (Dr J.), Observatoire de Paris, 61, avenue de l'Observatoire, 75-Paris 14e, France

BOURGEOIS (Prof. Dr P. E.), Directeur Honoraire, Observatoire Royal de Belgique, rue Paul Hankar, 31, B-1180 Bruxelles, Belgium

BOURY (Dr A. J. J. L.), Institut d'Astrophysique, Université de Liège, rue Chinrue, 9, B-4870, Theux, Belgium

BOUŠKA (Dr J.), Dept. of Astronomy and Astrophysics, Charles University, Švédská 8, Prague 5-Smíchov, Czechoslovakia

BOUVIER (Prof. P. B.), Observatoire de Genève, 1290 Sauverny (GE), Switzerland

BOWEN (Dr E. G.), CSIRO, Division of Radiophysics, Post Office Box 76, Epping, N.S.W., 2121, Australia

BOWEN (Dr I. S.), Mount Wilson and Palomar Observatories, 813 Santa Barbara Street, Pasadena 4, California, U.S.A.

* BOWYER (Dr C. S.), Astronomy Department, University of California, Berkeley, California 94720, U.S.A.

BOYARCHUK (Dr A. A.), Crimean Astrophysical Observatory, USSR Academy of Sciences, P/O Nauchny, Crimea, U.S.S.R.

BOYCE (Dr P. B.), P.O. Box 1269, Flagstaff, Arizona 86001, U.S.A.

BOYD (Prof. R. L. F.), University College, Gower Street, London, U.K.

BOZIS (Dr G.), Astronomical Department, University of Thessaloniki, Thessaloniki, Greece

BRACCESI (A.), Laboratorio di Radioastronomia, Instituto di Fisica, Bologna, Italy

BRACEWELL (Prof. R. N.), Radio Astronomy Institute, Stanford University, Stanford, California 94305, U.S.A.

BRAES (Dr L. L. E.), Sterrewacht, Leiden, The Netherlands

BRAHDE (R.), Institute of Theoretical Astrophysics, University of Oslo, Blindern, Oslo 3, Norway

BRANDT (Dr J. C.), Code 680, Laboratory for Solar Physics, Goddard Space Flight Center, Greenbelt, Maryland 20771, U.S.A.

BRANDT (Dr V. E.), Sternberg Astronomical Institute, Moscow, U.S.S.R.

BRANSCOMB (Dr L. M.), National Bureau of Standards, Washington 25, D.C., U.S.A.

* BRANSON (Dr N. J. B. A.), Mullard Radio Astronomy Observatory, Cavendish Laboratory, Free School Lane, Cambridge, U.K.

* BRAUDE (Dr S. Ya.), Institute of Radiophysics and Electronics, Academy of Sciences of the Ukrainian SSR, Khar'kov, U.S.S.R.

BRAY (Dr R. J.), National Standards Laboratory, CSIRO, University Grounds, Sydney, N.S.W., Australia

BREALEY (G. A.), Dominion Astrophysical Observatory, R.R. 7 Victoria, B.C., Canada

* BREJDO (Mrs Dr I. I.), Main Astronomical Observatory, USSR Academy of Sciences, Pulkovo, U.S.S.R.

* BRINI (Prof. D.), Laboratorio T.E.S.R.E., via Castagnoli 1, Bologna, Italy

BRKIĆ (Z. M.), Astronomska Observatorija, Veliki Vracar, Beograd, Yugoslavia

* BROADFOOT (Dr L.), Kitt Peak National Observatory, 950 North Cherry, Tucson, Arizona 85721, U.S.A.

BROGLIA (Dr P.), Osservatorio Astronomico, Merate (Como), Italy

* BROSCHE (Dr P.), Astronomisches Rechen-Institut, Mönchhofstrasse 12–14, 6900 Heidelberg, F.R.G.

BROSTERHUS (Dr E. B. F.), Dominion Astrophysical Observatory, R.R. 7, Victoria, B.C., Canada

BROTEN (N. W.), Astrophysics Branch, National Research Council, Ottawa, Ontario, Canada

* BROUCKE (Dr R.), 1106 East Orange Grove, Pasadena, California 91103, U.S.A.

* BROUW (W. N.), Leiden Observatory, Leiden, The Netherlands

BROWN (Dr Harrison) Division of Geological Sciences, California Institute of Technology, Pasadena 4, California, U.S.A.

BROWN (P. Lancaster), Karlsvik, 30 Enghams Wood Road, Beaconsfield, Bucks., U.K.

BROWN (Dr R. Hanbury), Chatterton Astronomy Department, School of Physics, University of Sydney, Sydney, N.S.W. 2006, Australia

BROWNLEE (Dr R. R.), P.O. Box 1663, Los Alamos, New Mexico, U.S.A.

BRÜCK (Prof. H. A.), Astronomer Royal for Scotland, Royal Observatory, Blackford Hill, Edinburgh 9, Scotland, U.K.

BRÜCK (Mrs Dr M. T.), Royal Observatory, Blackford Hill, Edinburgh 9, Scotland, U.K.

BRÜCKNER (Dr G.), U.S. Naval Research Laboratory, Plasma Physics Section, Code 7700, Washington D.C., 20390, U.S.A.

BRUMBERG (Dr V. A.), Institute of Theoretical Astronomy, USSR Academy of Sciences, Leningrad, U.S.S.R.

BRUNK (W.), Lunar and Planetary Programs, NASA Headquarters, Washington, D.C., 20546, U.S.A.

BRUWER (J. A.), Republic Observatory, Johannesburg, South Africa

BRUZEK (Dr A.), Fraunhofer Institut, Schöneckstrasse 6, Freiburg i.Br., F.R.G.

BUCHAR (Prof. Dr E.), Astronomical Institute, Technical University, Karlovo náměstí 13, Prague
2, Czechoslovakia

BULLARD (Sir Edward C.), University of Cambridge, Department of Geodesy and Geophysics, Madingley Road, Cambridge, U.K.

BULLEN (Prof. K. E.), Department of Applied Mathematics, University of Sydney, Australia

BUMBA (Dr V.), Astronomical Institute, Czechoslovak Academy of Sciences, Observatory Ondřejov, Czechoslovakia

BURBIDGE (Mrs Dr E. M.), University of California, La Jolla, California 92038, U.S.A.

BURBIDGE (Dr G. R.), University of California, La Jolla, California 92038, U.S.A.

BURCH (Dr C. R.), H.H. Royal Fort Wills Physics Laboratory, Bristol 8, U.K.

BURGESS (Dr A.), University of Cambridge, Department of Applied Mathematics and Theoretical Physics, Silver Street, Cambridge, U.K.

* BURGESS (Dr D. D.), Physics Department, Imperial College of Science and Technology, London S.W. 7, U.K.

BURKE (Dr B. F.), Department of Terrestrial Magnetism, Room 26–459, Carnegie Institution of Washington, Washington D.C., U.S.A.

* BURKE (Prof. J. A.), Department of Physics, University of Victoria, Victoria B.C., Canada

BUSBRIDGE (Miss Dr I. W.), 'Haremere' Westerham Road, Keston, Kent, BR 2 6HH, U.K.

BUSCOMBE (Dr W.), Astronomy Department, North-Western University, Evanston, Illinois 60201, U.S.A.

BUTLER (Dr H. E.), Royal Observatory, Blackford Hill, Edinburgh 9, Scotland, U.K.

* BYKOV (Dr M. F.), Astronomical Institute of the Usbek SSR Academy of Sciences, Tashkent, U.S.S.R.

* BYSTROV (Dr N.F.), Main Astronomical Observatory, USSR Academy of Sciences, Pulkovo, U.S.S.R.

* BYSTROVA (Mrs Dr N.V.), Main Astronomical Observatory, USSR Academy of Sciences, Pulkovo, U.S.S.R.

* CACCIANI (Dr A.), Osservatorio Astronomico di Roma, Rome, Italy

CALAMAI (Prof. G.), Osservatorio astrofisico di Arcetri, Firenze, Italy

CAMERON (Dr A. G. W.), Belfer Graduate School of Science, Yeshiva University, Amsterdam Ave & 186th Street, New York, N.Y. 10033, U.S.A.

CAMERON (Dr R. C.), NASA, Goddard Space Flight Center, Greenbelt, Maryland 20771, U.S.A.

CAMERON (Mrs W.), NASA, Goddard Space Flight Center, Greenbelt, Maryland 20771, U.S.A.

CAMICHEL (Dr H.), Observatoire de Toulouse, 31-Toulouse, France

CAMM (Dr G. L.), Department of Mathematics, The University, Manchester 13, England

CANAVAGGIA (Miss Dr R.), Observatoire de Paris, 61, avenue de l'Observatoire, Paris 14e, France

CANDY (M. P.), Perth Observatory, Bickley, Western Australia 6076

CAPRIOLI (Dr G.), Osservatorio astronomico di Monte Mario, Via Trionfale 204, Roma, Italy

CAPRIOTTI (Dr E.R.), Astronomy Department, The Ohio State University, 174 West 18th Avenue Columbus, Ohio 43210, U.S.A.

CARDÚS (J. O.), Astronome de l'Observatorio del Ebro, Roquetas, Tarragona, Spain

* CARLQVIST (P. A.), Royal Institute of Technology, Division of Plasma Physics, Stockholm 70, Sweden

CARPENTER (Dr L. H.), NASA, Goddard Space Flight Center, Greenbelt, Maryland 20771, U.S.A.

* CAROUBALOS (Dr C.), Observatoire de Meudon, 92-Meudon, France

CARR (Dr T. D.), Department of Physics, University of Florida, Gainesville, Florida, U.S.A.

CARRASCO (Dr R.), Director, Observatorio Astronómico Nacional de Madrid, Alfonso XII, 3 Madrid (7), Spain

CARROL (Sir John A.), 14, Belveder Grove, Wimbledon, London, S.W. 19, U.K.

* CARROLL (Prof. P. K.), Department of Physics, University College, Belfield, Dublin 4, Ireland

CARSON (Dr T. R.), University Observatory, Buchanan Gardens, St Andrews, Fife, Scotland, U.K.

CARVER (Prof. J. H.), Department of Physics, University of Adelaide, Adelaide, Australia

* CASTELLANI (Dr V.), Laboratorio di Astrofisica, Casella Postale 67, Frascati (Roma), Italy

* CATALANO (Dr S.), Osservatorio Astrofisico Città Universitaria, Viale A. Doria, 95123 Catania, Sicilia, Italy

* CAUGHLAN (Mrs Dr G.), Department of Physics, Montana State University, Bozeman, Montana 59715, U.S.A.

CAYREL (Dr R.), Observatoire de Paris, Section d'Astrophysique, 92-Meudon, France

CAYREL-DE-STROBEL (Mrs Dr G.), Institut d'Astrophysique, 98bis Boulevard Arago, Paris 14e, France

CECCARELLI (Prof. M.), Instituto di Fisica Generale, Università di Bologna, Italy

CECCHINI (Prof. G.), Direttore, Osservatorio Astronomico di Torino in Pino Torinese, Italy

CEPLECHA (Dr Z.), Astronomical Institute, Czechoslovak Academy of Sciences, Observatory Ondřejov, Czechoslovakia

CESCO (Dr C. U.), Observatorio Félix Aguilar, Facultad de Ingeniería, Universidad Nacional de Cuyo, San Juan, Argentina

CESCO (Dr R. P.), Observatorio Astronomico, La Plata, Argentina

CESTER (Prof. Dr B.), Osservatorio Astronomico, Via G.B. Tiepolo, 15, Trieste (409), Italy

CHALONGE (Dr D.), Institut d'Astrophysique, 98bis Boulevard Arago, Paris 14e, France

CHAMBERLAIN (Dr J. W.), Director, The Lunar Science Institute, 3303 NASA Road 1, Houston, Texas 77058, U.S.A.

CHANDRA (S.), General Electric Company, Room M 1325, P.O. 8555, Philadelphia, Pennsylvania, U.S.A.

CHANDRASEKHAR (Prof. S.), Laboratory for Astrophysics & Space Research, 933 East 56th Street, Chicago, Illinois 60637, U.S.A.

* CHAPMAN (Dr R. D.), Solar Physics Group, Goddard Space Flight Center, Greenbelt, Maryland 20771, U.S.A.

* CHAPRONT (Dr J.), Bureau des Longitudes, 3, Rue Mazarine, Paris 8e, France

CHARVIN (P.), Observatoire de Meudon, 5, place Janssen, 92-Meudon, France

CHAVIRA (E.), Observatorios de Tonantzintla y Tacubaya, Apdo Postal 25264, C. Universitaria, México 20, D.F., Mexico

CHEBOTAREV (Prof. G. A.), Institute of Theoretical Astronomy, Leningrad B-164, U.S.S.R.

CHEKIRDA (Dr A. T.), Khar'kov Observatory, Khar'kov, U.S.S.R.

* CHEN (Dr Kwan-Yu), Astronomy Department, Rosemary Hill Observatory, University of Florida, Gainesville, Florida 32601, U.S.A.

CHENAKAL (Dr V. L.), Lomonosov Museum of the Academy of Sciences of the USSR, Leningrad, U.S.S.R.

* CHEREDNICHENKO (Dr V. I.), Kiev Polytechnical Institute, Kiev 56, U.S.S.R.

* CHEREPASHCHUK (Dr A. M.), Sternberg Astronomical Institute, Moscow, U.S.S.R.

CHERNEGA (Dr N. A.), Astronomical Observatory, Kiev University, Kiev, U.S.S.R.

* CHERTOPRUD (Dr V. E.), Astronomical Council, USSR Academy of Sciences, Vavilov Str. 34, Moscow V-312, U.S.S.R.

CHINCARINI (G.), Mc Donald Observatory at Mount Locke, P.O. Box 1337, Fort Davis, Texas 79734, U.S.A.

CHIS (Prof. G.), Observatoire Astronomique, Rue de la République 109, Cluj, Roumania

* CHITRE (Dr S. N.), Tata Institute of Fundamental Research, Homi Bhabha Road, Colaba, Bombay-5, India

CHIU (Dr H. Y.), Institute for Space Studies, 475 Riverside Drive, New York, New York 10027, U.S.A.

CHIUDERI (Dr F. D.), Osservatorio Astrofisico di Arcetri, Largo E. Fermi, 5, Firenze, Italy

CHOPINET (Miss M.), Observatoire de l'Université de Bordeaux, 33-Floirac, France

CHRISTIANSEN (Prof. W. N.), School of Electrical Engineering, The University of Sydney, Sydney, N.S.W., Australia, 2006

CHRISTY (Prof. R. F.), Physics Department, California Institute of Technology, Pasadena, California, U.S.A.

CHUBB (Dr T. A.), 5023 N 38th Street, Arlington 7, Virginia, U.S.A.

CHUDOVICHEVA (Mrs Dr N. A.), Engelhardt Observatory, Kazan, U.S.S.R.

CHUGAJNOV (Dr P. F.), Crimean Astrophysical Observatory, USSR Academy of Sciences, Crimea, U.S.S.R.

* CHURMS (J.), Royal Observatory, Observatory, Cape, South Africa

CHUVAEV (Dr K. K.), Crimean Astrophysical Observatory, USSR Academy of Sciences, P/O Nauchny, Crimea, U.S.S.R.

CHVOJKOVÁ (Dr E.), Astronomical Institute, Czechoslovak Academy of Sciences, Budečská 6, Prague 2, Czechoslovakia

* CID PALACIOS (Dr R.), Facultad de Ciencias, Ciudad Universitaria, Zaragoza, Spain

CILLIÉ (Prof. G. G.), Department of Mathematics, University of Stellenbosch, Cape Province, South Africa

CIMINO (Prof. Dr M.), Direttore, Osservatorio Astronomico su Monte Mario, Rome, Italy

CLARK (Dr B. G.), National Radio Astronomy Observatory, P.O. Box 2, Green Bank, West Virginia 24944, U.S.A.

CLARK (Dr G.), Physics Department, Massachusetts Institute of Technology, Cambridge, Massachusetts 02139, U.S.A.

CLAYTON (Dr D. D.), Space Science Department, Rice University, Houston, Texas, U.S.A.

CLEMENCE (Dr G. M.), Yale University Observatory, Box 2023, Yale Station, New Haven, Connecticut, U.S.A.

* CLEMENT (Prof. Dr M. J.), David Dunlap Observatory, Richmond Hill, Ontario, Canada

CLIMENHAGA (Dr J. L.), Department of Physics, University of Victoria, Victoria, B.C., Canada

* CLUBE (Dr S. V. M.), Royal Greenwich Observatory, Herstmonceux Castle, Hailsham, Sussex, U.K.

* COCKE (Dr J.), Steward Observatory, University of Arizona, Tucson, Arizona 85721, U.S.A.

CODE (A. D.), 6524 Sterling Hall, University of Wisconsin, Madison, Wisconsin, U.S.A.

* COELHO BALSA (Dr M. C.), Observatório Astronómico de Universidade de Coimbra, Coimbra, Portugal

* COHEN (Dr J. M.), Institute for Advanced Study, School of Natural Sciences, Princeton, New Jersey 08540, U.S.A.

COHEN (Prof. M. H.), Department of Astronomy, California Institute of Technology, Pasadena, California 91109, U.S.A.

* COLBURN (Dr D. S.), Ames Research Center, Moffet Field, California 94035, U.S.A.

COLGATE (Dr S.), New Mexico Institute of Mining and Technology, Socorro, New Mexico, U.S.A.

COLLINDER (Dr P.), Odensgatan 5 B, S-752 22 Uppsala, Sweden

COLLINS II (Dr G. W.), Department of Astronomy, Ohio State University, Columbus, Ohio 43210, U.S.A.

COLLINSON (E. H.), Culpho End, Culpho, Ipswich, Suffolk, U.K.

* COLOMBO (Prof. Dr G.), Instituto Meccanica Applicata, Universita di Padova, via F. Marzolo 9, Padova, Italy

CONDON (Dr E. U.), Joint Institute for Laboratory Astrophysics, University of Colorado, Boulder, Colorado 80310, U.S.A.

CONNES (Mrs J.), Service de Calcul Numérique, Observatoire de Meudon, 5, place Janssen, 92-Meudon, France

CONNES (P.), Laboratoire du C.N.R.S., Bellevue, France

CONTI (Dr P. S.), Lick Observatory, University of California, Santa Cruz, California 95060, U.S.A.

CONTOPOULOS (Prof. Dr G.), Astronomical Department, University of Thessaloniki, Thessaloniki, Greece. Assistant General Secretary of the Union.

CONWAY (Dr R. G.), Nuffield Radio Astronomy Laboratories, Jodrell Bank, Macclesfield, Cheshire, U.K.

COOK (Dr A. F.), Smithsonian Astrophysical Observatory, Cambridge, Massachusetts 02138, U.S.A.

COOK (Dr A. H.), Department of Geophysics, University of Edinburgh, 6 South Oswald Road, Edinburgh, EH9 2HX, U.K.

* COOKE (Dr B. A.), X-Ray Astronomy Group, Physics Department, Leicester University, Leicester, U.K.

* COOPER (B.F.C.), CSIRO, Division of Radiophysics, P.O. Box 76, Epping 2121, Australia

CORLISS (Dr C. H.), National Bureau of Standards, Washington, D.C. 20234, U.S.A.

COSTAIN (Dr C. H.), Dominion Radio Astrophysical Observatory, P.O. Box 248, Penticton, B.C., Canada

COUDER (Dr A.), Observatoire de Paris, 61 Avenue de l'Observatoire, Paris 14e, France

COUDERC (Dr P.), Observatoire de Paris, 61 Avenue de l'Observatoire, Paris 14e, France

COURTÈS (Dr G.), Observatoire de Marseille, 2 Place Le Verrier, 23 Marseille-IV, France

COUSINS (Dr A. W. J.), Royal Observatory, Observatory, C.P., South Africa

COUTEAU (Dr P. C.), Observatoire de Nice, Le Mont-Gros, 06-Nice, France

COUTREZ (Prof. Dr R. A. J.), Directeur, Institut d'Astronomie et Astrophysique, Université Libre de Bruxelles, avenue F.D. Roosevelt, 50, B-1050 Bruxelles, Belgium

COVINGTON (A. E.), Astrophysics Branch, National Research Council, Ottawa, Ontario, Canada

COWLEY (Mrs Dr A. P.), Astronomy Department, University of Michigan, Ann Arbor, Michigan 48104, U.S.A.

* COWLEY (Dr Ch.), Astronomy Department, University of Michigan, Ann Arbor, Michigan 48104, U.S.A.

COWLING (Prof. T. G.), Department of Mathematics, The University, Leeds 2, U.K.

Cox (Dr A. N.), Box 1663, Los Alamos, New Mexico, U.S.A.

Cox (Prof. Dr J. F. J. G.), Recteur Honoraire, Université Libre de Bruxelles, avenue Louise, 351, B-1050 Bruxelles, Belgium

Cox (Dr J. P.), Joint Institute for Laboratory Astrophysics, 1511 University Avenue, Boulder, Colorado, U.S.A.

* COYNE (Dr G. V., S.J.), Specola Vaticana, Vatican City State

* CRAMPTON (Dr D.), Dominion Astrophysical Observatory, R.R. 7, Victoria B.C., Canada

CRAWFORD (Dr D. L.), Kitt Peak National Observatory, 950 North Cherry Avenue, Tucson, Arizona, U.S.A.

CRISTALDI (Dr S.), Piazza Vaccarini 11, Catania, Italy

CRISTESCU (Dr C.), Observatoire Astronomique, rue Cutitul de Argint 5, Bucarest 28, Roumania

* CRUVELLIER (Dr P.), Observatoire de Marseille, Place le Verrier, 13-Marseille (4), France

CSADA (Dr I. K.), Konkoly Observatory, Konkoly Thege U. 13–17, Budapest, XII, Hungary

CUDABACK (Dr D. D.), Radio Astronomy Laboratory, 633 Campbell Hall, University of California, Berkeley, California 94720, U.S.A.

CUFFEY (J.), New Mexico State University, Department of Earth Sciences and Astronomy, University Park, New Mexico 88001, U.S.A.

* CULHANE (Dr J. L.), Palo Alto Research Laboratory, 3241 Hanover Street, Palo Alto, California 94304, U.S.A.

CUNNINGHAM (Dr L. E.), Department of Astronomy, University of California, Berkeley 4, California 94720, U.S.A.

* CUNY (Miss Dr Y.), Observatoire de Meudon, 92-Meudon, France

* CUPERMAN (Dr S.), Department of Physics and Astronomy, Tel-Aviv University, Ramat Aviv, Israel

CUREA (Prof. Dr I.), Recteur de l'Université de Timisoara, Timisoara, Roumania

CZYZAK (Prof. S. J.), Astronomy Department, The Ohio State University, Columbus, Ohio 43210, U.S.A.

* DACHS (Dr J.), Astronomisches Institut der Ruhr-Universität, Postfach 2148, 4630 Bochum-Querenburg, F.R.G.

DADAEV (Dr A. N.), Pulkovo Observatory, Leningrad, U.S.S.R.

DADIĆ (Dr Ž.), Institut za povijest nauka Jazu, Demetrova 18, Zagreb 3, Yugoslavia

DAENE (Dr H.), Heinrich-Hertz-Institut für Solar-Terrestrische Physik, Berlin-Adlershof, Rudower Chaussee, G.D.R.

* DAINTREE (Dr E. J.), Nuffield Radio Astronomy Laboratories, Jodrell Bank, Macclesfield, Cheshire, U.K.

* DALGARNO (Prof. A.), Smithsonian Astrophysical Observatory, 60 Garden Street, Cambridge, Massachusetts 02138, U.S.A.

DALLAPORTA (Prof. N.), Instituto di Fisica Teorica, Università di Padova, Italy

DAMBARA (Dr T.), Geographical Survey Institute, 7-1000 Kamimeguro, Meguro-ku, Tokyo, Japan

DANBY (Dr J. M. A.), Yale University Observatory, Box 2023, Yale Station, New Haven, Connecticut, U.S.A.

DANIELSON (Dr R. E.), Princeton University Observatory, Peyton Hall, Princeton, N.J. 08540, U.S.A.

DANZIGER (Dr I. J.), Harvard College Observatory, 60 Garden Street, Cambridge, Massachusetts 02138, U.S.A.

DA SILVA (Dr A. V. C. S.), Observatorio Astronomico da Universidade Santa Clara, Coimbra, Portugal

DA SILVA (Prof. Dr I. F.), Faculdade de Ciencias, Lisbon, Portugal

DAUBE-KURZEMNIECE (Mrs Dr I. A.), Astrophysical Laboratory of the Academy of Sciences of the Latvian SSR, Riga, U.S.S.R.

* DAUTCOURT (Dr G.), Zentralinstitut für Astrophysik, Telegrafenberg, 15 Potsdam, G.D.R.

* DAVIDSON (Prof. W.), University of Otago, P.O. Box 56, Dunedin, New Zealand

DAVIES (Dr J. G.), Nuffield Radio Astronomy Laboratories, Jodrell Bank, Macclesfield, Cheshire, U.K.

DAVIES (Dr R. D.), Nuffield Radio Astronomy Laboratories, Jodrell Bank, Macclesfield, Cheshire, U.K.

DAVIS (Dr J.), Department of Physics, University of Sydney, Sydney 2006 N.S.W., Australia

DAVIS JR (Prof. L.), Physics Department, California Institute of Technology, Pasadena, California 91109, U.S.A.

* DAVIS (Dr M. M.), NRAO, Edgemont Road, Charlottesville, Va. 22901, U.S.A.

DAVIS (Prof. M. S.), The University of North Carolina, Department of Physics, Chapel Hill, N.C. 27514, U.S.A.

* DAVIS (Dr R. J.), Smithsonian Astrophysical Observatory, 60 Garden Street, Cambridge, Massachusetts 02138, U.S.A.

* DAWE (Dr A. J.), Department of Astronomy, Leicester University, Leicester, LE1 7RH, U.K.

DÉBARBAT (Miss S.), Observatoire de Paris, 61, avenue de l'Observatoire, Paris 14e, France

DEBEHOGNE (Dr H.), Observatoire Royal de Belgique, avenue Circulaire, 3, B-1180 Bruxelles, Belgium

DECAUX (B.), Bureau International de l'Heure, Observatoire de Paris, 61, avenue de l'Observatoire, Paris 14e, France

DEEMING (Dr T. J.), Astronomy Department, 404 Physics Building, University of Texas, Austin, Texas 78712, U.S.A.

De FEITER (Dr L.D.), Astronomical Observatory 'Sonnenborgh', Zonnenburg 2, Utrecht, The Netherlands

* DE GRAAF (Dr T.), Institute for Theoretical Physics, Westersingel 34, Groningen, The Netherlands

DE GRAAFF (Dr W.), Laboratorium voor Ruimte-Onderzoek, Beneluxlaan 21, Utrecht, The Netherlands

* DE GROOT (M.J.H.), European Southern Observatory, Bergedorfer Strasse 131, 205 Hamburg 80, F.R.G.

DE GROOT (Dr T.), Sterrewacht 'Sonnenborgh', Servaas Bolwerk 13, Utrecht, The Netherlands

DEINZER (Dr W.), Universitäts-Sternwarte, 34 Göttingen, Geismarlandstrasse 11, F.R.G.

DE JAGER (Prof. Dr C.), Director, The Astronomical Institute, Space Research Laboratory, Beneluxlaan 21, Utrecht, The Netherlands. General Secretary of the Union.

* DE JONGE (Dr J. K.), Department of Astronomy, University of Pittsburgh, Riverview Park, Pittsburgh, Pennsylvania 15214, U.S.A.

DE KOCK (R. P.), Royal Observatory, Observatory, C.P., South Africa

DE KORT (S. J.) (Rev. Dr J.), Department of Astronomy, Faculty of Science, Catholic University, Driehuizerweg 200, Nijmegen, The Netherlands

* DELACHE (Dr Ph.), Observatoire de Nice, Le Mont-Gros, 06-Nice, France

DELANNOY (J.), Observatoire de Paris, Section d'Astrophysique, 92-Meudon, France

DELBOUILLE (Prof. Dr L.), Institut d'Astrophysique, Université de Liège, avenue de Cointe, 5 B-4200 Cointe-Ougrée, Belgium

DELHAYE (Prof. J.), Observatoire de Paris, 61, avenue de l'Observatoire, Paris-14e, France

* DE LOORE (Dr C. W. H.), Astrofysisch Instituut, Vrije Universiteit Brussel, F.D. Rooseveltlaan, 50, B-1050 Brussel, Belgium

DELSEMME (Prof. A. H.), Department of Physics & Astronomy, The University of Toledo, Toledo, Ohio 43606, U.S.A.

DE MARCUS (Prof. W. C.), Department of Physics, University of Kentucky, Lexington, Kentucky, U.S.A.

DEMARQUE (Prof. P. R.), Yale University Observatory, Box 2023, Yale Station, New Haven, Connecticut 06520, U.S.A.

* DEMENKO (Mrs Dr A. A.), Astronomical Observatory of the University, Ukrainian SSR, Kiev, U.S.S.R.

* DEMERS (Prof. Dr S.), Institute of Astronomy, Laurentian University, Sudbury, Ontario, Canada

DE MORAES (Dr A.), Caixa Postal 30627, São Paulo, SP, Brazil

DE MOTTONI (Dr G.), Via Fratelli Rosselli 15/23, Genova, Italy

DENISSE (Dr J. F.), Directeur de l'Observatoire de Paris, 61, avenue de l'Observatoire, Paris-14e, France

DENNISON (Dr E. W.), Mount Wilson and Palomar Observatories, 813 Santa Barbara Street, Pasadena, California, U.S.A.

DENT (Dr W. A.), Physics and Astronomy Department, University of Massachusetts, Amherst, Massachusetts 01002, U.S.A.

DEPRIT (Dr A.), Professor at the University of Louvain; on leave at Boeing Scientific Research Laboratories, P.O. Box 3981, Seattle, Washington 98124, U.S.A.

* DE SABBATA (Prof. Dr V.), Instituto di Fisica, Università di Bologna, via Irnerio 46, Bologna, Italy

* DE SOUSA PEREIRA OSÓRIO (Dr J. J.), Observatório Astronómico Universidade de Porto, Monte da Virgem, Vila Nova de Gaia, Portugal

DESSY (Dr J. Landi), Observatorio Nacional Argentino, Cordoba, Argentina

DETRE (Dr L.), Director, Konkoly Observatory, Szabadsághegy, Budapest XII, Hungary

* DEUBNER (Dr F. L.), Fraunhofer Institut, Schöneckstrasse 6, 7800 Freiburg, F.R.G.

DEUTSCH (Prof. Dr A. N.), Pulkovo Observatory, Leningrad-M 140, U.S.S.R.

DE VAUCOULEURS (Prof. G.), Department of Astronomy, University of Texas, Austin 12, Texas 78412, U.S.A.

* DE VEGT (Dr Ch.), Sternwarte Bergedorf, Gojenbergsweg 112, 205 Hamburg 80, F.R.G.

DEWHIRST (Dr D. W.), The Observatories, Madingley Road, Cambridge, U.K.

DE WITT (Dr J. H.), Arthur J. Dyer Observatory, Vanderbilt University, Nashville 5, Tennessee, U.S.A.

DEZSÖ (Prof. L.), Heliophysical Observatory, Hungarian Academy of Sciences, Debrecen 10, Hungary

DIBAY (Dr E. A.), Sternberg Astronomical Institute, Moscow, U.S.S.R.

DICK (Prof. Dr J.), Rote Kreuz-Strasse 7, Potsdam-Babelsberg 2, G.D.R.

* DICKE (Dr R. H.), P.O. Box 708, Jadwin Hall, Princeton University, Princeton, New Jersey 08540, U.S.A.

* DICKEL (Mrs Dr H. R.), Department of Astronomy, University of Illinois Observatory, Urbana, Illinois 61801, U.S.A.

DICKEL (Dr J. R.), University of Illinois Observatory, Urbana, Illinois 61801, U.S.A.

* DICKENS (R. J.), Royal Greenwich Observatory, Herstmonceux Castle, Hailsham, Sussex, England

DIECKVOSS (Prof. Dr W.), Hamburger Sternwarte, Hamburg-Bergedorf, F.R.G.

DIETER (Mrs Dr N. H.), Radio Astronomy Laboratory, University of California, Berkeley, California 94720, U.S.A.

DIMITROFF (Prof. G. Z.), R.F.D. n° 1, Hartland, Vermont 05048, U.S.A.

DIMOV (Dr N. A.), Crimean Astrophysical Observatory, USSR Academy of Sciences, Crimea, U.S.S.R.

* DINESCU (Dr A.), Observatorul Astronomic, Strada Cutítul de Argint 5, Bucuresti 28, Roumania

DINGENS (Prof. Dr P. S. A.), Sterrenkundig Instituut, Rijksuniversiteit Gent, J. Plateaustraat, 22, B-9000 Gent, Belgium

DINGLE (Prof. H.), 104, Downs Court Road, Purley, Surrey, U.K.

DINULESCU (Prof. N.), Rue Corbeni, 30, Secteur 2, Bucarest, Roumania

* DIRIKIS (Dr M. A.), Astronomical Observatory of the Latvian University, Latvian SSR, Riga, U.S.S.R.

DIVAN (Miss Dr L.), Institut d'Astrophysique, 98bis Boulevard Arago, Paris-14e, France

DIVARI (Dr N. B.), Odessa Politechnical Institute, Odessa, U.S.S.R.

DIXON (Dr M. E.), Department of Astronomy, University of Edinburgh, Edinburgh, U.K.

DIZER (Prof. Dr M.), Astronomical Observatory, Kandilli, Çengelköy, Istanbul, Turkey

DJURKOVIĆ (P. M.), Astronomical Observatory, Volgina 7, Beograd, Yugoslavia

* DLUZHNEVSKAYA (Mrs Dr O. B.), Astronomical Council, USSR Academy of Sciences, Vavilov Street 34, Moscow V-312, U.S.S.R.

* DOAN (Dr N. H.), Observatoire de Lyon, 69, St. Denis Laval, France

DOAZAN (Mrs V.), Observatoire de Paris, 61, avenue de l'Observatoire, 75-Paris-14e, France

* DOBRITSCHEV (V. M.), Department of Astronomy, '7th November' Street 1, Sofia, Bulgaria

DOBRONRAVIN (Dr P. P.), Crimean Astrophysical Observatory, USSR Academy of Sciences, P/O Nauchny, Crimea, U.S.S.R.

DOBROVOL'SKIJ (Dr O. V.), Institute of Astrophysics, Dushanbe, U.S.S.R.

DOBRZYCKI (Dr J.), Institute of the History of Science and Technics, Nowy Swiat 72, Warszawa, Poland

DODSON-PRINCE (Mrs Dr H. W.), McMath Hulbert Observatory, University of Michigan, Pontiac 4, Michigan, U.S.A.

DOĞAN (Dr N.), Faculty of Science, Department of Astronomy, Ankara, Turkey

* DOHERTY (L. H.), Astrophysics Branch National Research Council, Ottawa 7, Ontario, Canada

DOHERTY (Dr L. R.), Astronomy Department, Sterling Hall, University of Wisconsin, Madison, Wisconsin 53706, U.S.A.

* DOLGINOV (Prof. Dr A. Z.), Physical-Technical Institute, USSR Academy of Sciences, Leningrad, U.S.S.R.

DOLIDZE (Dr M. V.), Abastumani Observatory, Abastumani, Georgia, U.S.S.R.

DOLLFUS (Dr A.), Observatoire de Paris, Section d'Astrophysique, 92-Meudon, France

DOMBROVSKIJ (Dr V. A.), Leningrad University Observatory, Leningrad, U.S.S.R.

DOMINKO (Prof. Dr F.), Saranovićeva 11, Ljubljana, Yugoslavia

DOMMANGET (Dr J.), Chef du Département d'Astrométrie et de Mécanique Céleste, Observatoire Royal de Belgique, avenue Circulaire, 3, B-1180 Bruxelles, Belgium

DONITCH (N.), address unknown

DONN (Dr B. D.), Code 613, Goddard Space Flight Center, Greenbelt, Maryland, U.S.A.

* DOROSHKEVICH (Dr A. G.), Institute of Applied Mathematics, USSR Academy of Sciences, Moscow, U.S.S.R.

* DORSCHNER (Dr J.), University Observatory, Schillergässchen 2, 69 Jena, G.D.R.

DOS REIS (Prof. M.), Director of the Observatory, Coimbra, Portugal

DOSSIN (Dr F.), European Southern Observatory, Am Bahnhof 21, 205 Hamburg 80, F.R.G.

* DOUGHTY (Dr N. A.), Department of Physics, University of Canterbury, Private Bag, Christchurch, New Zealand

DOUGLAS (Dr A. E.), Division of Physics, National Research Council, Ottawa, Ontario, Canada

DOUGLAS (Miss Dr A. V.), 127 King St. W., Kingston, Ontario, Canada

DOUGLAS (Prof. J. N.), Department of Astronomy, The University of Texas, Austin, Texas, U.S.A.

DRAKE (Prof. F. D.), Department of Astronomy, Cornell University, Ithaca, New York 14850, U.S.A.

DRÂMBĂ (Prof. Dr C.), Observatoire de Bucarest, 5 Rue Cutitul de Argint, Bucarest 28, Roumania

DRESSLER (Dr K.), Labor. Phys. Chem. E.T.H., Zürich, Switzerland

DROFA (Dr V. K.), Astronomical Observatory, Kiev University, Kiev, U.S.S.R.

* DUBINSKIJ (Dr B. A.), Scientific Council for 'Radio Astronomy', USSR Academy of Sciences, Moscow, U.S.S.R.

DUBOSHIN (Prof. Dr G. N.), Sternberg State Astronomical Institute, Moscow, U.S.S.R.

DUBOV (Dr E. E.), Crimean Astrophysical Observatory, USSR Academy of Sciences, P/O Nauchny, Crimea, U.S.S.R.

DUCHESNE (Dr M.), Observatoire de Paris, 61, Avenue de l'Observatoire, Paris-14e, France

DUFAY (M.), Institut de Physique Générale, Faculté des Sciences de Lyon, 18 quai Claude Bernard, Lyon, France

DUFLOT (Mrs Dr M.), Observatoire de Marseille, 2, Place Le Verrier, Marseille-4e, France

DUFLOT (R.), Chargée de Recherches CNRS, Observatoire de Marseille, 2, place Le Verrier, 13-Marseille-4e, France

* DULK (Dr G. A.), Department of Astro-Geophysics, University of Colorado, Boulder, Colorado 80302, U.S.A.

DUMONT (R.), Observatoire de Bordeaux, 33-Floriac, France

* DUMONT (Miss Dr S.), Observatoire de Meudon, 92-Meudon, France

DUNCOMBE (Dr R. L.), Director, Nautical Almanac Office, U.S. Naval Observatory, Washington D.C. 20390, U.S.A.

DUNHAM JR (Dr T.), Department of Physics, University of Tasmania, Hobart, Tasmania, Australia

* DUNKELMAN (Dr L.), Astrophysics Branch, Code 613, Goddard Space Flight Center, Greenbelt, Maryland 20771, U.S.A.

DUNN (Dr R. B.), Sacramento Peak Observatory, Sunspot, New Mexico, U.S.A.

* DURNEY (Dr B.), High Altitude Observatory, Boulder, Colorado 80302, U.S.A.

* DVORYASHIN (Dr A. S.), Crimean Astrophysical Observatory, USSR Academy of Sciences, Crimea, U.S.S.R.

DYER (Dr E. R.), National Academy of Sciences, 2101 Constitution Avenue, N.W., Washington 25, D.C., 20037, U.S.A.

* DYSON (Dr F. J.), Institute for Advanced Studies, Princeton, New Jersey 08540, U.S.A.

* DYSON (Dr J. E.), Department of Astronomy, The University, Manchester, M13 9PL, U.K.

DZHAPIASHVILI (Dr V. P.), Vice-Director, Abastumani Astrophysical Observatory, Abastumani, Georgia, U.S.S.R.

DZIGVASHVILI (Dr R. M.), Abastumani Astrophysical Observatory, Georgian Academy of Sciences, Abastumani, U.S.S.R.

EBBIGHAUSEN (Prof. E. G.), Department of Physics, University of Oregon, Eugene, Oregon, U.S.A.

ECKERT (Dr W. J.), 216 Leonia Avenue, Leonia, New Jersey 07605, U.S.A.

EDDY (Dr J.), High Altitude Observatory, Boulder, Colorado 80302, U.S.A.

EDLÉN (Prof. B.), Institute of Physics, Sölvegatan 14, S-223 62 Lund, Sweden

EDMONDS JR (Dr F. N.), Department of Astronomy, The University of Texas, Austin, Texas 78712, U.S.A.

EDMONDSON (Dr F.K.), Swain Hall West 319A, Indiana University, Bloomington, Indiana 47401, U.S.A.

EELSALU (Dr H.), W. Struve Astrophysical Observatory, Tartu, Tôravere, Estonia, U.S.S.R.

EFREMOV (Dr Ju. I.), Astronomical Council, USSR Academy of Sciences, Vavilov Str. 34, Moscow V-312, U.S.S.R.

EGGEN (Dr O. J.), Director, Mount Stromlo Observatory, Canberra, A.C.T., Australia

* EGGLETON (Dr P. P.), Institute of Theoretical Astronomy, Madingley Road, Cambridge, U.K.

EICHHORN-VON WURMB (Prof. H. K.), Department of Astronomy, University of South Florida, Tampa, Florida 33549, U.S.A.

EINASTO (Dr J.), W. Struve Astrophysical Observatory, Tartu, Tôravere, Estonia, U.S.S.R.

ELFORD (Dr W. G.), Department of Physics, University of Adelaide, Adelaide, Australia

ELGARÖY (Dr Ö.), Institute of Theoretical Astrophysics, University of Oslo, Blindern, Oslo 3, Norway

ELLIOTT (Dr I.), Dunsink Observatory, Castleknock, Co. Dublin, Ireland

* ELLIS (Dr G. F. R.), Department of Applied Mathematics and Theoretical Physics, Silver Street, Cambridge, U.K.

ELLIS (Prof. G. R. A.), University of Tasmania, P.O. Box 2520 Hobart, Tasmania, Australia

ELLYETT (Prof. C. D.), Department of Physics, University of Newcastle, Newcastle, Australia

ELSÄSSER (Prof. Dr H.), Max-Planck-Institut für Astronomie und Landessternwarte, 6900 Heidelberg-Königstuhl, F.R.G.

ELSMORE (B.), Mullard Radio Observatory, Cavendish Laboratory, Free School Lane, Cambridge, U.K.

ELSTE (Dr G.), Department of Astronomy, University of Michigan, Ann Arbor, Michigan 48104, U.S.A.

* ELSTON (Prof. W. E.), University of New Mexico, Department of Geology, Albuquerque, New Mexico 87106, U.S.A.

ELVEY (Dr C. T.), 5359 E. Hawthorne Street, Tucson, Arizona 85711, U.S.A.

ELVIUS (Mrs Dr A. M.), Stockholm Observatory, S-13300 Saltsjöbaden, Sweden

ELVIUS (Prof. T.), Director, Astronomical Observatory, S-22224 Lund, Sweden

ELWERT (Prof. Dr G.), Astronomisches Institut der Universität, Gmelinstrasse 6, Tübingen, F.R.G.

* EL'YASBERG (Prof. Dr P. E.), Institute of Cosmical Research, USSR Academy of Sciences, Moscow, U.S.S.R.

EMINZADE (Dr T. A.), Shemakha Astrophysical Observatory, Azerbajdzhan Academy of Sciences, Baku, U.S.S.R.

ENGELHARD (Dr E.), Physikalisch-Technische Bundesanstalt (PTB), Bundesallee 100, 33 Braunschweig, F.R.G.

* ENGIN (Dr S.), Faculty of Science, Department of Astronomy, Ankara, Turkey

* ENGVOLD (O.), Institute of Theoretical Astrophysics, University of Oslo, P.O. Box 1029, Blindern, Oslo 3, Norway

ENSLIN (Dr H.), Deutsches Hydrographisches Institut, 2 Hamburg 4, F.R.G.

EPSTEIN (Dr E. E.), Aerospace Corporation, P.O. Box 95085, Los Angeles, California 90045, U.S.A.

EPSTEIN (Dr I.), Rutherfurd Observatory, Columbia University, New York 27, N.Y., U.S.A.

ERICKSON (Dr W. C.), University of Maryland, Department of Physics and Astronomy, College Park, Maryland, U.S.A.

ERIKSEN (G.), Institute of Theoretical Astrophysics, University of Oslo, Blindern, Oslo 3, Norway

ERPYLEV (Dr N. P.), Astronomical Council, USSR Academy of Sciences, Vavilov Str. 34, Moscow V-312, U.S.S.R.

* ERUSHEV (Dr N. N.), Crimean Astrophysical Observatory, USSR Academy of Sciences, Crimea, U.S.S.R.

ESHLEMAN (Prof. V. R.), Radioscience Laboratory, Stanford University, Stanford, California, U.S.A.
* ESIPOV (Dr V. F.), Sternberg Astronomical Institute, Moscow, U.S.S.R.
ESKIOĞLU (Dr A. N.), P. K. 7, Fener–Istanbul, Turkey
ESSEN (Dr L.), National Physical Laboratory, Teddington, Middlesex, U.K.
EVANS (Prof. D. S.), Department of Astronomy, University of Texas, Austin, Texas 78712, U.S.A.
* EVANS (Dr J. V.), MIT Lincoln Laboratory, Millstone/Haystack Observatory, Lexington, Massachusetts 02173, U.S.A.
EVANS (Dr J. W.), Director, Sacramento Peak Observatory, Sunspot, New Mexico, U.S.A.
EVDOKIMOV (Dr Yu. V.), Astronomical Observatory, Kazan University, Kazan, U.S.S.R.
* EVERHART (Dr E.), Department of Physics, University of Denver, 2115 S. University Blvd, Denver, Colorado 80210, U.S.A.
EWEN (Dr H. I.), Ewen Knight Corporation, Oak and Pine Streets, East Natick, Massachusetts, U.S.A.
EZER (Mrs Dr D.), Research Associate, Institute for Space Studies, 475 Riverside Drive, New York 27, N.Y., U.S.A.
EZERSKIJ (Dr V. I.), Astronomical Observatory, Kharkov University, Kharkov, U.S.S.R.

FABRE (Dr H.), Observatoire de Nice, Le Mont Gros, 06-Nice, France
* FALCIANI (Dr R.), Osservatorio Astronomico di Capodimonte, Via Moiariello 16, 80131 Naples, Italy
* FANTI (Dr R.), Physical Institute of the University of Bologna, via Irnerio 46, Bologna, Italy
* FARAGGIANA (Miss Dr R.), Astronomical Observatory Trieste, via G.B. Tiepolo 11. Trieste, Italy
FATCHIKHIN (Dr N. V.), Pulkovo Observatory, Leningrad, U.S.S.R.
FAULKNER (Dr J.), Lick Observatory, University of California, Santa Cruz, California 95060, U.S.A.
FAZIO (Dr G. G.), Smithsonian Astrophysical Observatory, 60 Garden Street, Cambridge, Massachusetts 02138, U.S.A.
FEAST (Dr M. W.), Radcliffe Observatory, P.O. Box 373, Pretoria, South Africa
FEDOROV (Prof. E. P.), The Main Astronomical Observatory of the Ukrainian Academy of Sciences, Kiev-127, Goloseevo, U.S.S.R.
FEDYNSKIJ (Prof. Dr V. V.), Astronomical Council, USSR Academy of Sciences, Vavilov Str. 34, Moscow V-312, U.S.S.R.
FEHRENBACH (Prof. Ch.), Directeur de l'Observatoire de Marseille, 2, place le Verrier, 13-Marseille-4e, France
FEINSTEIN (Dr A.), Observatorio Astronómico, La Plata, Argentina
* FEIX (Dr G.), Astronomisches Institut der Ruhr-Universität, Postfach 2148, 4630 Bochum-Querenburg, F.R.G.
* FELDMAN (Dr U.), Department of Physics and Astronomy, Tel-Aviv University, Ramat-Aviv, Israel
FELENBOK (P.), Observatoire de Paris, Section d'Astrophysique, 92-Meudon, France
FELLGETT (Prof. P. B.), Department of Applied Physical Sciences, University of Reading, Whiteknights Park, Reading, Berks., U.K.
FELTEN (Dr J. E.), Steward Observatory, University of Arizona, Tucson, Arizona 85721, U.S.A.
FENKART (Dr R. P.), Astronomisches Institut der Universität Basel, Venusstrasse 7, CH-4102 Binningen, Switzerland
FERNIE (Dr J.D.), David Dunlap Observatory, Richmond Hill, Ontario, Canada
FERRARI D'OCCHIEPPO (Prof. Dr K.), Universitäts-Sternwarte, Türkenschanzstrasse 17, Wien 18/110, Austria
FERRARO (Prof. V. C. A.), Queen Mary College, University of London, Mile End Road, London E.1, U.K.

FERRAZ-MELLO (Dr S.), Instituto Tecnológico de Aeronáutica, Observatório, Sao José dos Campos, São Paulo, Brazil

FESENKOV (Prof. Dr V. G.), Meteorite Committee, USSR Academy of Sciences, Stroiteley Street 3, Building 1, Doorway 2, Moscow V-313, U.S.S.R.

FESENKOVA (Mrs Prof. Dr E. V.), Astronomical Council, Vavilov Str. 34, USSR Academy of Sciences, Moscow V-312, U.S.S.R.

FICHERA (Prof. E.), Osservatorio Astronomico di Capodimonte, Napoli, Italy

FICHTEL (Dr C. E.), NASA, Goddard Space Flight Center, Greenbelt, Maryland 20771, U.S.A.

FIELD (Dr G. B.), Department of Astronomy, University of California, Berkeley, California 94720, U.S.A.

FIELDER (Dr G.), Lunar and Planetary Unit, Department of Environmental Sciences, University of Lancaster, Bailrigg, Lancaster, U.K.

FILLIOZAT (Dr J.), Professeur au Collège de France, 11, place Marcelin Berthelot, 75-Paris-5e, France

FINDLAY (Dr J. W.), National Radio Astronomy Observatory, Edgemont Dairy Road, Charlottesville, Virginia, U.S.A.

* FINK (Dr U.), Lunar and Planetary Laboratory, University of Arizona, Tucson, Arizona 85721, U.S.A.

* FINN (Dr G. D.), Institute for Astronomy, 2525 Correa Road, Honolulu, Hawaii 96822, U.S.A.

FINSEN (Dr W. S.), P.O. Box 4204, Johannesburg, South Africa

FIREMAN (Dr E. L.), Smithsonian Astrophysical Observatory, 60 Garden Street, Cambridge, Massachusetts 02138, U.S.A.

FIROR (Dr J. W.), High Altitude Observatory of the University of Colorado, Boulder, Colorado, U.S.A.

FISCHEL (Dr D.), Code 613, NASA, Goddard Space Flight Center, Greenbelt, Maryland 20771, U.S.A.

* FISCHER (Prof. Dr P. L.), Universitäts-Sternwarte Wien, A-1180 Wien, Türkenschantzstrasse 17, Austria

FISHER (P. C.), Lockheed Palo Alto Research Laboratory, 3251 Hanover Street, Palo Alto, California 94304, U.S.A.

* FISHER (Dr R. R.), Institute for Astronomy, University of Hawaii, 2525 Correa Road, Honolulu, Hawaii 96822, U.S.A.

* FISHKOVA (Mrs Dr L. M.), Abastumani Astrophysical Observatory, Abastumani, Georgia, U.S.S.R.

FITCH (Dr W. S.), Steward Observatory, University of Arizona, Tucson, Arizona, U.S.A.

FLECKENSTEIN (Prof. Dr J. O.), Rebgasse 32, 4000 Basel, Switzerland

FLEISCHER (Dr R.), 1733 Church Street, N.W., Washington, D.C. 20036, U.S.A.

* FLETCHER (J. M.), Dominion Astrophysical Observatory, R.R. 7, Victoria B.C., Canada

* FLORENSKIJ (Dr K. P.), Institute of Cosmical Research, USSR Academy of Sciences, Moscow, U.S.S.R.

FOKKER (Dr A. D.), Sterrewacht Sonnenborgh, Servaas Bolwerk 13, Utrecht, The Netherlands

* FOMALONT (Dr E. B.), National Radio Astronomy Observatory, Green Bank, W.Va 24944, Box 2, U.S.A.

* FORBES (Dr J. E.), Washburn Observatory, University of Wisconsin, 475 North Charter Street, Madison, Wisconsin 53706, U.S.A.

FORD JR (Dr W. K.), Department of Terrestrial Magnetism, Carnegie Institution of Washington, Washington, D.C. 30015, U.S.A.

FORTINI (Prof. T.), Osservatorio Astronomico, Rome, Italy

* FOSTER (Mrs Dr P. R.), Department of Natural Philosophy, University of Aberdeen, Aberdeen, AB9 2UE, U.K.

FOWLER (Prof. W. A.), California Institute of Technology, 1201 E. California Blvd., Pasadena, California, U.S.A.

Fox (W. E.), 40 Windsor Road, Newark, Nottinghamshire, U.K.

* Fracassini (Prof. Dr M.), Osservatorio Astronomico di Milano, via Brera 28, Milano, Italy

Fracastoro (Prof. M. G.), Direttore, Osservatorio Astronomico di Torino, Pino Torinese, Torino, 10025, Italy

Franklin (Dr F. A.), Smithsonian Institution Astrophysical Observatory, 60 Garden Street, Cambridge, Massachusetts 02138, U.S.A.

* Frantsman (Dr Yu. L.), Radio Astrophysical Observatory, Latvian Academy of Sciences, Riga, Latvian SSR, U.S.S.R.

Franz (Dr O. G.), Lowell Observatory, Flagstaff, Arizona 86002, U.S.A.

Fredga (Miss Dr K.), Institute of Plasma Physics, Royal Institute of Technology, S-100 44 Stockholm 70, Sweden

Fredrick (Prof. L. W.), Leander McCormick Observatory, University of Virginia, Charlottesville, Virginia 22903, U.S.A.

* Freeman (Dr K. C.), Mount Stromlo and Siding Spring Observatories, Research School of Physical Sciences, The Australian National University, Private Bag, Canberra 2600, Australia

* Freiesleben (Dr H. C.), Hydrografisches Institut der Universität, Alt-Oberweg 24, 2000 Hamburg 4, F.R.G.

Freitas-Mourão (Dr R. R. de), Observatorio Nacional, Rua General Bruce 586, São Cristóvão, Rio de Janeiro, Brasil

Fresa (Prof. A.), Via Porto 23, Salerno, Italy

* Fricke (Dr K. J.), Universitätssternwarte, Geismarlandstr. 11, 3400 Göttingen, F.R.G.

Fricke (Prof. Dr W.), Direktor des Astronomischen Rechen-Instituts, Mönchhofstrasse 12–14, 69 Heidelberg, F.R.G.

Friedemann (Dr C.), Universitätssternwarte und Astrophysikalisches Institut der Friedrich-Schiller-Universität, Jena, G.D.R.

* Friedlander (Dr M. W.), Washington University, Department of Physics, St. Louis, Missouri 63130, U.S.A.

Friedman (Dr H.), Code 7100, U.S. Naval Research Laboratory, Washington, D.C. 20390, U.S.A.

Fringant (Miss Dr A. M.), Institut d'Astrophysique, 98 bis Boulevard Arago, Paris 14e, France

* Frisch (Dr U.), Institut d'Astrophysique, 98 bis Boulevard Arago, Paris 14e, France

Fritzová-Švestková (Mrs Dr L.), Astronomical Institute, Czechoslovak Academy of Sciences, Observatory Ondřejov, Czechoslovakia

* Frolov (Dr M. S.), Astronomical Council, USSR Academy of Sciences, Vavilov Str. 34, V-312, U.S.S.R.

Fuchs (Prof. Dr J.), Universitäts-Sternwarte, Universitätsstrasse 4, Innsbruck, Austria

* Fujimoto (Dr M.), Department of Physics, Faculty of Science, Nagoya University, Furocho, Chikusa-ku, Nagoya, Japan

Fujita (Dr Y.), Department of Astronomy, Faculty of Science, University of Tokyo, Bunkyo-ku, Tokyo, Japan

* Fursenko (Mrs Dr M. A.), Institute of Theoretical Astronomy, USSR Academy of Sciences, Leningrad, U.S.S.R.

* Gabriel (Dr A. H.), Astrophysics Research Unit, Culham Laboratory, Abingdon, Berkshire, U.K.

Gabriel (Dr M. R. L.), Institut d'Astrophysique, Université de Liège, avenue de Cointe, 5, B-4200 Cointe-Ougrée, Belgium

Gaide (A.), Observatoire Cantonal de Genève, Genève, Sauverny, 1290, Switzerland

Gaizauskas (Dr V.), Astrophysics Branch, National Research Council, M-50, Ottawa 7, Ontario, Canada

Galibina (Mrs Dr I. V.), Institute of Theoretical Astronomy, USSR Academy of Sciences, Leningrad, U.S.S.R.

* GALKIN (Dr L. S.), Crimean Astrophysical Observatory, USSR Academy of Sciences, Crimea, U.S.S.R.

GALLET (R. M.), National Bureau of Standards, Boulder, Colorado, U.S.A.

GALLOUËT (Dr L.), Observatoire de Paris, 61, av. de l'Observatoire, 75-Paris 14e, France

GALPERIN (Dr G. I.), Institute of Physics and of the Atmosphere, B. Gruzinskaya 10, Moscow, U.S.S.R.

GALT (Dr J. A.), Dominion Radio Astrophysical Observatory, Box 248, Penticton, B.C., Canada

GAMA (Dr L. I.), Observatorio Nacional, R. General Bruce 586, S. Cristovâo, Rio de Janeiro, Brazil

GAMOW (Prof. G.), University of Colorado, Boulder, Colorado 80302, U.S.A.

GAPOSCHKIN (Dr S.), Harvard College Observatory, Cambridge, Mass. 02138, U.S.A.

GARDNER (Dr F. F.), CSIRO Division of Radiophysics, Post Office Box 76, Epping, N.S.W., 2121, Australia

GARFINKEL (Dr B.), Department of Astronomy, Yale University, New Haven, Connecticut 06520, U.S.A.

* GARMIRE (Dr G. P.), 328 Downs, California Institute of Technology, Pasadena, California 91109, U.S.A.

* GARRISON (Prof. R. F.), Department of Astronomy, University of Toronto, Toronto 5, Ontario, Canada

GARSTANG (Dr R. H.), Joint Institute for Laboratory Astrophysics, University of Colorado, Boulder, Colorado 80304, U.S.A.

* GARTON (Prof. W. R. S.), Department of Physics, Imperial College, London S.W. 7, U.K.

GASCOIGNE (Dr S. C. B.), Mount Stromlo Observatory, Canberra, A.C.T., Australia

* GASKA (Dr S.), Astronomical Institute, N. Copernicus University, Sienkiewicza 30, Torun, Poland

* GAUSTAD (Dr J. E.), Astronomy Department, University of California, Berkeley, California 94720, U.S.A.

* GAUTIER (Dr D.), Observatoire de Meudon, 92-Meudon, France

GAVIOLA (Dr E.), Instituto Fisico Balseiro, Bariloche, Rio Negro, Argentina

* GAVRILOV (Dr I. V.), Main Astronomical Observatory, Ukrainian Academy of Sciences, Kiev, U.S.S.R.

GEAKE (Dr J. E.), Faculty of Technology, University of Manchester, Manchester 1, U.K.

GEBBIE (Dr K.), Joint Institute for Laboratory Astrophysics, University of Colorado, Boulder, Colorado 80302, U.S.A.

GEHRELS (Dr T.), Space Sciences Building, University of Arizona, Tucson, Arizona 85721, U.S.A.

GEL'FREJKH (Dr G. B.), Main Astronomical Observatory, USSR Academy of Sciences, Pulkovo, U.S.S.R.

GENT (H.), Royal Radar Establishment, St Andrews Road, Great Malvern, Worcestershire, U.K.

GENTILI DE GIUSEPPE (M.), Observatoire du Pic-du-Midi, 65-Bagnères-de-Bigorre (Hautes-Pyrénées), France

* GEORGELIN (Dr Y.), Observatoire de Marseille, Place le Verrier, 13-Marseille 4e, France

GERSHBERG (Dr R. E.), Crimean Astrophysical Observatory, USSR Academy of Sciences, Crimea, U.S.S.R.

GEYER (Dr E. H.), Observatorium Hoher List (Universität Bonn), 5568 Daun/Eifel, F.R.G.

GHABRUS (Dr Roushdy Azer), Helwan Observatory, Helwan, near Cairo, U.A.R.

GIACAGLIA (Prof. G. E. O.), 227 Taylor Hall, The University of Texas at Austin, Austin, Texas 78712, U.S.A.

GIACCONI (Dr R.), American Science and Engineering Inc., 11 Carleton Street, Cambridge, Mass. 02142, U.S.A.

GIANNONE (Dr P.), Osservatorio Astronomico, Viale Parco Mellini 84, 00185 Roma, Italy

GIANNUZZI (Prof. M. A.), Corso Matteotti 190, Albano, Rome, Italy

GICLAS (H. L.), Lowell Observatory, Flagstaff, Arizona, U.S.A.

GIESE (Dr R. H.), Ruhr-Universität Bochum, Bereich Extraterrestrische Physik, NA Gebäude, 463 Bochum, F.R.G.

GINGERICH (Dr O.J.), Smithsonian Astrophysical Observatory, 60 Garden Street, Cambridge, Mass. 02138, U.S.A.

GINZBURG (Prof. V.L.), Corresponding Member of the USSR Academy of Sciences, Physical Institute, Leninsky Prospect 53, Moscow, U.S.S.R.

GIOVANELLI (Dr R.G.), National Standards Laboratory, CSIRO, University Grounds, Sydney, N.S.W., Australia

GJELLESTAD (Miss G.), Universitetet i Bergen, Geofysisk Institutt, Avd. C., Bergen, Norway

* GLAGOLEVSKIJ (Dr Ju. V.), Special Astrophysical Observatory, USSR Academy of Sciences, St. Zelenchukskaya, Stavropolsky Kraj, U.S.S.R.

* GLASER (Dr H.), Solar Physics Program, Physics and Astronomy Programs NASA, Office of Space Science and Applications, Washington, D.C. 20546, U.S.A.

* GLEDHILL (Dr J.A.), Department of Physics, Rhodes University, P.O. Box 94, Grahamstown, South Africa

GLEISSBERG (Prof. Dr W.), Senckenberganlage 23, 6000 Frankfurt, F.R.G.

GLIESE (Dr W.), Astronomisches Rechen-Institut, Mönchhofstrasse 12–14, 69 Heidelberg, F.R.G.

GNEVYSHEV (Dr M.N.), Astronomical Observatory, Leningrad-M 140, U.S.S.R.

GNEVYSHEVA (Mrs Dr R.S.), Main Astronomical Observatory, USSR Academy of Sciences, Pulkovo, U.S.S.R.

GODART (Prof. Dr O.), Université de Louvain, De Croylaan, 27, B-3030 Heverlee, Belgium

GODOLI (Prof. Dr G.), Direttore, Osservatorio Astrofisico, Catania 95123, Italy

GÖKDOĞAN (Mrs Prof. Dr N.), Director, University Observatory, Beyazit, Istanbul, Turkey

GÖKMEN (Dr T.), Kandilli Observatory, Çengelköy, Istanbul, Turkey

GOLAY (Prof. Dr M.), Directeur de l'Observatoire de Genève, 1290 Sauverny (GE), Switzerland

GOLD (Prof. T.), Space Science Bldg., Cornell University, Ithaca, N.Y., 14850, U.S.A.

GOLDBERG (Prof. L.), Harvard College Observatory, Cambridge, Mass. 02138, U.S.A.

GOLDREICH (Dr P.), California Institute of Technology, Pasadena, California 91109, U.S.A.

* GOLDSTEIN (Dr R.M.), Jet Propulsion Laboratory, California Institute of Technology, 4800 Oak Grove Drive, Pasadena, California 91103, U.S.A.

GOLDSTEIN (Dr S.J.), Leander McCormick Observatory, University of Virginia, Charlottesville, Virginia 22903, U.S.A.

GOLDSWORTHY (Prof. F.A.), Department of Mathematics, University of Leeds, Leeds 2, Yorkshire

* GOLDWIRE (Dr H.C. Jr), Space Science Department, Rice University, Houston, Texas 77001, U.S.A.

GOLLNOW (Dr H.), Mount Stromlo Observatory, Canberra, A.C.T., Australia

GOMES (Dr A.M.), University of Brazil, Rua Ipiranga, 25-40 Andar, Rio de Janeiro, ZC-01, Brazil

GONDOLATSCH (Prof. Dr F.), Astronomisches Rechen-Institut, Mönchhofstrasse 12-14, 69 Heidelberg, F.R.G.

GONZALÉZ (G.), Observatorio Astronomico, Universidad Nacional de Mexico, Tonantzintla, Mexico

* GONZE (Ir. R.F.J.), Observatoire Royal de Belgique, avenue Circulaire, 3, B-1180 Bruxelles, Belgium

GOODY (Dr E.M.), Director, Blue Hill Observatory, Harvard University, Cambridge, Mass. 02138, U.S.A.

* GOPALA RAO (U.V.), Astrophysical Observatory, Kodaikanal-3, India

GOPASYUK (Dr S.I.), Crimean Astrophysical Observatory, USSR Academy of Sciences, Crimea, U.S.S.R.

GORBATSKIJ (Dr V.G.), Leningrad University Observatory, Leningrad, U.S.S.R.

* GORDON (Dr I.M.), Institute of Radiophysics and Electronics, Academy of Sciences of the Ukrainian SSR, Khar'kov 85, U.S.S.R.

GORDON (Dr J.E.), Director of the Nikolaiev Observatory, Nikolaiev, U.S.S.R.

GORGOLEWSKI (Dr S.), Astronomical Observatory, N. Copernicus University, Sienkiewicza 30, Torun, Poland

GORSHKOV (Prof. Dr P.M.), State University, Leningrad, U.S.S.R.

GORYNYA (Dr A. A.), Kiev University Observatory, Kiev, U.S.S.R.

* GOSACHINSKIJ (Dr I. V.), Main Astronomical Observatory, USSR Academy of Sciences, Pulkovo, U.S.S.R.

* GOSS (Dr W. M.), CSIRO, Division of Radiophysics, P.O. Box 76, Epping, N.S.W., Australia

GOSSNER (Mrs S. D.), U.S. Naval Observatory, Washington 25, D.C., U.S.A.

GOTTLIEB (K.), Mount Stromlo Observatory, Canberra, A.C.T., Australia

GOUDAS (Dr C. L.), University of Patras, Patras, Greece

GOUGENHEIM (Prof. A.), Directeur du Service Hydrographique de la Marine, 30, Boulevard Flandrin, 75-Paris 16e, France

* GOUGH (Dr D. O.), Institute of Theoretical Astronomy, Madingley Road, Cambridge, U.K.

GOULD (Dr R.), Physics Department, University of California-San Diego, La Jolla, California, U.S.A.

* GOWER (Mrs A. Ch.), University of British Columbia, Department of Physics, Vancouver 8, B.C., Canada

* GOWER (Prof. Dr J. F. R.), Department of Physics, University of British Columbia, Vancouver 8, B.C., Canada

GOY (G.), Observatoire Cantonal de Genève, Genève, Sauverny 1290, Switzerland

GOYAL (A. N.), Department of Mathematics, University of Rajasthan, Jaipur, India

* GRADSZTAJN (Dr E.), Institut de Physique Nucléaire d'Orsay, 91-Gif s/Yvette, France

GRAHAM (Dr J. A.), Cerro Tololo Inter-American Observatory, Casilla 63-D, La Serena, Chile

GRAHL (Dr B. H.), Universitäts-Sternwarte Bonn/Rhein, Poppelsdorfer Allee 49, 53 Bonn, F.R.G.

GRAINGER (Dr J.), University of Manchester, Institute of Science & Technology, Sackville Street, Manchester 1, U.K.

GRANDON (Prof. R.), Carlos Edwards 1270, Santiago de Chile, Chile

GRANT (Dr I. P.), Pembroke College, Oxford, U.K.

GRATTON (Prof. L.), Laboratorio di Astrofisica, Casella Postale 67, 00044 Frascati (Roma), Italy. Vice-President of the Union.

* GRAY (Dr D. F.), Department of Astronomy, The University of Western Ontario, London 72, Ontario, Canada

GREEN (Dr J.), Research Scientist, Advanced Research Laboratory, Douglas Aircraft Company, Inc., 5251 Bolsa Avenue, Huntington Beach, California 92646, U.S.A.

GREEN (Prof. L. C.), Strawbridge Observatory, Haverford College, Haverford, Pennsylvania 19041, U.S.A.

GREEN (Dr R. M.), Department of Astronomy, The University, Glasgow, W.2, U.K.

GREENBERG (Dr J. M.), Department of Physics, Rensselaer Polytechnic Institute, Troy, N.Y., U.S.A.

GREENSTEIN (Prof. J. L.), Hale Observatories, California Institute of Technology, 1201 E. California Street, Pasadena, California, 91109, U.S.A.

* GREGUL (Mrs Dr A. J.), Astronomical Observatory of the University, Kiev, Ukrainian SSR, U.S.S.R.

* GREISEN (Dr K. I.), Cornell University, Physics Department, Clark Hall, Ithaca, New York 14850, U.S.A.

* GREVESSE (Dr N.), Institut d'Astrophysique, Université de Liège, avenue de Cointe, 5, B-4200 Cointe-Ougrée, Belgium

* GREWING (Dr M.), Institut für Astrophysik und extraterr. Forschung, Poppelsdorfer Allee 49, 5300 Bonn, F.R.G.

* GREYBER (Dr H. D.), Martin-Marietta Corp. (S-2000), Denver, Col. 80201, U.S.A.

GRIFFIN (Dr R. E. M.), The Observatories, Madingley Road, Cambridge, U.K.

GRIFFIN (Dr R. F.), The Observatories, Madingley Road, Cambridge, U.K.

GRIGOREVSKIJ (Dr V. M.), Kishinev State University, Kishinev, U.S.S.R.

* GRIGORYAN (Dr K. A.), Byuranskaya Astrophysical Observatory, Armenian Academy of Sciences, Byurakan, U.S.S.R.

GROTH (Dr H. G.), Universitäts-Sternwarte München, Sternwartstrasse 23, 8 Munich 27, F.R.G.

* GRUBISSICH (Prof. Dr C.), via Aosta 34/5, 35100 Padova, Italy

GRUSHINSKIJ (Prof. Dr N. P.), Sternberg Astronomical Institute, Moscow, U.S.S.R.

GRYGAR (Dr J.), Astronomical Institute, Czechoslovak Academy of Sciences, Observatory Ondřejov, Czechoslovakia

GRZEDZIELSKI (Dr S.), Astronomical Observatory of Warsaw University, Al. Ujazdowskie 4, Warsaw, Poland

GUÉRIN (Dr P.), Institut d'Astrophysique, 98bis, Bld Arago, Paris-14e, France

* GUEST (Dr J. E.), University of London Observatory, Mill Hill Park, London N.W. 7, U.K.

GUIGAY (Dr G.), Observatoire de Marseille, 2, Place le Verrier, 13-Marseille-4e, France

GUINOT (Dr B.), Observatoire de Paris, 61 avenue de l'Observatoire, Paris-14e, France

* GULKIS (Dr S.), Jet Propulsion Laboratory, 1836–365, 4800 Oak Grove Drive, Pasadena, California 91103, U.S.A.

* GUNN (Dr J. E.), Hale Observatory, California Institute of Technology, Pasadena, California 91109, U.S.A.

GÜNTHER (Dr O.), Zentralinstitut für Astrophysik, DDR 15 Potsdam, Telegrafenberg, G.D.R.

GÜNTZEL-LINGNER (Dr U.), Astronomisches Rechen-Institut, Mönchhofstrasse 12–14, 69 Heidelberg, F.R.G.

GURTOVENKO (Dr E. A.), Main Astronomical Observatory, Ukrainian Academy of Sciences, Kiev, U.S.S.R.

* GURSKY (Dr H.), American Science and Engineering, Inc., 11 Carleton Street, Cambridge, Massachusetts 02142, U.S.A.

GURZADIAN (Dr G. A.), Byurakan Astrophysical Observatory, Armenian SSR, Erevan, Armenia, U.S.S.R.

* GUSEJNOV (Dr R. E.), Shemaha Astrophysical Observatory, Academy of Sciences of Azerbaidzhan SSR, Shemada, U.S.S.R.

GUSSMANN (Dr E.-A.), Zentralinstitut für Astrophysik, DDR 15 Potsdam, Telegrafenberg, G.D.R.

GUTH (Dr V.), Astronomical Institute, Czechoslovak Academy of Sciences, Observatory Ondřejov, Czechoslovakia

GUTHRIE (B. N. G.), The Royal Observatory, Edinburgh 9, U.K.

GUTIÉRREZ-MORENO (Mrs Dr A.), Observatorio Astronómico de la Universidad de Chile, Casilla 36-D, Santiago de Chile, Chile

GYLDENKERNE (K.), Universitets Observatoriet, Brorfelde, pr. 4340 Tølløse, Denmark

HAAS (Prof. Dr J.), Lipschitz-Str. 1, 53 Bonn, F.R.G.

HABIBULLIN (Prof. Dr Sh. T.), Director, Astronomical Observatory of State University, Lenina 18, Kazan, U.S.S.R.

* HABING (Dr H. J.), Kapteyn Laboratorium, Postbus 800, Groningen, The Netherlands

HACHENBERG (Prof. Dr O.), Radiosternwarte Bonn, Poppelsdorfer Allee 49, 53 Bonn, F.R.G.

HACK (Prof. M.), Director, Astronomical Observatory, Via Tiepolo 11, 34131 Trieste I, Italy

HADDOCK (Dr F.), The Observatory, University of Michigan, Ann Arbor, Michigan, U.S.A.

HADJIDEMETRIOU (Dr J. D.), Astronomical Department, University of Thessaloniki, Thessaloniki, Greece

HAERENDEL (Dr G.), Institut für extraterrestrische Physik des Max-Planck-Instituts für Physik und Astrophysik, 8046 Garching b. München, F.R.G.

HAFFNER (Prof. H.), Astronomisches Institut der Universität, 87 Würzburg, Büttnerstrasse 72, F.R.G.

HAGEN (Dr J. P.), Professor of Radio Astronomy, The Pennsylvania State University, University Park, Pennsylvania, U.S.A.

* HAGFORS (Dr T.), MIT Lincoln Laboratory, Millstone, Haystack, Lexington, Massachusetts 02173, U.S.A.

HAGIHARA (Prof. Y.), Department of Astronomy, Faculty of Science, University of Tokyo, Bunkyo-ku, Tokyo, Japan

* HAJDUK (Dr A.), Astronomical Institute, Slovak Academy of Sciences, Dúbravská Cesta, Bratislava, Czechoslovakia

HALL (Dr J. S.), Director, Lowell Observatory, Flagstaff, Arizona, U.S.A.

HALL (Dr R. G.), U.S. Naval Observatory, Washington, D.C. 20390, U.S.A.

HALLAM (Dr K. L.), Laboratory for Optical Astronomy, Code 273, Goddard Space Flight Center, Greenbelt, Maryland 20771, U.S.A.

HALLIDAY (Dr I.), Upper Atmosphere Research Section, Astrophysics Branch, National Research Council, Ottawa 7, Ontario, Canada

* HAMADA (Prof. T.), Department of Physics, Ibaraki University, Mito 310, Japan

HÄMEEN-ANTTILA (Prof. Dr A.), The Astronomical Institute of the Oulu University, Oulu, Finland

HAMID (Dr S. El-Din), Department of Astronomy, Faculty of Science, University Fouad, Giza, Cairo, U.A.R.

HANASZ (Dr J.), Astronomical Observatory, Toruń-Piwnice, Poland

* HANSEN (Dr C. J.), Department of Astrophysics and Physics, University of Colorado, Boulder, Colorado, U.S.A.

* HANSEN (Dr R.), High Altitude Observatory, Boulder, Colorado 80302, U.S.A.

HANSSON (N.), Lund Observatory, S-222 24 Lund, Sweden

* HARDEBECK (Mrs Dr E. G.), California Institute of Technology, Pasadena, California 91109, U.S.A.

HARDIE (Dr R. H.), Dyer Observatory, Vanderbilt University, Nashville 5, Tennessee, U.S.A.

HARDING (G. A.), Royal Greenwich Observatory, Herstmonceux Castle, Hailsham, Sussex, U.K.

HARDORP (Prof. Dr J.), Department of Earth and Space Sciences, State University of New York, Stony Brook, New York 11790, U.S.A.

HARO (Dr G.), Observatorio Astronómico Nacional, Apartado Postal 70–264, Ciudad Universitaria, Mexico 20, D.F., Mexico

HARRIS (B. J.), Perth Observatory, Bickley, 6076 Western Australia

HARRISON (Prof. E. R.), Department of Physics and Astronomy, University of Massachusetts, Amherst, Massachusetts 01002, U.S.A.

HARROWER (Prof. G. A.), Department of Physics, Queen's University, Kingston, Ontario, Canada

HARTE (Dr E. A.), Department of Applied Mathematics, University College of Swansea, Singleton Park, Swansea, Wales, U.K.

* HARTNER (Dr W.), Institut für Geschichte der Naturwissenschaften, Johann Wolfgang Goethe Universität, Frankfurt am Main, F.R.G.

* HARTWICK (Prof. Dr F. D. A.), Department of Physics, University of Victoria, Victoria, B.C., Canada

HARTZ (Dr T. R.), Communications Research Centre, Dept. of Communications P.O. Box 490, Terminal A, Ottawa 2, Ontario, Canada

* HARVEY (Dr Ch.), Observatoire de Meudon, 92-Meudon, France

HARWIT (Dr M.), Cornell University, Center for Radiophysics and Space Research, Clark Hall, Ithaca, New York 14850, U.S.A.

HARWOOD (Miss M.), Harvard College Observatory, Cambridge, Massachusetts 02138, U.S.A.

HASER (Dr L.), Max-Planck-Institut für Physik und Astrophysik, Institut für Extraterrestrische Physik, 8046 Garching bei München, F.R.G.

HATTORI (Dr A.), Kwasan Observatory, University of Kyoto, Yamashina, Kyoto, Japan

HAUCK (B.), Observatoire Cantonal de Genève, Genève, Sauverny 1290, Switzerland

HAUG (Dr U.), Hamburger Sternwarte, 205 Hamburg-Bergedorf, F.R.G.

* HAUGE (Ö.), Institute of Theoretical Astrophysics, University of Oslo, P.O. Box 1029, Blindern, Oslo 3, Norway

HAUPT (Prof. Dr H.), Director, Universitäts-Sternwarte, Universitätsplatz 5, A-8010 Graz, Austria

HAUPT (R.), Assistant Director of the American Ephemeris and Nautical Almanac, U.S. Naval Observatory, Washington, D.C., U.S.A.

HAUPT (R.F.), U.S. Naval Observatory, Washington, D.C., 20390, U.S.A.

* HAWKING (Dr S.W.), Institute of Theoretical Astronomy, Madingley Road, Cambridge, U.K.

HAWKINS (Dr G.S.), Harvard College Observatory, Cambridge, Mass. 02138, U.S.A.

* HAYAKAWA (Prof. Dr S.), Department of Physics, Faculty of Science, Nagoya University, Furocho, Chikusa-ku, Nagoya, Japan

HAYASHI (Prof. C.), Department of Nuclear Science, Faculty of Science, Kyoto University, Kyoto, Japan

* HAYLI (Dr A.), Institut d'Astrophysique, 98bis, Boulevard Arago, Paris-14e France

* HAZARD (Dr C.), Institute of Theoretical Astronomy, Madingley Road, Cambridge, U.K.

* HAZER (Dr S.), Faculty of Science, Department of Astronomy, P.K. 21, Bornova, Izmir, Turkey

HAZLEHURST (Dr J.), The Astronomy Centre, University of Sussex, Falmer, near Brighton, Sussex, U.K.

HEARD (Dr J.F.), David Dunlap Observatory, Richmond Hill, Ontario, Canada

HEARN (Dr A.G.), Department of Physics, University of Queensland, St. Lucia, Brisbane 4067, Australia

HECKMANN (Prof. Dr O.), Schmiedesberg 2b, 2057 Reinbek, F.R.G. Adviser to the Executive Committee.

HEDDLE (Dr D.W.O.), Department of Physics, University of York, Heslington, York, U.K.

HEDEMAN (Miss E.R.), McMath-Hulbert Observatory, Lake Angelus Road, Pontiac, Michigan, U.S.A.

HEESCHEN (Dr D.S.), National Radio Astronomy Observatory, Edgemont Dairy Road, Charlottes-ville, Virginia 22901, U.S.A.

HEIDMANN (Dr J.), Observatoire de Paris, Section d'Astrophysique, 92-Meudon, France

* HEILES (Dr C.E.), Astronomy Department, University of California, Berkeley, California 94720, U.S.A.

HEINTZ (Prof. W.D.), Sproul Observatory, Swarthmore, Pa. 19081, U.S.A.

HEINTZE (Dr J.R.W.), Sterrewacht 'Sonnenborgh', Servaas Bolwerk 13, Utrecht, The Nether-lands

* HEISER (Dr A.), A. J. Dyer Observatory, Box 1803, Vanderbilt University, Nashville, Tenn. 37203, U.S.A.

HEISKANEN (Prof. V.A.), Geodetic Institute Helsinki, Finland

HELFER (Dr H.L.), Department of Physics and Astronomy, University of Rochester, Rochester, N.Y., 14628, U.S.A.

HELLMAN (Prof. C.D.), Dept. of History, Queens College of the City University of New York, Flushing, New York 11367, U.S.A.

* HELMKEN (Dr H.), Smithsonian Astrophysical Observatory, 60 Garden Street, Cambridge, Mas-sachusetts 02138, U.S.A.

HEMENWAY (Dr C.L.), 100 Fuller Road, Albany, New York 12205, U.S.A.

HEMMLEB (Dipl. Ing. G.), Zentralinstitut Physik der Erde, DDR 15, Potsdam, Telegrafenberg, G.D.R.

HENIZE (Dr K.G.), Astronaut Office, NASA Manned Spacecraft Center, 2101 Webster-Seabrook Road, Houston, Texas 77058, U.S.A.

HENN (Miss F.), Astronomisches Rechen-Institut, Mönchhofstrasse 12–14, 69 Heidelberg, F.R.G.

HÉNON (M.), Observatoire de Nice, Le Mont-Gros, 06-Nice, France

* HENRIKSEN (Prof. Dr R.N.), Astronomy Group, Department of Physics, Queens University, Kingston, Ontario, Canada

* HENRY (Dr R.C.), E. O. Hulbert Center for Space Research, Naval Research Laboratory, Washington, D.C. 20390, U.S.A.

HERBIG (Dr G.H.), Lick Observatory, University of California, Santa Cruz, California 95060, U.S.A.

HERCZEG (Dr T.), Hamburger Sternwarte, 205 Hamburg-Bergedorf, F.R.G.

HERGET (Dr P.), Cincinnati Observatory, Observatory Place, Cincinnati, Ohio 45208, U.S.A.

HERMAN (Mrs Dr R.), Observatoire de Paris, Section d'Astrophysique, 92-Meudon, France

* HERR (Dr R. B.), Physics Department, University of Delaware, Newark, Delaware 19711, U.S.A.

HERRICK (Dr S.), University of California, Los Angeles 24, California 90024, U.S.A.

HERS (J.), Republic Observatory, Johannesburg, South Africa

HERTZ (Dr H. G.), 2112 Florida Ave., N.W., Washington, D.C. 20008, U.S.A.

HERZBERG (Dr G.), Distinguished Research Scientist, National Research Council, Ottawa, Ontario, Canada

HERZBERG (Mrs Dr L. H.), Communications Research Centre, Dept. of Communications, Box 490, Terminal A, Ottawa 2, Ontario, Canada

* HESSER (Dr J. E.), Cerro Tololo Inter-American Observatory, Casilla 63-D, La Serena, Chile

HEWISH (Dr A.), Cavendish Laboratory, Free School Lane, Cambridge, U.K.

HEY (Dr J. S.), 4 Shortlands Close, Willingdon, Sussex, U.K.

HEYDEN, S.J. (Dr F. J.), Georgetown College Observatory, Washington 7, D.C., U.S.A.

HIBBS (A. R.), Jet Propulsion Laboratory, California Institute of Technology, Pasadena, California 91109, U.S.A.

* HIDAJAT (Dr B.), Acting Director, Bosscha Observatory, Lembang, Java, Indonesia

HIEI (Dr E.), Tokyo Astronomical Observatory, University of Tokyo, Mitaka, Tokyo, Japan

HIGGS (Dr L. A.), Astrophysical Branch, National Research Council, Ottawa, Ontario, Canada

HILL (E. R.), CSIRO, Division of Radiophysics, Post Office Box 76, Epping, N.S.W., 2121, Australia

* HILL (Dr G.), Dominion Astrophysical Observatory, R.R. 7, Victoria, B.C., Canada

HILL (Dr P. W.), University Observatory, St Andrews, Fife, U.K.

* HILLIARD (Dr R.), Steward Observatory, University of Arizona, Tucson, Arizona 85721, U.S.A.

HILTNER (Dr W. A.), Yerkes Observatory, Williams Bay, Wisconsin, U.S.A.

HINDMAN (J. V.), Sinding Spring Observatory, Coonabarabran, N.S.W., Australia

HINDMARSH (Prof. W. R.), Physics Department, The University, Newcastle upon Tyne, 1, U.K.

HINTEREGGER (Dr H. E.), Hq. AFCRL (CRUU) Stop 30, L. G. Hanscom Field, Bedford, Massachusetts 01731, U.S.A.

HIRAYAMA (Dr T.), Tokyo Astronomical Observatory, University of Tokyo, Mitaka, Tokyo, Japan

HIROSE (Dr H.), Tokyo Astronomical Observatory, Mitaka, Tokyo, Japan

HIRST (P. W.), Water's Edge, Greenbanks Road, Rondebosch, Cape Province, South Africa

HITOTUYANAGI (Prof. Dr Z.), Department of Astronomy, Faculty of Science, Tohoku University, Sendai, Japan

* HJELMING (Dr R. M.), National Radio Astronomy Observatory, Edgemont Road, Charlottesville, Va. 22901, U.S.A.

HOAG (Dr A. A.), Kitt Peak National Observatory, P.O. Box 4130, Tucson, Arizona 85717, U.S.A.

* HOBBS (Dr L. M.), Yerkes Observatory, Williams Bay, Wisconsin 53191, U.S.A.

HOBBS (Dr R. W.), NASA/Goddard Space Flight Center, Code 614, Solar Physics Branch, Laboratory for Space Sciences, Greenbelt, Maryland 20771, U.S.A.

HODGE (Prof. P. W.), Astronomy Department, University of Washington, Seattle, Washington 98105, U.S.A.

HOFFLEIT (Miss Dr E. D.), Yale University Observatory, Box 2023 Yale Station, New Haven, Conn., U.S.A.

HØG (E.), Hamburger Sternwarte, 205 Hamburg 80, F.R.G.

HÖGBOM (Dr J. A.), Sterrewacht, Leiden, The Netherlands

HOGG (Dr D. E.), National Radio Astronomy Observatory, Edgemont Dairy Road, Charlottesville, Virginia 22901, U.S.A.

HÖGLUND (Dr B.), Onsala Space Observatory, 43034 Onsala, Sweden

* HOHL (Dr F.), NASA Hampton, NASA Langley Research Center, Virginia, U.S.A.

HOLDEN (F.), Astronomer-in-Charge, Lamont-Hussey Observatory, P.O. Box 268, Bloemfontein, O.F.S., Republic of South Africa

HOLMBERG (Prof. E. B.), Astronomical Observatory, Box 515, S-751 20 Uppsala 1, Sweden

* HOLWEGER (Dr H.), Institut für Theoretische Physik, Oldhausenstr., 2300 Kiel, F.R.G.

HOOGHOUDT (B. G.), University Observatory, Sterrewacht, Leiden, The Netherlands

HOPMANN (Prof. Dr J.), Meckenheimer Allee 153, D 53 Bonn, F.R.G.

HOPPE (Prof. Dr J. A.), Heinrich-Hertz Institut für Solar-Terrestrische Physik, Berlin-Adlershof, Rudower Chaussee, G.D.R.

* HORÁK (Dr T. B.), Astronomical Institute, Czechoslovak Academy of Sciences, Budečská 6, Prague 2, Czechoslovakia

* HORÁK (Dr Z.), Technical University, Technická 4, Prague 6, Czechoslovakia

HORI (Dr G.), Yale University Observatory, Box 2023, Yale Station, New Haven, Connecticut, U.S.A.

* HOSHI (Dr R.), Department of Physics, Kyoto University, Kyoto, Japan

* HOSKIN (Dr M. A.), Churchill College, Cambridge, U.K.

HOSOKAWA (Dr Y.), Yamagata University, Koshirakawa-cho, Yamagata-shi, Yamagata, Japan

HOTINLI (Dr M.), University Observatory, Beyazit, Istanbul, Turkey

HOUCK (Dr T. E.), Washburn Observatory, University of Wisconsin, Madison, Wisconsin 53706, U.S.A.

HOUSE (Dr L. L.), High Altitude Observatory, Boulder, Colorado 80302, U.S.A.

HOUTGAST (Dr J.), Astronomical Observatory 'Sonneborgh', Zonnenburg 2, Utrecht, The Netherlands

HOUZIAUX (Prof. Dr L. N.), Faculté des Sciences, Département d'Astrophysique, Plaine de Nimy, B-7000 Mons, Belgium

* HOVENIER (J. W.), The Observatory Leiden, Leiden, The Netherlands

HOWARD (Dr R. F.), Mount Wilson and Palomar Observatories, 813 Santa Barbara Street, Pasadena, California 91106, U.S.A.

HOWARD (Dr W. E. III), National Radio Astronomy Observatory, Edgemont Road, Charlottesville, Virginia 22901, U.S.A.

HOYLE (Prof. F.), St. John's College, Cambridge, U.K.

HUANG (Dr Su-Shu), Dearborn Observatory, Northwestern University, Evanston, Illinois 60201, U.S.A.

* HUBE (Dr D. P.), Department of Physics, University of Alberta, Edmonton 7, Alberta, Canada

HUBENET (Dr H.), Sterrewacht 'Sonnenborgh', Servaas Bolwerk 13, Utrecht, The Netherlands

HUFFER (Dr C. M.), 5059 Campanile Drive, San Diego, California 92115, U.S.A.

HUGHES (Dr J. A.), U.S. Naval Observatory, Washington, D.C. 20390, U.S.A.

HUGHES (Dr V. A.), Department of Physics and Astronomy, Queen's University, Kingston, Ontario, Canada

HUGUENIN (Dr G. R.), Dept. of Physics and Astronomy, University of Massachusetts, Amherst, Mass. 01002, U.S.A.

HUMASON (Dr M. L.), P.O. Box 165, Mendocino, California, U.S.A.

HUMMER (Dr D.), Joint Institute for Laboratory Astrophysics, University of Colorado, Boulder, Colorado, U.S.A.

HUMPHREYS (C. J.), Williamsburg on the Wabash, 400 N. River Road (APT. 1122), W. Lafayette, Indiana 47906, U.S.A.

HUNAERTS (Dr J. J.), Observatoire Royal de Belgique, avenue Circulaire, 3, B-1180 Bruxelles, Belgium

HUNGER (Prof. Dr K.), Lehrstuhl für Astrophysik, Technische Universität Berlin, 1 Berlin 10, Ernst-Reuter-Platz 7, F.R.G.

HUNTEN (Dr D. M.), Kitt Peak National Observatory, Box 4130, Tucson, Arizona 85717, U.S.A.

HUNTER (Dr A.), Royal Greenwich Observatory, Herstmonceux Castle, Hailsham, Sussex, U.K.

HUNTER (Dr C.), Department of Mathematics, The Florida State University, Tallahasse, Florida 32306, U.S.A.

HUNTER (Dr J. H.), Yale University Observatory, Box 2023, Yale Station, New Haven, Connecticut 06520, U.S.A.

HURUHATA (Prof. M.), Tokyo Astronomical Observatory, Mitaka, Tokyo, Japan

* HUTCHINGS (Dr J. B.), Dominion Astrophysical Observatory, R.R. 7, Victoria, B.C., Canada

HYDER (Dr C. L.), Sacramento Peak Observatory, Sunspot, New Mexico 88349, U.S.A.

HYNEK (Dr J. A.), Dearborn Observatory, Northwestern University, Evanston, Illinois 60201, U.S.A.

IANINI (Dr G.), Observatorio Astronómico, Cordoba, Argentina

IBEN (Prof. I. Jr), Room 6–203, Physics Department, Massachusetts Institute of Technology, Cambridge, Mass. 02139, U.S.A.

IDLIS (Dr G. M.), Astrophysical Institute, Alma-Ata, U.S.S.R.

IIJIMA (Dr S.), Tokyo Astronomical Observatory, Mitaka, Tokyo, Japan

IKEDA (Dr T.), International Latitude Observatory, Mizusawa-shi, Iwate-Ken, Japan

* IKHSANOV (Dr R. N.), Main Astronomical Observatory, USSR Academy of Sciences, Pulkovo, U.S.S.R.

IKHSANOVA (Mrs Dr V. N.), Main Astronomical Observatory, USSR Academy of Sciences, Pulkovo, U.S.S.R.

* IMSHENNIK (Dr V. S.), Institute of Applied Mathematics, USSR Academy of Sciences, Moscow, U.S.S.R.

INGHAM (Dr M. F.), Department of Astrophysics, University Observatory, South Parks Road, Oxford, U.K.

INGRAO (H. C.), 58 Hundreds Road, Wellesley Hills, Massachusetts 02181, U.S.A.

* INNANEN (Dr K. A.), Physics Department and Centre for Research in Experimental Space Science, York University, 4700 Keele St., Downsview, Toronto, Canada

IONNISIANI (Dr B. K.), Main Astronomical Observatory, Academy of Sciences, Pulkovo, U.S.S.R.

IOSHPA (Dr B. A.), Institute of Terrestrial Magnetism, Ionosphere, and Propagation of Radiowaves, USSR Academy of Sciences, Moscow, U.S.S.R.

IRELAND (Dr J. G.), Royal Observatory, Edinburgh 9, U.K.

IRIARTE (B.), Tonantzintla and Tacubaya Observatories, Mexico

IRVINE (Dr W. M.), Department of Physics and Astronomy, University of Massachusetts, Amherst, Massachusetts 01002, U.S.A.

IRWIN (Dr J. B.), Department of Earth and Planetary Sciences, Newark State College, Union, N. J. 07083, U.S.A.

* ISHIDA (Dr K.), Tokyo Astronomical Observatory, The University of Tokyo, 2–21–1 Osawa, Mitaka, Tokyo, Japan

ITZIGSOHN (M.), Observatorio Astronómico, La Plata, Argentina

* IVANCHUK (Dr V. I.), Kiev University, Observatornaya 3, Kiev 53, Ukrainian SSR, U.S.S.R.

IVANOV (Dr V. V.), Astronomical Observatory, Leningrad University, Leningrad, U.S.S.R.

IVANOV-KHOLODNY (Dr G. S.), Institute of Applied Geophysics, Academy of Sciences of the USSR, Moscow, U.S.S.R.

IWANISZEWSKA-LUBIENSKA (Mrs Dr C.), Astronomical Observatory, N. Copernicus University, Sienkiewiecza 30, Torun, Poland

IWANOWSKA (Miss Prof. W.), Astronomical Observatory, N. Copernicus University, Sienkiewicza 30, Torun, Poland

* IZVEKOV (Dr V. A.), Institute of Theoretical Astronomy, USSR Academy of Sciences, Leningrad, U.S.S.R.

JACCHIA (Dr L.), Smithsonian Astrophysical Observatory, 60 Garden Street, Cambridge, Mass. 02138, U.S.A.

JACKSON (C.), Yale-Columbia Station, Mount Stromlo Observatory, Canberra, A.C.T., Australia

JACKSON (Dr P.), Universitäts-Sternwarte Wien, Türkenschanzstrasse 17, A 1180-Vienna, Austria

JACOBSEN (Prof. T. S.), University of Washington, Seattle 5, Washington 98115, U.S.A.

JACQUINOT (P.), Laboratoires de Bellevue, 1 Place Aristide-Briand, Bellevue, 92-Meudon, France

JÄGER (Prof. Dr F. W.), Heinrich-Hertz-Institut für Solar-Terrestrische Physik, DDR 15 Potsdam, Telegrafenberg, G.D.R.

JAKIMIEC (Dr J.), Wrocław University Astronomical Institute, Kopernika 11, Wrocław 9, Poland

JAMES (Dr R. A.), Department of Astronomy, The University, Manchester 13, U.K.

JÁNOSSY (Prof. L.), Director of the Central Physical Institute of Budapest, Budapest, Hungary

JÄRNEFELT (Prof. G. J.), Ohjaajantie 3 B 18, Helsinki 40, Finland

JARRETT (Prof. A. H.), Professor-Director of the Boyden Observatory, Boyden Observatory, P.O. Box 334, Bloemfontein, Republic of South Africa

JARZEBOWSKI (Dr T.), Astronomical Observatory, Kopernika 11, Wrocław, Poland

JASCHEK (Dr C. O. R.), Observatorio Astronómico, La Plata, Argentina

JASCHEK (Mrs Dr M.), Observatorio Astronómico, La Plata, Argentina

JASTROW (Dr R.), NASA, Institute for Space Studies, 2880 Broadway, New York, N.Y. 10025, U.S.A.

* JAUNCEY (Dr D. L.), Center for Radiophysics and Space Research, Cornell University, Ithaca, New York 14850, U.S.A.

JAVET (Prof. Dr P.), Chantemerle 19, 1010 Lausanne, Switzerland

* JAYARAJAN (A. P.), Astrophysical Observatory, Kodaikanal-3, India

JEFFERIES (Dr J. T.), Institute for Astronomy, University of Hawaii, 2840 Kolowalu Street, Honolulu, Hawaii 96822, U.S.A.

JEFFERS (Dr H. M.), Lick Observatory, Mount Hamilton, California, U.S.A.

JEFFREYS (Lady), Girton College, Cambridge, U.K.

JEFFREYS (Sir Harold), 160 Huntingdon Road, Cambridge, U.K.

* JEFFREYS (Dr W.), University of Texas, Department of Astronomy, Austin, Texas 78712, U.S.A.

JELLEY (Dr J. V.), Nuclear Physics Division, Building 8, Atomic Energy Research Establishment, Harwell, Didcot, Berkshire, U.K.

* JENKINS (Dr E. B.), Princeton University Observatory, Princeton, New Jersey 08540, U.S.A.

JENKINS (Miss L. F.), Yale University Observatory, Box 2023 Yale Station, New Haven, Conn., U.S.A.

JENNISON (Prof. R. C.), The Electronics Laboratory, University of Kent, Canterbury, Kent, U.K.

JENSCH (A.), Carl Zeiss Observatory, Jena, G.D.R.

JENSEN (Prof. E.), Director, Institute of Theoretical Astrophysics, University of Oslo, Blindern, Oslo 3, Norway

* JERZYKIEWICZ (Dr M.), Astronomical Institute, University of Wrocław, Ul. Kopernika 11, Wrocław 9, Poland

* JOHANSEN (Dr K. T.), University Observatory, Brorfelde, DK-4340, Tølløse, Denmark

JOHNSON (Dr H. L.), Lunar and Planetary Laboratory, University of Arizona, Tucson, Arizona, U.S.A.

JOHNSON (Dr H. M.), Dept. 52–14, Bldg. 201, Lockheed Missiles & Space Co., 3251 Hanover Street, Palo Alto, California 94304, U.S.A.

JOHNSON (Prof. H. R.), University of Indiana, Department of Astronomy, Swain Hall West 319, Bloomington, Indiana 47401, U.S.A.

JOHNSON (Dr M. C.), Department of Physics, University, Edgbaston, Birmingham 15, U.K.

* JOKIPII (Dr J. R.), 405–47 Downs Lab., California Institute of Technology, Pasadena, California 91109, U.S.A.

JONES (A. F.), 14 Main Road, Tahunanui, Nelson, New Zealand

* JONES (Dr D. H. P.), Mount Stromlo Observatory, Canberra 2600, A.C.T. Australia

JORDAN (Dr C.), The Culham Laboratories, Abingdon, Berkshire, U.K.

JORDAN (Dr H. L.), Director, am Institut für Plasmaphysik der Kernforschungsanlage, Postfach 365, 517 Jülich 1, F.R.G.

* JØRGENSEN (Dr H. E.), University Observatory, Øster Voldgade 3, DK-1350, Copenhagen K., Denmark

JOSHI (M. N.), Tata Institute of Fundamental Research, Bombay, India

* JOSHI (S. C.), Uttar Pradesh State Observatory, Manora Peak, Naini Tal, India

JOY (Dr A. H.), Mount Wilson and Palomar Observatories, 813 Santa Barbara Street, Pasadena, California, U.S.A.

JUGAKU (Dr J.), Tokyo Astronomical Observatory, Mitaka, Tokyo, Japan

* JUNG (Dr J.), Observatoire de Paris, 61, Avenue de l'Observatoire, Paris 14e, France

JUNKES S. J. (Dr J.), Specola Vaticana, Castel Gandolfo (Rome), Vatican City State

JURKEVICH (Dr I.), Department of Astronomy, David Rittenhouse Laboratory, University of Pennsylvania, Philadelphia, Pennsylvania 19104, U.S.A.

KABURAKI (Prof. M.), Department of Astronomy, Faculty of Science, University of Tokyo, Yayoicho, Bunkyo-ku, Tokyo, Japan

KADLA (Mrs Dr Z.), Pulkovo Observatory, Leningrad M 140, U.S.S.R.

* KAFTAN-KASSIM (Dr M. A.), State University of New York, 1400 Washington Avenue, Albany, New York 12203, U.S.A.

KAHN (Dr F. D.), Department of Astronomy, The University, Manchester 13, U.K.

KAI (Dr K.), Tokyo Astronomical Observatory, University of Tokyo, Mitaka, Tokyo, Japan

KAIDANOVSKIJ (Dr N. L.), Main Astronomical Observatory, Academy of Sciences, Pulkovo, U.S.S.R.

KAISER (Prof. T. R.), Department of Physics, The University, Sheffield 10, U.K.

KAKINUMA (Dr T.), Research Institute of Atmospherics, Nagoya University, Toyokawa, Aichi, Japan

* KAKUTA (Dr Ch.), International Latitude Observatory of Mizusawa, Mizusawa, Japan

KALANDADZE (Mrs Dr N. B.), Astrophysical Observatory, Abastumani, Georgia, U.S.S.R.

KALER (Dr J. B.), University of Illinois Observatory, Urbana, Illinois 61801, U.S.A.

KALINIAK (Dr A. A.), Main Astronomical Observatory, Academy of Sciences, Pulkovo, U.S.S.R.

KALINKOV (M.), Section d'Astronomie de l'Académie Bulgare des Sciences, Rue 7 Novembre No. 1, Sofia, Bulgaria

KALKOFEN (Dr W.), Smithsonian Astrophysical Observatory, 60 Garden Street, Cambridge, Massachusetts 02138, U.S.A.

* KALLOGLJAN (Dr A. T.), Byurakan Astrophysical Observatory, Armenian Academy of Sciences, Byurakan, U.S.S.R.

KALMYKOV (Dr A. M.), Tashkent Astronomical Observatory, Uzbek Academy of Sciences, Tashkent, U.S.S.R.

* KALNAJS (Dr A.), Harvard College Observatory, Cambridge, Massachusetts 02138, U.S.A.

KAMIEŃSKI (Prof. Dr M.), Kopernik's str. N. 8/18 apart. 2, Warsaw, Poland

KAMIJO (Dr F.), Department of Astronomy, University of Tokyo, Yayoi, Bunkyo-ku, Tokyo, Japan

* KAMINISHI (K.), Department of Physics, Kumamoto University, Kumamoto, Japan

* KANDEL (Dr R. S.), Department of Astronomy, Boston University, Boston, Massachusetts 02215, U.S.A.

KANNO (Dr M.), Hida Observatory, University of Kyoto, Kamitakara, Gifu-ken, Japan

KAO (Dr Ping-tse), 12, Lane 65, Ling-I Str., Taipei, Taiwan

KAPLAN (Dr J.), Department of Physics, University of California at Los Angeles, Los Angeles, California, U.S.A.

KAPLAN (L.), Jet Propulsion Laboratory, 4800 Oak Grove Drive, Pasadena, California, U.S.A.

KAPLAN (Prof. S. A.), Radio Physical Research Institute of Gorki University, Gorki, Liadova 25114, U.S.S.R.

KARANDIKAR (Dr R. V.), Director, Nizamiah Observatory, Begumpet, Hyderabad-16, India

KARDASHEV (Dr N. S.), Sternberg Astronomical Institute, Moscow, U.S.S.R.

* KARLSSON (Dr B. A. R.), Institute of Astronomy, S-222 24 Lund, Sweden

* KARPINSKIJ (Dr V. N.), Main Astronomical Observatory, USSR Academy of Sciences, Pulkovo, U.S.S.R.

KARYAGINA (Miss Dr Z. V.), Astrophysical Institute, Alma-Ata, U.S.S.R.

* KASHSCHEEV (Prof. Dr B. L.), Khar'kov Polytechnical Institute, Khar'kov, U.S.S.R.

KASTLER (Prof. A.), Ecole Normale Supérieure, 45 Rue d'Ulm, Paris-14e, France

KATASEV (Dr L. A.), Commission on Comets and Meteors, Astronomical Council, USSR Academy of Sciences, Vavilov Str. 34, Moscow V-312, U.S.S.R.

KATO (Dr S.), Department of Astronomy, University of Tokyo, Yayoi, Bunkyo-ku, Tokyo, Japan

KATSIS (Dr D. N.), 12, rue Varnis, Nea Smyrne, Athens, Greece

KAUFMANN (Prof. P.), Centro de Rádio-Astronomia e Astrofísica, Universidade Mackenzie, C.P. 8792, São Paulo, Brazil

KAULA (Dr W. M.), Institute of Geophysics and Planetary Physics, University of California, Los Angeles, California 90024, U.S.A.

KAWABATA (Dr Kin-aki), Tokyo Astronomical Observatory, University of Tokyo, Mitaka, Tokyo, Japan

KAWAGUCHI (Dr I.), Kwasan Observatory, University of Kyoto, Yamashina, Kyoto, Japan

KAZES (Dr I.), Observatoire de Paris, Section d'Astrophysique, 92-Meudon, France

KAZIMIRCHAK-POLONSKAYA (Mrs Dr H. I.), Institute of Theoretical Astronomy, USSR Academy of Sciences, Leningrad V-164, U.S.S.R.

KEAY (Dr C. S. L.), University of Newcastle, New South Wales, Australia

KEENAN (Dr P. C.), Perkins Observatory, Delaware, Ohio 43015, U.S.A.

KELLER (Dr G.), Program Director for Astronomy, National Science Foundation, Washington 25, D.C., U.S.A.

* KELLER (Dr H. U.), Sternwarte und Planetarium der Stadt Bochum, Castroper Strasse 67, D-4630 Bochum, F.R.G.

KELLERMAN (Dr K. I.), National Radio Astronomical Observatory, Box 2, Green Bank, West Virginia 24944, U.S.A.

KELLOG (Dr W. W.), National Center for Atmospheric Research, Boulder, Colorado 80302, U.S.A.

KENDERDINE (Dr S.), Mullard Radio Astronomy Observatory, Cavendish Laboratory, Free School Lane, Cambridge, U.K.

* KENNEDY (Prof. J. E.), Department of Physics, University of Saskatchewan, Saskatoon, Canada

KERES (Prof. H.), University of Tartu, L. Koidula 6, Tartu, Estonia, U.S.S.R.

KERR (Dr F. J.), Astronomy Program, University of Maryland, College Park, Maryland 20742, U.S.A.

KESSLER (Dr K. G.), Atomic Physics Division, National Bureau of Standards 13.00, Washington, D.C. 20234, U.S.A.

KHACHIKYAN (Dr E. E.), Byurakan Astrophysical Observatory, Armenian Academy of Sciences, Byurakan, U.S.S.R.

KHARADZE (Prof. Dr E. K.), Director, Astrophysical Observatory, Abastumani, Georgia, U.S.S.R.

KHAVTASI (Dr J. Sh.), Institute Cybernetics, Academy of Sciences, Chitadze str. 6, Tbilisi, Georgia, U.S.S.R.

* KHETSURIANI (Dr T. S.), Abastumani Astrophysical Observatory, Abastumani, Georgia, U.S.S.R.

KHOKHLOVA (Mrs Dr V. L.), Astronomical Council, USSR Academy of Sciences, Vavilov Str. 34, Moscow, U.S.S.R.

KHOLOPOV (Dr P. N.), Astronomical Council of the USSR Academy of Sciences, Vavilov Str. 34, Moscow, U.S.S.R.

* KHROMOV (Dr G. S.), Sternberg Astronomical Institute, Moscow, U.S.S.R.

KIANG (Dr T.), Dunsink Observatory, Castleknock, Co. Dublin, Ireland

KIENLE (Prof. Dr H.), Ziegelhäuser Landstrasse 31, 6900 Heidelberg, F.R.G.

KIEPENHEUER (Prof. Dr K. O.), Direktor des Fraunhofer Institutes, Schöneckstrasse 6, Freiburg i.Br., F.R.G.

* KIKUCHI (Dr S.), Astronomical Institute, University, Sendai, Japan

KILADZE (Dr R. I.), Abastumani Astrophysical Observatory, Georgian Academy of Sciences, Abastumani, U.S.S.R.

* KIM (Y. H.), Pyongyang Astronomical Observatory, Academy of Sciences, Pyongyang, Korea, P.D.R.

* KING (Dr D. S.), Department of Astronomy, University of New Mexico, Albuquerque, New Mexico 87106, U.S.A.

KING (Dr H. C.), McLaughlin Planetarium, Royal Ontario Museum, University of Toronto, 100 Queen's Park, Toronto 5, Canada

KING (Dr I. R.), Berkeley Astronomy Department, University of California, Berkeley, California, 94720, U.S.A.

KING (Dr R. B.), Department of Physics, California Institute of Technology, Pasadena, California, 91109, U.S.A.

KINMAN (Dr T. D.), Kitt Peak National Observatory, P.O. Box 4130, 950 North Cherry Avenue, Tucson, Arizona 85717, U.S.A.

KIPPENHAHN (Prof. R.), Universitäts-Sternwarte Göttingen, Geismarlandstrasse 11, 34 Götingen, F.R.G.

KIPPER (Prof. A.), W. Struve Astrophysical Observatory, Tartu, Tôravere, Estonia, U.S.S.R.

KIRAL (Dr A.), University Observatory, Beyazit, Istanbul, Turkey

KISELEV (Dr A. A.), Main Astronomical Observatory, USSR Academy of Sciences, Pulkovo, U.S.S.R.

KISLYAKOV (Dr A. G.), Radiophysical Research Institute, Gorkij, U.S.S.R.

KISSELL (Dr K. E.), Director, General Physics Research Laboratory, Aerospace Research Laboratories (ARP), Wright-Patterson Air Force Base, Ohio 45433, U.S.A.

KITAMURA (Dr M.), Tokyo Astronomical Observatory, Mitaka, Tokyo, Japan

KIZILIRMAK (Prof. A.), Ege University Observatory, P.K. 21, Bornova, Izmir, Turkey

* KJELDSETH MOE (Dr O.), Institute of Theoretical Astrophysics, University of Oslo, P.O. Box 1029, Blindern, Oslo 3, Norway

* KLARE (Dr G.), Bad. Landessternwarte, 6900 Heidelberg-Königstuhl, F.R.G.

* KLARMAN (Dr J.), Washington University, Department of Physics, Saint Louis, Missouri 63130, U.S.A.

KLECZEK (Dr J.), Astronomical Institute, Czechoslovak Academy of Sciences, Observatory Ondřejov, Czechoslovakia

KLEMOLA (Dr A. R.), Lick Observatory, University of California, Santa Cruz, California 95060, U.S.A.

* KLEPCZYNSKI (Dr W. J.), U.S. Naval Observatory, Washington, D.C. 20390, U.S.A.

KLIMSHIN (Dr I. A.), Astronomical Observatory, Lvov University, Lvov, U.S.S.R.

KLOCK (Dr B. L.), U.S. Naval Observatory, Department of Navy, Washington, D.C. 20390, U.S.A.

KNIPE (G. F. G.), Republic Observatory, Gill Street, Johannesburg, South Africa

KO (Dr Hsien-Ching), Department of Electrical Engineering, Ohio State University, Columbus, Ohio 43210, U.S.A.

* KOBRIN (Dr M. M.), Scientific Research Radiophysical Institute, Gorki University, Gorki U.S.S.R.

KOCH (Dr R. H.), Flower and Cook Observatory, University of Pennsylvania, Philadelphia, Pennsylvania 19104, U.S.A.

* KODAIRA (Dr K.), Tokyo Astronomical Observatory, Mitaka, Tokyo, Japan

KOEBCKE (Dr F.), Astronomical Observatory, Sloneczna 36, Poznan, Poland

* KOECKELENBERGH (DrA.), Observatoire Royal de Belgique, avenue Circulaire, 3, B-1180 Bruxelles, Belgium

* KOEHLER (Prof.DrJ.A.R.), Department of Physics, University of Saskatchewan, Saskatoon, Canada

KOELBLOED (DrD.), Astronomical Institute, Roetersstraat 1a, Amsterdam, The Netherlands

KOGURE (DrT.), Department of Physics, Faculty of Science, Ibaraki University, Mito (310), Japan

KÖHLER (Prof.DrH.), Sauerbruchstrasse 6, Heidenheim a.d. Brenz, F.R.G.

KOHOUTEK (DrL.), Hamburger Sternwarte, 205 Hamburg-Bergedorf, F.R.G.

KOLCHINSKIJ (DrI.G.), Main Astronomical Observatory, Ukrainian Academy of Sciences, Kiev, U.S.S.R.

KOLESOV (DrA.K.), Astronomical Observatory, Leningrad, U.S.S.R.

* KONDO (DrY.), Chief Astronomy Branch Tg4, Manned Spacecraft Center, Houston, Texas 77058, U.S.A.

KÖNIG (DrA.), Bahnhofstr. 46b, 6903 Neckargemünd, F.R.G.

* KONIN (DrV.V.), Nikolaev Observatory, Nikolaev, U.S.S.R.

KONOPLEVA (Dr V.P.), Main Astronomical Observatory, Academy of Sciences, Ukrainian SSR, Kiev 127, U.S.S.R.

KONOVOVICH (DrE.V.), Sternberg Astronomical Institute, Moscow V-234, U.S.S.R.

KOPAL (Prof.Z.), Department of Astronomy, The University, Manchester 13, U.K.

KOPECKÝ (DrM.), Astronomical Institute, Czechoslovak Academy of Sciences, Observatory Ondřejov, Czechoslovakia

KOPYLOV (DrI.M.), Director, Special Astrophysical Observatory, USSR Academy of Sciences, St. Zelenchukskaya, Stavropolsky Kraj, U.S.S.R.

* KORCHAK (DrA.A.), Institute of Terrestrial Magnetism, Ionosphere and Propagation of Radio Waves, USSR Academy of Sciences, Moscow, U.S.S.R.

KORDYLEWSKI (DrK.), Astronomical Observatory, Kopernika 27, Cracow, Poland

KOROL' (DrA.K.), Main Astronomical Observatory, Ukrainian Academy of Sciences, Kiev, U.S.S.R.

KOROL'KOV (DrD.V.), Main Astronomical Observatory, USSR Academy of Sciences, Pulkovo, U.S.S.R.

KOSIN (DrG.S.), Main Astronomical Observatory, USSR Academy of Sciences, Pulkovo, U.S.S.R.

* KOSTINA (MrsDrL.D.), Main Astronomical Observatory, USSR Academy of Sciences, Pulkovo, U.S.S.R.

* KOSTYAKOVA (MrsDrE.B.), Sternberg Astronomical Institute, Moscow, U.S.S.R.

KOSTYLEV (DrK.V.), Engelhardt Observatory, Kazan, U.S.S.R.

KOTEL'NIKOV (Acad.V.A.), Commission for Radioastronomy, USSR Academy of Sciences, Moscow, U.S.S.R.

KOTHARI (DrD.S.), Department of Physics, University of Delhi, New Delhi, India

KOTSAKIS (Prof.DrD.), University of Athens, Director of Astronomical Institute, National Observatory, 189, Hippocrates Str., Athens (708), Greece

KOURGANOFF (Prof.V.), Faculté des Sciences, Service d'Astronomie, 91-Orsay, France

KOVAL' (DrI.K.), Main Astronomical Observatory, Ukrainian Academy of Sciences, Kiev, U.S.S.R.

KOVALEVSKY (DrJ.), Bureau des Longitudes, 3, Rue Mazarine, Paris-6e, France

* KOVAR (MrsDrN.S.), Associate Professor, Physics Department, University of Houston, Houston, Texas 77004, U.S.A.

* KOVAR (DrR.P.), NASA Manned Spacecraft Center, Houston, Texas 77058, U.S.A.

* KOVETZ (DrA.), Department of Physics and Astronomy, Tel-Aviv University, Ramat-Aviv, Israel

KOWATCHEW (B.J.), Section d'Astronomie de l'Académie Bulgare des Sciences, Rue 7 Novembre No. 1, Sofia, Bulgaria

Kox (Dr H.), Hamburger Sternwarte, 205 Hamburg-Bergedorf, F.R.G.

Kozai (Dr Y.), Tokyo Astronomical Observatory, Mitaka, Tokyo, Japan

* Kozhevnikov (Dr N. I.), Sternberg Astronomical Institute, Moscow, U.S.S.R.

Kozieł (Prof. Dr K.), Director Jagellonian University Observatory, Kopernika 27/3, Cracow, Poland

* Kozlovsky (Dr B. Z.), Department of Physics and Astronomy, Tel-Aviv University, Ramat Aviv, Israel

Kozyrev (Dr N. A.), Pulkovo Observatory, Leningrad, U.S.S.R.

Kraft (Dr R. P.), Mount Wilson and Palomar Observatories, 813 Santa Barbara Street, Pasadena, California, U.S.A.

Kramer (Dr Kh. N.), Astronomical Observatory, Odessa University, Odessa, U.S.S.R.

Kranjc (Dr A.), Osservatorio Astronomico di Bologna, Italy

Krasovskij (Prof. Dr I.), Box 1269, Moscow K 9, U.S.S.R.

Krat (Prof. V. A.), Pulkovo Observatory, Leningrad M-140, U.S.S.R.

Kraus (Dr J. D.), Radio Observatory, Ohio State University, Columbus, Ohio, U.S.A.

Krause (Dr F.), Zentralinstitut für Astrophysik, DDR 15 Potsdam, Telegrafenberg, G.D.R.

Kraushaar (Prof. W. L.), Dept. of Physics, University of Wisconsin, Sterling Hall, Madison, Wisconsin 53706, U.S.A.

Kresák (Dr L.), Astronomical Institute, Slovak Academy of Sciences, Dúbravská cesta A/11, Bratislava, Czechoslovakia

Kresáková (Mrs Dr M.), Astronomical Institute, Slovak Academy of Sciences, Dúbravská cesta A/11, Bratislava, Czechoslovakia

Krinov (Dr E. L.), Committee on Meteorites of the Academy of Sciences of the USSR, ul. M. Ul'ianovoy 3, korpus 1, Moscow W-313, U.S.S.R.

Krishna Swamy (Dr K. S.), Theoretical Astrophysics Group, Tata Institute of Fundamental Research, Colaba, Bombay 5, India

Krishnan (T.), ASTRO Research Corporation, 1330 Cacique, Santa Barbara, California 93103, P.O. Box 4128, U.S.A.

Kristenson (Dr H.), Arkitektvägen 6, S-245 00 Staffanstorp, Sweden

Kristian (Dr J.), Mount Wilson and Palomar Observatories, Carnegie Institution of Washington, California Institute of Technology, 813 Santa Barbara Street, Pasadena, California 91106, U.S.A.

Křivský (Dr L.), Astronomical Institute, Czechoslovak Academy of Sciences, Observatory Ondřejov, Czechoslovakia

* Kříž (Dr S.), Astronomical Institute, Czechoslovak Academy of Sciences, Observatory Ondřejov, Czechoslovakia

Krogdahl (Prof. W. S.), Department of Astronomy, University of Kentucky, Lexington, Kentucky, U.S.A.

Kron (Dr G. E.), U.S. Naval Observatory, Flagstaff Station, Flagstaff, Arizona 86001, U.S.A.

Kron (Mrs K. G.), P.O. Box 217, Flagstaff, Arizona 86001, U.S.A.

Krook (Dr M.), Harvard College Observatory, Cambridge, Massachusetts 02138, U.S.A.

Krüger (Dr A.), Heinrich-Hertz-Institut für Solar-Terrestrische Physik, Berlin-Adlershof, Rudower Chaussee, G.D.R.

Krüger (Prof. E.), Via Mauro Macchi 65, Milano, Italy

Kruszewski (Dr A.) Astronomical Observatory of Warsaw University, Al. Ujazdowskie 4, Warsaw, Poland

Krzeminski (Dr W.), Zakład Astronomii P.A.N., Aleje Ujazdowskie 4, Warszawa, Poland

* Kubota (Dr J.), Kwasan Observatory, Yamashina, Kyoto, Japan

Kuhi (Dr L. V.), Berkeley Astronomy Department, University of California, Berkeley, California 94720, U.S.A.

Kuiper (Dr G. P.), Lunar and Planetary Laboratory, University of Arizona, Tucson, Arizona, 85721, U.S.A.

KUKARKIN (Prof. Dr B. V.), Sternberg Astronomical Institute, Moscow V-234, U.S.S.R.

KUKLIN (Dr G. V.), Post Office Meget, Sibizmiran, Irkutsk, U.S.S.R.

KULIKOV (Prof. Dr K. A.), Sternberg Astronomical Institute, Moscow V-234, U.S.S.R.

KULIKOVSKIJ (Dr P. G.), Sternberg State Astronomical Institute, Moscow V-234, U.S.S.R.

* KULKARNI (Dr P. V.), Physical Research Laboratory, Navrangpura, Ahmedabad-9, India

KUMAR (Prof. S. S.), Leander McCormick Observatory, University of Virginia, Charlottesville, Virginia, U.S.A.

KUMSISHVILI (Dr J. I.), Abastumani Astrophysical Observatory, Georgian Academy of Sciences, Abastumani, U.S.S.R.

KUNDU (Dr M. R.), Astronomy Program, University of Maryland, College Park, Maryland 20742, U.S.A.

* KUNKEL (Dr W. E.), Cerro Tololo Inter-American Observatory, Casilla 63-D, La Serena, Chile

KÜNZEL (H.), Heinrich-Hertz-Institut für Solar-Terrestrische Physik, DDR 15 Potsdam, Telegrafenberg, G.D.R.

KUPERUS (Dr M.), Astronomical Observatory 'Sonnenborgh', Servaas Bolwerk 13, Utrecht, The Netherlands

KUPPERIAN (Dr J. E. Jr), Astrophysics Branch, Goddard Space Flight Center, NASA, Greenbelt, Maryland, U.S.A.

KUPREVICH (Dr N. F.), Pulkovo Observatory, Leningrad, U.S.S.R.

* KURIL'CHIK (Dr V. N.), Sternberg Astronomical Institute, Moscow, U.S.S.R.

KURT (Dr V. G.), Sternberg Astronomical Institute, Moscow V-234, U.S.S.R.

KURTH (Dr R.), Professor of Mathematics, Georgia Institute of Technology, Atlanta, Ga., U.S.A.

KUSHWAHA (Prof. R. S.), Department of Mathematics, University of Jodhpur, Jodhpur (Raj), India

KUSTAANHEIMO (Prof. P.), Klaukkala, Finland

* KUTUZOV (Dr S. A.), The Astronomical Observatory of the Leningrad University, Leningrad, 199178, U.S.S.R.

* KUZMENKO (Mrs Dr K. N.), Astronomical Observatory of the University, Odessa, U.S.S.R.

* KUZ'MIN (Dr A. D.), Physical Institute, USSR Academy of Sciences, Moscow, U.S.S.R.

KUZ'MIN (Dr G. G.), W. Struve Astrophysical Observatory, Tartu, Tôravere, Estonia, U.S.S.R.

KWEE (Dr K. K.), Sterrewacht Leiden, The Netherlands

* LABEYRIE (Dr L.), G.E.A. B.P. n° 2, 91-Gif s/Yvette, France

LABORDE (Dr G.), Observatoire de Paris, Section d'Astrophysique, 92-Meudon, France

LABS (Dr D.), Landessternwarte, Königstuhl, 69 Heidelberg, F.R.G.

LACCHINI (Prof. G. B.), Borgo Durbecco, Via A. Cicognani No. 10, Faenza (Ravenna), Italy

LACOMBE (Dr C. G.), Observatorio Nacional, Rua General Bruce, 586, São Cristovão (ZC-08), Rio de Janeiro, Gb., Brazil

LACROUTE (Prof. P.), Directeur de l'Observatoire de Strasbourg, 11, rue de l'Université, 67-Strasbourg (Bas-Rhin), France

LAFFINEUR (Dr M.), Institut d'Astrophysique, 98bis Boulevard Arago, Paris-14e, France

LAGERQVIST (Prof. A.), Institute of Physics, Stockholm University, Vanadisvägen 9, S-113 46 Stockholm VA, Sweden

LAHIRI (Sir N. C.), Officer-in-charge, Nautical Almanac Division, Meteorological Department, Alipore, Calcutta, India

LALLEMAND (Dr A.), Observatoire de Paris, 61, avenue de l'Observatoire, Paris-14e, France

LAMBERT (Dr D.), Department of Astrophysics, University Observatory, South Parks Rd, Oxford, U.K.

LAMBRECHT (Prof. Dr H.), Direktor der Universitäts-Sternwarte, Schillergässchen 2, 69 Jena, G.D.R.

LAMLA (Dr E.), Astronomische Institute der Universität, 53 Bonn, Poppelsdorfer Allee 49, F.R.G.

LANDINI (M.), Osservatorio Astrofisico di Arcetri, Firenze, Italy

LANDOLT (Dr A.), Department of Astronomy and Physics, Louisiana State University, Baton Rouge, Louisiana 70803, U.S.A.

LA PAZ (Dr L.), Director, Institute of Meteoritics, University of New Mexico, Albuquerque, New Mexico, U.S.A.

LA POINTE (Dr S. M.), Department of Physics, University of Montreal, Montreal, Quebec, Canada

* LAPUSHKA (Dr K. K.), Astronomical Observatory, Latvian University, Riga, Latvian SSR, U.S.S.R.

LARGE (Dr M. I.), Chatterton Astronomy Department, University of Sydney, Sydney, Australia

LARINK (Prof. Dr J.), Hamburger Sternwarte, 205 Hamburg-Bergedorf, F.R.G.

LARSSON-LEANDER (Dr G.), Lund Observatory, S-222 24 Lund, Sweden

* LASKER (Dr B. M.), Cerro Tololo Inter-American Observatory, Casilla 63-D, La Serena, Chile

* LATYPOV (Dr A. A.), Astronomical Institute of the Usbek SSR, Academy of Sciences, Tashkent, U.S.S.R.

LAUSTSEN (Dr S.), European Southern Observatory, ETP Division, CERN, Ch-1211 Geneva 23, Switzerland

LAUTMAN (Dr D. A.), Smithsonian Astrophysical Observatory, 60 Garden Street, Cambridge, Massachusetts 02138, U.S.A.

LAVDOVSKIJ (Dr V. V.), Pulkovo Observatory, Leningrad, U.S.S.R.

LAVROV (Dr M. I.), Engelhardt Astronomical Observatory, Kazan University, Kazan, U.S.S.R.

* LAVRUKHINA (Mrs Prof. Dr A. K.), Institute of Geochemistry and Analytic Chemistry, USSR Academy of Sciences, Moscow, U.S.S.R.

LAWRENCE, Douglas Advanced Research Laboratory, 5251 Bolsa Avenue, Huntington Beach, California 92646, U.S.A.

LAYZER (Prof. D.), Harvard College Observatory, 60 Garden Street, Cambridge, Massachusetts 02138, U.S.A.

* LEBEDINETS (Dr V. N.), Commission on Comets and Meteors, Astronomical Council, USSR Academy of Sciences, Moscow, U.S.S.R.

LEBOVITZ (Dr N.), Department of Mathematics, Eckhart Hall, Room 309, University of Chicago, 1118-32 East 58th Street, Chicago 60637, U.S.A.

LECAR (Dr M.), Smithsonian Astrophysical Observatory, 60 Garden Street, Cambridge, Massachusetts 02138, U.S.A.

* LE CONTEL (J. M.), Observatoire de Paris, 61, avenue de l'Observatoire, Paris-14e, France

LEDERLE (Dr T.), Astronomisches Rechen-Institut, Mönchhofstrasse 12–14, 69 Heidelberg, F.R.G.

LEDOUX (Prof. Dr P.), Institut d'Astrophysique, Université de Liège, avenue de Cointe, 5, B-4200 Cointe-Ougrée, Belgium

* LEFÈVRE (Dr J.), Observatoire de Nice, Le Mont Gros, 06-Nice, France

LEHNERT (Dr B. P.), Royal Institute of Technology, Department of Electronics, S-10044 Stockholm 70, Sweden

* LEGG (Dr T. H.), Astrophysics Branch National Research Council, Ottawa 7, Ontario, Canada

LEIGHTON (Prof. R. B.), California Institute of Technology, Pasadena, California 91109, U.S.A.

LEIKIN (Dr G. A.), Astronomical Council of the USSR Academy of Sciences, Vavilov Str. 34, Moscow V-312, U.S.S.R.

* LENA (Dr P.), Observatoire de Meudon, 92-Meudon, France

LENGAUER (Dr G. G.), Main Astronomical Observatory, USSR Academy of Sciences, Pulkovo, U.S.S.R.

LEQUEUX (J.), Observatoire de Paris, Section d'Astrophysique, 92-Meudon, France

LEROY (Dr J.-L.), Observatoire du Pic-du-Midi, 65-Bagnères-de Bigorre, Hautes Pyrénées, France

LE SQUEREN (Mrs A. M.), Observatoire de Paris, Section d'Astrophysique, 92-Meudon, France

LETFUS (Dr V.), Astronomical Institute, Czechoslovak Academy of Sciences, Observatory Ondřejov, Czechoslovakia

* LEUNG (Dr Kam Ching), Institute for Space Studies, Goddard Space Flight Center, New York 10025, U.S.A.

LEVALLOIS (J.J.), Institut Geographique National, 2, av. Pasteur, 94-Saint-Mandé, France

LEVIN (Dr B.J.), O. Schmidt Institute of Physics of the Earth, Academy of Sciences of the USSR, B. Gruzinskaya 10, Moscow D-242, U.S.S.R.

LÉVY (Dr J.R.), Observatoire de Paris, 61, avenue de l'Observatoire, Paris-14e, France

* LI (J.Y.), Pyongyang Astronomical Observatory, Academy of Sciences, Pyongyang, Korea, P.D.R.

LIDDELL (U.), National Aeronautics and Space Administration, Lunar and Planetary Program, Office of Space Science and Applications, Washington, D.C. 20546, U.S.A.

LILLER (Dr M.), Harvard College Observatory, 60 Garden Street, Cambridge, Massachusetts 02138, U.S.A.

LILLER (Prof. W.), Harvard College Observatory, 60 Garden Street, Cambridge, Massachusetts 02138, U.S.A.

LILLEY (Prof. A. E.), Harvard College Observatory, Cambridge, Massachusetts 02138, U.S.A.

LIMBER (Dr D. N.), Yerkes Observatory, Williams Bay, Wisconsin, U.S.A.

LIN (Dr C. C.), Department of Mathematics, Massachusetts Institute of Technology, Cambridge, Massachusetts, U.S.A.

LIN (Rong-an), 64, Kung Yuen Road, Taipei, Taiwan

LINDBLAD (Dr B. A.), Lund Observatory, S-222 24 Lund, Sweden

LINDBLAD (Dr P. O.), Stockholm Observatory, S-133 00 Saltsjöbaden, Sweden

LINDLEY (W. M.), Pentonwarra, Trevone, Padstow, Cornwall, U.K.

* LINDOFF (Dr U. I. G. S.), Institute of Astronomy, S-222 24 Lund, Sweden

LINDSAY (Dr E. M.), Director, Armagh Observatory, Armagh, North Ireland, U.K.

LINES (Dr A. W.), Royal Aircraft Establishment, Farnborough, Hants., U.K.

LINFOOT (Dr E. H.), The Observatories, Madingley Road, Cambridge, U.K.

LING (Dr Chih-bing), Institute of Mathematics, Academia Sinica, P.O. Box No. 143, Taipei, Taiwan

LINK (Dr F.), Institut d'Astrophysique, 98bis, Boul. Arago, Paris-14e, France

LINNELL (Dr A. P.), Department of Astronomy, Michigan State University, East Lansing, Michigan 48823, U.S.A.

LINNIK (Prof. Dr V. P.), Pulkovo Observatory, Leningrad, U.S.S.R.

LIPPINCOTT (S. L.), Sproul Observatory, Swarthmore College, Swarthmore, Pa., U.S.A.

LIPSKIJ (Dr Y. N.), Sternberg Astronomical Institute, Moscow, U.S.S.R.

LISZKA (Dr L.), Geophysical Observatory, S-981 00 Kiruna, Sweden

LITTLE (A. G.), Chatterton Astronomy Department, School of Physics, University of Sydney, Sydney, N.S.W., Australia

LITTLE (C. Gordon), Chief, Division 87, National Bureau of Standards, Room 3001, Radio Bldg., Boulder, Colorado, U.S.A.

LITTLEFIELD (Dr T. A.), Department of Physics, King's College, Newcastle upon Tyne, U.K.

LIVINGSTON (Dr W. C.), Kitt Peak National Observatory, P.O. Box 4130, Tucson, Arizona 85717, U.S.A.

* LIVSHITS (Dr M. A.), Institute of Terrestrial Magnetism, Ionosphere and Propagation of Radio Waves, USSR Academy of Sciences, Moscow, U.S.S.R.

LOCHTE-HOLTGREVEN (Prof. Dr W.), Institut für Experimentalphysik der Universität, Kiel, F.R.G.

LOCKE (Dr J. L.), Astrophysics Branch National Research Council, Ottawa 2, Ontario, Canada

LODÉN (Mrs Dr K.), Stockholm Observatory, S-133 00 Saltsjöbaden, Sweden

LODÉN (Dr L. O.), Astronomiska Observatoriet, Box 515, 751 20 Uppsala, Sweden

LOHMANN (Prof. Dr W.), Astronomisches Rechen-Institut, Mönchhofstrasse 12–14, 69 Heidelberg, F.R.G.

* LONGAIR (Dr M. S.), Mullard Radio Astronomy Observatory, Cavendish Laboratory, Cambridge, U.K.

LORÓN (Dr M. M.), Observatorio Astronomico, Alfonso XII, No. 3, Madrid, Spain

LORTET-ZUCKERMANN (Mrs M. C.), Institut d'Astrophysique, 98bis, Boulevard Arago, Paris-14e, France

* LOTOVA (Mrs Dr N. A.), Physical Institute, USSR Academy of Sciences, Moscow, U.S.S.R.

LOUGHHEAD (R. E.), National Standards Laboratory, CSIRO, University Grounds, Sydney, N.S.W., Australia

* LOUISE (Dr R.), L.A.S.-Traverse du Siphon, Les 3 Lucs, 13-Marseille 12e, France

LOURENS (J. v. B.), Royal Observatory, Observatory, Cape Province, South Africa

* LOVAS (M.), Konkoly Observatory of the Hungarian Academy of Sciences, Konkoly Thege Miklós u 13/17, Budapest XII, Hungary

LOW (Dr F.J.), Rice University, Department of Space Science, Houston, Texas 77001, U.S.A.

LOVELL (Sir Bernard), Nuffield Radio Astronomy Laboratories, Jodrell Bank, Macclesfield, Cheshire, U.K., Vice-President of the Union.

LOZINSKIJ (Dr A. M.), Astronomical Council of the USSR Academy of Sciences, Vavilov Str. 34, Moscow V-312, U.S.S.R.

LÜBECK (K.), Hamburger Sternwarte, 205 Hamburg-Bergedorf, F.R.G.

* LUKATSKAJA (Mrs Dr F. I.), Main Astronomical Observatory, Ukrainian Academy of Sciences, Kiev, U.S.S.R.

LUNDQUIST (Dr C. A.), 218 Parker Street, Newton, Massachusetts, U.S.A.

LUNEL (Mrs M.), Observatoire de Lyon, 69-Saint-Genis-Laval, France

LUPLAU JANSSEN (Dr C. E.), Urania Observatory, Dr Olgas Vej 25, Copenhagen F, Denmark

LÜST (Prof. Dr R.), Max-Planck-Institut für extraterrestrische Physik, 8046 Garching b. München, F.R.G.

LÜST (Dr Rhea), Max-Planck-Institut für Physik und Astrophysik, Föhringer Ring 6, 8000 München 23, F.R.G.

* LUUD (Dr L.S.), W. Struve Astrophysical Observatory, Estonian Academy of Sciences, Tartu, Estonian SSR, U.S.S.R.

LUYTEN (Prof. W.J.), The Observatory, University of Minnesota, Minneapolis 14, Minnesota, U.S.A.

LYNDEN-BELL (Dr D.), Royal Greenwich Observatory, Herstmonceux Castle, Hailsham, Sussex, U.K.

LYNDS (Dr B.T.), Steward Observatory, University of Arizona, Tucson, Arizona 85721, U.S.A.

LYNDS (Dr C.R.), Kitt Peak National Observatory, 950 North Cherry Avenue, P.O. Box 4130, Tucson, Arizona 85717, U.S.A.

LYNGÅ (Dr G.), Institutionen för astronomi, 222 24 Lund, Sweden

LYTTKENS (Dr E.), Skolgatan 33 B, S-752 21 Uppsala, Sweden

LYTTLETON (Dr R. A.), St. John's College, Cambridge, U.K.

McCARTHY, S.J. (Dr M. F.), Specola Vaticana, Castel Gandolfo, Vatican City State

McCLAIN (E. F.), 225 Maple Road, Morningside, Maryland 20023, U.S.A.

* McCLURE (Dr R. H.), Yale University, Department of Astronomy, New Haven, Connecticut 07520, U.S.A.

* McCRACKEN (Prof. K. G.), Department of Physics, University of Adelaide, Adelaide 5000, Australia

McCREA (Prof. W. H.), The University of Sussex, School of Mathematical and Physical Sciences, Falmer, Sussex, U.K.

McCROSKY (Dr R. E.), Smithsonian Astrophysical Observatory, 60 Garden Street, Cambridge, Massachusetts 02138, U.S.A.

McCUSKEY (Dr S. W.), Warner and Swasey Observatory, Case Institute of Technology, East Cleveland, Ohio 44112, U.S.A.

McDONALD (Dr F.B.), NASA, Goddard Space Flight Center, Greenbelt, Maryland 20771, U.S.A.

McDONALD (Mrs Dr J. K. Petrie), Dominion Astrophysical Observatory, Victoria, B.C., Canada

McELROY (Dr M. B.), Kitt Peak National Observatory, 950 North Cherry, Tucson, Arizona, U.S.A.

McGEE (Prof. J. D.), Physics Department, Imperial College of Science and Technology, Prince Consort Road, London S.W. 7, U.K.

McGEE (Dr R. X.), CSIRO, Division of Radiophysics, Post Office Box 76, Epping, N.S.W. 2121, Australia

McINTOSH (Dr B. A.), Astrophysics Branch, National Research Council, Ottawa 7, Ontario, Canada

McKENNA-LAWLOR (S. M. P.), Vishnu House, Blanchardstown, Dublin, Ireland

McKINLEY (Dr D. W. R.), Executive Director (Laboratories), National Research Council, Ottawa 7, Ontario, Canada

* McLEAN (Dr D.), CSIRO, Division of Radiophysics, P.O. Box 76, Epping 2121, Australia

* McMULLAN (Dr D.), Royal Greenwich Observatory, Herstmonceux Castle, Hailsham, Sussex, U.K.

McNALLY (Dr D.), University of London Observatory, Mill Hill Park, London, N.W. 7, U.K.

McNAMARA (Dr D. H.), Department of Physics, Brigham Young University, Provo, Utah, U.S.A.

McNARRY (L.R.), Astrophysics Branch, National Research Council, Ottawa, Ontario, Canada

McVITTIE (Prof. G. C.), University of Illinois Observatory, Urbana, Illinois 61801, U.S.A.

* McWHIRTER (Dr R. W. P.), Science Research Council, Astrophysics Research Unit, Culham Laboratory, Abingdon, Berkshire, U.K.

* MacCONNELL (Dr D. J.), Department of Astronomy, University of Michigan, Ann Arbor, Michigan 48104, U.S.A.

* MacGARROLL (Dr R.), Observatoire de Meudon, 92-Meudon, France

* MacLEOD (Dr J. M.), Astrophysics Branch, National Research Council, Ottawa 7, Canada

MacRAE (Dr D. A.), Director of the David Dunlap Observatory, Richmond Hill, Ontario, Canada

MACHADO (J. M. A. Braz), Serviços Radioelectricos dos C.T.T., Lisbon, Portugal

MACRIS (Dr C. J.), Director, Research Center for Astronomy and Applied Mathematics, Academy of Athens, 14 Anagnostopoulou Street, Athens (136), Greece

MADEIRA (Dr J. A.), Observatorio Astronomico, Tapada, Lisbon, Portugal

MADWAR (Prof. M. R.), 15 Rue Pahlewi, Dokki, Cairo, Egypt, U.A.R.

MAFFEI (Dr P.), Laboratorio di Astrofisica, Casella Postale, 67-Frascati (Rome), Italy

MAGALASHVILI (Dr N. L.), Abastumani Observatory, Abastumani, Georgia, U.S.S.R.

* MAGNARADZE (Mrs Dr N. G.), State University, Tbilisi, Georgia, U.S.S.R.

MAÎTRE (Dr V.), Observatoire, 25-Besançon, France

* MAKAROVA (Mrs Dr E. A.), Sternberg Astronomical Institute, Moscow, U.S.S.R.

MAKITA (Dr M.), Tokyo Astronomical Observatory, University of Tokyo, Mitaka, Tokyo, Japan

MALACARA (Dr D.), Observatorio Astronómico Nacional, Apdo. Postal 70–264, Ciudad Universitaria, Mexico 20, D.F., Mexico

* MALAISE (Dr D. J.), Institut d'Astrophysique, Université de Liège, avenue de Cointe, 5, B-4200 Cointe-Ougrée, Belgium

MALITSON (Mrs H. H.), Code 615, Radio Astronomy Section, Ionospheric and Radio Physics Branch, Laboratory for Space Sciences, NASA-Goddard Space Flight Center, Greenbelt, Maryland 20771, U.S.A.

* MALLIA (Dr E. A.), Department of Astrophysics, South Parks Road, Oxford OXI 3RQ, U.K.

MALMQUIST (Prof. K. G.), S:t Olofsgatan 10 A, S-753 21 Uppsala, Sweden

MALTBY (Dr P.), Institute of Theoretical Astrophysics, University of Oslo, Blindern, Oslo 3, Norway

MALVILLE (Dr J. M.), High Altitude Observatory, Boulder, Colorado, U.S.A.

MAMEDBEJLI (Dr G. D.), Shemakha Astrophysical Observatory, Azerbajdzhan Academy of Sciences, Baku, U.S.S.R.

MAMMAMO (A.), Osservatorio Astrofisico, Asiago (Vicenza), Italy

* MANARA (Dr A.), Osservatorio Astronomico di Milano, via Brera 28, Milano, Italy

MANDEL'SHTAM (Dr S. L.), Laboratory of Spectroscopy, Lebedev Institute, Academy of Sciences of the USSR, Moscow, U.S.S.R.

MANGENEY (A.), Observatoire de Meudon, 5, place Janssen, 92-Meudon, France

MANNINO (Prof. G.), Osservatorio Astronomico Universitario, Via Zamboni 33, Bologna, Italy

MARAN (Dr S. P.), Solar Physics Branch, Code 614, Goddard Space Flight Center, Greenbelt, Maryland 20771, U.S.A.

MARCUS (Prof. E.), Observatoire de Bucarest, 5 rue Cutitul de Argint, Bucarest, Roumania

MARGONI (R.), Osservatorio Astrofisico, Asiago (Vicenza), Italy

* MARIK (Dr M.), Astronomical Institute of the Eötvös Loránd University, Kun Béla tér 2, Budapest, Hungary

MARKARYAN (Dr B. E.), Byurakan Observatory, Erevan, Armenia, U.S.S.R.

MARKOWITZ (Prof. Wm.), Department of Physics, Marquette University, Milwaukee, Wisconsin 53233, U.S.A.

MARSDEN (Dr B. G.), Smithsonian Astrophysical Observatory, 60 Garden Street, Cambridge, Massachusetts 02138, U.S.A.

MARTEL-CHOSSAT (Mrs Dr M. T.), Observatoire de Lyon, Saint-Genis-Laval, (Rhône), France

MARTIN (N.), Astronome Adjoint, Observatoire de Marseille, 2, place Le Verrier, 13-Marseille-4e, France

MARTIN, JR (W. C.), Chief, Spectroscopy Section, National Bureau of Standards, Physics Building A-165, Washington D.C. 20234, U.S.A.

MARTINET (L.), Observatoire Cantonal de Genève, Genève, Sauverny 1290, Switzerland

MARTRES (Mrs M. J.), Observatoire de Meudon, Place Jules Janssen, 92-Meudon, France

MARTYNOV (Prof. Dr D. Ya.), Director of the Sternberg Astronomical Institute, Moscow University, Leninskije Gory, Moscow-V234, U.S.S.R.

MARVIN (Dr U. B.), Smithsonian Astrophysical Observatory, Cambridge, Massachusetts 02138, U.S.A.

MARX (Prof. G.), Roland Eötvös University, Puskin u. 5–7, Budapest VIII, Hungary

MASANI (Prof. A.), Osservatorio Astronomico di Brera, Via Brera 28, Milano, Italy

MASSEVICH (Mrs Dr A. G.), Astronomical Council of the USSR Academy of Sciences, Vavilov Str. 34, Moscow V-312, U.S.S.R.

MATHEWS (Dr W. G.), Physics Department, University of California-San Diego, La Jolla, California 92038, U.S.A.

MATHEWSON (Dr D. S.), CSIRO, Division of Radiophysics, University Grounds, Chippendale, N.S.W., Australia

MATHIS (Dr J. S.), Astronomy Department, University of Wisconsin, Madison, Wisconsin 53706, U.S.A.

* MATSUNAMI (Dr N.), Tokyo Astronomical Observatory, Mitaka, Tokyo, Japan

* MATSUOKA (Dr M.), Institute of Space and Aeronautical Science, University of Tokyo, Komaba, Meguro-Ku, Tokyo, Japan

MATSUSHIMA (Prof. Dr S.), Department of Astronomy, The Pennsylvania State University, University Park, Pennsylvania 16802, U.S.A.

MATTHEWS (Dr T. A.), Department of Physics and Astronomy, University of Maryland, College Park, Md. 20740, U.S.A.

MATTIG (Dr W.), Fraunhofer Institut, Schöneckstrasse 6, 78 Freiburg i. Breisgau, F.R.G.

* MATVEYENKO (Dr L. I.), Institute for Space Research, Profsojuznaja 88, Moscow V 485, U.S.S.R.

* MAUDER (Dr H.), Remeis-Sternwarte, Sternwartstr. 7, 8600 Bamberg, F.R.G.

MAVRIDIS (Prof. Dr L. N.), Department of Geodetic Astronomy, University of Thessaloniki, Thessaloniki, Greece

MAXWELL (Dr A.), Harvard Radio Astronomy Station, Fort Davis, Texas, U.S.A.

MAYALL (Mrs M. W.), 187 Concord Avenue, Cambridge, Massachusetts 02138, U.S.A.

MAYALL (Dr N. U.), Kitt Peak National Observatory, 950 North Cherry Avenue, Tucson, Arizona, U.S.A.

MAYER (C.H.), Code 7131, U.S. Naval Research Laboratory, Washington 25, D.C., U.S.A.

MAYER (Dr P.), Dept. of Astronomy and Astrophysics, Charles University, Švédská 8, Prague 5-Smíchov, Czechoslovakia

MAYER (Dr U.), Astronomisches Institut der Universität, Waldhäuser Str. 64, 74 Tübingen, F.R.G.

* MAYFIELD (Dr E.B.), Solar Physics Department, The Aerospace Corporation, P.O. Box 95085, Los Angeles, California 90045, U.S.A.

MEABURN (Dr J.), Department of Astronomy, The University, Manchester 13, U.K.

* MEADOWS (Dr A.J.), Astronomy Department, University of Leicester, University Road, Leicester, U.K.

MEDD (W.J.), Astrophysics Branch, National Research Council, Ottawa, Ontario, Canada

MEEKS (Dr M.L.), Millstone Hill Field Station, MIT-Lincoln Laboratory, Lexington, Massachusetts, U.S.A.

MEFFROY (J.), Chef de Travaux, Faculté des Sciences, 34-Montpellier, France

MEGRELISHVILI (Mrs Dr T.G.), Abastumani Astrophysical Observatory, Georgian Academy of Sciences, Abastumani, U.S.S.R.

MEHLTRETTER (Dr J.P.), Fraunhofer Institut, Schöneckstrasse 6, 78 Freiburg, Breisgau, F.R.G.

MEIN (P.), Observatoire de Meudon, 5, place Janssen, 92-Meudon, France

MEINEL (Dr A.B.), Kitt Peak National Observatory, 950 North Cherry Avenue, Tucson, Arizona, U.S.A.

MEINIG (Dr M.), Zentralinstitut Physik der Erde, DDR 15 Potsdam, Telegrafenberg, G.D.R.

MELCHIOR (Prof. Dr P.), Chef du Département d'Astronomie de Position et de Géodynamique, Observatoire Royal de Belgique, avenue Circulaire 3, B-1180 Bruxelles, Belgium

MEL'NIKOV (Prof. Dr O.A.), Pulkovo Observatory, Leningrad, U.S.S.R.

* MEN' (Dr A.V.), Institute of Radiophysics and Electronics, Academy of Sciences, Ukrainian SSR, Khar'kov, U.S.S.R.

MENDEZ (Dr M.), Observatorio Astronómico Nacional, Apdo. Postal 70–264, Ciudad Universitaria, México 20, D.F., Mexico

MENDOZA V. (Dr E.), Observatorio Astronomico Nacional, 20 Piso Torre de Ciencias, Ciudad Universitaria, Mexico 20, D.F., Mexico

MENON (Dr T.K.), Tata Institute of Fundamental Research, Homi Bhabha Road, Colaba, Bombay 5, India

MENZEL (Dr D.H.), Harvard College Observatory, 60 Garden Street, Cambridge, Massachusetts 02138, U.S.A.

MERGENTALER (Prof. Dr J.), Astronomical Observatory, Kopernika 11, Wroclaw, Poland

MERMAN (Dr G.A.), Institute of Theoretical Astronomy, Leningrad, U.S.S.R.

MERMAN (Mrs Dr N.V.), Main Astronomical Observatory, USSR Academy of Sciences, Pulkovo, U.S.S.R.

MERRILL (Dr J.E.), University of Florida Observatories, Department of Physics and Astronomy, University of Florida, Gainesville, Florida 32601, U.S.A.

MERTON (Dr G.), 17 Holywell, Oxford, U.K.

MESSAGE (Dr P.J.), Mathematics Department, University of Liverpool, Liverpool 3, U.K.

MESTEL (Prof. L.), Department of Mathematics, The University, Manchester 13, U.K.

MEURERS (Prof. Dr J.), Direktor der Universitätssternwarte, Türkenschanzstrasse 17, A-1180 Vienna, Austria

MEYER (Dr F.), Max-Planck-Institut für Physik und Astrophysik, Föhringer Ring 6, 8000 München 23, F.R.G.

* MEYER-HOFMEISTER (Mrs Dr E.), Max-Planck-Institut für Physik und Astrophysik, Föhringer Ring 6, 8000 München 23, F.R.G.

MEZGER (Dr P.G.), Max-Planck-Institut für Radioastronomie, Argelanderstrasse 3, 53 Bonn, F.R.G.

MIANES (P.), Observatoire de Lyon, 69-Saint-Genis-Laval, France

MICHARD (Dr R.), Observatoire de Paris, Section d'Astrophysique, 92-Meudon, France

* MICHEL (Dr E.C.), Department of Space Sciences, Rice University, Houston, Texas 77001, U.S.A.

MICHEL (Ingénieur H.), rue Ten Bosch, 54, B-1050 Bruxelles, Belgium

MICHKOVITCH (Prof. V. V.), Serbian Academy of Sciences and Arts, Knez Mihaïlova 35, Belgrade, Yugoslavia

MICZAIKA (Dr G.R.), TRW Systems, 1 Space Park, Redendo Beach, California, U.S.A.

MIDDLEHURST (Miss B. M.), c/o Royal Astronomical Society, Burlington House, Piccadilly, London, W.C. 1, U.K.

MIGEOTTE (Prof. Dr M.), Institut d'Astrophysique, Université de Liège, avenue de Cointe 5, B-4200 Cointe-Ougrée, Belgium

* MIHAILA (Dr I.), Observatorul Astronomic, Strada Cutitul de Argint 5, Bucuresti 28, Roumania

MIHALAS (Dr D.), Department of Astronomy and Astrophysics, University of Chicago, Chicago, Illinois 60611, U.S.A.

* MIKAIL (Dr J. S.), Helwan Observatory, Near Cairo, Egypt, U.A.R.

MIKESELL (Dr A. H.), U.S. Naval Observatory, Washington, D.C. 20390, U.S.A.

MIKHAILOV (Prof. Dr A. A.), Director of the Pulkovo Observatory, Leningrad M 140, U.S.S.R.

MIKHEL'SON (Dr N. N.), Main Astronomical Observatory, Academy of Sciences, Pulkovo, U.S.S.R.

MILET (B.), Aide Astronome, Observatoire de Nice, Le Mont-Gros, 06-Nice, France

MILFORD (Dr S. N.), Astrophysics and Geophysics Research, Grumman Aircraft Engineering Corporation, Bethpage, Long Island, New York, U.S.A.

MILLER (Dr F. D.), Department of Astronomy, The University of Michigan, Physics-Astronomy Building, Ann Arbor, Michigan 48104, U.S.A.

* MILLER (Dr J.S.), Lick Observatory, University of California, Santa Cruz, California 95060, U.S.A.

MILLER (Dr R. H.), Institute for Computer Research, University of Chicago, 5640 Ellis Avenue, Chicago, Illinois 60637, U.S.A.

MILLER, S.J. (Rev. W.J.), Director, Fordham University Astronomical Laboratory, Box 88, Fordham University, New York, N.Y. 10458, U.S.A.

MILLIGAN (J. E.), Astrophysics Branch, Goddard Space Flight Center, Greenbelt, Maryland, U.S.A.

MILLMAN (Dr P. M.), Astrophysics Branch, National Research Council, Ottawa 7, Ontario, Canada

MILLS (Dr B. Y.), School of Physics, University of Sydney, Sydney, N.S.W., Australia

* MILOVANOVIĆ (Dr V.), Service de Variation de latitude et du service de l'heure à l'Observatoire Astronomique de Belgrade, Belgrade, Yugoslavia

MININ (Dr I. N.), Leningrad University Observatory, Leningrad, U.S.S.R.

MINKOWSKI (Dr R. L.), Radio Astronomy Laboratory, University of California, Berkeley, California 94720, U.S.A.

* MINNET (H.C.), Anglo-Australian Telescope Project Office, c/o Department of Education and Science, P.O. Box 826, Canberra 2601, Australia

MIRZABEKYAN (Dr E. G.), Astrophysical Observatory, Byurakan, Armenia, U.S.S.R.

MIRZOYAN (Dr L. V.), Byurakan Astrophysical Observatory, Erevan, Armenia, U.S.S.R.

* MISNER (Dr Ch. W.), Department of Astronomy, University of Maryland, College Park, Maryland 20742, U.S.A.

MISSANA (Prof. N.), Osservatorio Astronomico di Pino Torinese, (Torino), Italy

* MITCHELL (R.), Steward Observatory, University of Arizona, Tucson, Arizona 85721, U.S.A.

MITCHELL, JR. (Dr W. E.), Perkins Observatory, Delaware, Ohio, U.S.A.

MITIĆ (L. A.), Astronomical Observatory, Volgina 7, Belgrade, Yugoslavia

MITRA (Dr A. P.), Assistant Director, National Physical Laboratory, New Delhi, India

MITROFANOVA (Miss Dr L. A.), Main Astronomical Observatory, USSR Academy of Sciences, Pulkovo, U.S.S.R.

MIYADI (Dr M.), 3–6 Senkawa, Chôfu-shi, Tokyo, Japan

MIYAMOTO (Prof. Dr S.), Director, Kwasan Observatory, Faculty of Science, University of Kyoto, Yamashina, Kyoto, Japan

MOERDIJK (Dr W. G.), Sterrenkundig Instituut, Rijksuniversiteit Gent, J. Plateaustraat 22, B-9000 Gent, Belgium

MOFFET (Dr A. T.), Radio Astronomy Laboratory, California Institute of Technology, Pasadena, California 91109, U.S.A.

MOGILEVSKIJ (Prof. E. I.), Institute of Terrestrial Magnetism, Ionosphere and Propagation of Radiowaves, USSR Academy of Sciences, Moscow, U.S.S.R.

MOHLER (Prof. O. C.), McMath-Hulbert Observatory, 895 Lake Angelus Road North, Pontiac 4, Michigan, U.S.A.

MOHR (Prof. J. M.), Dept. of Astronomy and Astrophysics, Charles University, Švédská 8, Prague 5-Smíchov, Czechoslovakia

MOISEEV (Dr I. G.), Crimean Astrophysical Observatory, USSR Academy of Sciences, P/O Nauchny, Crimea, U.S.S.R.

MOLCHANOV (Prof. A. P.), Radio Astronomical Laboratory, Institute of Physics, Leningrad State University, V-164, Leningrad, U.S.S.R.

MØLLER (O.), Ole Rømer Observatory, Arhus C, Denmark

* MONAGHAN (Dr J. J.), Department of Mathematics, Monash University, Clayton 3168, Australia

MONFILS (Prof. Dr A.), Institut d'Astrophysique, Université de Liège, avenue de Cointe 5, B-4200 Cointe-Ougrée, Belgium

MONIN (Dr G. A.), Crimean Astrophysical Observatory of the Academy of Sciences of the USSR, Pochtovoe, Crimea, U.S.S.R.

* MONNETT (Dr G.), Observatoire de Marseille, Place le Verrier, Marseille, France

MOORE, O.B.E. (P.), Farthings, 39 West Street, Selsey, Sussex, U.K.

MOORE-SITTERLY (Mrs Dr Ch.), National Bureau of Standards 151.00, Washington, D.C. 20234, U.S.A.

MORANDO (B.), Bureau des Longitudes, 3, rue Mazarine, 75-Paris-6e, France

MOREAU (Dr F.), Astronome Honoraire, Observatoire Royal de Belgique, avenue Georges Bergmann, 115, B-1050 Bruxelles, Belgium

MORENO (Prof. H.), Observatório Astronómico Nacional, Casilla 36-D, Santiago, Chile

MORETON (G. E.), Division of Physics, CSIRO, University Grounds, Chippendale, N.S.W., Australia

MORGAN (Dr W. W.), Yerkes Observatory, Williams Bay, Wisconsin 53191, U.S.A.

MORIMOTO (Dr M.), Tokyo Astronomical Observatory, University of Tokyo, Mitaka, Tokyo, Japan

MORIYAMA (Dr F.), Tokyo Astronomical Observatory, Mitaka, Tokyo, Japan

MOROZ (Dr V. I.), Sternberg Astronomical Institute, Moscow, U.S.S.R.

MORRIS (Dr S. C.), Dominion Astrophysical Observatory, R.R. 7 Victoria, B.C., Canada

MORRISON (L. V.), Royal Greenwich Observatory, Herstmonceux Castle, Hailsham, Sussex, U.K.

MORRISON (Dr P.), Department of Physics, Massachusetts Institute of Technology, Cambridge, Massachusetts 02139, U.S.A.

MORTON (Dr D. C.), Princeton University Observatory, Peyton Hall, Princeton, New Jersey 08540, U.S.A.

MORTON (Dr G. A.), R.C.A. Laboratories, Princeton, New Jersey, U.S.A.

MOTZ (Dr L.), Rutherfure Observatory, Columbia University, Box 57 Pupin Hall, New York, New York 10027, U.S.A.

MOURADIAN (Z.), Observatoire de Meudon, 5 place Janssen, 92-Meudon, France

* MOUTSOULAS (Dr M.), Department of Astronomy, The University, Manchester, M 13 9PL, U.K.

* MRKOS (A.), Dept. of Astronomy and Astrophysics, Charles University, Švédská 8, Prague 5-Smíchov, Czechoslovakia

MUGGLESTONE (Dr D.), University of Queensland, Brisbane, Queensland, Australia

MULDERS (Dr G. F. W.), c/o Federation of Amer. Soc. Experimental Biology, 9650 Rockville Pike, 20014 Bethesda, (Maryland) U.S.A.

* MULHOLLAND (Dr J. D.), P.O. Box 631, La Canada, California 91011, U.S.A.

MULLALY (Dr R. F.), School of Electrical Engineering, University of Sydney, Sydney, Australia

MÜLLER (Miss Prof. E. A.), Observatoire de Genève, 1290 Sauverny (GE), Switzerland

MÜLLER (Prof. Dr R.), Postfach 8, 8204 Degerndorf/Inn, F.R.G.

MULLER (Dr A. B.), Kapteyn Astronomical Laboratory, Broerstraat 7, Groningen, The Netherlands

MULLER (Prof. Ir. C. A.), Radio Observatory, Bosrand 25, Dwingeloo, The Netherlands

MULLER (Dr P.), Observatoire de Paris, Section d'Astrophysique, 92-Meudon, France

MUMFORD (G. S.), Department of Mathematics, Tufts University, Medford, Massachusetts 02155, U.S.A.

MÜNCH (Prof. G.), California Institute of Technology, 1201 E. California Street, Pasadena, California, U.S.A.

MURRAY (C. A.), Royal Greenwich Observatory, Herstmonceux Castle, Hailsham, Sussex, U.K.

MUSEN (Dr P.), National Aeronautics and Space Administration, Goddard Space Flight Center, Greenbelt, Maryland, U.S.A.

* MUSMAN (Dr S.), Sacramento Peak Observatory, Sunspot, New Mexico 88349, U.S.A.

MUSTEL (Prof. Dr E. R.), Astronomical Council, USSR Academy of Sciences, Vavilov Str. 34, Moscow V-312, U.S.S.R. Vice-President of the Union

MUTSCHLECNER (Dr P.), Department of Astronomy, Indiana University, Bloomington, Indiana 47401, U.S.A.

* MYACHIN (Dr V. F.), Institute of Theoretical Astronomy, USSR Academy of Sciences, Leningrad, U.S.S.R.

* MYERSCOUGH (Mrs Dr V. P.), Department of Mathematics, Queen Mary College, Mile End Road, London E.1, U.K.

* NACOZY (Dr P. E.), Department of Aerospace Engineering, The University of Texas, Austin, Texas 78712, U.S.A.

NADOLSKI (Prof. Dr V.), Université de Iasi, Rue M. Sadoveanu 5, Iasi, Roumania

NAGASAWA (Dr S.), Tokyo Astronomical Observatory, Mitaka, Tokyo, Japan

NAHON (Prof. F.), Observatoire de Paris, 61 Avenue de l'Observatoire, 75-Paris-14e, France

* NAKAGAWA (Dr Y.), High Altitude Observatory, Boulder, Colorado 80302, U.S.A.

NAKANO (S.), 47, Myogadani, Bunkyo-ku, Tokyo, Japan

NAMBA (Dr O.), Astronomical Observatory 'Sonnenborgh', Zonnenburg 2, Utrecht, The Netherlands

NANDY (Dr K.), Royal Observatory, Edinburgh 9, U.K.

NAQVI (Dr A. M.), Office of Research, Aerospace Corporation, San Bernardino, California 92402, U.S.A.

NARAYAN (Dr A. L.), Maharaja's College, Vizianagram, India

NARAYANA (J. V.), Meteorologist, Astrophysical Observatory, Kodaikanal 3, India

* NARIAI (Dr H.), Research Institute for Theoretical Physics, Hiroshima University, Takehara, Hiroshima-Ken, Japan

* NARIAI (Dr K.), Tokyo Astronomical Observatory, Mitaka, Tokyo, Japan

NARLIKAR (Dr J. V.), King's College, Cambridge, U.K.

NAZAROVA (Dr T. N.), Institute of Geochemistry and Analytic Chemistry, USSR Academy of Sciences, Moscow, U.S.S.R.

NECKEL (Dr H.), Hamburger Sternwarte, 205 Hamburg-Bergedorf, F.R.G.

* NECKEL (Dr Th.), Max-Planck-Institut für Astronomie, 6900 Heidelberg-Königstuhl, F.R.G.

* NE'EMAN (Dr Y.), Department of Physics and Astronomy, Tel-Aviv University, Ramat Aviv, Israel

NEFED'EV (Dr A. A.), Director, Engelhardt Observatory, Kazan, U.S.S.R.

NEFED'EVA (Dr A. I.), Engelhardt Observatory, Kazan, U.S.S.R.

* NEFF (Dr J. S.), Department of Physics and Astronomy, State University of Iowa, Iowa City, Iowa 52240, U.S.A.

NELSON (Prof. B.), Department of Astronomy, San Diego State College, San Diego 15, California, U.S.A.

NEMIRO (Dr A. A.), Pulkovo Observatory, Leningrad, U.S.S.R.

* NESMYANOVICH (Dr A. T.), Kiev University, Kiev, Ukrainian SSR, U.S.S.R.

NESS (Dr N. F.), NASA, Goddard Space Flight Center, Greenbelt, Maryland 20771, U.S.A.

* NEUGEBAUER (Dr G.), California Institute of Technology, Department of Astrophysics, Pasadena, California 91109, U.S.A.

NEUGEBAUER (Dr G.), Physics Department, California Institute of Technology, 1201 East California Street, Pasadena, California 91109, U.S.A.

NEUPERT (Dr W. M.), Solar Physics Branch, Goddard Space Flight Center, Greenbelt, Maryland, U.S.A.

NEUŽIL (Dr L.), Astronomical Institute, Czechoslovak Academy of Sciences, Observatory Ondřejov, Czechoslovakia

NEVEN (Dr L.), Chef du Département d'Astrophysique, Observatoire Royal de Belgique, avenue Circulaire, 3, B-1180 Bruxelles, Belgium

NEVIN (Prov. T. E.), University College, Dublin, Ireland

NEWKIRK JR. (Dr G. A.), High Altitude Observatory of the University of Colorado, Boulder, Colorado 80302, U.S.A.

NEWTON (H. W.), 'Alandale', Dean Road, Seaford, Sussex, U.K.

NEWTON (Dr R. R.), Applied Physics Laboratory, Johns Hopkins University, 8621 Georgia Avenue, Silver Spring, Maryland, U.S.A.

NEYMAN (Prof. J.), Statistical Laboratory, University of California, Berkeley 4, California, U.S.A.

NICHOLLS (Prof. R. W.), Department of Physics, York University, Keele Street, Toronto, Ontario, Canada

NICHOLSON (W.), Royal Greenwich Observatory, Herstmonceux Castle, Hailsham, Sussex, U.K.

NICOLET (Prof. Dr M.), Directeur de l'Institut d'Aéronomie Spatiale de Belgique, avenue Circulaire 3, B-1180 Bruxelles, Belgium

NICOLINI (Prof. T.), Direttore, Osservatorio Astronomico di Capodimonte, Napoli, Italy

NICOLOV (N. S.), Université de Sofia, Faculté Physique, Sofia, Bulgaria

NICOLSON (G. D.), c/o National Institute for Telecommunications Research, P.O. Box 3718, Johannesburg, South Africa

NIKITIN (Dr A. A.), Observatory, State University, Leningrad, U.S.S.R.

NIKOLSKY (Dr G. M.), Institute of Terrestrial Magnetism, Ionosphere & Radio Wave Propagation, Academy of Sciences of the USSR, Moscow, U.S.S.R.

NIKONOV (Dr V. B.), Crimean Astrophysical Observatory of the Academy of Sciences of the USSR, P/O Nauchny, Crimea, U.S.S.R.

NILSSON (Dr C.), Smithsonian Astrophysics Observatory, 60 Garden Street, Cambridge, Massachusetts 02138, U.S.A.

NINGER-KOSIBOWA (Dr S.), Astronomical Observatory, Kopernika 11, Wrocław, Poland

NININGER (Dr H. H.), Sedona's Meteorite Museum, P.O. Box 146, Sedona, Arizona, U.S.A.

NISHI (Dr K.), Tokyo Astronomical Observatory, University of Tokyo, Mitaka, Tokyo, Japan

* NISHIDA (Prof. M.), Department of Physics, Kyoto University, Kitashirakawa, Kyoto, Japan

* NISHIMURA (Dr S.), Tokyo Astronomical Observatory, Mitaka, Tokyo, Japan

NISSEN (Prof. J. J.), Observatorio Astronomico, San Juan, Argentina

* NISSEN (Dr P. E.), Ole Roemer Observatory, DK-8000 Aarhus C, Denmark

NOCI (Dr G.), Osservatorio astrofisico di Arcetri, Firenze, Italy

* NOERDLINGER (Dr P. D.), New Mexico Institute of Mining and Technology, Department of Physics and Geophysics, Socorro, New Mexico 87801, U.S.A.

NORDSTRÖM (Dr H.), Gjörwellsgatan 15, S-112 60 Stockholm, Sweden

NORLIND (Dr W.), Lund Observatory, S-222 24 Lund, Sweden

NØRLUND (Prof. N. E.), Director, Geodetic Institute, Malmögade 6, Copenhagen, Denmark

NOTNI (Dr P.), Zentralinstitut für Astrophysik, Sternwarte Babelsberg, DDR 1502 Potsdam-Babelsberg, Rosa-Luxemburg-Strasse 17a, G.D.R.

NOTUKI (Dr N.), 1–23 Higashida, Suginami-ku, Tokyo, Japan

NOVIKOV (Dr I. D.), Mathematical Institute of the USSR Academy of Sciences, Moscow, U.S.S.R.

NOVOPASHENNYJ (Dr B. V.), Astronomical Observatory, Odessa University, Park Chevchenko, Odessa, U.S.S.R.

NOVOSELOV (Prof. Dr V. S.), Astronomical Observatory, Leningrad University, Leningrad, U.S.S.R.

NOWACKI (Dr H.), Astronomisches Rechen-Institut, Franz-Knauff Str. 18, 69 Heidelberg, F.R.G.

NOYES (Dr R. W.), Smithsonian Institution, 60 Garden Street, Cambridge, Massachusetts 02138, U.S.A.

* NUSSBAUMER (Dr H.), JILA, University of Colorado, Boulder, Colorado 80302, U.S.A.

* OBASHEV (S. O.), Astrophysical Institute, Kazakh SSR Academy of Sciences, Alma-Ata 20, U.S.S.R.

OBI (Dr S.), Department of Earth Science, College of General Education, University of Tokyo, Komaba, Shibuya, Tokyo, Japan

OBRIDKO (Dr V. N.), Institute of Terrestrial Magnetism, Ionosphere, and Propagation of Radiowaves, USSR Academy of Sciences, Moscow, U.S.S.R.

OCCHIONERO (Dr F.), Laboratorio de Astrofisica, Casella Postale 67, 00044 Frascati (Roma), Italy

O'CONNELL, S. J. (Dr J. K.), Borgo S. Spirito 5, 00193 Roma, Italy

* O'CONNELL (Dr R. F.), Department of Physics and Astronomy, Louisiana State University, Baton Rouge, Louisiana 70803, U.S.A.

* ODA (Dr M.), Institute of Space and Aeronautical Science, University of Tokyo, Komaba, Meguro-Ku, Tokyo, Japan

O'DELL (Dr R. C.), Yerkes Observatory, Williams Bay, Wisconsin 53191, U.S.A.

ODGERS (Dr G. J.), Dominion Astrophysical Observatory R.R. 7, Victoria, B.C., Canada

OETKEN (Dr L.), Zentralinstitut für Astrophysik, DDR 15 Potsdam, Telegrafenberg, G.D.R.

OGORODNIKOV (Prof. Dr K. F.), University, Dept. of Astronomy, Leningrad, U.S.S.R.

* O'HANDLEY (Dr D.), California Institute of Technology, Jet Propulsion Laboratory, 4800 Oak Grove Drive, Pasadena, California 91103, U.S.A.

ÖHMAN (Prof. Y.), Stockholm Observatory, S-133 00 Saltsjöbaden, Sweden

* OHYAMA (Prof. N.), 1–22–26, Hirisawa, Hamamatsu, Japan

OJA (Dr T.), Kvistaberg Observatory, S-19051 Bro, Sweden

OKE (Dr J. B.), Mount Wilson and Palomar Observatories, 1201 E. California Street, Pasadena 4, California 91104, U.S.A.

O'KEEFE (J. A.), 3712 Thornapple Street, Chevy Chase 15, Maryland, U.S.A.

* OKI (Prof. Dr T.), Department of Earth Science, Faculty of Education, Fukushima University, Hamada-cho 12–23, Fukushima, Japan

* OKUDA (Prof. Dr H.), Department of Physics, Kyoto University, Kitashirakawa, Sakyo-Ku, Japan

OKUDA (Dr T.), Director of International Latitude Observatory of Mizusawa, Mizusawa-shi, Iwate-ken, Japan

ÖLANDER (Prof. V. R.), Geodeettinen Laitos, Geodetic Institute, Hämeentie 31, Helsinki 50, Finland

OLEAK (Dr H.), Zentralinstitut für Astrophysik, Sternwarte Babelsberg, DDR 1502 Potsdam-Babelsberg, Rosa-Luxemburg-Str. 17a, G.D.R.

OLIVIER (Dr C. P.), American Meteor Society, 521 N. Wynnewood Avenue, Narberth, Pennsylvania, U.S.A.

OLLONGREN (Dr A.), Centraal Rekeninstituut, Stationsweg 46, Leiden, The Netherlands

OLSEN (Dr K. H.), Los Alamos Scientific Laboratory, Box 1663 Group J-15, Los Alamos, New Mexico, 87544, U.S.A.

OLSON (Dr E. C.), University of Illinois Observatory, Urbana, Illinois 61801, U.S.A.

OMAROV (Dr T. B.), Astrophysical Institute of the Kazakh Academy of Sciences, Alma-Ata, U.S.S.R.

OMER JR (Dr G. C.), Department of Physical Sciences, University of Florida, Gainesville, Florida, U.S.A.

ONDERLIČKA (Dr B.), Department of Astronomy, Purkyně University, Brno, Czechoslovakia

* ONEGINA (Mrs Dr A. B.), Main Astronomical Observatory, Ukrainian Academy of Sciences, Kiev, U.S.S.R.

ŌNO (Dr Y.), Department of Physics, Hokkaido University, Sapporo, Hokkaido, Japan

OORT (Prof. Dr J. H.), President Kennedylaan 169, Oegstgeest, The Netherlands

OOSTERHOFF (Prof. Dr Th.), Sterrewacht Leiden, The Netherlands

OPALSKI (Dr W.), Institut Astronomique de l'École Polytechnique, Koszykowa 75, Warsaw, Poland

ÖPIK (Dr E. J.), Armagh Observatory, Armagh, Northern Ireland, U.K.

OPOLSKI (Dr A.), Astronomical Observatory, Kopernika 11, Wrocław, Poland

ORLIN (Dr H.), Coast & Geodetic Survey, Rockville, Maryland 20852, U.S.A.

ORLOV (Dr A. A.), Sternberg Astronomical Institute, Moscow, U.S.S.R.

ORLOVA (Mrs Dr N. S.), The Pulkovo Observatory, Leningrad M-140, U.S.S.R.

ORRALL (Prof. Dr F. Q.), Institute for Astronomy, 2840 Koluwalu St., University of Hawaii, Honolulu, Hawaii 96822, U.S.A.

ORTE LLEDO (D. A.), Observatorio de Marina, San Fernando, Cádiz, Spain

ORÚS (Pr. Dr. J. J. de), Barcelona University, Barcelona, Spain

ŌSAKI (Dr T.), Ryukoku University, Shimo-gyo-ku, Kyoto, Japan

* ŌSAKI (Dr Y.), Department of Astronomy, Faculty of Science, University of Tokyo, Bunkyo-Ku, Tokyo, Japan

OSAWA (Dr K.), Tokyo Astronomical Observatory, Mitaka, Tokyo, Japan

OSKANYAN (V.), Astrophysical Observatory, Byurakan, Armenian SSR, U.S.S.R.

OSTER (Prof. L.), Joint Institute for Laboratory Astrophysics, University of Colorado, Boulder, Colorado, U.S.A.

OSTERBROCK (Prof. D. E.), Washburn Observatory, University of Wisconsin, 475 North Charter Street, Madison, Wisconsin 53706, U.S.A.

OSTRIKER (Dr J. P.), Princeton University Observatory, Peyton Hall, Princeton, New Jersey 08540, U.S.A.

OTERMA (Dr L.), Sirkkalank 31, Turku 6, Finland

OTTELET (Dr I. J. G. J.), Institut d'Astrophysique, Université de Liège, avenue de Cointe, 5, B-4200 Cointe-Ougrée, Belgium

OVENDEN (Prof. Dr M. W.), Professor of Astronomy, c/o Department of Geophysics, The University of British Columbia, Vancouver 8, B.C., Canada

OWAKI (Dr N.), Marine Research Laboratory, Hydrographic Office, Tsukili, Chuo-ku, Tokyo, Japan

OWEN (Dr T.), IIT Research Institute (ASC), 10 West 35th Street, Chicago, Illinois 60616, U.S.A.

OWREN (Dr L.), Department of Physics, Div. B., University of Bergen Allegaten 53–55, 5000 Bergen, Norway

ÖZEMRE (Dr K.), University Observatory, Beyazit, Istanbul, Turkey

OZERNOJ (L. M.), Lebedev Physical Institute, Moscow, U.S.S.R.

OZSVATH (Prof. I.), Earth & Planetary Sciences Laboratory, Graduate Research Center, P.O. Box 30365, Dallas, Texas 75230, U.S.A.

PACHNER (Prof. J.), Physics Department, University of Saskatchewan, Regina Campus, Regina, Sask., Canada

PACHOLCZYK (Prof. Dr A. G.), Steward Observatory, University of Arizona, Tucson, Arizona, U.S.A.

* PACINI (Dr F.), Laboratorio di Astrofisica, Casella Postale 67, Frascati (Roma), Italy

PACZYNSKI (Dr B.), Astronomical Observatory, Al. Ujazdowskie 4, Warszawa, Poland

PAGE (Prof. T. L.), Director, Van Vleck Observatory, Wesleyan University, Middletown, Conn., 06457, U.S.A.

PAGEL (Dr B. E. J.), Royal Greenwich Observatory, Herstmonceux Castle, Hailsham, Sussex, U.K.

PAJDUŠÁKOVÁ (Mrs Dr L.), Director, Skalnaté Pleso Observatory, Slovak Academy of Sciences, Tatranská Lomnica, Czechoslovakia

PAL (Dr A.), Université, Cluj, Roumania

* PALMER (D. R.), Royal Greenwich Observatory, Herstmonceux Castle, Hailsham, Sussex, U.K.

PALMER (Dr H. P.), Nuffield Radio Astronomy Laboratories, Jodrell Bank, Macclesfield, Cheshire, U.K.

* PALMER (Dr P. E.), Department of Astronomy and Astrophysics, Ryerson Laboratory 162, 1100–14 East 58th Street, Chicago, Illinois 60637, U.S.A.

PANAJOTOV (Dr L. A.), Pulkovo Observatory, Leningrad, U.S.S.R.

PANDE (M. C.), U.P. State Observatory, Nainital, India

* PAPAGIANNIS (Dr M.), Department of Astronomy, Boston University, Boston, Massachusetts 02215, U.S.A.

* PAQUET (Dr P. E. G.), Observatoire Royal de Belgique, avenue Circulaire, 3, B-1180 Bruxelles, Belgium

PARIJSKIJ (Dr Yu. N.), Main Astronomical Observatory, USSR Academy of Sciences, Pulkovo, U.S.S.R.

PARIJSKIJ (Dr N. N.), O. Schmidt Institute of Physics of the Earth, USSR Academy of Sciences, B. Gruzinskaya 10, Moscow, U.S.S.R.

PARKER (Dr E. N.), Institute for Nuclear Studies, University of Chicago, Chicago, Illinois 60637, U.S.A.

PARKER (Dr R. A. R.), Washburn Observatory, University of Wisconsin, Madison, Wisconsin 53705, U.S.A.

PARKINSON (Dr W. H.), Harvard College Observatory, 60 Garden Street, Cambridge 38, Massachusetts 02138, U.S.A.

PASINETTI (L.), Osservatorio Astronomico di Merate (Como), Italy

PATON (J.), Department of Natural Philosophy, The University, Edinburgh 8, U.K.

* PAULINY-TOTH (Dr I. I. K.), NRAO, Edgemont Road, Charlottesville, Virginia 22901, U.S.A.

PAVLOV (Prof. Dr N. N.), Pulkovo Observatory, Leningrad, U.S.S.R.

* PAVLOVSKAYA (Mrs Dr E. D.), Sternberg Astronomical Institute, Moscow, U.S.S.R.

PAYNE-GAPOSCHKIN (Mrs Dr C. H.), Harvard College Observatory, Cambridge, Massachusetts 02138, U.S.A.

PEACH (Dr G.), Department of Physics, University College, Gower Street, London W.C. 1, U.K.

* PEACH (Dr J. V.), Department of Astrophysics, South Parks Road, Oxford, U.K.

PEALE (Dr S. J.), Department of Physics, University of California, Santa Barbara, California 93106, U.S.A.

PEARSE (Prof. R. W. B.), Imperial College of Science and Technology, Prince Consort Road, London, S.W. 7, U.K.

PEAT (Dr D. W.), The Observatories, Madingley Road, Cambridge, U.K.

PECKER (Prof. J.-C.), Observatoire de Meudon, place Jules Janssen, 92-Meudon, France

PECKER-WIMEL (Mrs Prof. Ch.), Institut d'Astrophysique, 98 bis Boulevard Arago, Paris 14e, France

PEDOUSSAUT (Dr A.), Observatoire de Toulouse, 31-Toulouse (Haute Garonne), France

* PEEBLES (Dr P. J. E.), Department of Physics, Princeton University, Princeton, New Jersey 08540, U.S.A.

PEERY (Dr B. F.), Department of Astronomy, Indiana University, Bloomington, Indiana 47405, U.S.A.

* PEIMBERT-SIERRA (Dr M.), Instituto de Astronomía U.N.A.M., Apartado Postal 70–264, México D.F., Mexico

PEKERIS (Prof. Ch. L.), The Weizmann Institute of Science, Department of Mathematics, Rehovoth, Israel

* PELLAS (Dr P.), Muséum d'Histoire Naturelle-Minéralogie, 61, rue de Buffon, Paris, France

PELS-KLUYVER (Mrs Dr H. A.), Sterrewacht Leiden, The Netherlands

PELSENEER (Prof. Dr J.), Université Libre de Bruxelles, avenue des Grenadiers 76, B-1050 Bruxelles, Belgium

PELTIER (L. C.), 327 S. Bredeick Street, Delphos, Ohio, U.S.A.

PENSADO (J.), Observatoire, Alfonso XII, 3, Madrid, Spain

* PENZIAS (Dr A.), Bell Telephone Laboratories, Crawford Hill Laboratory, Box 400, Holmdel, New Jersey 07733, U.S.A.

PEREK (Dr L.), Astronomical Institute, Czechoslovak Academy of Sciences, Budečská 6, Prague 2, Czechoslovakia. Adviser to the Executive Committee

* PERINOTTO (Dr M.), Osservatorio Astrofisico di Asiago, Vicenza, Italy

* PEROLA (Dr G. C.), Instituto di Scienze Fisiche, Via Celoria 16, 20133, Milano, Italy

PERRY (C. L.), Louisiana State University, Baton Rouge, Louisiana 70803, U.S.A.

PESCH (Prof. P.), Warner & Swasey Observatory, Case Institute of Technology, Taylor and Brunswick Roads, East Cleveland, Ohio 44112, U.S.A.

PETERSEN (J. Otzen), University Observatory, Øster Voldgade 3, Copenhagen K, Denmark

PETERSON (Dr L. E.), Physics Department, Revelle College, University of California-San Diego, La Jolla, California 92038, U.S.A.

PETFORD (Dr A. D.), University Observatory, South Park Road, Oxford, U.K.

PETRI (Dr W.), Unterleiten 2, Post-Box 106, 8162 Schliersee (OBB), F.R.G.

* PETROSIAN (Dr V.), Institute of Plasma Physics, Stanford University, Stanford, California 94305, U.S.A.

* PETROV (Prof. Dr G. I.), Institute of Cosmical Research, USSR Academy of Sciences, Moscow, U.S.S.R.

PETROV (Dr G. M.), Nikolaev Department of the Main Astronomical Observatory, USSR Academy of Sciences, Nikolaev, Ukrainian SSR, U.S.S.R.

PETROVSKAYA (Mrs Dr M. S.), Institute of Theoretical Astronomy, USSR Academy of Sciences, Leningrad, U.S.S.R.

PETTENGILL (Prof. G. H.), Department of Earth and Planetary Sciences, Massachusetts Institute of Technology, Cambridge, Massachusetts 02139, U.S.A.

PETTIT (Dr E.), Mount Wilson and Palomar Observatories, 813 Santa Barbara Street, Pasadena 4, California, U.S.A.

PETTIT (Mrs H. B. Kniflich), Boeing Scientific Research Laboratories, P.O. Box 3981, Seattle, Washington 98124, U.S.A.

PEYTURAUX (Dr R.), Institut d'Astrophysique, 98 bis, Boulevard Arago, Paris-14e, France

PFLUG (Dr K.), Heinrich-Hertz-Institut für Solar-Terrestrische Physik, Berlin-Adlershof, Rudower Chaussee, G.D.R.

PHILIP (Dr A. G. Davis), Dudley Observatory, 100 Fuller Road, Albany, New York 12208, U.S.A.

PHILLIPS (Prof. J. G.), Department of Astronomy, University of California, Berkeley, California, U.S.A.

PIASKOVSKY (Prof. D. V.), University Astronomical Observatory, Kiev, U.S.S.R.

PICK-GUTMAN (Mrs M.), Observatoire de Paris, Section d'Astrophysique, 92-Meudon, France

PIDDINGTON (Dr J. H.), CSIRO, National Standards Laboratory, Division of Physics, University Grounds, City Road, Chippendale, N.S.W., Australia 2008

PIERCE (Dr A. K.), Kitt Peak National Observatory, 950 North Cherry Avenue, P.O. Box 4130, Tucson, Arizona 85717, U.S.A.

PIKEL'NER (Prof. S. B.), Sternberg Astronomical Institute, Moscow V-234, U.S.S.R.

PILOWSKI (Prof. Dr K.), Geodätisches Institut der Technischen Hochschule, Nienburger Strasse 1, Hannover, F.R.G.

PINTO (Dr G.), Osservatorio Astronomico, Padova, Italy

PIOTROWSKI (Dr S.), Astronomical Observatory, Ujazdowskie 4, Warsaw, Poland

PISHMISH DE RECILLAS (Mrs Dr P.), Observatorio Astronómico Nacional, Universidad Nacional de Mexico, Ciudad Universitaria, Apartado Postal 70264, Mexico 20, D.F., Mexico

PLAKIDIS (Prof. Dr S.), 4B Eridanou Street, Athens (612), Greece

PLASKETT (Prof. H. H.), 48 Blenheim Drive, Oxford, U.K.

PLASSARD (Dr J.), Director, Ksara Observatory, Ksara, Lebanon

PLATZECK (Dr R.), Comision Nacional de la Energia Atomica, San Carlos de Bariloche, Argentina

PLAUT (Dr L.), Kapteyn Astronomical Laboratory, Broerstraat 7, Groningen, The Netherlands

PLAVCOVÁ (Mrs Dr Z.), Department of Astronomy, University of California, 405 Hilgard Avenue, Los Angeles, California 90024, U.S.A.

PLAVEC (Dr M.), Department of Astronomy, University of California, 405 Hilgard Avenue, Los Angeles, California 90024, U.S.A.

* PNEUMAN (Dr G.), High Altitude Observatory, Boulder, Colorado 80302, U.S.A.

PODOBED (Dr V. V.), Sternberg Astronomical Institute, Moscow V-3234, U.S.S.R.

POGO (Dr A.), Mount Wilson and Palomar Observatories, 813 Santa Barbara Street, Pasadena 4, California, U.S.A.

* POHL (Dr E.), Sternwarte Nürnberg, Lützowstr. 10, 8500 Nürnberg, F.R.G.

* POLETTO (Mrs Dr G.), Osservatorio Astrofisico Firenze, Largo E. Fermi 5, Firenze, Italy

POLOSKOV (Dr S. M.), Publishing House of Foreign Literature, Novo-Aleksejevskaya 52, Moscow, U.S.S.R.

POLOZHENTSEV (Dr D. D.), Pulkovo Observatory, Leningrad M-140, U.S.S.R.

POLUPAN (Dr P. N.), Astronomical Observatory, Kiev University, Kiev, U.S.S.R.

* PONSONBY (Dr J. E. B.), Nuffield Radio Astronomy Laboratories, Jodrell Bank, Macclesfield, Cheshire, U.K.

POPOV (Dr N. A.), Gravimetrical Observatory, Ukrainian Academy of Sciences, Poltava, U.S.S.R.

POPOVA (Mrs Dr M. D.), Section d'Astronomie de l'Académie Bulgare des Sciences, Rue 7 Novembre No. 1, Sofia, Bulgaria

POPOVIĆ (Prof. Dr B.), Ognjena Price 80, Beograd, Yugoslavia

POPOVICI (Prof. Dr C), Observatoire de Bucarest, 5 Rue Cutitul de Argint, Bucarest, Roumania

POPPER (Prof. D. M.), Department of Astronomy, University of California, Los Angeles 24, California, U.S.A.

PORFIR'EV (Dr V. V.), Main Astronomical Observatory, Ukrainian Academy of Sciences, Kiev, U.S.S.R.

PORTER (Dr J. G.), Whitestones, Hempstead Lane, Hailsham, Sussex, U.K.

* PORTER (Prof. N. A.), Department of Physics, University College, Belfield, Dublin 4, Ireland

POSTOIEV (Dr A.), Instituto Astronómico e Geofísico, Universidade de São Paulo, Caixa Postal 30627, São Paulo, Brazil

POTTASCH (Dr S. R.), Sterrekundig Laboratorium Kapteyn, Broerstraat 7, Groningen, The Netherlands

POTTER (Dr H. I.), Main Astronomical Observatory, Academy of Sciences, Pulkovo, U.S.S.R.

POUNDS (Dr K. A.), University of Leicester, Dept. of Physics, University Road, Leicester, LE 17 RH, U.K.

POVEDA (Dr A.), Observatorio Astronómico, Universidad Nacional de México, Apartado Postal 70264, México 20, D.F., Mexico

* PRADERIE (Mrs Dr F.), Observatoire de Meudon, 92-Meudon, France

PRADHAN (Dr), Nizamiah Observatory, Hyderabad, India

PRENDERGAST (Dr K. H.), 1402 Pupin, Columbia University, New York, U.S.A.

PRENTICE (J.P.M.), Star Ridge, Battisford, Needham Market, Suffolk, U.K.

PRESTON (Dr G.W.III), Lick Observatory, University of California, Santa Cruz, California 95060, U.S.A.

PRICE (Dr M.J.), Astro Sciences Center, IIT Research Institute, 10 West 35th Street, Chicago, Illinois 60616, U.S.A.

* PRICE (Dr R.M.), 26–463 M.I.T., Cambridge, Massachusetts 02139, U.S.A.

PRIESTER (Prof. W.), Universitäts-Sternwarte, Poppelsdorfer Allee 49, 53 Bonn, F.R.G.

PROBSTEIN (Dr R.F.), Department of Mechanical Engineering, Massachusetts Institute of Technology, Cambridge, Massachusetts 02139, U.S.A.

* PROCHAZKA (Dr F.V.), Observatory of the University of Vienna, Türkenschantzstrasse 17, A-1180 Vienna, Austria

PROKOF'EV (Prof. V.K.), Crimean Astrophysical Observatory of the Academy of Sciences of the USSR, P/O Nauchny, Crimea, U.S.S.R.

PROKOF'EVA (Mrs Dr I.A.), Main Astronomical Observatory, Academy of Sciences, Pulkovo, U.S.S.R.

PRONIK (Dr V.I.), Crimean Astrophysical Observatory, USSR Academy of Sciences, Crimea, U.S.S.R.

PROTHEROE (Dr W.M.), Department of Astronomy, 174 W. Eighteenth Street, Ohio State University, Columbus, Ohio 43210, U.S.A.

PROTITCH (M.B.), Astronomical Observatory, Veliki Vracar, Belgrade, Yugoslavia

PROVERBIO (Dr E.), Stazione Astronomica Internazionale di Latitudine, 09014 Carloforte (Cagliari), Italy

* PRYCE (Dr M.H.L.), Institute of Astronomy and Space Science, Department of Physics, University of British Columbia, Vancouver 8, B.C., Canada

PRZYBYLSKI (Dr A.), Mount Stromlo Observatory, Canberra, A.C.T., Australia

* PSKOVSKIJ (Dr Ju.P.), Sternberg Astronomical Institute, Moscow, U.S.S.R.

PURGATHOFER (Dr A.Th.), Universitäts-Sternwarte Wien, Türkenschanzstrasse 17, A-1180 Vienna, Austria

* PYPER (Mrs Dr D.M.), Department of Physics and Astronomy, Tel-Aviv University, Ramat Aviv, Israel

QUIJANO (L.), Observatoire, San Fernando, Spain

QVIST (Dr B.), Åbo Akademi, Åbo, Finland

RABBEN (Dr H.H.), Max-Planck-Institut für Physik und Astrophysik, Abt. Extraterrestrische Physik, Garching b. München, F.R.G.

RABE (Prof. Dr E.), Cincinnati Observatory, Observatory Place, Cincinnati, Ohio 45208, U.S.A.

* RACKHAM (Dr T.W.), Armagh Observatory and Planetarium, 46 Ashley Park, Armagh, Northern Ireland

RADHAKRISHNAN (V.), CSIRO, Division of Radiophysics, Post Office Box 76, Epping, N.S.W. 2121, Australia

* RADLOVA (Mrs Dr L.N.), Institute of Sciences and Technics Information, Department of Astronomy, USSR Academy of Sciences, Moscow, U.S.S.R.

RAGHAVAN (N.), Department of Physics, Indian Institute of Technology, Kanpur, India

* RAHE (Dr J.), Lehrstuhl für Astrophysik, T.U. Berlin, Strasse d. 17. Juni 135, 1000 Berlin, F.R.G.

RAHIM (Dr M.H.A.), Helwan Observatory, Helwan (Near Caïro), U.A.R.

RAIMOND (Dr E.), University Observatory, Sterrewacht, Leiden, The Netherlands

RAJCHL (Dr J.), Astronomical Institute, Czechoslovak Academy of Sciences, Observatory Ondřejov, Czechoslovakia

RAJU (Dr P.K.), Lehrstuhl für Theoretische Astrophysik, Hausserstrasse 64, 7400 Tübingen, F.R.G.

RAKAVY (Prof. G.), University of Jerusalem, Einstein Institute of Physics, Department of Theoretical Physics, The Hebrew University of Jerusalem, Jerusalem, Israel

RAKOS (Dr K. D.), Universitätssternwarte, Türkenschanzstrasse 17, A-1180 Wien, Austria

RAKSHIT (Prof. H.), Bengal Engineering College, Sibpore, Hewrak, India

* RAMATY (Dr R.), NASA Goddard Space Flight Center, Greenbelt, Maryland, U.S.A.

RAMBERG (Prof. Dr J. M.), European Southern Observatory, 131 Bergedorfer Strasse, 205 Hamburg 80, F.R.G.

RANDIĆ (Dr L.), Tehnićki Fakultet, Zagreb, p.p. 89, Yugoslavia

* RAO (Dr U. R.), Physical Research Laboratory, Navrangpura, Ahmedabad-9, India

RASMUSEN (Dr H. Q.), Vaerslevgaarden, 4400 Kalundborg, Denmark

RAYROLE (J.), Observatoire de Meudon, 5, place Janssen, 92-Meudon, France

* RAZIN (Dr V. A.), Scientific Research Radiophysical Institute, Gorki University, Gorki, U.S.S.R.

RAZMADZE (Dr T. S.), Institute Geophysics, Academy of Sciences, Cosmic Ray Station, Tbilisi, U.S.S.R.

REAVES (Dr G.), Astronomy Department, University of Southern California, Los Angeles, California 90007, U.S.A.

REBER (Dr G.), Research Corporation, 405 Lexington Avenue, New York 17, N.Y., U.S.A.

REDDISH (Dr V. C.), Royal Observatory, Edinburgh 9, U.K.

REDMAN (Prof. R. O.), Director of the Observatories, Madingley Road, Cambridge, U.K.

* REES (Dr M. J.), Institute of Theoretical Astronomy, Madingley Road, Cambridge, U.K.

REEVES (Dr E. M.), Harvard College Observatory, 60 Garden Street, Cambridge, Massachusetts 02138, U.S.A.

REEVES (Dr H.), SEP-Saclay-BP No. 2, Gif sur Yvette (91), France

* REFSDAL (Dr S.), Department of Physics, University of Nebraska, Lincoln, Nebraska, U.S.A.

REID (Dr J. H.), Lockheed Electronics Company, 16811 El Camino Real, Houston, Texas 77058, U.S.A.

REINMUTH (Dr K.), Häusserstrasse 61, Heidelberg, F.R.G.

REIZ (Prof. A.), University Observatory, Østervoldgade 3, Copenhagen-K, Denmark

REMY-BATTIAU (Mrs Dr L.), Conservateur, Institut d'Astrophysique, Université de Liège, avenue de Cointe, 5, B-4200 Cointe-Ougrée, Belgium

RENSE (W. A.), Physics Department, University of Colorado, Boulder, Colorado, U.S.A.

* RENZINI (Dr A.), Astronomical Observatory Bologna, via Zamboni 32, Bologna, Italy

REQUIÈME (Y.), Observatoire de Bordeaux, 33-Floirac, France

REUNING (Dr E.), Department of Physics and Astronomy, University of Georgia, Athens, Georgia 30601, U.S.A.

REYNOLDS (Dr J. H.), Department of Physics, University of California, Berkeley, California, U.S.A.

RICE (D. A.), Coast and Geodetic Survey, Rockville, Maryland 20852, U.S.A.

RICHARDSON (Dr E. H.), Dominion Astrophysical Observatory, R.R. 7, Victoria, B.C., Canada

RICHARDSON (R. S.), Griffith Observatory, P.O. Box 27787, Los Felix Station, Los Angeles 27, California, U.S.A.

RICHTER (Dr G.), Zentralinstitut für Astrophysik, Sternwarte Sonneberg, DDR 64 Sonneberg, G.D.R.

RICHTER (Dr N.), Zentralinstitut für Astrophysik, Karl-Schwarzschild-Observatorium, DDR 6901 Tautenburg, G.D.R.

* RIEGEL (Dr K. W.), Astronomy-UCLA, Los Angeles, California 90024, U.S.A.

* RIEU (Dr N. Q.), Observatoire de Meudon, 92-Meudon, France

* RIGHINI (Dr A.), Osservatorio Astrofisico di Catania, Città Universitaria, Catania, Sicilia, Italy

RIGHINI (Prof. G.), Direttore, Osservatorio Astrofisico di Arcetri, Largo E. Fermi 5, Firenze, 50125, Italy

RIGUTTI (Prof. M.), Director, Osservatorio Astronomico di Capodimonte, via Moiariello 16, 80131 Napoli, Italy

RIIHIMAA (Dr J. J.), c/o Prof. Tiuri, Technical University, Department of Electrotechnics, Albert-Street 40–42, Helsinki, Finland

RIIVES (Dr V.), Tartu State University, Department of Theoretical Physics, Tartu, Estonia, U.S.S.R.

RINDLER (Prof. W.), Graduate Research Center, P.O. Box 30365, Dallas, Texas 75230, U.S.A.

RINEHART (Dr J. S.), Senior Research Fellow, Institutes for Environmental Research, ESSA, Boulder, Colorado 80302, U.S.A.

RING (Prof. J.), Department of Applied Physics, The University, Hull, U.K.

RINGNES (Dr T. S.), Institute of Theoretical Astrophysics, University of Oslo, Blindern, Oslo 3, Norway

RINGUELET (Dr A. E.), Observatorio Astronomico, Universidad Nacional de la Plata, La Plata, Argentina

ROACH (Dr F. E.), 2987 Kalakaua Avenue (602), Honolulu, Hawaii 96815, U.S.A.

* ROBE (Dr H. A. Gh.), Institut d'Astrophysique, Université de Liège, avenue de Cointe, 5, B-4200 Cointe-Ougrée, Belgium

ROBERTS (Dr J. A.), CSIRO, Radiophysics Laboratory, P.O. Box 76, Epping, N.S.W. 2121, Australia

ROBERTS (Dr M. S.), National Radio Astronomy Observatory, Edgemont Road, Charlottesville, Virginia, U.S.A.

ROBERTS (Dr W. O.), Director, National Center for Atmospheric Research, Boulder, Colorado, U.S.A.

ROBERTSON (W. H.), Sydney Observatory, Sydney, N.S.W., Australia

ROBINSON (Dr B. J.), CSIRO, Division of Radiophysics, Post Office Box 76, Epping, N.S.W. 2121, Australia

ROBINSON (Prof. I.), Graduate Research Center, P.O. Box 30365, Dallas, Texas 75230, U.S.A.

ROBLEY (Dr B.), Observatoire du Pic-du-Midi, Bagnères-de-Bigorre, Hautes-Pyrénées, France

RODDIER (F.), Faculté des Sciences de Nice, Avenue Valrose, 06-Nice, France

RODGERS (Dr A. W.), Mount Stromlo Observatory, Canberra, A.C.T., Australia

* RODONO (Dr M.), Osservatorio Astrofisico, Città Universitaria, viale A. Doria, 95123 Catania, Sicilia, Italy

RODRIGUEZ (M.), Director del Instituto y Observatorio de Marina, San Fernando (Cádiz), Spain

ROEDER (Dr R. C.), David Dunlap Observatory, Richmond Hill, Ontario, Canada

ROEMER (Miss Dr E.), Lunar and Planetary Laboratory, University of Arizona, Tucson, Arizona 85721, U.S.A.

* ROGER (Dr R. S.), Dominion Radio Astrophysical Observatory, Box 248, Penticton, B.C., Canada

* ROGERS (Dr A. E. E.), Massachusetts Institute of Technology, Lincoln Laboratory, Lexington, Massachusetts 02173, U.S.A.

ROGERSON (Prof. J. B.), Princeton University Observatory, Peyton Hall, Princeton, New Jersey 08540, U.S.A.

* ROGSTAD (Dr D. H.), Kapteyn Laboratory, Broerstraat 7, Groningen, The Netherlands

* ROHLFS (Prof. K.), Max-Planck-Institut für Radioastronomie, Argelanderstr. 3, 5300 Bonn, F.R.G.

ROLAND (Miss Dr G.), Institut d'Astrophysique, Université de Liège, avenue de Cointe 5, B-4200 Cointe-Ougrée, Belgium

ROMAN (Miss Dr N. G.), NASA Headquarters, Washington, D.C. 20546, U.S.A.

ROMAÑA (Dr A.), Directeur, Observatorio de Ebro, Roquetas, Tarragona, Spain

ROMANO (Dr G.), Viale S. Francesco 7, Treviso, Italy

ROMPOLT (Dr B.), Astronomical Institute of Polish Academy of Sciences, Kopernika 11, Wrocław, Poland

* RONAN (C. A.), Ballards Place, Cowlinge, Newmarket, Suffolk, U.K.

* ROOD (Dr H. J.), Van Vleck Observatory, Wesleyan University, Middletown, Connecticut 06457, U.S.A.

RÖSCH (Prof. J.), Directeur de l'Observatoire du Pic-du-Midi, 65 Bagnères-de-Bigorre, (Hautes-Pyrénées), France

Rose (DrW.K.), Physics Department, Massachusetts Institute of Technology, Cambridge, Massachusetts, U.S.A.

Rosen (Prof.DrB.), Université de Liège, rue du Stade 19, B-4200 Ougrée, Belgium

* Rosen (DrE.), The City College of the City University of New York, Department of History, New York, N.Y. 10031, U.S.A.

Rosino (Prof.L.), Director of the Astrophysical Observatory, Asiago (Vicenza), Italy

Roslund (DrC.), Lund Observatory, S-222 24 Lund, Sweden

Rosseland (Prof.S.), Institute of Theoretical Astrophysics, University of Oslo, Blindern, Oslo 3, Norway

Rossi (DrB.), Department of Physics, Massachusetts Institute of Technology, Cambridge, Massachusetts 02139, U.S.A.

* Rostas (DrF.), Observatoire de Meudon, 92-Meudon, France

Rottenberg (DrJ.A.), 124 Chestnut Ave, Pointe Claire 720, P.Q., Canada

* Rountree Lesh (MrsDrJ.), Observatoire de Meudon, 92-Meudon, France

* Rouse (DrC.A.), Gulf General Atomic Inc., P.O. Box 608, San Diego, California 92112, U.S.A.

Routly (Dr P.M.), Princeton University Observatory, Peyton Hall, Princeton, New Jersey 08540, U.S.A.

Roxburgh (DrI.W.), Mathematics Department, Queen Mary College, Mile End Road, London E.1., U.K.

Roy (DrA.E.), Department of Astronomy, University of Glasgow, Glasgow W.2, U.K.

Rozhkovski (DrD.A.), Astrophysical Institute, Alma Ata, U.S.S.R.

Rubashev (DrB.M.), Pulkovo Observatory, Leningrad, U.S.S.R.

Ruben (DrG.), Zentralinstitut für Astrophysik, DDR 15 Potsdam, Telegrafenberg, G.D.R.

Rubin (MrsDrV.C.), Department of Terrestrial Magnetism, Carnegie Institution of Washington, 5241 Broad Branch Road N.W., Washington, D.C., U.S.A.

* Rublev (DrS.V.), Special Astrophysical Observatory, USSR Academy of Sciences, St. Zelenchukskaya, Stavropolsky Kraj, U.S.S.R.

Rudkjøbing (Prof.M.), Director of the Ole Rømer Observatory, Aarhus C, Denmark

Rudnicki (DrK.), Jagiellonian University Observatory, ul. Kopernika 27, Kraków, Poland

Rufener (DrF.-G.), Observatoire de Genève, 1290, Sauverny (GE), Switzerland

Rule (B.R.), Mount Wilson and Palomar Observatories, 813 Santa Barbara Street, Pasadena, California, U.S.A.

Runcorn (Prof.S.K.), Physics Department, The University, Newcastle-upon-Tyne, U.K.

Ruprecht (DrJ.), Astronomical Institute, Czechoslovak Academy of Sciences, Budečská 6, Prague 2, Czechoslovakia

Ruskol (MrsDrE.L.), Institute of Physics of the Earth, USSR Academy of Sciences, Moscow, U.S.S.R.

Russell (DrJ.A.), Astronomy Department, University of Southern California, University Park, Los Angeles, California 90007, U.S.A.

* Rusu (MrsDrL.), Observatorul Astronomic, Strada Cutitul de Argint 5, Bucuresti 28, Roumania

Rutllant (Prof.F.), Casilla 933, Viña del Mar, Chile

* Růžičková-Topolová (MrsDrB.), Astronomical Institute, Czechoslovak Academy of Sciences, Observatory Ondřejov, Czechoslovakia

Ryabov (Dr Ju.A.), Sternberg Astronomical Institute, Moscow, U.S.S.R.

* Rybicki (DrG.B.), Smithsonian Astrophysical Observatory, 60 Garden Street, Cambridge, Massachusetts 02138, U.S.A.

Rybka (Prof.DrE.), Szopena 20/3, Kraków, Poland

Rybka (DrP.), Astronomical Institute of the Wrocław University, Dembowskiego 19, Wrocław 9, Poland

Rydbeck (DrO.E.H.), Institute of Electronphysics I, Chalmers Technical University, Fack, S-402 20 Göteborg 5, Sweden

Ryle (SirMartin), Cavendish Laboratory, Free School Lane, Cambridge, U.K.

* RYTER (Dr Ch.), CEN SACLAY-SEP, B.P. N° 2, 91-Gif sur Yvette, France
* RZHIGA (Dr O. N.), Institute of Radio and Electronics, USSR Academy of Sciences, Moscow, U.S.S.R.

SACK (Dr N.), Einstein Institute of Physics, Dept. of Theoretical Physics, The Hebrew University of Jerusalem, Jerusalem, Israel
* SADEH (Dr D.), Department of Physics and Astronomy, Tel-Aviv University, Ramat Aviv, Israel
SADLER (D. H.), 8 Collington Rise, Bexhill-on-Sea, Sussex, U.K.
SADLER (Mrs F. M. McBain), H.M. Nautical Almanac Office, Royal Greenwich Observatory, Herstmonceux Castle, Hailsham, Sussex, U.K.
SAFRONOV (Dr V. S.), O. J. Schmidt Institute of the Physics of the Earth, USSR Academy of Sciences, B. Gruzinskaya 10, Moscow 242, U.S.S.R.
SAGAN (Prof. C.), Center for Radiophysics and Space Research, Laboratory for Planetary Studies, Cornell University, Ithaca, New York 14850, U.S.A.
* SAGGION (Prof. Dr A.), Instituto di Fisica 'G. Galilei', Via Marzolo 8, 35100 Padova, Italy
SAGITOV (Dr M. U.), Sternberg Astronomical Institute, Moscow, U.S.S.R.
* SAGNIER (Dr J. L.), Bureau des Longitudes, 3, Rue Mazarine, Paris-8e, France
SAHADE (Dr J.), 47–848 (9°A), La Plata, Argentina. Vice-President of the Union.
SAITO (Dr K.), Tokyo Astronomical Observatory, Mitaka, Tokyo, Japan
SAITO (Dr S.), Kwasan Observatory, University of Kyoto, Yamashina, Kyoto, Japan
* SAKASHITA (Dr S.), Department of Physics, Hokkaido University, Sapporo, Japan
SAKHAROV (Dr V. I.), Main Astronomical Observatory, USSR Academy of Sciences, Pulkovo, U.S.S.R.
SAKURAI (Dr K.), Radio Astronomy Branch, NASA Goddard Space Flight Center, Greenbelt, Maryland 20771, U.S.A.
* SAKURAI (Prof. Dr T.), Department of Aeronautical Engineering, Faculty of Engineering, Kyoto University, Kyoto, Japan
* SALIN (Dr A.), Observatoire de Meudon, 92-Meudon, France
* SALISBURY (Dr J. W.), Sacramento Peak Observatory, Sunspot, New Mexico 88349, U.S.A.
* SALOMONOVICH (Dr A. E.), Physical Institute, USSR Academy of Sciences, Moscow, U.S.S.R.
SALPETER (Dr E. E.), Laboratory of Nuclear Studies, Cornell University, Ithaca, N.Y., U.S.A.
SALPETER S.J. (Dr E. W.), Specola Vaticana, Castel Gandolfo, Vatican City State
SALUKVADZE (Dr G. N.), Abastumani Astrophysical Observatory, Georgian Academy of Sciences, Abastumani, U.S.S.R.
SAMAHA (Prof. A. H. M.), 5 Wadi El-Nile Street, Maadi, Cairo, U.A.R.
SANAMIAN (Dr V. A.), Byurakan Astrophysical Observatory, Erevan, Armenia, U.S.S.R.
* SANCHEZ-MARTINEZ (Dr F.), Observatorio de Teide, Universidad de la Laguna, Tenerife, Spain
* SANCISI (R.), Kapteyn Laboratory, Groningen, The Netherlands
SANDAGE (A.), Mount Wilson and Palomar Observatories, 813 Santa Barbara Street, Pasadena 4, California, U.S.A.
SANDAKOVA (Dr E. V.), Kiev University Observatory, Kiev, U.S.S.R.
SANDERS (Dr P.), Observatoire Royal de Belgique, avenue Circulaire, 3, B-1180 Bruxelles, Belgium
SANDIG (Prof. H.-U.), Observatory of the Technical University, Lohrmann Institut, Dresden 8027, G.D.R.
* SANDULEAK (Dr N.), Warner and Swasey Observatory, Taylor and Brunswick Roads, East Cleveland, Ohio 44112, U.S.A.
SANTOS (Dr A. J. Baptista dos), Observatorio Astronomico, Tapada, Lisbon, Portugal
SANWAL (Dr N. B.), Department of Astronomy, Osmania University, Hyderabad-7 (A.P.), India
SAPAR (Dr A.), W. Struve Astrophysical Observatory, Tartu, Tôravere, Estonia, U.S.S.R.
SARABHAI (V. A.), Physical Research Laboratory, Navrangpura, Ahmedabad, India
SARGENT (Dr W. L.), California Institute of Technology, Pasadena, California 91109, U.S.A.

SARMA (M.B.K.), Department of Astronomy, Osmania University, Hyderabad, India

* SASLAW (DrW.C.), Astronomy Department, University of California, Berkeley, California 94720, U.S.A.

SASTRY (Ch.V.), Astrophysical Observatory, Kodaikanal-3, India

* SATO (DrH.), Department of Physics, Kyoto University, Kyoto, Japan

SATO (DrY.), Tokyo Astronomical Observatory, Mitaka, Tokyo, Japan

* SAVAGE (DrB.), Washburn Observatory, University of Wisconsin, 475 North Charter Street, Madison, Wisconsin 53706, U.S.A.

SAVEDOFF (Prof.M.P.), C.F. Kenneth Mees Observatory University of Rochester, River Campus, Rochester, N.Y. 14627, U.S.A.

SAWYER HOGG (MrsProf.H.B.), David Dunlap Observatory, Richmond Hill, Ontario, Canada

* SAXENA (DrP.P.), Uttar Pradesh State Observatory, Manora Peak, Naini Tal, India

* SCARFE (DrC.D.), Department of Physics, University of Victoria, Victoria, B.C., Canada

* SCARGLE (DrJ.D.), Lick Observatory, University of California, Santa Cruz, California 94060, U.S.A.

SCHADEE (DrA.), Astronomical Observatory 'Sonnenborgh', Servaas Bolwerk 13, Utrecht, The Netherlands

SCHAIFERS (DrK.), Landessternwarte, Heidelberg-Königstuhl, F.R.G.

SCHALÉN (Prof.C.), Lund Observatory, S-222 24 Lund, Sweden

SCHATZMAN (Prof.E.L.), Institut d'Astrophysique, 98bis, Boulevard Arago, Paris-14e, France

SCHEEPMAKER (DrA.C.), Boerhaavestraat 59, Voorhout, The Netherlands

SCHEFFLER (DrH.), Landessternwarte, 69 Heidelberg-Königstuhl, F.R.G.

SCHEUER (DrP.A.G.), Mullard Radio Astronomy Observatory, Cavendish Laboratory, Free School Lane, Cambridge, U.K.

* SCHILD (DrR.E.), Smithsonian Astrophysical Observatory, 60 Garden Street, Cambridge, Massachusetts 02138, U.S.A.

SCHILLER (Prof.DrK.), Pirschweg 6, 6079 Buchschlag über Sprendlingen, F.R.G.

SCHILT (Prof.J.), 481 Greenville Avenue, Centerdale 11, Rhode Island, U.S.A.

* SCHKODROV (DrV.G.), Department of Astronomy, '7th November' Street 1, Sofia, Bulgaria

SCHLÜTER (Prof.DrA.), Max-Planck-Institut für Physik und Astrophysik, Institut für Astrophysik, Aumeisterstrasse 6, München 23, F.R.G.

* SCHLÜTER (DrD.), Sternwarte Bergedorf, Gojenbergsweg 112, 205 Hamburg 80, F.R.G.

* SCHMAHL (DrG.), Universitätssternwarte, Geismarlandstr. 11, 3400 Göttingen, F.R.G.

SCHMALBERGER (DrD.C.), State University of New York at Albany, ES 314, Albany, New York 12203, U.S.A.

SCHMEIDLER (Prof.DrF.), Universitäts-Sternwarte, Sternwartstrasse 23, München 27, F.R.G.

SCHMIDT (Prof.DrH.), Direktor der Universitäts-Sternwarte, Poppelsdorfer Allee 49, 5300 Bonn, F.R.G.

SCHMIDT (DrH.U.), Max-Planck-Institut für Physik und Astrophysik, Föhringer Ring 6, 8 München 23, F.R.G.

SCHMIDT (DrK.-H.), Zentralinstitut für Astrophysik, Sternwarte Babelsberg, DDR 1502 Potsdam-Babelsberg, Rosa-Luxemburg-Strasse 17a, G.D.R.

SCHMIDT (Prof.M.), Mount Wilson and Palomar Observatories, 1201 East California Blvd., Pasadena, California 91109, U.S.A.

SCHMIDT (DrTh.), Landessternwarte Heidelberg-Königstuhl, 69 Heidelberg, F.R.G.

SCHMIDT-KALER (Prof.DrTh.), 581 Witten, Steinhügel 105, F.R.G.

SCHMITT (A.), Observatoire de Strasbourg, Strasbourg (Bas-Rhin), France

* SCHMITTER Y MARTIN DEL CAMPO (DrE.F.), Institute of Astronomy U.N.A.M., México D.F., Mexico

* SCHNEIDER (DrM.), Observatoire de Nice, Le Mont Gros, 06-Nice, France

* SCHNELL (MrsDrA.), Universitäts-Sternwarte, Türkenschantzstrasse 17, A-1180 Vienna, Austria

* SCHOMBERT (DrJ.L.), U.S. Naval Observatory, Washington, D.C. 20390, U.S.A.

* SCHÖNEICH (Dr W.), Forschungsgruppenleiter am Zentralinstitut für Astrophysik (Bereich Stern-physik), Potsdam, G.D.R.
* SCHROEDER (Dr D. J.), Beloit College, Beloit, Wisconsin 53511, U.S.A.
SCHRÖTER (Dr E. H.), Universitäts-Sternwarte, Geismarlandstr. 11, 34 Göttingen, F.R.G.
SCHRUTKA-RECHTENSTAMM (Prof. Dr G.), Universitäts-Sternwarte, Türkenschanzstrasse 17, A-1180 Vienna, Austria
SCHUBART (Dr J.), Astronomisches Rechen-Institut, Mönchhofstrasse 12–14, 69 Heidelberg, F.R.G.
SCHÜCKING (Dr E.), Astronomy Department, University of Texas, Austin, Texas, U.S.A.
* SCHULER (Dr W.), Observatory of Neuchâtel, 2000 Neuchâtel, Switzerland
SCHULTE (Dr D. H.), Itek Corporation, 10 Maguire Road, Lexington, Massachusetts, U.S.A.
SCHÜRER (Prof. Dr M.), Astronomisches Institut der Universität Bern, Sidlerstrasse 5, Bern, Switzerland
SCHWARZ (Dr U. J.), Kapteyn Astronomical Laboratory, Broerstraat 7, Groningen, The Netherlands
SCHWARZSCHILD (Prof. M.), Princeton University Observatory, Peyton Hall, Princeton, New Jersey 08540, U.S.A.
SCIAMA (Prof. D. W.), Department of Astrophysics, South Parks Road, Oxford, OX1 3 RQ, U.K.
SCONZO (Dr P.), Federal Systems Division I.B.M., 1730, Cambridge Street, Cambridge, Massachusetts 02138, U.S.A.
SCOTT (Prof. E. L.), Statistical Laboratory, University of California, Berkeley, CA 94720, U.S.A.
SCOTT (F. P.), U.S. Naval Observatory, Washington, D.C. 20390, U.S.A.
SCOTT (Dr P. F.), Mullard Radio Astronomy Observatory, Cavendish Laboratory, Free School Lane, Cambridge, U.K.
* SEAQUIST (Dr E. R.), David Dunlap Observatory, Richmond Hill, Ontario, Canada
SEARLE (Dr L.), Hale Observatories, 813 Santa Barbara Street, Pasadena, California 91106, U.S.A.
SEARS (Dr R. L.), Astronomy Department, University of Michigan, Ann Arbor, Michigan 48104, U.S.A.
SEATON (Prof. M. J.), Department of Physics, University College London, Gower Street, London, W.C.1, U.K.
SEDDON (H.), The Royal Observatory, Edinburgh 9, U.K.
SEEGER (Prof. Ch. L.), Dept. of Earth Sciences and Astronomy, New Mexico State University, P.O. Box 3 AB, Las Cruces, N.M. 88001, U.S.A.
* SEGGEWISS (Dr W.), Universitätssternwarte, Poppelsdorfer Allee 49, 5300 Bonn, F.R.G.
SEHNAL (Dr L.), Astronomical Institute, Czechoslovak Academy of Sciences, Observatory Ondřejov, Czechoslovakia
* SEIDELMANN (Dr P. K.), U.S. Naval Observatory, Washington, D.C. 20390, U.S.A.
SEIELSTAD (Dr G. A.), California Institute of Technology, Pasadena, California 91109, U.S.A.
SEITTER (Dr W. C.), Universitäts-Sternwarte Bonn, Observatorium Hoher List, 5568 Daun/Eifel, F.R.G.
SEKANINA (Dr Z.), Smithsonian Astrophysical Observatory, 60 Garden Street, Cambridge, Massachusetts 02138, U.S.A.
SEKIGUCHI (Dr N.), Tokyo Astronomical Observatory, University of Tokyo, Mitaka, Tokyo, Japan
* SEMEL (Dr M.), Observatoire de Meudon, 92-Meudon, France
SÉMIROT (Prof. Dr P.), Directeur de l'Observatoire de Bordeaux, 33-Floirac, France
SEN (Dr H. K.), High Temperature Section, Thermal Radiation Laboratory, Air Force Cambridge Research Centre, Bedford, Massachusetts, U.S.A.
* V. SENGBUSCH (Dr K.), Max-Planck-Institut für Physik und Astrophysik, Föhringer Ring 6, 8000 München 23, F.R.G.
SERKOWSKI (Dr K.), Lunar and Planetary Laboratory, University of Arizona, Tucson, Arizona 85721, U.S.A.
SÉRSIC (Dr J. L.), Observatorio Astronómico, Córdoba, Argentina
SERVAJEAN (Dr R.), Observatoire de Paris, Section d'Astrophysique, 92-Meudon, France

SETTI (G.), Laboratorio di Radioastronomia, Instituto di Fisica, Bologna, Italy

ŠEVARLIĆ (Dr B. M.), Assistant Professor, Astronomska Observatorija, Veliki Vracar, Belgrade, Yugoslavia

SEVERNYJ (Prof. Dr A. B.), Director, Crimean Astrophysical Observatory of the USSR Academy of Sciences, P/O Nauchny, Crimea, U.S.S.R.

* SHAH (Dr G. A.), Tata Institute of Fundamental Research, Homi Bhabha Road, Colaba, Bombay-5, India

SHAKESHAFT (Dr J. R.), Mullard Radio Astronomy Observatory, Cavendish Laboratory, Free School Lane, Cambridge, U.K.

* SHAKHOVSKOJ (Dr N. M.), Crimean Astrophysical Observatory, USSR Academy of Sciences, Crimea, U.S.S.R.

SHANE (Dr C. D.), P.O. Box 582, Santa Cruz, California 95060, U.S.A.

SHANE (W. W.), Sterrewacht, Leiden 2401, The Netherlands

SHAO (Mr Cheng-yuan), Harvard College Observatory, 60 Garden Street, Cambridge, Massachusetts 02138, U.S.A.

SHAPIRO (Dr I. I.), Building 54-622, Massachusetts Institute of Technology, Cambridge, Mass. 02139, U.S.A.

* SHAPIRO (Dr M. M.), Laboratory for Cosmic Ray Physics, Code 7020, Naval Research Laboratory, Washington, D.C. 20390, U.S.A.

SHAPLEY (A. H.), CRPL, National Bureau of Standards, Boulder, Colorado, U.S.A.

SHAPLEY (Prof. Dr H.), Sharon Cross Road, Peterboro, New Hampshire, U.S.A.

SHAPLEY (Mrs M. B.), Sharon Cross Road, Peterboro, New Hampshire, U.S.A.

SHARAF (Mrs Dr Sh. G.), Institute of Theoretical Astronomy, USSR Academy of Sciences, Leningrad, U.S.S.R.

SHAROV (Dr A. S.), Sternberg Astronomical Institute, Moscow, U.S.S.R.

SHARPLESS (Dr S. L.), Department of Physics and Astronomy, The University of Rochester, Rochester, New York 14627, U.S.A.

* SHAVIV (Dr G.), Department of Physics and Astronomy, Tel-Aviv University, Ramat Aviv, Israel

SHAW (Prof. J. H.), Department of Physics and Astronomy, Ohio State University, 174 W. 18th Ave, Columbus, Ohio, U.S.A.

SHAW (Prof. R. W.), Clark Hall of Science, Cornell University, Ithaca, N.Y., U.S.A.

SHCHEGLOV (Dr P. V.), Sternberg Astronomical Institute, Moscow, U.S.S.R.

SHCHEGLOV (Prof. Dr V. P.), Director of the Astronomical Observatory, Tashkent, U.S.S.R.

SHCHEGOLEV (Dr D. E.), Pulkovo Observatory, Leningrad, U.S.S.R.

* SHCHERBINA-SAMOJLOVA (Mrs Dr I. S.), Institute of Science and Technics Information, Department of Astronomy, USSR Academy of Sciences, Moscow, U.S.S.R.

SHCHIGOLEV (Prof. B. M.), Sternberg Astronomical Institute, Moscow, U.S.S.R.

SHEELEY JR (Dr N. R.), Kitt Peak Observatory, 950 North Cherry Avenue, Tucson, Arizona 85717, U.S.A.

SHEN (Dr B. S. P.), Flower and Cook Observatory, University of Pennsylvania, Philadelphia, Pennsylvania 19104, U.S.A.

SHEN (Dr Zee), Department of Mathematics, National Taiwan University, Taipei, Taiwan

* SHEPTUNOV (Dr G. S.), Main Astronomical Observatory, USSR Academy of Sciences, Pulkovo, U.S.S.R.

* SHER (Dr D.), Louisiana State University, Baton Rouge, Louisiana 70803, U.S.A.

SHERIDAN (K. V.), CSIRO, Div. of Radiophysics, Post Office Box 76, Epping, N.S.W., 2121, Australia

* SHIMIZU (Prof. Dr M.), The Institute of Space and Aeronautical Science, The University of Tokyo, Tokyo, Japan

SHIMIZU (Dr T.), Institute of Astrophysics, College of Science, University of Kyoto, Kyoto, Japan

* SHIMODA (Dr M.), Department of Astronomy, Faculty of Science, University of Tokyo, Bunkyo-ku, Tokyo, Japan

SHIRYAEV (Dr A. V.), Astronomical Observatory, Leningrad University, Leningrad, U.S.S.R.

SHKLOVSKY (Dr I. S.), Sternberg Astronomical Institute, Moscow, Prospect 13, U.S.S.R.

SHOEMAKER (E.), U.S. Geological Survey, Menlo Park, California, U.S.A.

SHORE (Dr B. W.), Dept. of Physics and Astronomy, Cardwell Hall, Kansas State University, Manhattan, Kansas 66502, U.S.A.

SHTEINS (Dr K. A.), Latvian University, Riga, U.S.S.R.

* SHU (Dr F.), Department of Earth and Space Sciences, State University of New York, Stony Brook, New York 11790, U.S.A.

* SHUKLA (K. S.), Reader, Department of Mathematics and Astronomy, Lucknow University, Lucknow, U.P., India

SHUL'BERG (Dr A. R.), Astronomical Observatory, Odessa University, Odessa, U.S.S.R.

* SHUTER (Dr W. L. H.), Department of Physics, University of British Columbia, Vancouver 8, B.C., Canada

SIBAHARA (Dr R.), n° 30, Kawashima-Sakurazonocho, Ukyo-ku, Kyoto, Japan

* SIDA (Prof. D. W.), Carleton University, Ottawa, Ontario, Canada

SIEGEL (C. L.), Göttingen University, Bunsenstrasse 3–5, Göttingen, F.R.G.

* ŠIMEK (Dr M.), Astronomical Institute, Czechoslovak Academy of Sciences, Observatory Ondřejov, Czechoslovakia

SIMON (Dr G. W.), Sacramento Peak Observatory, Sunspot, New Mexico, U.S.A.

* SIMON (Dr M.), State University of New York, Stony Brook, New York 11790, U.S.A.

SIMON (Dr P.), Observatoire de Paris, Section d'Astrophysique, 92-Meudon, France

SIMON (Prof. Dr R. L. E.), Institut d'Astrophysique, Université de Liège, avenue de Cointe, 5, B-4200 Cointe-Ougrée, Belgium

SIMOVLJEVIĆ (Dr J.), Prirodno matem. fakultet, katedra za astronomiju Beograd, Studentski trg 16, Beograd, Yugoslavia

SINNERSTAD (Dr U.), Stockholm Observatory, S-133 00 Saltsjöbaden, Sweden

SINTON (Dr W. M.), Institute for Astronomy, 2840 Koluwalu St., University of Hawaii, Honolulu, Hawaii 96822, U.S.A.

SINVHAL (Dr S. D.), Director, Uttar Pradesh State Observatory, Naini Tal, India

SINZI (Dr A. M.), Hydrographic Office, 5 chome, Tsukiji, Chuo-ku, Tokyo, Japan

SISSON (G. M.), Planetrees, Wall, Hexham, Northumberland, U.K.

SITARSKI (Dr G.), Astronomical Institute of Polish Academy of Sciences, Al. Ujazdowskie 4, Warszawa, Poland

SITNIK (Dr G. F.), Sternberg Astronomical Institute, Moscow, U.S.S.R.

SIVARAMAN (K. R.), Astrophysical Observatory, Kodaikanal-3, India

* SKALAFURIS (Dr A.), Department of Physics, Bartol Research Foundation of the Franklin Institute, Swarthmore, Pennsylvania 19081, U.S.A.

SKUMANICH (Dr A.), High Altitude Observatory, Boulder, Colorado 80301, U.S.A.

SLAUCITAJS (Prof. Dr S.), Observatorio Astronómico, La Plata, Argentina

SLAVENAS (Prof. Dr P. V.), Kestucio gatve 17–3, Vilnius 4, Lithuanian SSR, U.S.S.R.

SLEBARSKI (T. B.), University Observatory, St Andrews, U.K.

SLEE (O. B.), CSIRO, Division of Radiophysics, Post Office Box 76, Epping, N.S.W. 2121, Australia

SLETTEBAK (Prof. Dr A.), Perkins Observatory, Delaware, Ohio, U.S.A.

SLONIM (Mrs Dr E. M.), Astronomical Observatory, Tashkent, U.S.S.R.

* SLYSH (Dr V. I.), Institute of Cosmical Research, USSR Academy of Sciences, Moscow, U.S.S.R.

SMAK (Dr J.), Astronomical Observatory of Warsaw University, Al. Ujazdowskie 4, Warsaw, Poland

SMART (Prof. W. M.), Westbourne House, Westbourne Road, Lancaster, U.K.

SMERD (Dr S. F.), CSIRO, Division of Radiophysics, Post Office Box 76, Epping, N.S.W., 2121, Australia

SMEYERS (Abbé Dr P.), Astronomisch Instituut, Katholieke Universiteit Leuven, Naamsestraat, 61, B-3000 Leuven, Belgique

SMILEY (Dr C. H.), Ladd Observatory, Brown University, Providence, R.I., U.S.A.

SMIT (Prof. J. A.), Physical Laboratory of the University, Bijlhouwerstraat 6, Utrecht, The Netherlands

SMITH (Dr A. G.), Department of Physics, University of Florida, Gainesville, Florida, U.S.A.

SMITH (B. A.), The Observatory, New Mexico State University, Las Cruces, New Mexico 88001, U.S.A.

SMITH (Mrs Dr E. v. P.), Department of Physics and Astronomy, University of Maryland, College Park, Maryland 20742, U.S.A.

SMITH (Prof. F. Graham), Nuffield Radio Astronomy Laboratories, Jodrell Bank, Macclesfield, Cheshire, U.K.

SMITH (Dr G.), Department of Astrophysics, University Observatory, South Parks Road, Oxford, U.K.

SMITH (Prof. Harlan J.), Director, McDonald Observatory, Astronomy Department, Room 407 Physics Bld., The University of Texas, Austin, Texas 78712, U.S.A

SMITH (Dr Henry J.), National Aeronautics and Space Administration, Code SG, Washington 25, D.C., U.S.A.

SMITH (H. M.), Royal Greenwich Observatory, Herstmonceux Castle, Hailsham, Sussex, U.K.

* SMITH (Dr M. G.), Kitt Peak National Observatory, 950 North Cherry, Tucson, Arizona 85721, U.S.A.

* SMOLUCHOWSKI (Dr R.), Solid State Sciences, Princeton University, Princeton, New Jersey 08540, U.S.A.

SMYTH (Dr M. J.), Royal Observatory, Blackford Hill, Edinburgh 9, U.K.

* SOBERMAN (Dr R. K.), General Electric Co., Valley Forge Space Technology Center, Box 8555, Philadelphia, Pa. 19101, U.S.A.

SOBIESKI (Dr S.), Astrophysics Branch, Code 613, NASA, Goddard Space Flight Center, Greenbelt, Maryland 20771, U.S.A.

SOBOLEV (Dr V. M.), Main Astronomical Observatory, USSR Academy of Sciences, Pulkovo, U.S.S.R.

SOBOLEV (Prof. Dr V. V.), The University, Leningrad, U.S.S.R.

SOBOLEVA (Mrs Dr N. S.), Main Astronomical Observatory, USSR Academy of Sciences, Pulkovo, U.S.S.R.

SOCHER (Dr H.), Universitäts-Sternwarte Wien, Türkenschanzstrasse 17, A-1180 Vienna, Austria

* SOFIA (Dr S.), University of South Florida, Department of Astronomy, Tampa, Florida 33620, U.S.A.

SOLOMON (Dr P. M.), Astronomy Department, Columbia University, Box 110 Pupin Hall, New York, New York 10027, U.S.A.

SOMERVILLE (Dr W. B.), University College, Gower Street, London, W.C.1, U.K.

* SONNETT (Dr Ch. P.), NASA Ames Research Center, Moffet Field, California 94035, U.S.A.

* SOROCHENKO (Dr R. L.), Physical Institute, USSR Academy of Sciences, Moscow, U.S.S.R.

SOUFFRIN (P.), Observatoire de Nice, Le Mont-Gros, 06-Nice, France

* SOUFFRIN (Mrs Dr S.), Institut d'Astrophysique, 98bis, Boulevard Arago, Paris-14e, France

SOUTHWORTH (Dr R. B.), Harvard College Observatory, 60 Garden Street, Cambridge, Massachusetts 02138, U.S.A.

* SPEER (Dr R. J.), Physics Department, Imperial College, Prince Consort Road, London S.W. 7, U.K.

SPIEGEL (Dr E.), Courant Institute of Mathematical Sciences, 251 Mercer Street, New York, New York 10003, U.S.A.

SPINRAD (Dr H.), Berkeley Astronomical Department, University of California, Berkeley, California, U.S.A.

SPITE (F.), Observatoire de Meudon, 5, place Janssen, 92-Meudon, France

* SPITE (Mrs N.), Observatoire de Meudon, 92-Meudon, France

SPITZER (Dr L. Jr), Director of the Princeton University Observatory, Peyton Hall, Princeton, New Jersey 08540, U.S.A.

* SREENIVASAN (Dr S. R.), Dept. of Physics, University of Calgary, Calgary, Alberta, Canada
* STABELL (R.), Institute of Theoretical Astrophysics, University of Oslo, P.O. Box 1029, Blindern, Oslo 3, Norway
  STAHR-CARPENTER (Mrs Dr M.), Leander McCormick Observatory, Box 3818, University Station, Charlottesville, Virginia 22903, U.S.A.
  STANILA (Dr G.), Observatoire Astronomique, str. Cutitul de Argint 5, Bucarest 28, Roumania
  STANLEY (G. J.), California Institute of Technology, Pasadena 4, California, U.S.A.
* STARYTSIN (Dr G. V.), Main Astronomical Observatory, USSR Academy of Sciences, Pulkovo, U.S.S.R.
  STAWIKOWSKI (Dr A.), Astronomical Observatory of Toruń University, Sienkiewicza 30, Toruń, Poland
  STEARNS (Dr C. L.), Van Vleck Observatory, Middletown, Connecticut, U.S.A.
  STECHER (T. P.), Code 672, Goddard Space Flight Center, Greenbelt, Maryland 20771, U.S.A.
  STEIGER (Prof. W. R.), Hawaii Institute of Geophysics, University of Hawaii, 2525 Correa Rd., Honolulu, Hawaii 96822, U.S.A.
* STEIN (Dr R. F.), Brandeis University, Waltham, Massachusetts, U.S.A.
  STEIN (Dr W. A.), University of California, Physics Department, La Jolla, California, U.S.A.
  STEINBERG (Dr J. L.), Service de Radioastronomie, Observatoire de Paris, Section d'Astrophysique, 92-Meudon, France
  STEINITZ (Dr R.), Institute of Planetary and Space Physics, Tel-Aviv University, Ramat-Aviv, Israel
  STEINLIN (Dr U. W.), Astronomisches Institut der Universität Basel, Venusstrasse 7, CH-4102 Binningen, Switzerland
* STENFLO (Dr J. O.), Institute of Astronomy, S-222 24 Lund, Sweden
  STEPANOV (Dr W. E.), Post Office 65, Sibizmiran, Irkutsk, U.S.S.R.
  STEPHENSON (Prof. C. B.), Warner and Swasey Observatory, Taylor and Brunswick Roads, East Cleveland, Ohio 44112, U.S.A.
  ŠTERNBERK (Dr B.), Astronomical Institute, Czechoslovak Academy of Sciences, Budečská 6, Prague 2, Czechoslovakia
  STERNE (Dr T. E.), Smithsonian Astrophysical Observatory, 60 Garden Street, Cambridge, Massachusetts 02138, U.S.A.
  STEWART (Dr J. Q.), Box 446, Sedona, Arizona 85336, U.S.A.
  STESHENKO (Dr N. V.), Crimean Astrophysical Observatory, USSR Academy of Sciences, P/O Nauchny, Crimea, U.S.S.R.
* STEWART (Dr P.), Department of Mathematics, The University, Manchester, M13 9 PL, U.K.
  STEYAERT (Dr H. L. C.), Sterrenkundig Instituut, Rijksuniversiteit Gent, J. Plateaustraat, 22, B-9000 Gent, Belgium
  STIBBS (Prof. D. W. N.), University Observatory, Buchanan Gardens, St. Andrews, Fife, U.K.
  STICKER (Prof. Dr B.), Direktor des Instituts für Geschichte der Naturwissenschaften, Hartungstrasse 5, Hamburg 13, F.R.G.
  STIEFEL (E.), Mathematische Abteilung, Eidgenösische Technische Hochschule, Zurich, Switzerland
* STOBIE (Dr R. S.), Institute of Theoretical Astronomy, Madingley Road, Cambridge, U.K.
  STOCK (Dr J.), Observatório Astronómico Nacional, Casilla 36-D, Santiago, Chile
  STODOŁKIEWICZ (Dr J. S.), Astronomical Observatory of Warsaw University, Warszawa, Al. Ujazdowskie 4, Poland
* ŠTOHL (Dr J.), Astronomical Institute, Slovak Academy of Sciences, Dúbravská cesta, Bratislava, Czechoslovakia
  STONE (Dr R. G.), Code 615, Radio Astronomy Section, Ionospheric and Radio Physics Branch, Laboratory for Space Sciences, NASA-Goddard Space Flight Center, Greenbelt, Maryland 20771, U.S.A.
  STOY (Dr R. H.), Royal Observatory, Edinburgh 9, U.K.

STOYKO (Mrs A.), Observatoire de Paris, 61, Avenue de l'Observatoire, Paris-14e, France

STOYKO (Dr N.), Observatoire de Paris, 61, Avenue de l'Observatoire, Paris-14e, France

STRAIŽYS (Dr V. L.), Astronomijos observatorija, Vilnius 31, Čiurlionio 29, Lithuania, U.S.S.R.

STRAND (Dr K. Aa.), U.S. Naval Observatory, Washington, D.C. 20390, U.S.A.

STRASSL (Prof. Dr H.), Direktor des Astronomischen Instituts der Universität, Steinfurterstrasse 107, Münster (Westfalen), F.R.G.

STRITTMATTER (P. A.), Peterhouse, Cambridge, U.K.

STROBEL (Dr W.), Astronomisches Rechen-Institut, Mönchhofstrasse 12–14, 69 Heidelberg, F.R.G.

STROHMEIER (Prof. Dr W.), Direktor der Remeis-Sternwarte, Sternwartstrasse 7, Bamberg, F.R.G.

STROM (Dr R. G.), Lunar and Planetary Laboratory, University of Arizona, Tucson, Arizona 75721, U.S.A.

STROM (Dr S. E.), Department of Earth and Space Sciences, State University of New York, Stony Brook, N.Y. 11790, U.S.A.

STRÖMGREN (Prof. B.), Observatoriet, Østervoldgade 3, Copenhagen K, Denmark. President of the Union.

STRONG (Prof. J. D.), Astronomy Research Facility, University of Massachusetts, Amherst, Massachusetts 01002, U.S.A.

STUMPFF (Dr P.), National Radio Astronomy Observatory, Edgemont Dairy Road, Charlottesville, Va., U.S.A.

* STURCH (Dr C.), University of Rochester, River Campus Station, Rochester, New York 14627, U.S.A.

STURROCK (Dr P. A.), Institute for Plasma Research, Stanford University, Stanford, California, U.S.A.

SUDA (Dr K.), Astronomical Institute, Tohoku University, Sendai, Japan

SUEMOTO (Dr Z.), Dept. of Astronomy, Faculty of Science, University of Tokyo, Bunkyo-ku, Tokyo, Japan

SUGAWA (Dr C.), International Latitude Observatory of Mizusawa, Mizusawa-shi, Iwate-ken, Japan

* SUGIMOTO (Dr D.), Institute of Earth Science and Astronomy, College of General Education, University of Tokyo, 3–8–1 Komaba, Meguro-ku, Tokyo 153, Japan

SUKHAREV (Dr L. A.), Pulkovo Observatory, Leningrad, U.S.S.R.

SULTANOV (Dr G. F.), Shemakha Astrophysical Observatory, Azerbajdzhan Academy of Sciences, Baku, U.S.S.R.

SUSLOV (Dr A. K.), Saltykov-Shtshedrin st.h. 12 fl. 28, Leningrad D-28, U.S.S.R.

* SVATOŠ (Dr J.), Dept. of Astronomy and Astrophysics, Charles University, Švédská 8, Prague 5-Smíchov, Czechoslovakia

* SVECHNIKOVA (Mrs Dr M. A.), Astronomical Observatory, Ural University, Sverdlovsk, U.S.S.R.

ŠVESTKA (Dr Z.), Astronomical Institute, Czechoslovak Academy of Sciences, Observatory Ondřejov, Czechoslovakia

SVOLOPOULOS (Prof. Dr S. N.), University of Ioannina, Ioannina, Greece

SWARUP (Dr G.), Tata Institute of Fundamental Research, Bombay, India

SWEET (Prof. P. A.), The University Observatory, Glasgow W.2, U.K.

* SWENSON (Dr G. W. Jr), Vermillion River Observatory, University of Illinois, Urbana, Illinois 61801, U.S.A.

SWENSSON (Dr J. W.), Institute of Physics, Sölvegatan 14, S-223 62 Lund, Sweden

SWIHART (Dr T. L.), Steward Observatory, University of Arizona, Tucson, Arizona, U.S.A.

* SWINGS (Dr J. P.), Institut d'Astrophysique, Université de Liège, avenue Léon Souguenet, 23, B-4050 Esneux, Belgium

SWINGS (Prof. Dr P.), Directeur de l'Institut d'Astrophysique, Université de Liège, avenue Léon Souguenet, 23, B-4050 Esneux, Belgium

SWOPE (Miss H. H.), Mount Wilson and Palomar Observatories, 813 Santa Barbara Street, Pasadena, California 91106, U.S.A.

SYKES (Dr J. B.), 20 Milton Lane, Steventon, Abingdon, Berkshire, U.K.

SYKES-HART (Mrs A. B.), University Observatory, Oxford, U.K.

SYMMS (L. S. T.), 9, The Grove, Bexhill on Sea, Sussex, U.K.

SYROVATSKIJ (S. I.), Lebedev Physical Institute, Moscow, U.S.S.R.

SYTINSKAYA (Mrs Dr N. N.), University Observatory, Leningrad, U.S.S.R.

* SYUNYAEV (Dr R. A.), Institute of Applied Mathematics, USSR Academy of Sciences, Moscow, U.S.S.R.

SZAFRANIEC (Miss Dr R.), Astronomical Observatory, Kopernika 27, Cracow, Poland

SZEBEHELY (Dr V.), Dept. of Aerospace Engineering, The University of Texas, 2315 Speedway, Austin, Texas 78712, U.S.A.

SZEIDL (Dr B.), Konkoly Observatory, P.O. Box 114, Box 67, Budapest XII, Hungary

TAFFARA (Prof. S.), Osservatorio Astronomico, Padova, Italy

* TAGLIAFERRI (Prof. Dr G.), Osservatorio Astrofisico di Arcetri, Largo Enrico Fermi 5, Florence, Italy

TAI (Dr Yuin-Kwei), Director, Institute of Geophysics, Central University, Taipei, Taiwan, Republic of China

TAKAGI (Dr S.), International Latitude Observatory of Mizusawa, Mizusawa-shi, Iwate-ken, Japan

TAKAKUBO (Dr K.), Astronomical Institute, Faculty of Science, Tohoku University, Sendai, Japan

TAKAKURA (Dr T.), Tokyo Astronomical Observatory, Mitaka, Tokyo, Japan

TAKASE (Dr B.), Tokyo Astronomical Observatory, Mitaka, Tokyo, Japan

TAKAYANAGI (K.), Institute of Space and Aeronautical Science, University of Tokyo, Komaba, Meguro-ku, Tokyo, Japan

TAKENOUCHI (Dr T.), Tokyo Astronomical Observatory, Mitaka, Tokyo, Japan

TALWAR (S. P.), Department of Physics, University of Delhi, Delhi, India

TAMMANN (Dr G. A.), Astronomisches Institut der Universität Basel, Venusstrasse 7, CH-4102 Binningen, Switzerland

TANABE (Dr H.), Tokyo Astronomical Observatory, University of Tokyo, Mitaka, Tokyo, Japan

TANAKA (Dr H.), Research Institute of Atmospherics, Nagoya University, Ichida-machi, Tokyokawa-shi, Aichi, Japan

TANAKA (Dr Y.), Werkgroep Kosmische Straling, Kamerlingh Onnes Laboratorium, Nieuwsteeg 18, Leiden, The Netherlands

TANDBERG-HANSSEN (Dr E.), High Altitude Observatory of the University of Colorado, Boulder, Colorado, U.S.A.

TANNAHILL (Dr T. R.), University Observatory, Glasgow W. 2, U.K.

TANNER (R. W.), Seismology Division, Earth Physics Branch, Energy Mines and Resources, Ottawa, Canada

TARDI (Prof. P.), 4, Villa de Segur, Paris-7e, France

TASSOUL (Dr J. L.), Département de Physique, Université de Montréal, P.O. Box 6128, Montréal 3, P.Q., Canada

* TATEVYAN (Dr S. K.), Astronomical Council, USSR Academy of Sciences, Vavilov Str. 34, Moscow V-312, U.S.S.R.

* TAUBER (Prof. G. E.), Department of Physics and Astronomy, Tel-Aviv University, Ramat Aviv, Israel

TAVARES (Dr J. T. L.), Avenida Dias da Silva, 173, R/C Esq., Coimbra, Portugal

TAVASTSHERNA (Dr K. N.), Main Astronomical Observatory, USSR Academy of Sciences, Pulkovo, Leningrad M-140, U.S.S.R.

TAYLER (Dr R. J.), Astronomy Centre, University of Sussex, School of Mathematical and Physical Sciences, Falmer, Sussex, U.K.

* TAYLOR (Dr D. J.), Stewart Observatory, University of Arizona, Tucson, Arizona 85721, U.S.A.

* TAYLOR (G. E.), H.M. Nautical Almanac Office, Royal Greenwich Observatory, Herstmonceux Castle, Hailsham, Sussex, U.K.

* Tech (Dr J.), U.S. Department of Commerce, National Bureau of Standards, Spectroscopy Section, Washington D.C. 20234, U.S.A.

Téhérany (D.), 83, avenue Rey, Teheran, Iran

Tejfel' (Dr V. G.), Astrophysical Institute of the Kazakh Academy of Sciences, Alma-Ata, U.S.S.R.

Teleki (Dr G.), Astronomical Observatory, Volgina 7, Belgrade, Yugoslavia

Temesváry (Dr S.), Department of Astronomy and Space Science, State University of New York, Albany, N.Y. 12203, U.S.A.

Tempesti (Prof. P.), Osservatorio Astronomico di Collurania, Teramo, Italy

* Teplitskaya (Mrs Dr R. B.), Sibizmir, USSR Academy of Sciences, Irkutsk, U.S.S.R.

Ter Haar (Dr D.), Magdalen College, Oxford, U.K.

Terrazas (L. R.), Tonantzintla and Tacubaya Observatories, Mexico

Terrien (Dr J.), Directeur, Bureau International des Poids et Mesures, Pavillon de Breteuil, 92-Sèvres (Seine-et-Oise), France

Terzan (Dr A.), Observatoire de Lyon, 69-St.-Genis-Laval, France

Terzian (Dr Y.), Center for Radiophysics and Space Research, Space Sciences Building, Cornell University, Ithaca, New York 14850, U.S.A.

Teske (Dr R. G.), Astronomy Department, University of Michigan, Ann Arbor, Michigan, U.S.A.

Texereau (J.), Observatoire de Paris, 61, avenue de l'Observatoire, 75-Paris-14e, France

Thackeray (Dr A. D.), Director of the Radcliffe Observatory, P.O. Box 373, Pretoria, South Africa

The (Dr Pik Sin), Astronomical Institute, University of Amsterdam, Roetersstraat 15, Amsterdam-C, The Netherlands

Thernöe (K. A.), University Observatory, Østervoldgade 3, Copenhagen-K, Denmark

Thiry (Prof. Y.), Prof. à la Faculté des Sciences de l'Université de Paris, 7–9 Quai St. Bernard, Paris-5e, France

Thomas (Dr D. V.), Royal Greenwich Observatory, Herstmonceux Castle, Hailsham, Sussex, U.K.

* Thomas (Dr H. C.), Max-Planck-Institut für Physik und Astrophysik, Föhringer Ring 6, 8000 München 23, F.R.G.

Thomas (Dr R. N.), Joint Institute for Laboratory Astrophysics, University of Colorado, Boulder, Colorado 80304, U.S.A.

Thompson (Dr A. R.), Radio Astronomy Institute, Stanford University, Stanford, California 94305, U.S.A.

Thompson (Dr G. I.), The Royal Observatory, Edinburgh 9, U.K.

Thomson (M. M.), Head, Time & Frequency Section, Physics Division, National Research Council, Montreal Road, Ottawa 7, Ontario, Canada

Thorne (Dr K.), Physics Department, California Institute of Technology, Pasadena, California 91109, U.S.A.

Tifft (Dr W. G.), Steward Observatory, University of Arizona, Tucson, Arizona 85721, U.S.A.

Tifrea (Dr E.), Observatoire Astronomique, Str. Cutitul de Argint 5, Bucarest 28, Roumania

Tilles (Dr D.), Oceanography Department, Oregon State University, Cornvallis, Oregon 97331, U.S.A.

Ting (Yeou-Tswen), Chief of Astronomical Department, Taiwan Provincial Weather Bureau, 64, Kung Yuen Road, Taipei, Taiwan, Republic of China

Tiuri (Prof. Dr M.), Technical University, Department of Electrotechnics, Albert-Street 40–42 Helsinki, Finland

Tlamicha (Dr A.), Astronomical Institute, Czechoslovak Academy of Sciences, Ondřejov, Czechoslovakia

Todoran (Dr I.), Observatoire, Université, Cluj, Roumania

Tolbert (Dr C. R.), Leander McCormick Observatory, University of Virginia, Charlottesville Virginia 22903, U.S.A.

TOMBAUGH (C.W.), New Mexico State University, Research Center, University Park, New Mexico, U.S.A.

TOOMRE (Dr A.), Department of Mathematics, Massachusetts Institute of Technology, Cambridge, Massachusetts 02139, U.S.A.

TORAO (M.), Tokyo Astronomical Observatory, Mitaka, Tokyo, Japan

TORGÅRD (Miss Dr I. H. M.), Nils Ericssonsgatan 28, 521 00, Falköping, Sweden

* TORRES-PEIMBERT (Dr S.), Instituto de Astronomía U.N.A.M., Apartado Postal 70–264, México D.F., Mexico

TORROJA (Prof. J. M.), Fac. de Ciencas, University of Madrid, Madrid 3, Spain

TOUSEY (Dr R.), U.S. Naval Research Laboratory, Washington 25, D.C., U.S.A.

TOVMASYAN (Dr G. M.), Byurakan Astrophysical Observatory, Armenian Academy of Sciences, Byurakan, U.S.S.R.

TOWNES (Dr C. H.), Physics Department, University of California, Berkeley, California 94720, U.S.A.

* TRAFTON (Dr L. M.), Department of Astronomy, University of Texas, Austin, Texas 78712, U.S.A.

TRAVING (Prof. Dr G.), Lehrstuhl für Theoretische Astrophysik der Universität Heidelberg, 69 Heidelberg, Berliner Strasse 19, F.R.G.

TREANOR, S.J. (Dr P. J.), Specola Vaticana, Castel Gandolfo, Vatican City State

TREDER (Prof. Dr H.-J.), Zentralinstitut für Astrophysik, Sternwarte Babelsberg, DDR 1502 Potsdam-Babelsberg, Rosa-Luxemburg-Strasse 17a, G.D.R.

TREFFTZ (Miss Dr E.), Max-Planck-Institut für Physik und Astrophysik, Aumeisterstrasse 6, München 23, F.R.G.

TREHAN (Dr S. K.), Department of Physics and Astrophysics, University of Delhi, Delhi, India

TRELLIS (M.), Détaché à l'Observatoire de Nice, Le Mont-Gros, 06-Nice, France

TREMKO (Dr J.), Skalnaté Pleso Observatory, Slovak Academy of Sciences, Tatranská Lomnica, Czechoslovakia

TREXLER (J.H.), U.S. Naval Research Laboratory, Washington 25, D.C., U.S.A.

TROITSKY (Dr V. S.), Scientific Radiophysical Institute, Gorky, U.S.S.R.

* TRURAN (Dr J. W.), Belfer Graduate School of Science, Yeshiva University, Amsterdam Ave. and 186th St., New York 10033, U.S.A.

* TRUTSE (Dr Yu. L.), Institute of Physics of Atmosphere, USSR Academy of Sciences, Moscow, U.S.S.R.

TS'AI (Chang-hsien), Taipei Observatory, Yuan Shan, Taipei, Taiwan, Republic of China

TSAO (Prof. Mo.), 47, 3rd Section, Hsin-I Road, Taipei, Taiwan, Republic of China

* TSAP (Dr T. T.), Crimean Astrophysical Observatory, USSR Academy of Sciences, Crimea, U.S.S.R.

* TSCHARNUTER (Dr W. M.), Institut für Theoretische Astronomie, University of Vienna, Türkenschantzstrasse 17, A-1180, Vienna, Austria

TSESEVICH (Prof. Dr V. P.), Director of the Astronomical Observatory, Odessa, U.S.S.R.

* TSUBAKI (Dr T.), Institute of Earth Science and Astrophysics, University of Shiga, Ohtsu, Japan

TSUBOKAWA (Dr I.), Professor of Tokyo University, Seismological Observatory Institute, No. 1 – 1 gou – 1 chome, Yayoimachi, Bunkuyo-ku, Tokyo, Japan

TSUCHIYA (A.), Tokyo Astronomical Observatory, University of Tokyo, Mitaka, Tokyo, Japan

* TSUJI (Dr T.), Department of Astronomy, Faculty of Science, University of Tokyo, Bunkyo-ku, Tokyo 113, Japan

* TSURUTA (Dr S.), Theoretical Studies Branch, Code 641, NASA Goddard Space Flight Center, Greenbelt, Maryland 20771, U.S.A.

TUCKER (R. H.), Royal Greenwich Observatory, Herstmonceux Castle, Hailsham, Sussex, U.K.

* TÜFEKÇIOĞLU (Dr Z.), ARGE, Ankara, Turkey

* TULL (Dr R. G.), Department of Astronomy, University of Texas, Austin, Texas 78712, U.S.A.

TUOMINEN (Prof. J. V.), Pihlajatie 49 B 20, Helsinki-Töölö, Finland

TURŁO (Dr Z.), Astronomical Observatory, Toruń-Piwnice, Poland

* TURNER (Dr K. C.), Department of Terrestrial Magnetism, Carnegie Institute of Washington, 5241 Broad Branch Road N.W., U.S.A.

* TURTLE (Dr A. J.), The Chatterton Astrophysics Department, School of Physics, University of Sydney 2006, Australia

* TUTUKOV (Dr A. V.), Astronomical Council, USSR Academy of Sciences, Vavilov Str. 34, Moscow V-312, U.S.S.R.

TUVE (Dr M. A.), Carnegie Institution of Washington, Department of Terrestrial Magnetism, 5241 Broad Branch Road, N.W., Washington 15, D.C., U.S.A.

TUZI (Dr K.), 2-2-20 Ôsawa, Mitaka-shi, Tokyo, Japan

TWISS (Dr R. Q.), National Physical Laboratory, Teddington, Middlesex, U.K.

* TYLER JR. (Dr G. L.), Center for Radar Astronomy, Stanford University, Stanford, California 94305, U.S.A.

* UCHIDA (Dr J.), Department of Technology, Tohoku-Gakuin University, Tagajo-machi, Miyagi-Ken, Japan

UCHIDA (Dr Y.), Tokyo Astronomical Observatory, University of Tokyo, Mitaka, Tokyo, Japan

* UDAL'TSOV (Dr V. A.), Physical Institute, USSR Academy of Sciences, Moscow, U.S.S.R.

UENO (Dr S.), Department of Astronomy, University of Southern California, University Park, Los Angeles, California 90007, U.S.A.

UESUGI (Dr A.), Yamagami-cho 2-8, Otsu, Shiga, Japan

UETA (Dr J.), Department of Astronomy, Faculty of Science, Kyoto University, Kyoto, Japan

* ULRICH (Dr B.), Astronomy Department, University of Texas, Austin, Texas 78712, U.S.A.

* ULRICH (Mrs Dr M. H.), Institut d'Astrophysique, 98 bis, Boulevard Arago, Paris-14e, France

UNDERHILL (Prof. Dr A. B.), Chief, Laboratory for Optical Astronomy, Goddard Space Flight Center, Greenbelt, Maryland 20771, U.S.A.

UNDERWOOD (J. H.), Goddard Space Flight Center, Greenbelt, Maryland 20771, U.S.A.

UNNO (Dr W.), Department of Astronomy, Faculty of Science, University of Tokyo, Bunkyo-ku Tokyo, Japan

UNSÖLD (Prof. Dr A.), Direktor des Instituts für Theoretische Physik, Olshausenstrasse, Neue Universität, Bau 13/I, Kiel, F.R.G.

UPGREN (Dr A.), Department of Astronomy, Wesleyan University, Middletown, Connecticut 06457, U.S.A.

UPTON (Dr E. K. L.), Astronomy Department, University of California, Los Angeles, California 90024, U.S.A.

URBARZ (Dr H.), Astronomisches Institut der Universität, Waldhaüserstr. 64, 74 Tübingen, F.R.G.

* URECHE (Dr V.), Observatorul Astronomic, Strada Cutitul de Argint 5, Bucuresti 28, Roumania

UREY (Prof. H. C.), University of California, La Jolla, California, U.S.A.

* USHER (Dr P. D.), Department of Astronomy, Pennsylvania State University, University Park, Pennsylvania 16802, U.S.A.

* VAINSTEIN (Dr L. A.), Physical Institute, USSR Academy of Sciences Moscow, U.S.S.R.

VÄISÄLÄ (Prof. Y.), Puolalanpuisto 1, Turku, Finland

VALNÍČEK (Dr B.), Astronomical Institute, Czechoslovak Academy of Sciences, Observatory Ondřejov, Czechoslovakia

* VAN AGT (Dr S. L. Th. J.), Astronomical Department of the University of Nijmegen, Nijmegen, The Netherlands

VAN ALBADA (Prof. Dr G. B.), Astronomical Institute, Roetersstraat 1a, Amsterdam-C, The Netherlands

* VAN ALBADA (Dr T. S.), Astronomical Institute, Broerstraat 7, Groningen, The Netherlands

* Van Allen (Dr J. A.), Department of Physics and Astronomy, University of Iowa, Iowa City, Iowa 52240, U.S.A.

* Van Altena (Dr W. F.), Yerkes Observatory, Astrometric Program, Williams Bay, Wisconsin 53191, U.S.A.

Van Biesbroeck (Prof. G.), Lunar & Planetary Laboratory, University of Arizona, Tucson, Arizona, U.S.A.

* Van Breda (Dr I. G.), University Observatory, Buchanan Gardens, St. Andrews, U.K.

Van Bueren (Prof. H. G.), Astronomical Observatory, 'Sonnenborgh', Zonneburg 2, Utrecht, The Netherlands

Van de Hulst (Prof. Dr H. C.), Sterrewacht, Leiden, The Netherlands

Van de Kamp (Prof. Dr P.), Director of the Sproul Observatory, Swarthmore College, Swarthmore, Pennsylvania 19081, U.S.A.

Vandekerkhove (Dr E.), Astronome Honoraire, Observatoire Royal de Belgique, avenue de Sumatra 7, B-1180 Bruxelles, Belgium

Van den Bergh (Prof. S.), David Dunlap Observatory, Richmond Hill, Ontario, Canada

Van den Bos (Dr W. H.), c/o Republic Observatory, Johannesburg, South Africa

* Van den Heuvel (E. P. J.), Astronomical Institute, Zonnenburg 2, Utrecht, The Netherlands

Van den Borght (Prof. R.), Department of Mathematics, Monash University, Melbourne, Australia

Van der Laan (Prof. Dr H.), Sterrewacht, Leiden, The Netherlands

Vanderlinden (Prof. ém. Dr H. L.), University Ghent, Hertogsdreef 15, B-1170 Brussels, Belgium

Vandervoort (Prof. P. O.), Ryerson Physical Laboratory, 1100–14 East 58th Street, Chicago, Illinois 60637, U.S.A.

Van Diggelen (Dr J.), Sterrewacht 'Sonnenborgh', Zonnenburg 2, Utrecht, The Netherlands

* Van Duinen (R. J.), University of Groningen, Astronomy Department, Space Research Division, Groningen, The Netherlands

* Van Flandern (Dr T. C.), U.S. Naval Observatory, Washington, D.C. 20390, U.S.A.

* Van Genderen (Dr A. M.), Leiden Southern Station, P.O. Box 13, Broederstroom, Transvaal, Republic of South Africa

Van Herk (Dr G.), Sterrewacht, Leiden, The Netherlands

Van Hoof (Prof. Dr A.), Astronomisch Instituut, Katholieke Universiteit Leuven, Naamsestraat 61, B-3000 Leuven, Belgium

* Van Horn (Dr H. M.), University of Rochester, River Campus Station, Rochester, New York 14627, U.S.A.

Van Houten (Dr C. J.), Sterrewacht, Leiden, The Netherlands

Van Houten-Groeneveld (Mrs Dr I.), Sterrewacht, Leiden, The Netherlands

* Van Leer (Dr B.), The Observatory Leiden, Leiden, The Netherlands

Van Regemorter (Dr H.), Observatoire de Paris, Section d'Astrophysique, 92-Meudon, France

Van Schewick (Dr H.), Universitäts-Sternwarte Bonn, Poppelsdorfer Allee 49, Bonn, F.R.G.

Van 't Veer (Mrs Dr C.), Institut d'Astrophysique, 98 bis Boulevard Arago, Paris-14e, France

Van 't Veer (Dr F.), Institut d'Astrophysique, 98 bis Boulevard Arago, Paris-14e, France

Van Woerden (Dr H.), Kapteyn Astronomical Laboratory, Hoogbouw W.S.N., Postbus 800, Groningen 8002, The Netherlands

Vanýsek (Prof. V.), Dept. of Astronomy and Astrophysics Charles University, Švédská 8, Prague 5-Smíchov, Czechoslovakia

Vardya (Dr M. S.), Astrophysics Division, Tata Institute of Fundamental Research, Colaba, Bombay 5, India

Varsavsky (Dr C. M.), Las Heras 1975 (6°A), Buenos Aires, Argentina

Vasilyev (Dr O. B.), Pulkovo h. 13 fl. 24, Leningrad M-140, U.S.S.R.

Vasil'ev (Dr V. M.), Main Astronomical Observatory, USSR Academy of Sciences, Pulkovo, U.S.S.R.

* VASIL'EVA (Mrs Dr G. J.), Main Astronomical Observatory, USSR Academy of Sciences, Pulkovo, U.S.S.R.

VASILEVSKIS (Dr S.), University of California, Santa Cruz, California 95060, U.S.A.

VAUGHAN JR. (Dr A. H.), Mount Wilson and Palomar Observatory, 813 Santa Barbara Street, Pasadena, California 91106, U.S.A.

VELGHE (Prof. Dr A. G.), Directeur de l'Observatoire Royal de Belgique, avenue Circulaire 3, B-1180 Bruxelles, Belgium

VERBAANDERT (Prof. Dr A. G.), Astronome Honoraire, Observatoire Royal de Belgique, avenue de l'Observatoire 90, B-1180 Bruxelles, Belgium

VERGNANO (Prof. A.), Osservatorio Astronomico, Pino Torinese, Italy

VERNIANI (Prof. F.), Instituto di Fisica, Via Irnerio 46, Bologna 40126, Italy

* VERON (Dr P.), Observatoire de Meudon, 92-Meudon, France

* VERSCHUUR (Dr C. L.), NRAO, Edgemont Road, Charlottesville, Va. 22901, U.S.A.

VETEŠNÍK (Dr M.), Astronomical Institute, Purkyně University, Kotlářská 2, Brno, Czechoslovakia

VICENTE (Prof. Dr R. O.), R. Mestre Aviz 30, R/c, Algés, Portugal

VIDAL (Dr J. M. C.), Avenida José Antonio 677, Barcelona, Spain

* VILA (Dr S. C.), Department of Astronomy, University of Pennsylvania, Philadelphia, Pa. 19104, U.S.A.

VINTI (Dr P. J.), Rm 37–340, M.I.T. Measurement Systems Laboratory, 70 Vassar Street, Cambridge, Massachusetts 02139, U.S.A.

VIRGOPIA (Dr N.), Osservatorio Astronomico, Via Trionfale 204, 00136 Rome, Italy

* VISVANATHAN (Dr N.), 787 Sawmill Brook Parkway, Newton Center, Massachusetts 02159, U.S.A.

VISWANATHAN (N.), Mount Wilson and Palomar Observatories, 813 Santa Barbara Street, Pasadena, California, U.S.A.

VITINSKIJ (Dr Yu. I.), Main Astronomical Observatory, USSR Academy of Sciences, Pulkovo, U.S.S.R.

VITKEVICH (Dr V. V.), Chief of the Department of Radio-Astronomy, Physical Institute, Moscow, U.S.S.R.

VOIGT (Prof. Dr H. H.), Direktor der Universitäts-Sternwarte, Geismarlandstrasse 11, 34 Göttingen, F.R.G.

VOLLAND (Dr H.), Astronomisches Institut der Universität, Poppelsdorfer Allee 49, 53 Bonn, F.R.G.

VON DER HEIDE (Dr J. E. B.), Hamburger Sternwarte, Hamburg-Bergedorf, F.R.G.

VON HOERNER (Dr S.), National Radio Astronomy Observatory, P.O. Box 2, Green Bank, West Virginia, U.S.A.

VON KLÜBER (Dr H.), The Observatories, Madingley Road, Cambridge, U.K.

VON WEIZSÄCKER (Prof. C. F.), Max-Planck-Institut, Riemerschmidstr. 7, 813 Starnberg, F.R.G.

VORONTSOV-VEL'YAMINOV (Prof. Dr B. A.), Sternberg Astronomical Institute, Moscow, V-234, U.S.S.R.

* VOROSHILOV (Dr V. I.), Main Astronomical Observatory, Ukrainian Academy of Sciences, Kiev, U.S.S.R.

VSEKHSVYATSKIJ (Prof. Dr S. K.), Astronomical Observatory, Kiev 53, U.S.S.R.

VYSSOTSKY (Prof. A. N.), 700 Melrose Drive, Apt. J2, Winter Park, Florida 32789, U.S.A.

WACHMANN (Prof. Dr A. A.), Augustastr. 2, 205 Hamburg 80, F.R.G.

WACKERNAGEL (Dr H. B.), University of Colorado, Colorado Springs, Colorado, U.S.A.

WADE (Dr C. M.), National Radio Astronomy Observatory, P.O. Box 2, Green Bank, West Virginia, U.S.A.

WAGMAN (Dr N. E.), Allegheny Observatory, Riverview Park, Pittsburgh, Pennsylvania 15214, U.S.A.

WALDMEIER (Prof. Dr M.), Director, Swiss Federal Observatory, Schmelzbergstrasse 25, 8006 Zürich, Switzerland

WALKER (Dr G. A. H.), Geophysics & Astronomy Department, University of British Columbia, Vancouver 8, B.C., Canada

WALKER (Dr M. F.), Lick Observatory, University of California, Santa Cruz, California 95060, U.S.A.

* WALKER (Dr R.), Washington University, St. Louis, Missouri 63130, U.S.A.

* WALKER (W. S. G.), Variable Star Section, RAS of New Zealand, 18 Pooles Road, Greerton, Tauranga, New Zealand

WALLACE (Dr L. V.), Kitt Peak National Observatory, 950 North Cherry Avenue, Tucson, Arizona, U.S.A.

WALLENQUIST (Prof. Dr Å. A. E.), Uppsala Observatory, Kvistaberg Station, S-190 51 Bro, Sweden

WALLERSTEIN (Dr G. N.), Department of Astronomy, Seattle, Washington, U.S.A.

WALRAVEN (Dr Th.), Leiden Southern Station, P.O. Box 13, Broederstroom, Transvaal, Republic of South Africa

WALSH (Prof. D.), Nuffield Radio Astronomy Laboratories, Jodrell Bank, Macclesfield, U.K.

WALTER (Prof. Dr K.), Astronomisches Institut der Universität, Waldhäuserstrasse 64, 74 Tübingen, F.R.G.

WAMPLER (Dr E. J.), Lick Observatory, Mount Hamilton, California 95140, U.S.A.

WAPSTRA (Prof. A.), Voltaplein 19, Amsterdam-O, The Netherlands

WARES (Dr G. W.), CRF Astrophysics, Space Physics Laboratory, Air Force Cambridge Research Laboratory, Laurence G. Hanscom Field, Bedford, Massachusetts 01730, U.S.A.

WARNER (Dr B.), Department of Astronomy, University of Texas, Austin, Texas 78712, U.S.A.

WARWICK (Mrs Dr C. S.), Sun-Earth Relationships Section, Ionosphere Research and Propagation Division, National Bureau of Standards, Boulder Laboratories, Boulder, Colorado, U.S.A.

WARWICK (Dr J. W.), High Altitude Observatory of the University of Colorado, Boulder, Colorado, U.S.A.

WARZEE (Dr J.), Astronome Honoraire, Observatoire Royal de Belgique, rue Mattot, 115, B-1410 Waterloo, Belgium

WATERFIELD (Dr R. L.), The Observatory, Woolston, Yeovil, Somerset, U.K.

* WATTENBERG (Prof. D.), Direktor der Archenhold-Sternwarte, Berlin-Treptow, G.D.R.

WATTS (Dr C. B.), U.S. Naval Observatory, Washington 25, D.C., U.S.A.

WAYMAN (Dr P. A.), Director, Dunsink Observatory, Castleknock, Co. Dublin, Ireland

WEAVER (Dr H. F.), Radio Astronomy Laboratory, University of California, Berkeley 4, California, U.S.A.

* WEBROVÁ (Mrs Dr L.), Astronomical Institute, Czechoslovak Academy of Sciences, Buděcská 6, Prague 2, Czechoslovakia

WEHLAU (Dr A.), Department of Astronomy, University of Western Ontario, London, Ontario, Canada

WEHLAU (Prof. Dr W. H.), Department of Astronomy, University of Western Ontario, London, Ontario, Canada

WEIDEMANN (Prof. Dr V.), Institut für Theoretische Physik und Sternwarte der Universität, 23 Kiel, Olshausenstrasse, Neue Universität, Haus C 4/1, F.R.G.

WEIGERT (Dr A.), Director, Hamburg Observatory, 205 Hamburg 80, F.R.G.

WEILL (G.), Institut d'Astrophysique, 98bis, boulevard Arago, 75-Paris-14e, France

WEIMER (Dr Th.), Observatoire de Paris, 61, avenue de l'Observatoire, Paris-14e, France

WEINBERG (Dr J. L.), The Dudley Observatory, 100 Fuller Road, Albany, New York 12205, U.S.A.

* WEISS (Dr N. O.), Department of Applied Mathematics and Theoretical Physics, Silver Street, Cambridge CB3 9EW, U.K.

* WELIACHEW (Dr L.), Observatoire de Meudon, 92-Meudon, France

WELLGATE (G. B.), Royal Greenwich Observatory, Herstmonceux Castle, Hailsham, Sussex, U.K.

WELLMANN (Prof. P.), Universitäts-Sternwarte, Sternwartstrasse 23, München 27, F.R.G.

WEMPE (Prof. Dr J.), Zentralinstitut für Astrophysik, DDR 15 Potsdam, Telegrafenberg, G.D.R.

* WENDKER (Dr H. J.), Max-Planck-Institut für Radioastronomie, Argelanderstr. 3, 5300 Bonn, F.R.G.

WENTZEL (Dr D. G.), Astronomy Program, University of Maryland, College Park, Md. 20740, U.S.A.

WENZEL (Dr W.), Zentralinstitut für Astrophysik, Sternwarte Sonneberg, DDR 64 Sonneberg, G.D.R.

WESSELINK (Dr A. J.), Yale University Observatory, Box 2023 Yale Station, New Haven, Connecticut 06520, U.S.A.

* WEST (Dr R. M.), European Southern Observatory, Bergedorfer Strasse 131, 205 Hamburg 80, F.R.G.

WESTERHOUT (Dr G.), Astronomy Program, University of Maryland, College Park, Maryland 20740, U.S.A.

WESTERLUND (Dr B. E.), European Southern Observatory, Bergedorfer Str. 131, 205 Hamburg 80, F.R.G.

* WESTFOLD (Prof. K. C.), Monash University, Clayton 3168, Australia

WEYMANN (Prof. R. J.), Steward Observatory, University of Arizona, Tucson, Arizona, U.S.A.

* WHEELER (Dr J. A.), Joseph Henry Laboratory, Princeton University, Princeton, New Jersey 08540, U.S.A.

WHIPPLE (Dr F. L.), Director, Smithsonian Astrophysical Observatory, 60 Garden Street, Cambridge, Massachusetts 02138, U.S.A.

WHITAKER (E. A.), Lunar and Planetary Laboratory, University of Arizona, Tucson, Arizona, U.S.A.

WHITE (Dr O. R.), Sacramento Peak Observatory, Sunspot, New Mexico 88349, U.S.A.

* WHITE (Dr R.), Steward Observatory, University of Arizona, Tucson, Arizona 85721, U.S.A.

WHITEOAK (Dr J. B.), CSIRO, Division of Radiophysics, Post Office Box 76, Epping, N.S.W., 2121, Australia

WHITFORD (Dr A. E.), Director of the Lick Observatory, University of California, Santa Cruz, California 95060, U.S.A.

WHITNEY (Prof. B. S.), University of Oklahoma Observatory, Norman, Oklahoma, U.S.A.

WHITNEY (Dr C. A.), Smithsonian Astrophysical Observatory, 60 Garden Street, Cambridge, Massachusetts 02138, U.S.A.

WHITROW (Dr G. J.), Department of Mathematics, Imperial College of Science and Technology, Prince Consort Road, London, S.W.7, U.K.

WICKRAMASINGHE (Dr N. C.), Jesus College, Cambridge, U.K.

WIDING (Dr K. D.), Code 7144, U.S. Naval Research Laboratory, Washington, D.C. 20390, U.S.A.

WIDORN (Dr T.), Universitäts-Sternwarte Wien, Türkenschanzstrasse 17, A-1180 Vienna, Austria

WIELDLING (Dr T.), Villavägen 15, S-611 00 Nyköping, Sweden

* WIELEBINSKI (Dr R.), Max-Planck-Institut für Radioastronomie, Argelanderstr. 3, 5300 Bonn, F.R.G.

* WIELEN (Dr R.), Astronomisches Recheninstitut, Mönchhofstr. 12–14, 6900 Heidelberg, F.R.G.

WIERZBIŃSKI (Prof. Dr St.), Astronomical Observatory, Kopernika 11, Wrocław, Poland

WIESE (W. L.), Chief, Plasma Spectroscopy Section, National Bureau of Standards, Physics Building A-149, Washington, D.C. 20234, U.S.A.

WIETH-KNUDSEN (Dr N. P.), Svend Trostsvej 12^{III}, Copenhagen-V, Denmark

WILCOCK (Dr W. L.), Department of Physics, University College of North Wales, Bangor, Caernarvonshire, U.K.

WILCOX (Dr J. M.), Space Sciences Laboratory, University of California, Berkeley, California, 94720, U.S.A.

WILD (Dr J. P.), CSIRO, Division of Radiophysics, Post Office Box 76, Epping, N.S.W., 2121, Australia

WILD (P.), Astronomisches Institut der Universität Bern, Sidlerstrasse 5, Bern, Switzerland

* WILD (Dr P. A. T.), Astronomy Department, University of Cape Town, Rondebosch, Cape, South Africa

* WILDEY (Dr R. L.), Center of Astrogeology, U.S. Geological Survey, 601 E.-Cedar Avenue, Flagstaff, Arizona 86001, U.S.A.

WILDT (Prof. R.), Yale University Observatory, Box 2023, Yale Station, New Haven, Connecticut 06520, U.S.A.

WILKENS (Prof. Dr A.), Oberföhringerstrasse 10, München 27, F.R.G.

WILKINS (Dr G. A.), Superintendent, H.M. Nautical Almanac Office, Royal Greenwich Observatory, Herstmonceux Castle, Hailsham, Sussex, U.K.

* WILLIAMS (Mrs Dr C. A.), Department of Astronomy, University of South Florida, Tampa, Florida 33620, U.S.A.

WILLIAMS (Dr D.), Radio-Astronomy Laboratory, University of California, Berkeley, California, U.S.A.

* WILLIAMS (Dr D. A.), Department of Mathematics, U.M.I.S.T., Manchester 1. U.K.

* WILLIAMS (Dr I. P.) Department of Mathematics, Queen Mary College, Mile End Road, London E.1, U.K.

WILLIAMS (Dr J. A.), Physics Department, Albion College, Albion, Michigan 49224, U.S.A.

* WILLIAMS (Dr R.), Steward Observatory, University of Arizona, Tucson 85721, U.S.A.

* WILLS (Dr D.), Department of Astronomy, University of Texas, Austin, Texas 78712, U.S.A.

WILLSTROP (Dr R. V.), University Observatories, Madingley Road, Cambridge, U.K.

WILSON (Dr A. G.), P.O. Box 113, Topanga, California, U.S.A.

* WILSON (Prof. B. G.), Deputy, President, Simon Fraser University, Burnaby, Canada

WILSON (Dr O. C.), Mount Wilson and Palomar Observatories, 813 Santa Barbara Street, Pasadena 4, California, U.S.A.

WILSON (Dr P. R.), Department of Applied Mathematics, University of Sydney, Sydney, 2006, N.S.W., Australia

WILSON (Dr R.), Science Research Council, Astrophysics Research Unit, Culham Laboratory, Abingdon, Berkshire, U.K.

* WILSON (Dr R. W.), Bell Laboratories, Crawford Hill Laboratory, Holmdel, New Jersey 07733, U.S.A.

* WINCKLER (Dr J. R.), School of Physics and Astronomy, University of Minnesota, Minneapolis, Minnesota 55455, U.S.A.

WINKLER (Dr G. M. R.), U.S. Naval Observatory, Department of the Navy, Washington, D.C., U.S.A.

* WISNIEWSKI (Dr W.), Astronomical Observatory, Kraków, Kopernika 27, Poland

* WITHBROE (Dr G.), Harvard College Observatory, 60 Garden St., Cambridge, Massachusetts 02138, U.S.A.

WITKOWSKI (Prof. Dr J.), Director of the Observatory of Poznań, Ul. Sloneczna 36, Poznań, Poland

* WITT (Dr A. N.), Department of Physics and Astronomy, University of Toledo, 2801 W. Bancroft St., Toledo, Ohio 43606, U.S.A.

WLÉRICK (Dr G.), Observatoire de Paris, Section d'Astrophysique, 92-Meudon, France

* WOLFF (Dr S. C.), Institute of Astronomy, University of Hawaii, 2525 Correa Road, Honolulu, Hawaii 96822, U.S.A.

WOLSTENCROFT (Dr R. D.), Institute for Astronomy, University of Hawaii, Honolulu, Hawaii 96822, U.S.A.

WOLTJER (Prof. L.), Department of Astronomy, Columbia University, New York 10027, N.Y., U.S.A.

WOOD (Prof. F. B.), University of Florida, Department of Physics and Astronomy, Gainesville, Florida 32601, U.S.A.

* WOOD (Dr H. J. III), Department of Astronomy, University of Virginia, Charlottesville, Va. 22903, U.S.A.

WOOD (Dr H. W.), Government Astronomer, The Observatory, Sydney, N.S.W., Australia

WOOD (Dr J. A.), Smithsonian Astrophysical Observatory, 60 Garden Street, Cambridge, Massachusetts 02138, U.S.A.

WOOLARD (E. W.), Director of the Nautical Almanac, U.S. Naval Observatory, Washington 25, D.C., U.S.A.

WOOLF (Dr N. J.), School of Physics and Astronomy, University of Minnesota, Minneapolis, Minnesota 55455, U.S.A.

WOOLLEY (Sir Richard), Astronomer Royal, Royal Greenwich Observatory, Herstmonceux Castle, Hailsham, Sussex, U.K.

WOOLSEY (E. G.), Earth Physics Branch, Dept. of Energy, Mines and Resources, Ottawa, Ontario, Canada

WORLEY (C. E.), U.S. Naval Observatory, Washington, D.C. 20390, U.S.A.

WORRALL (Dr G.), The Observatories, Madingley Road, Cambridge, U.K.

WOSZCYK (Dr A.), Astronomical Observatory of Toruń University, Sienkiewicza 30, Toruń, Poland

* WRAY (Dr J. D.), Astronomy Department, Northwestern University, Evanston, Illinois 60201, U.S.A.

WRIGHT (Miss Dr F. W.), Harvard College Observatory, Cambridge, Massachusetts 02138, U.S.A.

WRIGHT (Miss Dr H.), 6040 Boulevard E., West New York, New Jersey 07093, U.S.A.

WRIGHT (Dr J. P.), Smithsonian Astrophysical Observatory, 60 Garden Street, Cambridge, Massachusetts 02138, U.S.A.

WRIGHT (Dr K. O.), Director, Dominion Astrophysical Observatory, R.R.7, Victoria, B.C., Canada

WURM (Prof. Dr K.), Sonnenstr. 29, 8203 Oberaudorf/Inn, F.R.G.

WYATT JR (Dr S. P.), University of Illinois Observatory, Urbana, Illinois, U.S.A.

WYLLER (Dr A. A.), Bartol Research Foundations, Swarthmore, Pennsylvania, U.S.A.

* WYNNE (Prof. Ch. G.), Applied Optics Section, Department of Physics, Imperial College of Science and Technology, Prince Consort Road, London S.W.7, U.K.

XANTHAKIS (Prof. J.), Research and Computing Center, Academy of Sciences of Athens, Athens, Greece

* YABUSHITA (Dr S.), Department of Applied Mathematics and Physics, Kyoto University, Kyoto, Japan

YABUUTI (Dr K.), Research Institute of Humanistic Science, Kyoto University, Kyoto, Japan

YAKHONTOVA (Mrs Dr N. Samojlova), Institute of Theoretical Astronomy, Leningrad, U.S.S.R,

YAKOVKIN (Prof. Dr A. A.), Main Astronomical Observatory of the Academy of Sciences of the Ukrainian SSR, Kiev 41, U.S.S.R.

YAKOVKIN (Dr N. A.), Kiev University Observatory, Kiev, U.S.S.R.

YAMASHITA (Dr Y.), Department of Astronomy, University of Tokyo, Yayoi, Bunkyo-ku, Tokyo, Japan

YAPLEE (B. S.), Code 7134, U.S. Naval Research Laboratory, Washington, D.C. 20390, U.S.A.

* YAROV-YAROVOJ (Dr M. S.), Sternberg Astronomical Institute, Moscow, U.S.S.R.

YASUDA (Dr H.), Tokyo Astronomical Observatory, Mitaka, Tokyo, Japan

YAVNEL' (Dr A. A.), Meteorite Committee, USSR Academy of Sciences, Moscow, U.S.S.R.

YEN (Prof. J. L.), Department of Electrical Engineering, University of Toronto, Toronto, Ontario. Canada

YILMAZ (Dr F.), Observatoire de l'Université, Istanbul, Turkey

* YILMAZ (Dr N.), Faculty of Science, Department of Astronomy, Ankara, Turkey

YOSS (Dr K. M.), University of Illinois Observatory, Urbana, Illinois, U.S.A.

YOUNG (Dr A.), School of Physical Sciences, The New University of Ulster, Coleraine, Co. Londonderry, Northern Ireland, U.K.

YOUNG (Dr A. T.), 3760 Avenida Del Sol, Studio City, California 91604, U.S.A.

Yü (Dr Ching-Sung), Williams Observatory, Hood College, Frederick, Maryland, U.S.A.

Yumi (Dr S.), International Latitude Observatory of Mizusawa, Mizusawa-shi, Iwate-ken, Japan

Zabriskie (Prof. Dr F. R.), Department of Astronomy, 102 Whitmore Laboratory, Pennsylvania State University, University Park, Pennsylvania 16802, U.S.A.

Zacharov (Dr I.), Astronomical Institute, Czechoslovak Academy of Sciences, Observatory Ondřejov, Czechoslovakia

Zadunaisky (Prof. P. E.), Instituto 'Torcuato Di Tella', Florida 936, Buenos Aires, Argentina

Zagar (Prof. F.), Direttore dell' Osservatorio Astronomico, Via Brera 28, Milano, Italy

Zahn (J.-P.), Observatoire de Nice, Le Mont-Gros, 06-Nice, France

Zanstra (Prof. H.), Rustoord, Westhoutpark 34, Haarlem, The Netherlands

Zel'dovich (Acad. Ya. B.), Institute of Applied Mathematics, Miusscaya pl. 4, Moscow A-47, U.S.S.R.

Zel'manov (Dr A. L.), Sternberg Astronomical Institute, Moscow, U.S.S.R.

Zhelezniakov (Dr V. V.), Scientific Radiophysical Institute, Gorki, U.S.S.R.

Zhevakin (Dr S. A.), Scientific Radiophysical Institute, Gorki, U.S.S.R.

Zhongolovich (Prof. Dr I. D.), Institute of Theoretical Astronomy, Leningrad, U.S.S.R.

Zimmermann (Prof. Dr G. K.), Nikolaev Observatory, Nikolaev, U.S.S.R.

Zimmermann (Dr H.), University Observatory, Schillergässchen 2, Jena, G.D.R.

* Ziółkowski (Dr J.), Polish Academy of Sciences, Institute of Astronomy, Al. Ujazdowskie 4, Warsaw, Poland

Zirin (Prof. H.), Hale Observatories, California Institute of Technology, 1201 E. California Blvd., Pasadena, California 91109, U.S.A.

Zirker (Dr J. B.), Institute for Astronomy, 2840 Koluwalu St., University of Hawaii, Honolulu, Hawaii 96822, U.S.A.

Zonn (Prof. W.), Astronomical Observatory, Ujazdowskie 4, Warsaw, Poland

* Zuckermann (Dr B. M.), Department of Physics and Astronomy, University of Maryland, College Park, Maryland, U.S.A.

Zverev (Prof. Dr M. S.), Pulkovo Observatory, Leningrad M-140, U.S.S.R.

Zwaan (Dr C.), Astronomical Observatory 'Sonnenborgh', Zonnenburg 2, Utrecht, The Netherlands

Zwicky (Prof. Dr F.), California Institute of Technology, Pasadena, California 91109, U.S.A.

LIST OF COUNTRIES ADHERING TO THE UNION

The following is a list of the 46 countries that adhered to the Union in September 1970, giving also the year of admission, the approximate number of Members, and the Adhering Organizations.

Country	Year	Members	Adhering Organizations
Argentina	1927	16	Comité Nacional de Astronomía, La Plata.
Australia	1939	54	Australian National Committee for Astronomy, Sydney.
Austria	1955	17	Österreichische Akademie der Wissenschaften, Wien.
Belgium	1920	52	Académie Royale de Belgique, Bruxelles, Koninklijke Vlaamse Academie van België, Brussel
Brazil	1961	10	Conselho Nacional de Pesquisas, Rio de Janeiro.
Bulgaria	1957	9	Bulgarian Academy of Sciences '7 November', Sofia.
Canada	1920	89	National Committee of Astronomy for Canada, Dept. of Geophysics and Astronomy, University of British Columbia, Vancouver 8, B.C.
Chile	1947	5	Universidad de Chile, Santiago de Chile.
Colombia	1967	1	Academia Colombiana de Ciencias Exactas, Físicas y Naturales, Bogotá.
Cuba	1970	0	Instituto de Astronomía de la Academia de Ciencias de Cuba, La Habana
Czechoslovakia	1922	49	National Committee of Astronomy of Czechoslovakia, Czechoslovak Academy of Sciences, Praha.
Denmark	1922	19	Danish National Committee for Astronomy, det Kongelige Danske Videskabernes Selskab, København.
Finland	1948	10	Finnish Academy of Sciences and Letters, Helsinki.
France	1920	168	Comité National Français d'Astronomie, Académie des Sciences, Paris.
F.R.G.	1951	136	Rat Westdeutscher Sternwarten zu München.
G.D.R.	1951	33	Nationalkomitee für Astronomie der DDR, Deutsche Akademie der Wissenschaften zu Berlin.
Greece	1920	19	Greek National Committee for Astronomy, Academy of Athens, Athens.
Hungary	1947	12	Hungarian Academy of Sciences, Budapest.
India	1964	49	Indian National Committee for the International Astronomical Union, Indian National Science Academy, New Delhi.
Iran	1969	0	Tehran University, Tehran.
Ireland	1947	6	National Committee of Astronomy, Royal Irish Academy, Dublin.
Israel	1954	14	The Israel Academy of Sciences and Humanities, Jerusalem.
Italy	1921	77	Consiglio Nazionale delle Ricerche, Roma.
Japan	1920	105	National Committee of Astronomy of Japan, Science Council of Japan, Tokyo.

Country	Year	Members	Adhering Organizations
Korea (PDR)	1961	3	Academy of Sciences, Phyongyang.
Mexico	1921	13	Universidad Nacional de Mexico, Mexico D.F.
Netherlands	1922	64	Koninklijke Nederlandse Academie van Weten- schappen, Amsterdam.
New Zealand	1964	5	Royal Society of New Zealand, Wellington.
Norway	1922	14	Det Norske Videnskaps-Akademi, Oslo.
Poland	1922	42	Polska Akademia Nauk, Warsawa.
Portugal	1924	14	Seccao Portuguesa das Uniones Internacionais Astro- nómica, Geodésica e Geofisica, Lisboa.
Roumanie	1928	18	Conseil Scientifique de l'Observatoire de Bucarest, Bucarest.
South Africa	1938	13	Council for Scientific and Industrial Research, Pretoria.
Spain	1922	17	Comisión Nacional de Astronomía, Madrid.
Sweden	1925	42	Kungl. Vetenskaps-Akademien, Stockholm.
Switzerland	1923	21	Schweizerische Naturforschende Gesellschaft, Basel.
Taiwan (Republic of China)	1959	9	Academia Sinica, Nankang, Taipei.
Turkey	1961	18	Türk Astronomi Dernegi, Istanbul.
U.A.R.	1925	9	Astronomical Center at Helwan, Egyptian Region.
United Kingdom	1920	237	The Royal Society, London.
Uruguay	1970	0	Universidad de la República, Comité Nacional de Astronomía, Montevideo.
U.S.A.	1920	728	The National Academy of Sciences, Washington D.C.
U.S.S.R.	1920	362	Academy of Sciences of the U.S.S.R., Moscow.
Vatican City State	1932	7	Pontificia Academia delle Scienze, Città del Vaticano.
Venezuela	1953	2	Academia de Ciencias Físicas, Matemáticas y Natura- les, Caracas.
Yugoslavia	1935	14	Conseil Fédéral pour la Coordination de la Recherche Scientifique, Belgrade.